Lecture Notes in Con

Commenced Publication in 1973
Founding and Former Series Editoı
Gerhard Goos, Juris Hartmanis, and Jan van Leeuwen

Burkhard Monien   Ulf-Peter Schroeder (Eds.)

Algorithmic Game Theory

First International Symposium, SAGT 2008
Paderborn, Germany, April 30 – May 2, 2008
Proceedings

Volume Editors

Burkhard Monien
Ulf-Peter Schroeder
Universität Paderborn
Institut für Informatik
Fürstenallee 11, 33102 Paderborn, Germany
E-mail: {bm,ups}@upb.de

Library of Congress Control Number: 2008925094

CR Subject Classification (1998): H.4, K.4.4, J.1, J.4, G.1.2

LNCS Sublibrary: SL 3 – Information Systems and Application, incl. Internet/Web and HCI

ISSN 0302-9743
ISBN-10 3-540-79308-9 Springer Berlin Heidelberg New York
ISBN-13 978-3-540-79308-3 Springer Berlin Heidelberg New York

Springer is a part of Springer Science+Business Media

springer.com

Printed in Germany

Typesetting: Camera-ready by author, data conversion by Scientific Publishing Services, Chennai, India
Printed on acid-free paper SPIN: 12259106 06/3180 5 4 3 2 1 0

Preface

This volume contains the papers presented at the First International Symposium on Algorithmic Game Theory (SAGT 2008) held from April 30 to May 2 in Paderborn, Germany.

The purpose of SAGT is to bring together researchers from computer science, economics and mathematics to present and discuss original research at the intersection of algorithms and game theory. It is intended to cover all important areas of algorithmic game theory, such as: solution concepts in game theory; game classes (e.g., bimatrix, potential, Bayesian); exact and approximate computation of equilibria; convergence and learning in games; complexity classes in game theory; algorithmic aspects of fixed-point theorems; mechanisms, incentives and coalitions; cost-sharing algorithms and analysis; computational aspects of market equilibria; computational problems in economics, finance, decision theory and pricing; auction algorithms and analysis; price of anarchy and its relatives; representations of games and their complexity; economic aspects of distributed computing and the Internet; network formation on the Internet; congestion, routing and network design games; game-theoretic approaches to networking problems; Byzantine game theory.

There were 60 submissions. Each submission was reviewed by three Programme Committee members. The committee decided to accept 28 papers. The programme also included three invited talks from outstanding researchers Christos Papadimitriou, Nobel Memorial Prize winner Reinhard Selten and Paul Spirakis.

We would like to thank all the Programme Committee members and the external reviewers who assisted them in their work.

The members of the Organizing Committee as well as the developer of the EasyChair conference system deserve our gratitude for their contributions throughout the preparations.

We gratefully acknowledge support from the European Association for Theoretical Computer Science (EATCS), Integrated Project AEOLUS (IST-015964) of the European Union, University of Paderborn, and the City of Paderborn.

February 2008

Burkhard Monien
Ulf-Peter Schroeder

Organization

Programme Chair

Burkhard Monien	University of Paderborn, Germany

Programme Committee

Petra Berenbrink	Simon Fraser University, Canada
Xiaotie Deng	City University of Hong Kong, Hong Kong S.A.R.
Amos Fiat	Tel Aviv University, Israel
Dimitris Fotakis	University of the Aegean, Greece
Nicole Immorlica	CWI, Netherlands
Elias Koutsoupias	University of Athens, Greece
Stefano Leonardi	Università di Roma, Italy
Marios Mavronicolas	University of Cyprus, Cyprus
Vladimir V. Mazalov	Russian Academy of Sciences, Russia
Igal Milchtaich	Bar-Ilan University, Israel
Dov Monderer	Technion, Israel
Burkhard Monien	University of Paderborn, Germany (Chair)
Giuseppe Persiano	Università di Salerno, Italy
Tim Roughgarden	Stanford University, USA
Amin Saberi	Stanford University, USA
Bill Sandholm	University of Wisconsin, USA
Paul Spirakis	Computer Technology Institute, Greece
Bernhard von Stengel	London School of Economics, UK
Takashi Ui	Yokohama National University, Japan
Vijay V. Vazirani	Georgia Institute of Technology, USA
Berthold Vöcking	RWTH Aachen, Germany
Peter Widmayer	ETH Zurich, Switzerland

Local Organization

Bernard Bauer
Marion Rohloff
Ulf-Peter Schroeder
Thomas Thissen

External Reviewers

Arash Asadpour
Vincenzo Auletta
Luca Becchetti
Tianming Bu
Deeparnab Chakrabarty
Julia Chuyko
Artur Czumaj
Nikhil Devanur
Shahar Dobzinski
Ye Du
Amir Epstein
Michal Feldman
Gagan Goel
Alexis Kaporis
Spyros Kontogiannis
Aranyak Mehta
Luca Moscardelli
Hamid Nazerzadeh
Svetlana Olonotevsky
Panagiota Panagopoulou
Piotr Sankowski
Rahul Savani
Florian Schoppmann
Chris Shannon
Haralampos Tsaknakis
Carmine Ventre
Feng Wang
Ying Wang
Andreas Witzel

Table of Contents

Invited Talks

Session 1: Routing and Scheduling I

Session 2: Markets

Session 3: Routing and Scheduling II

Session 4: Mechanism Design

Session 5: Potpourri of Games

Session 6: Solution Concepts

Session 7: Cost Sharing

The Search for Equilibrium Concepts*

Christos H. Papadimitriou

Computer Science Division,
University of California at Berkeley, Berkeley, USA
christos@cs.berkeley.edu

Abstract. Game Theory is about predicting the behavior of groups of rational agents whose decisions affect each other's welfare, and such predictions are most often in the form of equilibrium concepts. There are several desiderata one might expect from an equilibrium concept: First and foremost it should be natural and convincing as a prediction of agent behavior. Then it should be *universal* — all games should have it, because otherwise it is an incomplete prediction. Since computer scientists became interested in Game Theory over the past decade, prompted by the advent of the Internet and its complex socioeconomic platform, another important question has been asked of an equilibrium concept: *Can it be computed efficiently?* Intractability makes an equilibrium concept problematic.

How do the major equilibrium concepts compare with respect to this battery of criteria (convincing, universal, efficient)? The pure Nash equilibrium is certainly natural and convincing, and can be computed efficiently, but is certainly non-universal (not just matching pennies: a random $n \times n$ game will fail to have a pure Nash equilibrium with probability asymptotic to $\frac{1}{e}$). In contrast, the mixed Nash equilibrium is famously universal by Nash's theorem [13], but was recently shown to be PPAD-complete, and thus presumably intractable [5]. (Whether it is convincing that agents will engage in precise randomization among equivalent alternatives just to keep others on their toes is an interesting question.) The correlated equilibrium is both efficient and universal, but assumes too much machinery to be truly convincing. Incidentally, the above account of the efficiency of equilibrium concepts applies to the normal form representation of games, and this form is inadequate for describing multiplayer games. In *succinctly representable games*, the situation is quite a bit different: Pure Nash equilibria can be NP-complete (graphical games) or PLS-complete (congestion games[10]), while for most succinct representations mixed Nash equilibria remain PPAD-complete [4] and correlated equilibria remain tractable [14]).

The recent proof of the intractability of finding a mixed Nash equilibrium has prompted researchers to look for new alternatives. The question of *approximate* mixed Nash equilibria was, naturally, one of the first to be pursued. We now

* Research supported by an NSF grant, a MICRO grant, and by a gift from Yahoo! Research.

B. Monien and U.-P. Schroeder (Eds.): SAGT 2008, LNCS 4997, pp. 1–3, 2008.

know that finding a FPTAS for this problem is also intractable [3] (but there is a PTAS for the very broad case of anonymous games with few strategies per player [8]), and that there are *pseudo*polynomial time approximation schemes when there are few players [12] *or* few strategies per player. However, progress towards better polynomial-time approximation algorithms has been rather slow [6,7,15], and this remains an interesting frontier.

In our search for novel equilibrium concepts we might as well consider ones that are a little more compelling in the context of the Internet, and this has led researchers to revisiting various sorts of *repeated games.* If agents react to each other's decisions by utility-optimizing (or just -improving) moves, the resulting dynamics ends up trapped in one of the sink strongly connected components of the directed graph of responses, and this has been called a *sink equilibrium* [11]. Or one can postulate that agents engage in arbitrary interactions by responding in consistent ways to what others are currently doing; this leads to a game in which pure strategies are *finite state automata* with the player's strategies as states, and the other players' current choices as input alphabet. Once such a choice of automata has been made, the play will quickly converge to a periodic behavior, whose average payoff then determines the players' utilities for this choice of automata. Now, if the automaton chosen by each player is the best possible response to the automata chosen by the others, then we have an interesting kind of equilibrium, which is called *unit recall equilibrium.* It turns out that, for a random game, such an equilibrium exists (and is in fact returned by an efficient algorithm) with probability asymptotic to one [9].

Nash equilibria had always been thought easier to find in the case of repeated games, because of a result known as *the Folk Theorem* [1] predicting a wealth of accessible equilibria based on a combination of periodic play and threats. It was recently shown [2] that the method for finding these equilibria suggested by the Folk Theorem is problematic from the complexity point of view, and that, in fact, finding a Nash equilibrium in a repeated game with three or more players is PPAD-complete — that is, no easier than finding one in the one-shot game.

References

1. Aumann, R., Shapley, L.: Long-term Competition: A Game-theoretic Analysis, mimeo. Hebrew University (1976)
2. Chayes, J., Borgs, C., Immorlica, N., Kalai, A., Mirokhni, V., Papadimitriou, C.: The Myth of the Folk Theorem. In: STOC (to appear, 2008)
3. Chen, X., Deng, X., Teng, S.: Computing Nash Equilibria: Approximation and Smoothed Complexity. In: FOCS (2006)
4. Daskalakis, C., Fabrikant, A., Papadimitriou, C.H.: The Game World Is Flat: The Complexity of Nash Equilibria in Succinct Games. In: Bugliesi, M., Preneel, B., Sassone, V., Wegener, I. (eds.) ICALP 2006. LNCS, vol. 4051, pp. 513–524. Springer, Heidelberg (2006)
5. Daskalakis, C., Goldberg, P., Papadimitriou, C.: The Complexity of Computing a Nash Equilibrium. In: STOC (2006)

6. Daskalakis, C., Mehta, A., Papadimitriou, C.: A Note on Approximate Nash Equilibria. In: Spirakis, P.G., Mavronicolas, M., Kontogiannis, S.C. (eds.) WINE 2006. LNCS, vol. 4286, pp. 297–306. Springer, Heidelberg (2006)
7. Daskalakis, C., Mehta, A., Papadimitriou, C.: Progress in Approximate Nash Equilibria. In: EC (2007)
8. Daskalakis, C., Papadimitriou, C.: Computing Equilibria in Anonymous Games. In: FOCS (2007)
9. Fabrikant, A., Papadimitriou, C.H.: The Complexity of Game Dynamics: BGP Oscillations, Sink Equlibria, and Beyond. In: SODA (2008)
10. Fabrikant, A., Papadimitriou, C.H., Talwar, K.: The Complexity of Pure Nash Equilibria. In: STOC (2004)
11. Goemans, M., Mirrokni, V., Vetta, A.: Sink Equilibria and Convergence. In: FOCS (2005)
12. Lipton, R., Markakis, E., Mehta, A.: Playing Large Games Using Simple Strategies. In: EC (2003)
13. Nash, J.: Noncooperative Games. Annals of Mathematics 54, 289–295 (1951)
14. Papadimitriou, C.H.: Computing Correlated Equilibria in Multiplayer Games. In: STOC (2005)
15. Tsaknakis, H., Spirakis, P.: An Optimization Approach for Approximate Nash Equilibria. In: WINE (2007)

Experimental Results on the Process of Goal Formation and Aspiration Adaptation

Reinhard Selten

Department of Economics,
Institute for Empirical Research in Economics,
University of Bonn, Adenauerallee 24-42, 53113 Bonn, Germany
rselten@uni-bonn.de

Abstract. We experimentally investigate how subjects deal with a multi-period planning and decision problem. The context is a profit maximization task in a computer-simulated monopoly market over fifty time periods. The subjects are provided with a computerized short-run planning tool allowing them to check feasibility of any aspiration level for any set of feedback variables of the respective planning period. Our results fall into two categories, first, regarding the selection of goal variables and, second, regarding the process of aspiration adaptation. As to the former category, we find that subjects with at least median success change their goal variables less frequently than those below median success. Relatedly, goal persistence, a measure of a subject's tendency to stick to the current goal system, is strongly positively correlated with success. As to the latter category, we find that aspiration levels tend to be changed in strong agreement with basic principles of Aspiration Adaptation Theory (Sauermann and Selten 1962, Selten 1998, 2001). In addition, we find that in many cases the process of aspiration adaptation leads into a nearly stationary situation in which the aspiration level does not significantly change over several periods. Those subjects who reach a nearly stationary situation tend to be more successful and more goal persistent than those who do not. Some subjects who reach a nearly stationary situation deviate from Aspiration Adaptation Theory in order to find a more profitable nearly stationary situation.

B. Monien and U.-P. Schroeder (Eds.): SAGT 2008, LNCS 4997, p. 4, 2008.

Approximate Equilibria for Strategic Two Person Games*

Paul G. Spirakis

Research Academic Computer Technology Institute, N. Kazantzaki Str., Patra University Campus, 26500 Rio-Patra, Greece
spirakis@cti.gr

Abstract. In view of the apparent intractability of constructing Nash Equilibria (NE in short) in polynomial time, even for bimatrix games, understanding the limitations of the approximability of the problem is an important challenge. The purpose of this document is to review a set of results, which have contributed significantly, and currently are the state-of-art with respect to the polynomial time construction of approximate Nash equilibria in bimatrix games. *Most of the results discussed here are joint work of the author and of the union of his coauthors in various papers, namely S. Kontogiannis, P. Panagopoulou and H. Tsaknakis.*

1 Introduction

One of the most appealing concepts in game theory is the notion of a Nash equilibrium: *A collection of strategies for the players from which no player has an incentive to unilaterally deviate from its own strategy.* The extremely nice thing about Nash equibria is that they always exist in any finite $k-$person game in normal form [20]. This is one of the most important reasons why Nash equilibria are considered to be the prevailing solution concept for finite games in normal form. The problem with Nash equilibria is that there can be exponentially many of them, of quite different characteristics, even for bimatrix games. Additionally, we do not know yet how to construct them in subexponential time. Therefore, $k-$NASH, the problem of computing an arbitrary Nash equilibrium of a finite $k-$person game in normal form, is a fundamental problem in algorithmic game theory and perhaps one of the most outstanding problems at the boundary of **P** [22]. Its complexity has been a long standing open problem, since the introduction of the pioneering (pivoting) algorithm of Lemke and Howson [18]. Unfortunately, it was recently shown by Savani and von Stengel [24] that this algorithm requires an exponential number of steps; moreover, it is also known that the smoothed complexity of the algorithm is likely to be superpolynomial [6]. Moreover, it is quite interesting that many (quite natural) refinements of $k-$NASH are known to be **NP**$-$complete problems [13,8].

* This work was partially supported by the 6th Framework Programme under contract 001907 (DELIS).

B. Monien and U.-P. Schroeder (Eds.): SAGT 2008, LNCS 4997, pp. 5–21, 2008.

A flurry of results in the last years has proved that $k-$NASH is indeed *complete problem* for the complexity class **PPAD** (introduced by Papadimitriou [21]), even for four players [9], three players [12], and two players [5]. In particular, the result of Chen and Deng [5], complemented by that of Abbott, Kane and Valiant [1], shows that 2−NASH is **PPAD**−complete even for win lose games.

Due to the apparent hardness even of 2−NASH, approximate solutions to Nash equilibria have lately attracted the attention of the research community. There are two different notions of approximate Nash equilibria: Those which require that each player gets the maximum possible payoff, within some additive constant ε (denoted here by ApproxNE), and those which require that each player is allowed to adopt **wpp**[1] only actions that are *approximate best responses* to the opponent's strategy, within an additive term ε (denoted here by SuppNE). ApproxNE seem to be the dominant notion of approximate equilibria in the literature, while SuppNE is a rather new notion (eg, see [6,7,10]).

1.1 Preliminaries

Mathematical Notation. For any integer $k \in \mathbb{N}$, let $[k] \equiv \{1, 2, \ldots, k\}$. We denote by $M \in F^{m \times n}$ any $m \times n$ matrix whose elements have values in some set F. We also denote by $(A, B) \in (F \times F)^{m \times n}$ any $m \times n$ matrix whose elements are *ordered pairs* of values from F. Equivalently, this structure can be seen as an ordered pair of $m \times n$ matrices $A, B \in F^{m \times n}$. Such a pair of matrices is called a **bimatrix**. A $k \times 1$ matrix is also considered to be an k-**vector**. Vectors are denoted by bold small letters (eg, $\mathbf{x}, \mathbf{y}$). A vector having a 1 in the i-th position and 0 everywhere else is denoted by $\mathbf{e_i}$. We denote by $\mathbf{1_k}$ ($\mathbf{0_k}$) the k-vector having 1s (0s) in all its coordinates. The $k \times k$ matrix $E = \mathbf{1_k} \cdot \mathbf{1_k}^T \in \{1\}^{k \times k}$ has value 1 in all its elements. For a pair of vectors $\mathbf{x}, \mathbf{y} \in \mathbb{R}^n$, we denote the *component–wise comparison* by $\mathbf{x} \geq \mathbf{y}$: $\forall i \in [n], x_i \geq y_i$. Matrices are denoted by capital letters (eg, $A, B, C, \ldots$), and bimatrices are denoted by ordered pairs of capital letters (eg, $(A, B), (R, C), \ldots$). For any $m \times n$ (bi)matrix M, M_j is its j-th column (as an $m \times 1$ vector), M^i is the i-th row (as a (transposed) $1 \times n$ vector) and $M_{i,j}$ is the (i, j)-th element. For any integer $k \geq 1$, we denote by $\Delta_k \equiv \{\mathbf{z} \in \mathbb{R}^k : \mathbf{z} \geq \mathbf{0};\ (\mathbf{1_k})^T \mathbf{z} = 1\}$ the $(k-1)$-simplex. For any point $\mathbf{z} \in \Delta_k$, its **support** $supp(\mathbf{z})$ is the subset of coordinates with positive value: $supp(\mathbf{z}) \equiv \{i \in [k] : z_i > 0\}$. Also, let $suppmax(\mathbf{z}) \equiv \{i \in [k] : z_i \geq z_j\ \forall j \in [k]\}$ be the subset of coordinates with maximum value, and denote by $max(\mathbf{z})$ the value of the maximum entry of $\mathbf{z}$. For a subset of coordinates $S \subseteq [k]$, let $max_S(\mathbf{z})$ be the value of the maximum entry of vector $\mathbf{v}$ within the subset S. We denote by $\overline{S}$ the complement of a subset of coordinates S, i.e. $\overline{S} = \{i \in [k], i \notin S\}$. For an arbitrary logical expression $\mathcal{E}$, we denote by $\mathbb{P}\{\mathcal{E}\}$ the probability of this expression being true, while $\mathbb{I}_{\{\mathcal{E}\}}$ is the indicator variable of whether this expression is true or false. For any random variable x, $\mathbb{E}\{x\}$ is its expected value (with respect to some probability measure).

[1] With positive probability.

Game Theoretic Definitions and Notation. An $m \times n$ **bimatrix game** $\langle A, B \rangle$ is a 2−person game in normal form, that is determined by the bimatrix $(A, B) \in (\mathbb{R} \times \mathbb{R})^{m \times n}$ as follows: The first player (called the **row player**) has an m−element *action set* $[m]$, and the second player (called the **column player**) has an n−element *action set* $[n]$. Each row (column) of the bimatrix corresponds to a different action of the row (column) player. The row and the column player's payoffs are determined by the $m \times n$ real matrices A and B respectively. In the special case that the payoff matrices have only rational entries, we refer to a **rational bimatrix game**. If both payoff matrices belong to $[0, 1]^{m \times n}$ then we have a $[0, 1]$−**bimatrix** (or else **normalized**) **game**. The special case of bimatrix games in which all elements of the bimatrix belong to $\{0, 1\} \times \{0, 1\}$, is called a $\{0, 1\}$−**bimatrix** (or else, **win lose**) **game**. A bimatrix game $\langle A, B \rangle$ is called **zero sum**, if it happens that $B = -A$. In that case the game is solvable in polynomial time, since the two players' optimization problems form a primal–dual linear programming pair. In all cases of bimatrix games we assume wlog[2] that $2 \leq m \leq n$.

Any probability distribution on the action set $[m]$ of the row player, ie, any point $\mathbf{x} \in \Delta_m$, is a **mixed strategy** for her. Ie, the row player determines her action independently from the column player, according to the probability distribution determined by $\mathbf{x}$. Similarly, any point $\mathbf{y} \in \Delta_n$ is a mixed strategy for the column player. Each extreme point $\mathbf{e_i} \in \Delta_m$ ($\mathbf{e_j} \in \Delta_n$) that enforces the use of the i-th row (j-th column) by the row (column) player, is called a **pure strategy** for her. Any element $(\mathbf{x}, \mathbf{y}) \in \Delta_m \times \Delta_n$ is a (mixed in general) **strategy profile** for the players. We now define the set of *approximate best responses* for the two players, that will help us simplify the forthcoming definitions:

Definition 1 (Approximate Best Response). *Fix arbitrary constant* $\varepsilon > 0$. *Given that the column player adopts a strategy* $\mathbf{y} \in \Delta_n$ *and the payoff matrix of the row player is* A, *the row player's set of* ε−**approximate (pure) best responses** *is:* $BR(\varepsilon, A, \mathbf{y}) \equiv \left\{\mathbf{x} \in \Delta_m : \mathbf{x}^T A\mathbf{y} \geq \mathbf{z}^T A\mathbf{y} - \varepsilon, \forall \mathbf{z} \in \Delta_m\right\}$ *and* $PBR(\varepsilon, A, \mathbf{y}) \equiv \left\{i \in [m] : A^i\mathbf{y} \geq A^r\mathbf{y} - \varepsilon, \forall r \in [m]\right\}$. *Similarly we define the column player's set of* ε−**approximate (pure) best responses**: $BR(\varepsilon, B^T, \mathbf{x}) \equiv \left\{\mathbf{y} \in \Delta_n : \mathbf{y}^T B^T\mathbf{x} \geq \mathbf{z}^T B^T\mathbf{x} - \varepsilon, \forall \mathbf{z} \in \Delta_n\right\}$ *and* $PBR(\varepsilon, B^T, \mathbf{x}) \equiv \left\{j \in [n] : B_j^T\mathbf{x} \geq B_r^T\mathbf{x} - \varepsilon, \forall r \in [n]\right\}$.

For the notion of Nash equilibria, originally introduced by Nash [20], we give the definition wrt[3] bimatrix games:

Definition 2 (Nash Equilibrium). *A strategy profile* $(\mathbf{x}, \mathbf{y})$ *is a Nash equilibrium for a bimatrix game* $\langle A, B \rangle$ *iff no player can improve her payoff by deviating (changing her strategy) unilaterally. More formally: For any bimatrix game* $\langle A, B \rangle$, *a profile* $(\mathbf{x}, \mathbf{y}) \in \Delta_m \times \Delta_n$ *is a* **Nash Equilibrium** *(NE in short) iff* $\mathbf{x} \in BR(0, A, \mathbf{y})$ *and* $\mathbf{y} \in BR(0, B^T, \mathbf{x})$. *Equivalently,* $(\mathbf{x}, \mathbf{y}) \in \Delta_m \times \Delta_n$ *is a NE of* $\langle A, B \rangle$ *iff supp*$(\mathbf{x}) \subseteq PBR(0, A, \mathbf{y})$ *and supp*$(\mathbf{y}) \subseteq PBR(0, B^T, \mathbf{x})$. *The set of profiles that are NE of* $\langle A, B \rangle$ *is denoted by* $NE(A, B)$.

[2] Without loss of generality.

[3] With respect to.

Due to the apparent difficulty in computing NE for arbitrary bimatrix games, the recent trend is to look for approximate equilibria. Two definitions of approximate equilibria that concern this paper are the following:

Definition 3 (Approximate Nash Equilibria). *For any* $\varepsilon > 0$ *and any bimatrix game* $\langle A, B \rangle$, $(\mathbf{x}, \mathbf{y}) \in \Delta_m \times \Delta_n$ *is an* $\varepsilon-$**approximate Nash Equilibrium** *(*$\varepsilon-$*ApproxNE) iff each player chooses an* $\varepsilon-$*approximate best response against the opponent:* $[\mathbf{x} \in BR(\varepsilon, A, \mathbf{y})] \wedge [\mathbf{y} \in BR(\varepsilon, B^T, \mathbf{x})]$ $(\mathbf{x}, \mathbf{y})$ *is an* $\varepsilon-$**well–supported Nash Equilibrium** *(*$\varepsilon-$*SuppNE) iff each player assigns positive probability only to* $\varepsilon-$*approximate* pure best responses *against the strategy of the opponent:* $\forall i \in [m], x_i > 0 \;\Rightarrow\; i \in PBR(\varepsilon, A, \mathbf{y})$ *and* $\forall j \in [n],\ y_j > 0 \;\Rightarrow\; j \in PBR(\varepsilon, B^T, \mathbf{x})$

It is not hard to see that any NE is both a $0-$ApproxNE and a $0-$SuppNE. Observe also that any $\varepsilon-$SuppNE is an $\varepsilon-$ApproxNE, but *not necessarily vice versa.* Indeed, the only thing we currently know towards this direction is that from an arbitrary $\frac{\varepsilon^2}{8n}-$ApproxNE one can construct an $\varepsilon-$SuppNE in polynomial time [6]. Note that both notions of approximate equilibria are defined wrt an *additive* error term ε. Although (exact) NE are known not to be affected by any positive scaling, it is important to mention that approximate notions of NE are indeed affected. Therefore, from now on we adopt the commonly used assumption in the literature (eg, [19,10,15,5,6]) that, when referring to $\varepsilon-$ApproxNE or $\varepsilon-$SuppNE, the bimatrix game is considered to be a $[0,1]-$bimatrix game. This is mainly done for sake of comparison of the results on approximate equilibria. Of particular importance are the uniform points of the $(k-1)-$simplex Δ_k:

Definition 4 (Uniform Profiles). *A point* $\mathbf{x} \in \Delta_r$ *is called a* $k-$**uniform strategy** *iff* $\mathbf{x} \in \Delta_r \cap \left\{0, \frac{1}{k}, \frac{2}{k}, \ldots, \frac{k-1}{k}, 1\right\}^r \equiv \Delta_r(k)$. *Ie,* $\mathbf{x}$ *assigns to each action a probability mass that is some multiple of* $\frac{1}{k}$. *In the special case that the only possibility for an action is to get either zero probability or* $\frac{1}{k}$, *we refer to a* **strict** $k-$**uniform strategy**. *We denote the space of strict* $k-$*uniform strategies by* $\hat{\Delta}_r(k) \equiv \Delta_r \cap \left\{0, \frac{1}{k}\right\}^r$. *A profile* $(\mathbf{x}, \mathbf{y}) \in \Delta_m \times \Delta_n$ *for which* $\mathbf{x}$ *is a (strict)* $k-$*uniform strategy and* $\mathbf{y}$ *is a (strict)* $\ell-$*uniform strategy, is called a* **(strict)** $(k, \ell)-$**uniform** *profile.*

We shall finally denote by $k-$NASH the problem of constructing an arbitrary NE for a finite $k-$player game in normal form.

1.2 Related Work and the Contribution of Our Team

The computability of NE in bimatrix games has been a long standing open problem for many years. The most popular algorithm of Lemke and Howson [18] for computing NE in these games, is an adaptation of Lemke's algorithm for finding solutions (if such exist) for arbitrary instances of the Linear Complementarity Problem (LCP). Unfortunately, it has been recently proved by Savani and von Stengel [24] that this pivoting algorithm may require an exponential number of steps before finding a NE, no matter which starting point is chosen. Moreover, even though the complexity of 2−NASH was unknown, it was well known that various (quite natural) restrictions of the problem (eg, uniqueness, bounds on support sizes, etc) lead to **NP**−hard problems [13,8].

A very recent series of research papers deal with the complexity of k−NASH. Initially [9,14] introduced a novel reduction technique and proved that 4−NASH is **PPAD**−complete. Consequently this result was extended to 3−player games [12]. Surprisingly, Chen and Deng [5] proved the same complexity result even 2−NASH. In view of all these hardness results for the k−NASH, understanding the limitations of the (in)approximability of the problem is quite important. To our knowledge, the first result that provides ε−ApproxNE within *subexponential* time, is the work of Lipton et al. [19]. In particular, for any *constant* $\varepsilon > 0$, they prove the existence of an ε−ApproxNE for arbitrary $n \times n$ bimatrix games, which additionally is a uniform profile that has supports of size at most $\lceil \log n/\varepsilon^2 \rceil$. This leads to a rather simple subexponential algorithm for constructing ε−ApproxNE for $[0,1]$−bimatrix games, simply by checking all possible profiles with support sizes at most $\lceil \log n/\varepsilon^2 \rceil$ for each strategy. This still remains the fastest strategy to date, for the general problem of providing ε−ApproxNE for any *constant* $\varepsilon > 0$. With respect to the tractability of a *Fully Polynomial Time Approximation Scheme* (FPTAS) for NE, [6] proved that providing a FPTAS for 2−NASH is also **PPAD**−complete. Namely, they proved that unless **PPAD** $\subseteq$ **P**, there is no algorithm that constructs ε−ApproxNE in time $poly(n, 1/\varepsilon)$, for any $\varepsilon = n^{-\Theta(1)}$. Moreover, they proved that unless **PPAD** $\subseteq$ **RP**, there is no algorithm that constructs a NE in time $poly(n, 1/\sigma)$, where σ is the size of the deviation of the elements of the bimatrix. This latter result essentially states that even the smoothed complexity of the algorithm of Lemke and Howson is not polynomial.

Two independent results [10,15] initiated the discussion of providing in polynomial time ε−ApproxNE for $[0,1]$−bimatrix games and some *constant* $1 > \varepsilon > 0$. In particular, [10] gave a simple $\frac{1}{2}$−ApproxNE for $[0,1]$−bimatrix games, involving only two strategies per player. In [15] our group presented a simple algorithm for computing a $\frac{3}{4}$−ApproxNE equilibrium for any bimatrix game in strongly polynomial time and we next showed how to extend this algorithm so as to obtain a (potentially stronger) parameterized approximation. Namely, we presented an algorithm that computes a $\frac{2+\lambda}{4}$−ApproxNE, where λ is the minimum, among all Nash equilibria, expected payoff of either player. The suggested algorithm runs in time polynomial in the number of strategies available to the players. Last year there was a series of results improving the constant for polynomial time constructions of ApproxNE. First [11] proposed an efficient construction of a 0.38−ApproxNE, and consequently [3] proposed a 0.36392−ApproxNE based on the solvability of zero sum bimatrix games (an idea that was borrowed by our group's work [16] for the efficient construction of SuppNE). Finally, [25] our group proposed a new methodology for determining ApproxNE of bimatrix games and based on that, we provided a polynomial time algorithm for computing 0.3393-ApproxNE. To our knowledge this is currently the best result for ApproxNE in bimatrix games.

As for the efficient approximation of SuppNE, [10] introduced the problem and proposed a quite simple algorithm, which, under a quite interesting graph theoretic conjecture, constructs in polynomial time some non–trivial SuppNE. Unfortunately, the status of this conjecture is still unkown (it is false for some small instances of graphs). [10] made also a quite interesting connection of the problem of constructing $\frac{1+\varepsilon}{2}$−SuppNE in an arbitrary $[0,1]$−bimatrix game, to that of constructing ε−SuppNE for a properly chosen win lose game of the same size. Our group continued this line of research and in [17] we studied the tractability of the more requiring notion of SuppNE. We demonstrated the existence of SuppNE with small supports and at the same time good quality. This directly implies a *subexponential time* algorithm for constructing SuppNE of arbitrary (constant) precision. An analogous result but for ApproxNE, was already known in [19]. We proved a much simpler, and slightly stronger result, as a corollary of Althöfer's approximation lemma [2]. We then proposed various algorithms for constructing SuppNE in win lose and normalized bimatrix games (ie, whose payoff matrices take values from $\{0,1\}$ and $[0,1]$ respectively). Our methodology for attacking the problem was based on two different approaches: The first [17] was graph theoretic, and we provided SuppNE whose quality is dependent on the *girth* of the Nash Dynamics graph in the win lose game, or a proper win lose image of the normalized game. In our second approach [16], based on the solvability of zero sum bimatrix games, we provided a 0.5−SuppNE for win lose games and a 0.658−SuppNE for normalized games. These are currently the best results for the stronger notion of SuppNE in bimatrix games.

Finally, concerning random $[0,1]$−bimatrix games, Bárány, Vempala and Vetta [4] considered the case where all the payoff values are (either uniform, or normal) **iid**[4] random variables in $[0,1]$. They analyzed a simple Las Vegas algorithm for finding a NE in such a game, by brute force on the support sizes, starting from smaller ones. The running time of their algorithm is $\mathcal{O}\big(m^2 n \log\log n + n^2 m \log\log m\big)$, **whp**[5]. In [17] we also studied random instances of bimatrix games and provided evidence for the efficient construction of SuppNE both in random normalized games and win lose games.

2 A First Remark

In this section we present one of the first very simple polynomial time algorithms for ε−ApproxNE, where ε is an absolute constant. For further details and more results we refer the interested reader to [15] and [10].

We will present a straightforward method for computing a $\frac{3}{4}$−ApproxNE for any positively normalized bimatrix game.

Lemma 1. *Consider any positively normalized* $m \times n$ *bimatrix game* $\Gamma = \langle A, B\rangle$ *and let* $A_{i_1,j_1} = \max_{i,j} A_{i,j}$ *and* $B_{i_2,j_2} = \max_{i,j} B_{i,j}$. *Then the pair of strategies* $(\hat{\mathbf{x}}, \hat{\mathbf{y}})$ *where* $\hat{\mathbf{x}} = \frac{1}{2}\mathbf{e_{i_1}} + \frac{1}{2}\mathbf{e_{i_2}}$ *and* $\hat{\mathbf{y}} = \frac{1}{2}\mathbf{e_{j_1}} + \frac{1}{2}\mathbf{e_{j_2}}$ *is a* $\frac{3}{4}$−*ApproxNE for* Γ.

[4] Independent, identically distributed.

[5] With high probability, ie, with probability $1 - m^{-c}$, for some constant $c > 0$.

Proof. First observe that $\hat{\mathbf{x}}^T A\hat{\mathbf{y}} = \sum_{i=1}^m \sum_{j=1}^n \hat{x}_i\hat{y}_j A_{i,j} = \frac{1}{4}(A_{i_1,j_1} + A_{i_1,j_2} + A_{i_2,j_1} + A_{i_2,j_2}) \geq \frac{1}{4}A_{i_1,j_1}$. Similarly, $\hat{\mathbf{x}}^T B\hat{\mathbf{y}} = \sum_{i=1}^m \sum_{j=1}^n \hat{x}_i\hat{y}_j B_{i,j} = \frac{1}{4}(B_{i_1,j_1} + B_{i_1,j_2} + B_{i_2,j_1} + B_{i_2,j_2}) \geq \frac{1}{4}B_{i_2,j_2}$. Now observe that, for any (mixed) strategies $\mathbf{x}$ and $\mathbf{y}$ of the row and column player respectively, $\mathbf{x}^T A\hat{\mathbf{y}} \leq A_{i_1,j_1}$ and $\hat{\mathbf{x}}^T B\mathbf{y} \leq B_{i_2,j_2}$ and recall that $A_{i,j}, B_{i,j} \in [0,1]$ for all $i \in [m]$, $j \in [n]$. Hence $\mathbf{x}^T A\hat{\mathbf{y}} \leq A_{i_1,j_1} = \frac{1}{4}A_{i_1,j_1} + \frac{3}{4}A_{i_1,j_1} \leq \hat{\mathbf{x}}^T A\hat{\mathbf{y}} + \frac{3}{4}$ and $\hat{\mathbf{x}}^T B\mathbf{y} \leq B_{i_2,j_2} = \frac{1}{4}B_{i_2,j_2} + \frac{3}{4}B_{i_2,j_2} \leq \hat{\mathbf{x}}^T B\hat{\mathbf{y}} + \frac{3}{4}$. Thus $(\hat{\mathbf{x}}, \hat{\mathbf{y}})$ is a $\frac{3}{4}-$ApproxNE equilibrium for Γ.

3 An Optimization Approach for ApproxNE

In this section we present an efficient algorithm that computes a $0.3393-$ApproxNE for any $[0,1]-$bimatrix game, the best approximation till now. The methodology is based on the formulation of an appropriate function of pairs of mixed strategies reflecting the maximum deviation of the players' payoffs from the best payoff each player could achieve given the strategy chosen by the other. We then seek to minimize such a function using descent procedures. For further details we refer the interested reader to [25].

Optimization formulation. Let $\Gamma = \langle A, B\rangle$ be an $m \times n$ positively normalized bimatrix game. Key to our approach is the definition of the following continuous function mapping $\Delta_m \times \Delta_n$ into $[0,1]$: $f(\mathbf{x},\mathbf{y}) = \max\left\{max(A\mathbf{y}) - \mathbf{x}^T A\mathbf{y}, max(B^T\mathbf{x}) - \mathbf{x}^T B\mathbf{y}\right\}$. It is evident that $f(\mathbf{x},\mathbf{y}) \geq 0$ for all $(\mathbf{x},\mathbf{y}) \in \Delta_m \times \Delta_n$ and that exact Nash equilibria of $\langle A, B\rangle$ correspond to pairs of strategies such that $f(x,\mathbf{y}) = 0$. Furthermore, $\varepsilon-$ApproxNE correspond to strategy pairs that satisfy $f(\mathbf{x},\mathbf{y}) \leq \varepsilon$. This function represents the maximum deviation of the players' payoffs from the best payoff each player could achieve given the strategy chosen by the other. An optimization formulation based on mixed integer programming methods was suggested in [23]. However, no approximation results were obtained there.

Remark: If each of A, B^T have at most δn nonzero entries then the fully mixed pair of strategies $\mathbf{x} = (1/m \ \cdots \ 1/m)^T$, $\mathbf{y} = (1/n \ \cdots \ 1/n)^T$ gives $f(\mathbf{x},\mathbf{y}) \leq \delta$.

The function $f(\mathbf{x},\mathbf{y})$ is not jointly convex with respect to both $\mathbf{x}$ and $\mathbf{y}$. However, it is convex in $\mathbf{x}$ alone, if $\mathbf{y}$ is kept fixed and vice versa. Let us define the two ingredients of the function $f(\mathbf{x},\mathbf{y})$ as follows: $f_A(\mathbf{x},\mathbf{y}) = \max(A\mathbf{y}) - \mathbf{x}^T A\mathbf{y}$ and $f_B(\mathbf{x},\mathbf{y}) = \max(B^T\mathbf{x}) - \mathbf{x}^T B\mathbf{y}$. From any point in $(\mathbf{x},\mathbf{y}) \in \Delta_m \times \Delta_n$ we consider variations of $f(\mathbf{x},\mathbf{y})$ along feasible directions in both players' strategy spaces of the following form:

$$(1-\varepsilon)\begin{bmatrix}\mathbf{x}\\ \mathbf{y}\end{bmatrix} + \varepsilon\begin{bmatrix}\mathbf{x}'\\ \mathbf{y}'\end{bmatrix}$$

where $0 \leq \varepsilon \leq 1$ and $(\mathbf{x}',\mathbf{y}') \in \Delta_m \times \Delta_n$ (the vectors in brackets are $(m+n)-$dimensional column vectors). The variation of the function along such a feasible direction is defined by the following relationship: $Df(\mathbf{x}, \mathbf{y}, \mathbf{x}', \mathbf{y}', \varepsilon) = f(\mathbf{x} + \varepsilon(\mathbf{x}' - \mathbf{x}), \mathbf{y} + \varepsilon(\mathbf{y}' - \mathbf{y})) - f(\mathbf{x}, \mathbf{y})$.

We have been able to derive an explicit formula for $Df(\mathbf{x}, \mathbf{y}, \mathbf{x}', \mathbf{y}', \varepsilon)$ which is a piecewise quadratic function of ε and the number of switches of the linear terms of the function is at most $m + n$. Therefore, for fixed $(\mathbf{x}', \mathbf{y}')$ this function can be minimized with respect to ε in polynomial time. Furthermore, there always exists a positive number, say $\varepsilon^\star$, such that for any $\varepsilon \leq \varepsilon^\star$ the coefficient of the linear term of this function of ε coincides with the gradient, as defined below. The number $\varepsilon^\star$ generally depends on both $(\mathbf{x}, \mathbf{y})$ and $(\mathbf{x}', \mathbf{y}')$. We define the gradient of f at the point $(\mathbf{x}, \mathbf{y})$ along an arbitrary feasible direction specified by another point $(\mathbf{x}', \mathbf{y}')$ as follows: $Df(\mathbf{x}, \mathbf{y}, \mathbf{x}', \mathbf{y}') = \lim_{\varepsilon \to 0} \frac{1}{\varepsilon} Df(\mathbf{x}, \mathbf{y}, \mathbf{x}', \mathbf{y}', \varepsilon)$.

The gradient $Df(\mathbf{x}, \mathbf{y}, \mathbf{x}', \mathbf{y}')$ of f at any point $(\mathbf{x}, \mathbf{y}) \in \Delta_m \times \Delta_n$ along a feasible direction (determined by another point $(\mathbf{x}', \mathbf{y}') \in \Delta_m \times \Delta_n$) provides the rate of decrease (or increase) of the function along that direction. For fixed $(\mathbf{x}, \mathbf{y})$, $Df(\mathbf{x}, \mathbf{y}, \mathbf{x}', \mathbf{y}')$ is a convex polyhedral function in $(\mathbf{x}', \mathbf{y}')$. In fact we have derived the explicit form of $Df(\mathbf{x}, \mathbf{y}, \mathbf{x}', \mathbf{y}')$ as the maximum of two linear forms in the $(\mathbf{x}', \mathbf{y}')$ space (see the derivations below). At any point $(\mathbf{x}, \mathbf{y})$ we wish to minimize the gradient function with respect to $(\mathbf{x}', \mathbf{y}')$ to find the steepest possible descent direction, or to determine that no such descent is possible.

Let us define the following subsets of coordinates: $S_A(\mathbf{y}) = suppmax(A\mathbf{y})$ and $S_B(\mathbf{x}) = suppmax(B^T\mathbf{x})$. By definition, $S_A(\mathbf{y}) \subset [m]$ and $S_B(\mathbf{x}) \subset [n]$. There are three cases:

(a) If $f_A(\mathbf{x}, \mathbf{y}) = f_B(\mathbf{x}, \mathbf{y})$ then
$$Df(\mathbf{x}, \mathbf{y}, \mathbf{x}', \mathbf{y}') = \max(T_1(\mathbf{x}, \mathbf{y}, \mathbf{x}', \mathbf{y}'), T_2(\mathbf{x}, \mathbf{y}, \mathbf{x}', \mathbf{y}')) - f(\mathbf{x}, \mathbf{y})$$
where $m_1(\mathbf{y}') = \max(A\mathbf{y}')$ over the subset $S_A(\mathbf{y})$, $m_2(\mathbf{x}') = \max(B^T\mathbf{x}')$ over the subset $S_B(\mathbf{x})$, $T_1(\mathbf{x}, \mathbf{y}, \mathbf{x}', \mathbf{y}') = m_1(\mathbf{y}') - \mathbf{x}^T A\mathbf{y}' - (\mathbf{x}')^T A\mathbf{y} + \mathbf{x}^T A\mathbf{y}$ and $T_2(\mathbf{x}, \mathbf{y}, \mathbf{x}', \mathbf{y}') = m_2(\mathbf{x}') - \mathbf{x}^T B\mathbf{y}' - (\mathbf{x}')^T B\mathbf{y} + \mathbf{x}^T B\mathbf{y}$.

(b) If $f_A(\mathbf{x}, \mathbf{y}) > f_B(\mathbf{x}, \mathbf{y})$ then $Df(\mathbf{x}, \mathbf{y}, \mathbf{x}', \mathbf{y}') = T_1(\mathbf{x}, \mathbf{y}, \mathbf{x}', \mathbf{y}') - f(\mathbf{x}, \mathbf{y})$.

(c) If $f_A(\mathbf{x}, \mathbf{y}) < f_B(\mathbf{x}, \mathbf{y})$ then $Df(\mathbf{x}, \mathbf{y}, \mathbf{x}', \mathbf{y}') = T_2(\mathbf{x}, \mathbf{y}, \mathbf{x}', \mathbf{y}') - f(\mathbf{x}, \mathbf{y})$.

The problem of finding $Df(\mathbf{x}, \mathbf{y})$ as the minimum over all $(\mathbf{x}', \mathbf{y}') \in \Delta_m \times \Delta_n$ of the function $Df(\mathbf{x}, \mathbf{y}, \mathbf{x}', \mathbf{y}')$, is a linear programming problem.

This problem can be equivalently expressed as the following minmax problem by introducing appropriate dual variables (we derive it for $(\mathbf{x}, \mathbf{y})$ such that $f_A(\mathbf{x}, \mathbf{y}) = f_B(\mathbf{x}, \mathbf{y})$ since this is the most interesting case and the cases where the two terms are different can be reduced to this by solving an LP, as we shall see below) as follows:

$$\text{minimize}_{(\mathbf{x}', \mathbf{y}') \in \Delta_m \times \Delta_n} \left\{ \max_{\mathbf{w}, \mathbf{z}, \rho} \left\{ [\rho \mathbf{w}^T, (1 - \rho)\mathbf{z}^T] G(\mathbf{x}, \mathbf{y}) \begin{bmatrix} \mathbf{y}' \\ \mathbf{x}' \end{bmatrix} \right\} \right\}$$

where:

(a) the maximum is taken with respect to dual variables $\mathbf{w}, \mathbf{z}, \rho$ such that $\mathbf{w} \in \Delta_m$, $supp(\mathbf{w}) \subset S_A(\mathbf{y})$ and $\mathbf{z} \in \Delta_n$, $supp(\mathbf{z}) \subset S_B(\mathbf{x})$ and $\rho \in [0, 1]$.

(b) The minimum is taken with respect to $(\mathbf{x}', \mathbf{y}') \in \Delta_m \times \Delta_n$.
(c) The matrix $G(\mathbf{x}, \mathbf{y})$ is the following $(m+n) \times (m+n)$ matrix:

$$G(\mathbf{x}, \mathbf{y}) = \begin{bmatrix} A - \mathbf{1_m x}^T A & -\mathbf{1_m y}^T A^T + \mathbf{1_m 1_m}^T \mathbf{x}^T A \mathbf{y} \\ -\mathbf{1_n x}^T B + \mathbf{1_n 1_n}^T \mathbf{x}^T B \mathbf{y} & B^T - \mathbf{1_n y}^T B^T \end{bmatrix}$$

The probability vectors $\mathbf{w}$ and $\mathbf{z}$ play the role of price vectors (or penalty vectors) for penalizing deviations from the support sets $S_A(\mathbf{y})$ and $S_B(\mathbf{x})$, and the parameter ρ plays the role of a trade-off parameter between the two parts of the function $f(\mathbf{x}, \mathbf{y})$. In fact, $\mathbf{w}, \mathbf{z}$ and ρ are not independent variables but they are taken all together to represent a single $(m+n)$−dimensional probability vector on the left hand side (the maximizing term) of the linear minmax problem.

Solving the above minmax problem we obtain $\mathbf{w}, \mathbf{z}, \rho, \mathbf{x}'$ and $\mathbf{y}'$ that are all functions of the point $(\mathbf{x}, \mathbf{y})$ and take values in their respective domains of definition. Let us denote by $V(\mathbf{x}, \mathbf{y})$ the value of the solution of the minmax problem at the point $(\mathbf{x}, \mathbf{y})$. The solution of this problem yields a feasible descent direction (as a matter of fact the steepest feasible descent direction) for the function $f(\mathbf{x}, \mathbf{y})$ if $Df(\mathbf{x}, \mathbf{y}) = V(\mathbf{x}, \mathbf{y}) - f(\mathbf{x}, \mathbf{y}) < 0$. Following such a descent direction we can perform an appropriate line search with respect to the parameter ε and find a new point that gives a lower value of the function $f(\mathbf{x}, \mathbf{y})$. Applying repeatedly such a descent procedure we will eventually reach a point where no further reduction is possible. Such a point is a stationary point that satisfies $Df(\mathbf{x}, \mathbf{y}) \geq 0$. In the next subsection we examine the approximation properties of stationary points. In fact, we prove that given any stationary point we can determine pairs of strategies such that at least one of them is a 0.3393-approximate Nash equilibrium.

Approximation properties of stationary points. Let us assume that we have a stationary point $(\mathbf{x}^\star, \mathbf{y}^\star)$ of the function $f(\mathbf{x}, \mathbf{y})$. Then, based on the above analysis and notation, the following relationship should be true: $Df(\mathbf{x}^\star, \mathbf{y}^\star) = V(\mathbf{x}^\star, \mathbf{y}^\star) - f(\mathbf{x}^\star, \mathbf{y}^\star) \geq 0$. Let $(\mathbf{w}^\star, \mathbf{z}^\star) \in \Delta_m \times \Delta_n, \rho^\star \in [0,1]$ be a solution of the linear minmax problem (with matrix $G(\mathbf{x}^\star, \mathbf{y}^\star)$) with respect to the dual variables corresponding to the pair $(\mathbf{x}^\star, \mathbf{y}^\star)$. Such a solution should satisfy the relations $supp(\mathbf{w}^\star) \subset S_A(\mathbf{y}^\star)$ and $supp(\mathbf{z}^\star) \subset S_B(\mathbf{x}^\star)$. Let us define the following quantities: $\lambda = \min_{\mathbf{y}':supp(\mathbf{y}') \subset S_B(\mathbf{x}^\star)} \{(\mathbf{w}^\star - \mathbf{x}^\star)^T A \mathbf{y}'\}$ and $\mu = \min_{\mathbf{x}':supp(\mathbf{x}') \subset S_A(\mathbf{y}^\star)} \{\mathbf{x}'^T B(\mathbf{z}^\star - \mathbf{y}^\star)\}$. From the fact that A, B are positively normalized it follows that both λ and μ are less than or equal to 1. At any point $(\mathbf{x}^\star, \mathbf{y}^\star)$ these quantities basically define the rates of decrease (or increase) of the function f along directions of the form $(1-\varepsilon)(\mathbf{x}^\star, \mathbf{y}^\star) + \epsilon(\mathbf{x}^\star, \mathbf{y}')$ and $(1-\epsilon)(\mathbf{x}^\star, \mathbf{y}^\star) + \epsilon(\mathbf{x}', \mathbf{y}^\star)$, i.e. the rates of decrease that are obtained when we keep one player's strategy fixed and move probability mass of the other player into his own maximum support, towards decreasing his own deviation from the maximum payoff he can achieve.

From the stationarity property of the point $(\mathbf{x}^\star, \mathbf{y}^\star)$ it follows that both λ and μ are nonnegative. Indeed, in the opposite case there would be a descent direction, which contradicts the stationarity condition. Let us define a pair of strategies $(\hat{\mathbf{x}}, \hat{\mathbf{y}}) \in \Delta_m \times \Delta_n$ as follows:

$$(\hat{\mathbf{x}}, \hat{\mathbf{y}}) = \begin{cases} (\mathbf{x}^\star, \mathbf{y}^\star) & \text{if } f(\mathbf{x}^\star, \mathbf{y}^\star) \le f(\tilde{\mathbf{x}}, \tilde{\mathbf{y}}) \\ (\tilde{\mathbf{x}}, \tilde{\mathbf{y}}) & \text{otherwise} \end{cases}$$

where

$$(\tilde{\mathbf{x}}, \tilde{\mathbf{y}}) = \begin{cases} \left(\frac{1}{1+\lambda-\mu}\mathbf{w}^\star + \frac{\lambda-\mu}{1+\lambda-\mu}\mathbf{x}^\star, \mathbf{z}^\star\right) & \text{if } \lambda \ge \mu \\ \left(\mathbf{w}^\star, \frac{1}{1+\mu-\lambda}\mathbf{z}^\star + \frac{\mu-\lambda}{1+\mu-\lambda}\mathbf{y}^\star\right) & \text{if } \lambda < \mu. \end{cases} .$$

We now express the main result of this section in the following theorem:

Theorem 1. *The pair of strategies* $(\hat{\mathbf{x}}, \hat{\mathbf{y}})$ *defined above, is a* 0.3393*-approximate Nash equilibrium.*

Proof. From the definition of $(\hat{\mathbf{x}}, \hat{\mathbf{y}})$ we have:

$$f(\hat{\mathbf{x}}, \hat{\mathbf{y}}) \le \min\{f(\mathbf{x}^\star, \mathbf{y}^\star), f(\tilde{\mathbf{x}}, \tilde{\mathbf{y}})\} \ . \tag{1}$$

Using the stationarity condition for $(\mathbf{x}^\star, \mathbf{y}^\star)$ we obtain that $f(\mathbf{x}^\star, \mathbf{y}^\star) \le V(\mathbf{x}^\star, \mathbf{y}^\star)$. But $V(\mathbf{x}^\star, \mathbf{y}^\star)$ is less than or equal to $\rho^\star E_1 + (1-\rho^\star)E_2$ where $E_1 = (\mathbf{w}^{\star T} A\mathbf{y}' - \mathbf{x}^{\star T} A\mathbf{y}' - \mathbf{x}'^T A\mathbf{y}^\star + \mathbf{x}^{\star T} A\mathbf{y}^\star)$ and $E_2 = (\mathbf{z}^{\star T} B^T\mathbf{x}' - \mathbf{x}^{\star T} B\mathbf{y}' - \mathbf{x}'^T B\mathbf{y}^\star + \mathbf{x}^{\star T} B\mathbf{y}^\star)$ and this holds $\forall(\mathbf{x}', \mathbf{y}') \in \Delta_m \times \Delta_n$.

Setting $\mathbf{x}' = \mathbf{x}^\star$ and $\mathbf{y}' : supp(\mathbf{y}') \subset S_B(\mathbf{x}^\star)$ in the above inequality we get:

$$f(\mathbf{x}^\star, \mathbf{y}^\star) \le \rho^\star \lambda \ . \tag{2}$$

Next, setting $\mathbf{y}' = \mathbf{y}^\star$ and $\mathbf{x}' : supp(\mathbf{x}') \subset S_A(\mathbf{y}^\star)$ in the same inequality, we get:

$$f(\mathbf{x}^\star, \mathbf{y}^\star) \le (1-\rho^\star)\mu \ . \tag{3}$$

Now using the definition of the strategy pair $(\tilde{\mathbf{x}}, \tilde{\mathbf{y}})$ above and exploiting the inequalities $(\mathbf{w}^\star - \mathbf{x}^\star)^T A\mathbf{z}^\star \ge \lambda$, since $supp(\mathbf{z}^\star) \subset S_B(\mathbf{x}^\star)$, and $\mathbf{w}^{*T} B(\mathbf{z}^\star - \mathbf{y}^\star) \ge \mu$, since $supp(\mathbf{w}^\star) \subset S_A(\mathbf{y}^\star)$, we obtain (assume $\lambda \ge \mu$):

$$\begin{aligned} f_A(\tilde{\mathbf{x}}, \tilde{\mathbf{y}}) = \max(A\tilde{\mathbf{y}}) - \tilde{\mathbf{x}}^T A\tilde{\mathbf{y}} &= \max(A\mathbf{z}^\star) - \left(\frac{1}{1+\lambda-\mu}\mathbf{w}^\star + \frac{\lambda-\mu}{1+\lambda-\mu}\mathbf{x}^\star\right)^T A\mathbf{z}^\star \\ &= \max(A\mathbf{z}^\star) - \frac{1}{1+\lambda-\mu}\mathbf{w}^{*T} A\mathbf{z}^\star - \frac{\lambda-\mu}{1+\lambda-\mu}\mathbf{x}^{*T} A\mathbf{z}^\star \\ &\le \max(A\mathbf{z}^\star) - \mathbf{x}^{*T} A\mathbf{z}^\star - \frac{\lambda}{1+\lambda-\mu} \le \frac{1-\mu}{1+\lambda-\mu} \ . \end{aligned}$$

Similarly, it can be shown that $f_B(\tilde{\mathbf{x}}, \tilde{\mathbf{y}}) \le \frac{1-\mu}{1+\lambda-\mu}$. From the above relationships we obtain:

$$f(\tilde{\mathbf{x}}, \tilde{\mathbf{y}}) \le \frac{1-\mu}{1+\lambda-\mu} \qquad \text{for } \lambda \ge \mu \tag{4}$$

A similar inequality can be obtained if $\lambda < \mu$ and we interchange λ and μ. In all cases, combining inequalities (2), (3), (4) and using the definition of $(\hat{\mathbf{x}}, \hat{\mathbf{y}})$ above, we get the following:

$$f(\hat{\mathbf{x}}, \hat{\mathbf{y}}) \le \min\left\{\rho^\star\lambda, (1-\rho^\star)\mu, \frac{1-\min\{\lambda,\mu\}}{1+\max\{\lambda,\mu\}-\min\{\lambda,\mu\}}\right\} . \tag{5}$$

We can prove that the quantity in (5) cannot exceed the number 0.3393 for any $\rho^\star, \lambda, \mu \in [0,1]$, and this concludes the proof of the theorem.

A stationary point of any general Linear Complementarity problem can be approximated arbitrarily close *in polynomial time* via the method of Ye [26]. In [25] we give an alternative approach, directly applicable to our problem; our method is an FPTAS with respect to approximating a stationary point and hence an approximate equilibrium of the stated quality.

Important Remarks

1. Our method can also derive the (simpler) inequality
$$f(\hat{\mathbf{x}}, \hat{\mathbf{y}}) \leq \min\{\rho^\star\lambda, (1-\rho^\star)\mu, 1-\min\{\lambda,\mu\}\} \tag{6}$$
which leads to the bound of [11].
2. For win-lose games, our descent procedure described in [25] leads actually to $\varepsilon = 0.25$.
3. For arbitrary imitation games (R, I) we get $\varepsilon = 4/13$, and for imitation games of the form
$$\left(\begin{bmatrix} \mathbf{0} & C^T \\ R & \mathbf{0} \end{bmatrix}, I\right)$$
we get $\varepsilon = 1/6$, by the same Optimization Approach.
4. Any stationary point is an exact equilibrium in constant sum games.

4 Existence of Uniform SuppNE

In this section we prove existence of SuppNE of arbitrary (constant) precision, with logarithmic (in the numbers of players' actions) support sizes. We also provide (to our knowledge) the *first polynomial time algorithms* for the construction of SuppNE in normalized and win lose bimatrix games, for some constant that is clearly *away* from the trivial bound of 1. For further details we refer the interested reader to [17].

The existence of uniform $\varepsilon-$ApproxNE with small support sizes is already known from [19]. In this section we prove a similar result but for SuppNE, based solely on the *Approximation Lemma* of Althöfer [2]:

Theorem 2 (Approximation Lemma [2]). *Assume any $m \times n$ real matrix $C \in [0,1]^{m\times n}$, any probability vector $\mathbf{p} \in \Delta_m$, and any constant $\varepsilon > 0$. Then, there exists another probability vector $\hat{\mathbf{p}} \in \Delta_m$ with $|supp(\hat{\mathbf{p}})| \leq k \equiv \lceil \log(2n)/(2\varepsilon^2) \rceil$, such that $|\mathbf{p}^T C_j - \hat{\mathbf{p}}^T C_j| \leq \varepsilon$, $\forall j \in [n]$. Moreover, $\hat{\mathbf{p}}$ is a $k-$uniform strategy, ie, $\hat{\mathbf{p}} \in \Delta_r(k)$.*

The following simple observation will be quite useful in our discussion:

Proposition 1. For any real matrix $C \in [0,1]^{m\times n}$ and any $\mathbf{p} \in \Delta_m$, for the empirical distribution $\hat{\mathbf{p}} \in \Delta_m$ produced by the Approximation Lemma it holds that positive probabilities only to rows whose indices belong to $supp(\hat{\mathbf{p}}) \subseteq supp(\mathbf{p})$.

We now demonstrate how the Approximation Lemma, along with the previous observation, guarantees the existence of a uniform profile which is also a $(2\varepsilon)-$SuppNE with support sizes at most $\lceil \log(2n)/(2\varepsilon^2) \rceil$, for any $\varepsilon > 0$:

Theorem 3. *Fix any positive constant $\varepsilon > 0$ and any $[0,1]-$bimatrix game $\langle A, B \rangle$. There is at least one $(k, \ell)-$uniform profile which is also a $(2\varepsilon)-$SuppNE for this game, where $k \leq \lceil \log(2n)/(2\varepsilon^2) \rceil$ and $\ell \leq \lceil \log(2m)/(2\varepsilon^2) \rceil$.*

Proof. Assume any profile $(\mathbf{p}, \mathbf{q}) \in NE(A, B)$, which we of course know to exist for any finite game in normal form [20]. We use the Approximation Lemma to assure the existence of some $k-$uniform strategy $\hat{\mathbf{p}} \in \Delta_m$ with $|supp(\hat{\mathbf{p}})| \leq k \equiv \lceil \log(2n)/(2\varepsilon^2) \rceil$, such that $|\mathbf{p}^T B_j - \hat{\mathbf{p}}^T B_j| \leq \varepsilon$, $\forall j \in [n]$. Similarly, we assume the existence of some $\ell-$uniform strategy $\hat{\mathbf{q}} \in \Delta_n$ with $|supp(\hat{\mathbf{q}})| \leq \ell \equiv \lceil \log(2m)/(2\varepsilon^2) \rceil$, such that $|A^i \mathbf{q} - A^i \hat{\mathbf{q}}| \leq \varepsilon$, $\forall i \in [m]$.

Observe now that, trivially, $\hat{\mathbf{p}}^T B - \mathbf{1}^T \cdot \varepsilon \leq \mathbf{p}^T B \leq \hat{\mathbf{p}}^T B + \mathbf{1}^T \cdot \varepsilon$. Similarly, $A \cdot \hat{\mathbf{q}} - \mathbf{1} \cdot \varepsilon \leq A \cdot \mathbf{q} \leq A \cdot \hat{\mathbf{q}} + \mathbf{1} \cdot \varepsilon$. Therefore (also exploiting the Nash Property of $(\mathbf{p}, \mathbf{q})$ and the fact that $supp(\hat{\mathbf{p}}) \subseteq supp(\mathbf{p})$) we have: $\forall i \in [m]$,

$$\hat{p}_i > 0 \overset{\text{/* Sampling */}}{\Longrightarrow} p_i > 0 \overset{\text{/* Nash Prop. */}}{\Longrightarrow} A^i \mathbf{q} \geq A^r \mathbf{q}, \ \forall r \in [m]$$

$$\overset{\text{/* Thm.2 */}}{\Longrightarrow} A^i \hat{\mathbf{q}} + \varepsilon \geq A^r \hat{\mathbf{q}} - \varepsilon, \ \forall r \in [m] \Longrightarrow A^i \hat{\mathbf{q}} \geq A^r \hat{\mathbf{q}} - 2\varepsilon, \ \forall r \in [m]$$

The argument for the column player is identical. Therefore, we conclude that $(\hat{\mathbf{p}}, \hat{\mathbf{q}})$ is a $(k, \ell)-$uniform profile that is also a $(2\varepsilon)-$SuppNE for $\langle A, B \rangle$.

5 A Linear Programming Approach for Constructing SuppNE

From now on we shall follow another line of attack, which exploits the efficiency of solving linear programs, plus the connection of zero sum games to linear programming. We start with a $0.5-$SuppNE for arbitrary win lose games and we consequently provide a $\left(\frac{\sqrt{11}}{2} - 1\right)-$SuppNE for any normalized game. For further details we refer the interested reader to [16].

5.1 Construction of a $\frac{1}{2}-$SuppNE for Win Lose Games

In this subsection we provide a $0.5-$SuppNE for win lose games, which directly translates to a $0.75-$SuppNE for arbitrary normalized games, if one exploits the nice observation of [10]. First, it can be shown that additive transformations (ie, shiftings) have no effect on well supported equilibria:

Lemma 2. *Fix arbitrary $[0,1]-$bimatrix game $\langle A, B \rangle$ and any real matrices $R, C \in \mathbb{R}^{m \times n}$, such that $\forall i \in [m], R^i = \mathbf{r}^T \in \mathbb{R}^n$ and $\forall j \in [n], C_j = \mathbf{c} \in \mathbb{R}^m$. Then, $\forall 1 > \varepsilon > 0, \forall (\mathbf{x}, \mathbf{y}) \in \Delta_m \times \Delta_n$, if $(\mathbf{x}, \mathbf{y})$ is an $\varepsilon-$SuppNE for $\langle A, B \rangle$ then it is also an $\varepsilon-$SuppNE for $\langle A + R, B + C \rangle$.*

Our next theorem tries to construct the "right" zero sum game that would stand between the two extreme zero sum games $\langle R, -R\rangle$ and $\langle -C, C\rangle$, wrt an arbitrary win lose bimatrix game $\langle R, C\rangle$.

Theorem 4. *For arbitrary win lose bimatrix game $\langle A, B\rangle$, there is a polynomial time constructible profile that is a $0.5-$SuppNE of the game.*

Proof. Consider arbitrary win lose game $\langle A, B\rangle \in \{(0,0),(0,1),(1,0)\}^{m\times n}$. We have excluded the $(1,1)-$elements because, as we already know, these are trivial PNE of the game. We transform the bimatrix (A, B) into a bimatrix (R, C) by subtracting $1/2$ from all the possible payoffs in the bimatrix: $R = A - \frac{1}{2}E$ and $C = B - \frac{1}{2}E$, where $E = \mathbf{1}\cdot\mathbf{1}^T$. We already know that this transformation does not affect the quality of a SuppNE (cf. Lemma 2).

We observe that the row player would never accept a payoff less than the one achieved by the (exact) Nash equilibrium $(\hat{\mathbf{x}}, \hat{\mathbf{y}})$ of the (zero sum) game $\langle R, -R\rangle$. This is because strategy $\hat{\mathbf{x}}$ is a maximin strategy for the row player, and thus the row player can achieve a payoff of at least $\hat{V}_I \equiv \hat{\mathbf{x}}^T R\hat{\mathbf{y}}$ by adopting $\hat{\mathbf{x}}$, for any possible column that the column player chooses **wpp**. Similarly, the column player would never accept a profile $(\mathbf{x}, \mathbf{y})$ with payoff for her less than $\tilde{V}_{II} \equiv \tilde{\mathbf{x}}^T C\tilde{\mathbf{y}}$, where $(\tilde{\mathbf{x}}, \tilde{\mathbf{y}})$ is the (exact) NE of the zero sum game $\langle -C, C\rangle$. So, we already know that any $0-$SuppNE for $\langle R, C\rangle$ should assure payoffs at least $\hat{V}_I$ and at least $\tilde{V}_{II}$ for the row and the column player respectively. Clearly, $(\hat{\mathbf{x}}, \tilde{\mathbf{y}})$ is a $\max\left\{\frac{1}{2} - \hat{V}_I, \frac{1}{2} - \tilde{V}_{II}\right\}-$ApproxNE of the game, but we cannot assure that it is a nontrivial SuppNE of $\langle R, C\rangle$. Nevertheless, inspired by this observation, we attempt to set up the right zero sum game that is somehow connected to $\langle R, C\rangle$, whose (exact) NE would provide a good SuppNE for $\langle R, C\rangle$. Therefore, we consider an arbitrary zero sum game $\langle D, -D\rangle$, for which it holds that $D = R + X \Leftrightarrow X = D - R$ and $-D = C + Y \Leftrightarrow Y = -(D + C)$ for some $m \times n$ bimatrix (X, Y). Let again $(\bar{\mathbf{x}}, \bar{\mathbf{y}}) \in NE(D, -D)$. Then clearly, by the definition of NE we have:

$$(\bar{\mathbf{x}}, \bar{\mathbf{y}}) \in NE(D, -D) = NE(R + X, C + Y) \Leftrightarrow$$

$$\begin{cases} \forall i, r \in [m],\ \bar{x}_i > 0 \Rightarrow (R+X)^i\bar{\mathbf{y}} \geq (R+X)^r\bar{\mathbf{y}} \Rightarrow R^i\bar{\mathbf{y}} \geq R^r\bar{\mathbf{y}} - [X^i - X^r]\bar{\mathbf{y}} \\ \forall j, s \in [n],\ \bar{y}_j > 0 \Rightarrow (C+Y)_j^T\bar{\mathbf{x}} \geq (C+Y)_s^T\bar{\mathbf{x}} \Rightarrow C_j^T\bar{\mathbf{x}} \geq C_s^T\bar{\mathbf{x}} - [Y_j - Y_s]^T\bar{\mathbf{x}} \end{cases}$$

Since $D = R + X = -(-D) = -(C + Y) \Leftrightarrow -Z \equiv R + C = -(X + Y)$, we can simply set $X = Y = \frac{1}{2}Z$, and then we conclude that:

$$(\bar{\mathbf{x}}, \bar{\mathbf{y}}) \in NE(D, -D) \Leftrightarrow \begin{cases} \forall i, r \in [m],\ \bar{x}_i > 0 \Rightarrow R^i\bar{\mathbf{y}} \geq R^r\bar{\mathbf{y}} - \frac{1}{2}\cdot[Z^i - Z^r]\bar{\mathbf{y}} \\ \forall j, s \in [n],\ \bar{y}_j > 0 \Rightarrow C_j^T\bar{\mathbf{x}} \geq C_s^T\bar{\mathbf{x}} - \frac{1}{2}\cdot[Z_j - Z_s]^T\bar{\mathbf{x}} \end{cases}$$

Observe now that, since $R, C \in \left\{\left(-\frac{1}{2}, -\frac{1}{2}\right), \left(-\frac{1}{2}, \frac{1}{2}\right), \left(\frac{1}{2}, -\frac{1}{2}\right)\right\}^{m\times n}$, any row of $Z = -(R + C)$ is a vector in $\{0,1\}^n$, and any column of Z is a vector in $\{0,1\}^m$. But it holds that $\forall\hat{\mathbf{z}}, \tilde{\mathbf{z}} \in \{0,1\}^k, \forall\mathbf{w} \in \Delta_k, (\hat{\mathbf{z}} - \tilde{\mathbf{z}})^T\mathbf{w} \leq \mathbf{1}^T\mathbf{w} = 1$. So we conclude that $\forall i, r \in [m], \forall\mathbf{y} \in \Delta_n, [Z^i - Z^r]\mathbf{y} \leq \mathbf{1}^T\mathbf{y} = 1$, and $\forall j, s \in [n], \forall\mathbf{x} \in \Delta_m, [Z_j - Z_s]^T\mathbf{x} \leq \mathbf{1}^T\mathbf{x} = 1$. Therefore we conclude that:

$$(\bar{\mathbf{x}}, \bar{\mathbf{y}}) \in NE\left(R + \frac{1}{2}Z, C + \frac{1}{2}Z\right) \Rightarrow \begin{cases} \forall i, r \in [m], \bar{x}_i > 0 \Rightarrow R^i\bar{\mathbf{y}} \geq R^r\bar{\mathbf{y}} - \frac{1}{2} \\ \forall j, s \in [n], \bar{y}_j > 0 \Rightarrow C_j^T\bar{\mathbf{x}} \geq C_s^T\bar{\mathbf{x}} - \frac{1}{2} \end{cases}$$
$$\Rightarrow (\bar{\mathbf{x}}, \bar{\mathbf{y}}) \in \tfrac{1}{2}\text{–SuppNE}(R, C)\,.$$

5.2 SuppNE for $[0,1]-$Bimatrix Games

Given our result on win lose games, applying a lemma of Daskalakis et al. [10, Lemma 4.6] for constructing $\frac{1+\varepsilon}{2}-$SuppNE of a $[0,1]-$bimatrix game $\langle A, B\rangle$ by any $\varepsilon-$SuppNE of a properly chosen win lose game of the same size, we could directly generalize our result on SuppNE for win lose games to SuppNE for any $[0,1]-$bimatrix game:

Corollary 1. For any $[0,1]-$bimatrix game $\langle A, B\rangle$, there is a 0.75–SuppNE that can be computed in polynomial time.

The question is whether we can do better than that. Indeed we can, but we first have to modify the rationale of the proof of Theorem 5. This way we shall get a weaker SuppNE for win lose games, which we can nevertheless extend to $[0,1]-$bimatrix games with only a small deterioration. The next theorem demonstrates the parameterized method for win lose games, which assures a $\phi-$SuppNE, where ϕ is the golden ratio.

Theorem 5. *For any win lose bimatrix game, there is a polynomial–time constructible $\varepsilon(\delta)-$SuppNE for any $0 < \delta < 1$, where $\varepsilon(\delta) \leq \max\left\{\delta, \frac{1-\delta}{\delta}\right\}$.*

Proof. Again we try to find a zero sum game that lies somehow between $\langle R, -R\rangle$ and $\langle -C, C\rangle$ and indeed provides a guaranteed SuppNE for $\langle R, C\rangle$. Therefore, we fix a constant $\delta \in (0,1)$ (to be determined later). Consequently, we consider the matrix $Z = -(R+C) \in \{0,1\}^{m\times n}$, indicating (with 1s) the elements of the bimatrix (R, C) that are $(-\frac{1}{2}, -\frac{1}{2})-$elements (ie, the $(0,0)-$elements of initial bimatrix (A, B)). All the other elements are 0s. We now consider the zero sum bimatrix game $\langle R+\delta Z, -(R+\delta Z)\rangle$, which is solvable in polynomial time (by use of linear programming). We denote with $(\bar{\mathbf{x}}, \bar{\mathbf{y}})$ the (exact) NE of this game. By the definition of NE, the row and the column player assign positive probability mass only to maximizing elements of the vectors $(R + \delta Z)\bar{\mathbf{y}}$ and $(-R - \delta Z)^T\bar{\mathbf{x}}$ respectively. That is:

$$(\bar{\mathbf{x}}, \bar{\mathbf{y}}) \in NE(R + \delta Z, -(R + \delta Z))$$
$$\Leftrightarrow \begin{cases} \forall i, r \in [m], \bar{x}_i > 0 \Rightarrow (R + \delta Z)^i\bar{\mathbf{y}} \geq (R + \delta Z)^r\bar{\mathbf{y}} \\ \forall j, s \in [n], \bar{y}_j > 0 \Rightarrow (-R - \delta Z)_j^T\bar{\mathbf{x}} \geq (-R - \delta Z)_s^T\bar{\mathbf{x}} \end{cases}$$
$$\Leftrightarrow \begin{cases} \forall i, r \in [m], \bar{x}_i > 0 \Rightarrow R^i\bar{\mathbf{y}} + \delta Z^i\bar{\mathbf{y}} \geq R^r\bar{\mathbf{y}} + \delta Z^r\bar{\mathbf{y}} \\ \forall j, s \in [n], \bar{y}_j > 0 \Rightarrow (1-\delta)R_j^T\bar{\mathbf{x}} + \delta(R + Z)_j^T\bar{\mathbf{x}} \leq (1-\delta)R_s^T + \delta(R + Z)_s^T\bar{\mathbf{x}} \end{cases}$$
$$\Leftrightarrow \begin{cases} \forall i, r \in [m], \bar{x}_i > 0 \Rightarrow R^i\bar{\mathbf{y}} + \delta Z^i\bar{\mathbf{y}} \geq R^r\bar{\mathbf{y}} + \delta Z^r\bar{\mathbf{y}} \\ \forall j, s \in [n], \bar{y}_j > 0 \Rightarrow (1-\delta)R_j^T\bar{\mathbf{x}} - \delta C_j^T\bar{\mathbf{x}} \leq (1-\delta)R_s^T\bar{\mathbf{x}} - \delta C_s^T\bar{\mathbf{x}} \end{cases}$$
$$\Leftrightarrow \begin{cases} \forall i, r \in [m], \bar{x}_i > 0 \Rightarrow R^i\bar{\mathbf{y}} \geq R^r\bar{\mathbf{y}} - \delta[Z^i - Z^r]\bar{\mathbf{y}} \geq R^r\bar{\mathbf{y}} - \varepsilon(\delta) \\ \forall j, s \in [n], \bar{y}_j > 0 \Rightarrow C_j^T\bar{\mathbf{x}} \geq C_s^T\bar{\mathbf{x}} - \frac{1-\delta}{\delta}\cdot[R_s^T - R_j^T]\bar{\mathbf{x}} \geq C_s^T\bar{\mathbf{x}} - \varepsilon(\delta) \end{cases}$$

where,

$$\varepsilon(\delta) \equiv \max_{i,r\in[m],j,s\in[n],\mathbf{x}\in\Delta_m,\mathbf{y}\in\Delta_n} \left\{ \delta \cdot \left[Z^i - Z^r\right]\mathbf{y}, \frac{1-\delta}{\delta} \cdot \left[R_s^T - R_j^T\right]\mathbf{x} \right\} \quad (7)$$

Obviously, for any $\delta \in (0,1]$ it holds that $(\bar{\mathbf{x}}, \bar{\mathbf{y}})$ is an $\varepsilon(\delta)-$SuppNE for $\langle R, C\rangle$.

We already proved that $\forall i, r \in [m], \forall \mathbf{y} \in \Delta_n, [Z^i - Z^r]\mathbf{y} \leq \mathbf{1}^T\mathbf{y} = 1$. Similarly, every column of R is a vector from $\left\{-\frac{1}{2}, \frac{1}{2}\right\}^m$. But the difference $\hat{\mathbf{u}} - \tilde{\mathbf{u}}$ of any vectors $\hat{\mathbf{u}}, \tilde{\mathbf{u}} \in \left\{-\frac{1}{2}, \frac{1}{2}\right\}^m$ is a vector from $\{-1, 0, 1\}^m$. Therefore, $\forall \hat{\mathbf{u}}, \tilde{\mathbf{u}} \in \left\{-\frac{1}{2}, \frac{1}{2}\right\}^m, \forall \mathbf{x} \in \Delta_m, (\hat{\mathbf{u}} - \tilde{\mathbf{u}})^T\mathbf{x} \leq \mathbf{1}^T\mathbf{x} = 1$. Therefore, we conclude that $\forall \delta \in (0,1], \varepsilon(\delta) \leq \max\left\{\delta, \frac{1-\delta}{\delta}\right\}$.

Remark: If we simply set $\delta = \frac{1-\delta}{\delta} = \frac{\sqrt{5}-1}{2}$, we conclude that $(\bar{\mathbf{x}}, \bar{\mathbf{y}})$ is a $\left(\frac{\sqrt{5}-1}{2}\right)-$SuppNE for $\langle R, C\rangle$, and therefore for $\langle A, B\rangle$. Of course, this golden ratio SuppNE is inferior to the previously constructed $0.5-$SuppNE for win lose games. But all we need is actually the bound of equation (7).

Now we can extend our technique for win lose games to a technique for arbitrary $[0,1]-$bimatrix games:

Theorem 6. *For any* $[0,1]-$*bimatrix game, a* $\left(\frac{\sqrt{11}}{2} - 1\right)-$*SuppNE is constructible in polynomial time.*

See [16] for a proof of the above theorem.

Remark: It is worth mentioning that if we had applied our technique to the first algorithm for computing $0.5-$SuppNE in win lose games, then this would lead to a $\frac{2}{3}-$SuppNE for the $[0,1]-$bimatrix game $\langle A, B\rangle$ which is strictly worse than our current result. Ie, equidistribution (between the two players) of the divergence from the zero sum game is not the right choice for the general algorithm.

6 Open Problems

In this work we have studied the tractability of ApproxNE and SuppNE, as well as the existence of SuppNE with small supports, both in normalized and win lose bimatrix games.

The important questions whether there exist polynomial time approximation schemes (PTAS) for the construction of $\varepsilon-$SuppNE or $\varepsilon-$ApproxNE, for any *positive constant* $1 > \varepsilon > 0$, still remain open. We only know that the construction of fully polynomial time approximation schemes (FPTAS) for the weaker notion of ApproxNE is as hard as the exact problem.

It would also be interesting to find polynomial time algorithms for constructing $\varepsilon-$SuppNE, for some constant $0 < \varepsilon < 0.5$ for win lose games and $0 < \varepsilon < 0.658$ for the general case. Similarly, for the case of other notion of approximate equilibria (ApproxNE), we do not currently know how to construct $\varepsilon-$ApproxNE for some precision $0 < \varepsilon < \frac{1}{3}$, or whether there is a matching lower bound on the approximability of ApproxNE.

References

1. Abbott, T., Kane, D., Valiant, P.: On the complexity of two-player win-lose games. In: Proc. of 46th IEEE Symp. on Found. of Comp. Sci. (FOCS 2005), pp. 113–122 (2005)
2. Althöfer, I.: On sparse approximations to randomized strategies and convex combinations. Linear Algebra and Applications 199, 339–355 (1994)
3. Bosse, H., Byrka, J., Markakis, E.: New Algorithms for Approximate Nash Equilibria in Bimatrix Games. In: Deng, X., Graham, F.C. (eds.) WINE 2007. LNCS, vol. 4858, Springer, Heidelberg (2007)
4. Bárány, I., Vempala, S., Vetta, A.: Nash equilibria in random games. In: Proc. of 46th IEEE Symp. on Found. of Comp. Sci. (FOCS 2005), pp. 123–131. IEEE Computer Society Press, Los Alamitos (2005)
5. Chen, X., Deng, X.: Settling the complexity of 2-player nash equilibrium. In: Proc. of 47th IEEE Symp on Found of Comp. Sci. (FOCS 2006), pp. 261–272. IEEE Computer Society Press, Los Alamitos (2006)
6. Chen, X., Deng, X., Teng, S.: Computing nash equilibria: Approximation and smoothed complexity. In: Proc. of 47th IEEE Symp on Found of Comp Sci (FOCS 2006), pp. 603–612. IEEE Computer Society Press, Los Alamitos (2006)
7. Chen, X., Deng, X., Teng, S.: Sparse Games Are Hard. In: Spirakis, P.G., Mavronicolas, M., Kontogiannis, S.C. (eds.) WINE 2006. LNCS, vol. 4286, pp. 262–273. Springer, Heidelberg (2006)
8. Conitzer, V., Sandholm, T.: Complexity results about nash equilibria. In: In Proc. of 18th Int Joint Conf on Art. Intel (IJCAI 2003), pp. 765–771. Morgan Kaufmann, San Francisco (2003)
9. Daskalakis, C., Goldberg, P., Papadimitriou, C.: The complexity of computing a nash equilibrium. In: Proc. of 38th ACM Symp on Th of Comp (STOC 2006), pp. 71–78 (2006)
10. Daskalakis, C., Mehta, A., Papadimitriou, C.: A Note on Approximate Nash Equilibria. In: Spirakis, P.G., Mavronicolas, M., Kontogiannis, S.C. (eds.) WINE 2006. LNCS, vol. 4286, pp. 297–306. Springer, Heidelberg (2006)
11. Daskalakis, C., Mehta, A., Papadimitriou, C.: Progress in approximate nash equilibrium. In: Proc. of 8th ACM Conf. on El. Comm. (EC 2007), pp. 355–358 (2007)
12. Daskalakis, C., Papadimitriou, C.: Three player games are hard. Technical Report TR05-139, Electr. Coll. on Comp. Compl. (ECCC) (2005)
13. Gilboa, I., Zemel, E.: Nash and correlated equilibria: Some complexity considerations. Games & Econ. Behavior 1, 80–93 (1989)
14. Goldberg, P., Papadimitriou, C.: Reducibility among equilibrium problems. In: Proc. of 38th ACM Symp. on Th. of Comp. (STOC 2006), pp. 61–70 (2006)
15. Kontogiannis, S.C., Panagopoulou, P., Spirakis, P.G.: Polynomial Algorithms for Approximating Nash Equilibria of Bimatrix Games. In: Spirakis, P.G., Mavronicolas, M., Kontogiannis, S.C. (eds.) WINE 2006. LNCS, vol. 4286, pp. 286–296. Springer, Heidelberg (2006)
16. Spirakis, P.G., Kontogiannis, S.C.: Efficient Algorithms for Constant Well Supported Approximate Equilibria in Bimatrix Games. In: Arge, L., Cachin, C., Jurdziński, T., Tarlecki, A. (eds.) ICALP 2007. LNCS, vol. 4596, pp. 595–606. Springer, Heidelberg (2007)
17. Kontogiannis, S.C., Spirakis, P.G.: Well Supported Approximate Equilibria in Bimatrix Games: A Graph Theoretic Approach. In: Kučera, L., Kučera, A. (eds.) MFCS 2007. LNCS, vol. 4708, pp. 596–608. Springer, Heidelberg (2007)

18. Lemke, C., Howson, J.T.: Equilibrium points of bimatrix games. Journal of the Society for Industrial and Applied Mathematics 12:413423, DELIS-TR-0487 (1964)
19. Lipton, R., Markakis, E., Mehta, A.: Playing large games using simple strategies. In: Proc. of 4th ACM Conf. on El. Comm. (EC 2003), pp. 36–41 (2003)
20. Nash, J.: Noncooperative games. Annals of Mathematics 54, 289–295 (1951)
21. Papadimitriou, C.: On the complexity of the parity argument and other inefficient proofs of existence. J. Comp. Sys. Sci. 48, 498–532 (1994)
22. Papadimitriou, C.: Algorithms, games and the internet. In: Proc. of 33rd ACM Symp. on Th. of Comp. (STOC 2001), pp. 749–753 (2001)
23. Sandholm, T., Gilpin, A., Conitzer, V.: Mixed-integer programming methods for finding Nash equilibria. In: Proc. of the 20th Nat. Conf. on Art. Intel (AAAI 2005), pp. 495–501 (2005)
24. Savani, R., von Stengel, B.: Hard-to-solve bimatrix games. Econometrica 74(2), 397–429 (2006)
25. Tsaknakis, H., Spirakis, P.: An optimization approach for approximate Nash equilibria. In: Deng, X., Graham, F.C. (eds.) WINE 2007. LNCS, vol. 4858, Springer, Heidelberg (2007)
26. Ye, Y.: A fully polynomial time approximation algorithm for computing a stationary point of the general linear complementarity problem. Mathematics of Operations Research 18, 334–345 (1993)

The Influence of Link Restrictions on (Random) Selfish Routing

Martin Hoefer[1,*] and Alexander Souza[2]

[1] Department of Computer Science, RWTH Aachen University, Germany
mhoefer@informatik.rwth-aachen.de
[2] Department of Computer Science, University of Freiburg, Germany
souza@informatik.uni-freiburg.de

Abstract. In this paper we consider the influence of link restrictions on the price of anarchy for several social cost functions in the following model of selfish routing. Each of n players in a network game seeks to send a message with a certain length by choosing one of m parallel links. Each player is restricted to transmit over a certain subset of links and desires to minimize his own transmission-time (latency). We study Nash equilibria of the game, in which no player can decrease his latency by unilaterally changing his link. Our analysis of this game captures two important aspects of network traffic: the dependency of the overall network performance on the total traffic t and fluctuations in the length of the respective message-lengths. For the latter we use a probabilistic model in which message lengths are random variables.

We evaluate the (expected) price of anarchy of the game for two social cost functions. For total latency cost, we show the tight result that the price of anarchy is essentially $\Theta\left(n\sqrt{m}/t\right)$. Hence, even for congested networks, when the traffic is linear in the number of players, Nash equilibria approximate the social optimum only by a factor of $\Theta\left(\sqrt{m}\right)$. This efficiency loss is caused by link restrictions and remains stable even under message fluctuations, which contrasts the unrestricted case where Nash equilibria achieve a constant factor approximation. For maximum latency the price of anarchy is at most $1+m^2/t$. In this case Nash equilibria can be (almost) optimal solutions for congested networks depending on the values for m and t. In addition, our analyses yield average-case analyses of a polynomial time algorithm for computing Nash equilibria in this model.

1 Introduction

Recently, there has been a lot of interest in considering network users as non-cooperative selfish players that unilaterally seek to optimize their experienced network latency. This serves to quantify the deterioration of the total system performance, and it builds a foundation to derive protocols taking possible selfish

* Supported by DFG Research Training Group "AlgoSyn" and by DFG Research Cluster "UMIC" within Excellence Initiative of the German government.

B. Monien and U.-P. Schroeder (Eds.): SAGT 2008, LNCS 4997, pp. 22–32, 2008.

defection into account. In their seminal work [13], Koutsoupias and Papadimitriou initiated this research direction by introducing the KP-model for selfish routing. Each of n players seeks to send a message with respective length t_j across a network consisting of m parallel capacitated links. The cost of a player j, called his *latency* ℓ_j, is the total length of messages on his chosen link i, scaled with the respective capacity. The latency corresponds to the duration of the transmission when the channel is shared by a set of players. Now each player strives to optimize his personally experienced latency by changing the chosen link for his message. He is satisfied with his link choice (also referred to as his strategy) if by unilaterally changing his link he cannot decrease his cost. If all players are satisfied, then the system is said to be in a stable state, called a *Nash equilibrium*.

In order to relate selfishly obtained stable solutions with those of an (imaginary) central authority, it is necessary to distinguish between the cost of the individual players and the *social cost* of the whole system caused by the community of all players. Naturally, depending on the choice of a social cost function selfish behavior is not always optimal. Consequently, the question arises how much worse a Nash equilibrium can be than the optimum. Koutsoupias and Papadimitriou [13] introduced the *price of anarchy* which is the ratio of the social cost of the worst Nash equilibrium and the optimum social cost, and proved initial bounds for special cases of the KP-model with maximum latency cost. Subsequently, generalized models with different latency functions, social cost functions and network topologies were considered for instance in [6,5,8,9,1,4]. For a recent survey on results related to network congestion games see [12].

In this paper we treat a generalization of the KP-model in which the link set a player i can choose from is a restricted subset of all links available. This model was treated before with maximum latency and polynomial load social cost. For maximum latency computing the social optimum solution is a special case of generalized assignment problems and the single source unsplittable flow problem [14,11]. Gairing et al. [7] gave a $(2 - 1/t_{\max})$-approximation algorithm for optimizing the social cost. They also showed how to compute in polynomial time a Nash equilibrium from any given starting solution without deteriorating the social cost, so the price of stability is 1. In [2] the price of anarchy for maximum latency was shown to be $O(\log m/\log\log m)$ and to decrease with the ratio $r = \text{cost}(s^*)/t_{\max}$. In particular, for $r = \Omega(\log m/\epsilon^2)$ it is $1+\epsilon$. Quadratic load social cost was recently studied in [3,18]. Suri et al. [18] show that the price of anarchy for identical machines is at least 2.012067, and Caragiannis et al. [3] provide a matching upper bound.

In contrast to previous work, we capture two important aspects of network traffic: the dependency of the overall network performance on the *total traffic* $t = \sum_j t_j$ and *fluctuations* in the length of the respective message-lengths t_j. In our model of fluctuation, the message-lengths are random variables T_j and the quality of equilibria is judged with the expected value of the price of anarchy, respectively stability. This idea of an *expected price of anarchy* was recently introduced by Mavronicolas et al. [15] (under the name diffuse price of anarchy) in the context the unrestricted KP-model with a cost-sharing mechanism. We considered the expected price of anarchy in [10], in which we were mostly concerned with the pure Nash equilibria of the unrestricted KP-model and the total latency $\sum_j \ell_j$ of all players. One main conclusion therein was that for highly congested networks, i.e., t being linear in n, Nash equilibria approximate the optimum solution within a constant factor.

In this paper, we characterize the loss of performance of Nash equilibria due to the presence of link restrictions. We show that the prices of stability and anarchy are essentially $\Theta(n\sqrt{m}/t)$ for total latency. Perhaps surprisingly this behaviour remains stable even in the stochastic counterpart. This means that – in contrast to other related average-case analyses, e.g., [17,16] – the averaging effects of fluctuations do not necessarily yield improved expected prices of the game.

Our results foster an interesting new research direction connecting game theory and average-case analysis in the context of traffic allocation and scheduling. We consider efficiency measures including randomness by presenting *tight* bounds on the (expected) price of anarchy. By capturing a notion of fluctuation, we bring a network game closer to practice. Secondly, our analysis yields an average-case analysis on the expected performance of a generic approximation algorithm for various scheduling problems. Most notably, our analysis holds under weak probabilistic assumptions. This extends previous work, e.g. [10,17,16] on average-case analyses of scheduling on identical *unrestricted* machines.

1.1 Model and Notation

We formulate the KP-model with scheduling terminology, where each link corresponds to one of m identical parallel machines. There are n players in the game, and each player seeks to assign a task to one of the machines. Each task j has a certain finite *length* t_j. We scale all task lengths by a positive factor without changing the approximation factors, i.e., we assume normalization $t_j \in [0,1]$, and w.l.o.g. $n \geq m$ throughout. With each player j we associate a set $A_j \neq \emptyset$ of *allowed machines*, and each player is restricted to assignment only to machines in the set A_j. The strategy of a player is the choice of one of the allowed machines. A *schedule* is a function s that maps each task j to a machine i obeying the restrictions A_j. The total length on machine i is its *load* $w_i = \sum_{k \text{ on } i} t_k$. Each machine i executes its assigned tasks in parallel and hence the finishing-time of a task j is proportional to the total length on the chosen link i, i.e., its *latency* is $\ell_j = \sum_{k \text{ on } i} t_k = w_i$. The disutility of each player is the latency of its task, i.e., the selfish incentive of every player is to minimize the individual latency.

A schedule s is said to be in a (pure) Nash equilibrium if no player can decrease his latency by unilaterally changing the machine his task is processed on. More formally, the schedule s has the property that for each task j

$$w_i + t_j \geq w_{s(j)} \quad \text{holds for every } i \in A_j. \tag{1}$$

This game is known to always admit pure Nash equilibria, see e.g. [7].

Schedules are valued with a certain *(social) cost* function cost : $\Sigma \to \mathbb{R}_+$, where Σ denotes the set of all schedules. A Nash equilibrium is simply a schedule that satisfies the stability criterion (1), whereas an *optimum* schedule minimizes the cost function over all possible schedules. Hence, it is natural to consider how much worse Nash equilibria can be compared to the optimum. The *price of anarchy* [13] relates the Nash equilibrium with highest social cost to the optimum, i.e, it compares the "worst" Nash equilibrium with the best possible solution. In contrast, the *price of stability* relates the Nash equilibrium with lowest social cost to the optimum, i.e, it compares the "best" Nash equilibrium with the best possible solution.

1.2 Our Concepts and Results

The main matter of this paper is to investigate the influence of link restrictions on Nash equilibria. We consider two different social cost functions: total latency $\sum_{j\in J} \ell_j$ and maximum latency $\max_{j\in J} \ell_j$.

Our focus lies on two important aspects of network traffic: the influence of the total traffic upon the quality of Nash equilibria and the question if fluctuations in the task-lengths have an positive averaging effect. In terms of fluctuations, we consider the following natural stochastic model. Throughout, upper-case letters denote random variables, lower-case letters their realisations, respectively constants.

Let the task-length T_j of a task j be a random variable over a bounded interval with expectation $\mathbb{E}[T_j]$. As before, a schedule is a *Nash equilibrium* if (1) holds, i.e., if the concrete *realisations* t_j of the random variables T_j satisfy the stability criterion. Consequently, the set of schedules that are Nash equilibria is a random variable itself. We define the *expected price of anarchy*

$$\mathrm{EPoA}(\Sigma) = \mathbb{E}\left[\max\left\{\frac{\mathrm{cost}(S)}{\mathrm{cost}(S^*)} \;:\; S \in \Sigma \text{ is a Nash equilibrium}\right\}\right].$$

The *expected price of stability* is obtained by replacing the maximum by the minimum in straightforward manner. Notice that each expected value is taken with respect to the random task-lengths T_j. This means that the expectation is accumulated by evaluating the prices for each outcome t_j of the random variables T_j and weighting with the respective probability.

In Section 2 we consider total latency $\sum_j \ell_j$, for which [10] shows that prices of stability and anarchy are $\Theta(n/t)$, i.e., they are both decreasing with t. Theorem 1, respectively Theorem 2, provide *tight* lower and upper bounds for the case with link restrictions: we show that the prices of anarchy, respectively stability are $\Theta(n\sqrt{m}/t)$. The question arises whether fluctuations in task-lengths help reducing this bound. Unfortunately, we show that the bounds remains stable. The expected prices of anarchy and stability are $\Theta(n\sqrt{m}/\mathbb{E}[T])$ under relatively weak assumptions on the probability distributions of the T_j.

For maximum latency $\max_j \ell_j$, it is already known (see [2]) that the price of stability is 1 and that the price of anarchy follows a tradeoff depending on the largest task length and the cost of the social optimum. We show a similar tradeoff in Theorem 4: the price of anarchy is at most $1+m^2/t$ and even in expectation it is at most $1+m^2/\mathbb{E}[T]$. Hence, Nash equilibria are almost optimal for congested networks even with link restrictions.

Moreover, there is an algorithm due to Gairing et al. [7] which computes pure Nash equilibria for our game in polynomial time. Our analyses of the expected prices of anarchy of these social cost functions provide average-case analyses of that algorithm, see, e.g. Theorem 3.

2 Total Latency Cost

In this section, we consider the social cost function *total latency* $\text{cost}(s) = \sum_j \ell_j$. Throughout, let p_i denote the number of *players* that use machine i, let $w_i = \sum_{j \text{ on } i} t_j$ be the *load* of machine i. Observe that we have the equality $\text{cost}(s) = \sum_j \ell_j = \sum_i p_i w_i$ for every feasible solution s. It will be convenient to denote $t = \sum_j t_j$ and $n = \sum_i p_i$ throughout. Recall that we normalize to $t_j \in [0,1]$. Before considering the general case, we restrict ourselves to games with so-called clustered restrictions.

Clustered Restrictions. We speak of *clustered restrictions* in the game if the set A of allowed machines can is characterized as follows. Let J denote the set of tasks and let $J_1, \ldots, J_k$ be a disjoint partition of the tasks in non-empty sets. Let M denote the set of machines and let $M_1, \ldots, M_k$ be a disjoint partition of the machines in non-empty sets. Let $j \in J_i$ for some $i \in \{1, \ldots, k\}$ then, the set of allowed machines for task j is $A_j = M_i$. This means that j is allowed to use exactly those machines in the class M_i, but no others.

Theorem 1. *For clustered restrictions $A = \{A_1, A_2, \ldots, A_n\}$ with task-partition $J_1, \ldots, J_k$ and machine-partition $M_1, \ldots, M_k$ we have:*

(1) Define $\varepsilon_1 = \varepsilon_1(n, m, t) = 2nm/t^2$. The prices of stability and anarchy of the game are

$$\frac{n\sqrt{m}}{4t}(1 - o(1)) \le \text{PoS}(\Sigma) \le \text{PoA}(\Sigma) \le \frac{n\sqrt{m}}{t} + \varepsilon_1. \tag{2}$$

The lower bound holds for $t \ge m$; the upper for $t \ge 2$.

(2) Define $\varepsilon_2 = \varepsilon_2(n, m, \mathbb{E}[T]) = 2nm/\mathbb{E}[T]^2$. Suppose $T = \sum_j T_j$ with $\mathbb{E}[T] = \omega\left(\sqrt{n \log n}\right)$, where the T_j are independent. Then the expected prices of stabilty and anarchy of the game are

$$\frac{n\sqrt{m}}{4\mathbb{E}[T]}(1 - o(1)) \le \text{EPoS}(\Sigma) \le \text{EPoA}(\Sigma) \le \left(\frac{n\sqrt{m}}{\mathbb{E}[T]} + \varepsilon_2\right)(1 + o(1)). \tag{3}$$

The lower bound holds with the additional assumption that $\mathbb{E}[T] \ge m$.

For the proof of an upper bound notice that the clustered restrictions divide the problem into a set of unrestricted problems corresponding to the aforementioned partition into task sets $J_1, \ldots, J_k$ and the machine sets $M_1, \ldots, M_k$. Define $c_i = \sum_{j \in J_i} t_j$ as the *load* of a cluster. Further let $m_i = |M_i|$ and $n_i = |J_i|$. For the next lemma define the vectors $\boldsymbol{n} = (n_1, \ldots, n_k)$, $\boldsymbol{c} = (c_1, \ldots, c_k)$ and $\boldsymbol{m} = (m_1, \ldots, m_k)$. Furthermore, let $\mathcal{F}(n, t, m) \subset \mathbb{N}^k \times \mathbb{R}^k \times \mathbb{N}^k$ denote the subspace of feasible $(\boldsymbol{n}, \boldsymbol{c}, \boldsymbol{m})$, which simultaneously satisfy all the following constraints:

$$n_i \geq c_i \qquad c_i > 0 \qquad m_i \geq 1 \qquad \sum_i n_i = n \qquad \sum_i c_i = t \qquad \sum_i m_i = m.$$

Lemma 1. *Define the function* $f(\boldsymbol{n}, \boldsymbol{c}, \boldsymbol{m}) = (\sum_{i=1}^k \frac{n_i c_i}{m_i})/(\sum_{i=1}^k \frac{c_i^2}{m_i})$. *We have that* $f(\boldsymbol{n}, \boldsymbol{c}, \boldsymbol{m}) \leq n\sqrt{m}/t$ *for* $(\boldsymbol{n}, \boldsymbol{c}, \boldsymbol{m}) \in \mathcal{F}(n, t, m)$.

Proof. For a geometric interpretation and intuition of the function f notice that for fixed n_i, the numerator is a hyperplane and the denominator is an elliptic paraboloid in the c_i. Therefore, f has a unique maximum, which can not be "very far" from the extremum of the elliptic paraboloid.

Without loss of generality, let $\frac{c_1}{m_1} \geq \frac{c_i}{m_i}$. Then for the numerator it is easy to see $\sum_{i=1}^k \frac{n_i c_i}{m_i} \leq n \frac{c_1}{m_1}$. This gives $f(\boldsymbol{n}, \boldsymbol{c}, \boldsymbol{m}) \leq (\frac{n c_1}{m_1})/(\sum_i \frac{c_i^2}{m_i})$. We strive to find the maximum value that this upper bound can attain. Hence, we try to maximize $f_1(\boldsymbol{c}, \boldsymbol{m}) = (\frac{c_1}{m_1})/(\sum_i \frac{c_i^2}{m_i})$ subject to $\frac{c_1}{m_1} \geq \frac{c_i}{m_i}$, $c_i > 0$, $t = \sum_i c_i$, $m_i \geq 1$ and $m = \sum_i m_i$ for all $i \leq k$.

How large can f_1 be? Let us fix values for m_1 and c_1. Then the denominator is minimized with the choice of $c_i = m_i(t - c_1)/(\sum_{\ell \geq 2} m_\ell)$ for the variables $c_2, \ldots, c_k$. Thus, we incorporate this assumption and get the remaining problem depending only on c_1 and m_1, which is to maximize $f_2(c_1, m_1) = (\frac{c_1}{m_1})/(\frac{c_1^2}{m_1} + \frac{(t-c_1)^2}{m-m_1})$ subject to $0 \leq c_1 \leq t$ and $1 \leq m_1 < m$.

Now assuming a fixed value for m_1, the best choice for c_1 is $c_1 = t\sqrt{\frac{m_1}{m}}$. Substitution and simplification reduces the problem to optimize only w.r.t. m_1, i.e. to maximize $f_3(m_1) = (\frac{1}{\sqrt{m m_1}})/(\frac{t}{m} + \frac{(1-\sqrt{m_1/m})^2 t}{m-m_1})$ subject to $1 \leq m_1 < m$. It is a technical, but straightforward, exercise to show that for the first derivative $f_3'(m_1) \leq 0$ for all $1 \leq m_1 < m$. Hence, $f_3(m_1)$ is monotonic decreasing and the maximum obtained with $m_1 = 1$:

$$f_3(m_1) \leq \frac{1/\sqrt{m}}{t/m + (1 - \sqrt{1/m})^2 t/(m-1)} \leq \frac{1/\sqrt{m}}{t/m} = \frac{\sqrt{m}}{t}.$$

We independently reduced the number of variables and finally derived $m_1 = 1$. A retrospective inspection shows that with our choices the constraints for $f_1(\boldsymbol{c}, \boldsymbol{m})$ and $\frac{c_1}{m_1} \geq \frac{c_i}{m_i}$ are satisfied. Thus, the upper bound for f_3 results in an upper bound for f_1, and finally in $f(\boldsymbol{n}, \boldsymbol{c}, \boldsymbol{m}) \leq n\sqrt{m}/t$. This proves the lemma. □

Finally, we need the following simple lemma, which is an adjustment from [10] to identical machines.

Lemma 2. *For every Nash equilibrium s for the selfish scheduling game without restrictions on identical machines* $\mathrm{cost}(s) \leq n(t+2m)/m$. *For an optimum schedule* s^* *for such a game we have that* $\mathrm{cost}(s^*) \geq t^2/m$.

Proof (Proof of Theorem 1.). For the upper bound in (2) we may apply Lemma 2 to the unrestricted problems given by task sets $J_1, \ldots, J_k$ and the machine sets $M_1, \ldots, M_k$. With Lemma 1 we obtain

$$\frac{\mathrm{cost}(s)}{\mathrm{cost}(s^*)} \leq \frac{\sum_{i=1}^k \frac{n_i(c_i+2m_i)}{m_i}}{\sum_{i=1}^k \frac{c_i^2}{m_i}} \leq \frac{n\sqrt{m}}{t} + \frac{2n}{\sum_{i=1}^k \frac{c_i^2}{m_i}} \leq \frac{n\sqrt{m}}{t} + \frac{2mn}{t^2}$$

To prove (3) we consider the probability that T deviates "much" from its expected value. Recall that $T = \sum_j T_j$ is a random variable. Let the random variables $S_0 = \mathbb{E}[T_1] + \cdots + \mathbb{E}[T_n]$ and $S_i = T_1 + \cdots + T_i + \mathbb{E}[T_{i+1}] + \cdots + \mathbb{E}[T_n]$ for $i = 1, \ldots, n$. The sequence $S_0, S_1, \ldots, S_n$ is a martingale, and differences are bounded by one: $|S_i - S_{i-1}| \leq 1$. Therefore we may apply the Azuma-Hoeffding inequality: $\Pr[|S_n - S_0| \geq \lambda] \leq 2\exp(-\lambda^2/2n)$. With the choice $\lambda = \sqrt{4n \log n}$ we have $\Pr\left[|T - \mathbb{E}[T]| \geq \sqrt{4n\log n}\right] \leq 2/n^2$. Clearly $\mathrm{PoA}(\Sigma) \leq n$ always holds because each task is counted at least once but at most n times. With $\mathbb{E}[T] = \omega\left(\sqrt{n \log n}\right)$ we find

$$\begin{aligned}
\mathrm{EPoA}(\Sigma) &\leq \mathbb{E}\left[\min\left\{n, \frac{n\sqrt{m}}{T} + \frac{nm}{T^2}\right\}\right] \\
&\leq \frac{n\sqrt{m}}{\mathbb{E}[T] - \sqrt{4n\log n}} + \frac{nm}{(\mathbb{E}[T] - \sqrt{4n\log n})^2} + n\frac{2}{n^2} \\
&= \left(\frac{n\sqrt{m}}{\mathbb{E}[T]} + \frac{nm}{\mathbb{E}[T]^2}\right)(1 + o(1)).
\end{aligned}$$

This proves the upper bounds. For the lower bounds we construct a deterministic task distribution and restrict the tasks to two sets of machines. We restrict the majority of tasks to a set of 2 machines, which creates a high price of stability similarly to the unrestricted case [10]. The remaining tasks on the remaining $m-2$ machines are used to account for the total load, and their presence reduces the price of stability to essentially $\Theta\left(n\sqrt{m}/t\right)$. Details appear in the full version. □

General Restrictions. We continue with general restrictions, i.e., the sets $A_j \neq \emptyset$ are not constrained in any further way. Our main result states that the price of anarchy for general restrictions behaves similarly as for clustered restrictions.

Theorem 2. *Under the assumptions of Theorem 1, the bounds stated therein remain valid if* ε_1 *and* ε_2 *are replaced by* $\varepsilon_1 = \frac{2nm^2}{t^2}$ *and* $\varepsilon_2 = \frac{2nm^2}{\mathbb{E}[T]^2}$.

We relate the price of anarchy with clustered restrictions to general restrictions. This requires an additional concept, which is closely related to clusters. Thus we use similar notation.

Definition 1. *For a Nash equilibrium, label machines in order of their loads* $w_1 \geq w_2 \geq \cdots \geq w_m$. *A partition of the set of machines into* groups $M_1, \ldots, M_k$ *has the property that for every group* $M_i = \{r_{i-1}+1, \ldots, r_i\}$ *the loads of machines* $w_{r_i} - w_{r_{i+1}} > t_{max}$ *and* $w_\ell - w_{\ell+1} \leq t_{\max}$ *for all* $\ell \in \{r_{i-1}+1, \ldots, r_i - 1\}$.

We denote by J_i the set of tasks that are on any of the machines in M_i, and by $c_i = \sum_{\ell \in M_i} w_\ell = \sum_{j \in J_i} t_j$ the *load* of group M_i. Intuitively, the groups have the shape of stairs. The load difference between two consecutive steps is at most $t_{\max}$, but the step between two consecutive groups is more than $t_{\max}$ high. In the chosen Nash equilibrium every task in a group would like to switch to a group with lower load. The reason it does not do so must be that the restrictions forbid the change.

Now consider the machines with their optimum loads $w_1^*, \ldots, w_m^*$. Let $M_1, \ldots, M_k$ be the groups induced by any chosen Nash equilibrium. The above observation implies that also in the optimum solution no task on any of the machines M_i can be on any of the machines in $M_{i+1}, \ldots, M_k$, because the restrictions forbid it. However, it is possible that certain tasks of M_i change to the groups $M_1, \ldots, M_{i-1}$. The following lemma quantifies the effect of such changes.

Lemma 3. *Let* s^* *be an optimum schedule for an instance of the restricted selfish scheduling game with arbitrary restrictions. Let* s *be a Nash equilibrium that induces groups* $M_1, \ldots, M_k$ *with* $m_1, \ldots, m_k$ *machines and loads* $c_1, \ldots, c_k$. *Then* $\text{cost}(s^*) \geq \sum_i c_i^2/m_i$.

Proof. Consider the optimum solution s^* with p_i^* players and w_i^* load on machine $i = 1, \ldots, m$. Group the machines into $M_1, \ldots, M_k$ as in the Nash equilibrium s. Define $c_i^* = \sum_{\ell \in M_i} w_\ell^*$ as the optimum load of the group M_i. Notice that $p_i^* \geq w_i^*$ because $t_j \leq 1$ for every task j. Clearly, the optimum cost of the group M_i is $\sum_{\ell \in M_i} p_\ell^* w_\ell^* \geq \sum_{\ell \in M_i} (w_\ell^*)^2 \geq (c_i^*)^2/m_i$. In order to prove the lower bound $\text{cost}(s^*) \geq \sum_i c_i^2/m_i$, we transform the profile of the Nash equilibrium $c_1, \ldots, c_k$ into the optimum profile $c_1^*, \ldots, c_k^*$ without decreasing its cost. Let $x_1, \ldots, x_k$ denote the *current load*, which is initially $x_1 = c_1, \ldots, x_k = c_k$ and finally $x_1^* = c_1^*, \ldots, x_k^* = c_k^*$. We say that a group i is currently underloaded if $x_i < c_i^*$, overloaded if $x_i > c_i^*$, and saturated if $x_i = c_i^*$.

Observe that – by the restrictions – load is only allowed to move from a group M_ℓ with index ℓ to a group M_j with smaller index j. Hence, if there is an overloaded group (at all), then there must be an underloaded group with smaller index. Conversely, if there is an underloaded group (at all), there must be an overloaded group with larger index, due to the same reason. This property suggests an intuitive algorithm to transform the load profiles with the invariant that whenever there is an overloaded group, there is also an underloaded group with smaller index (and vice versa).

We repeatedly find the overloaded group with largest index (denoted ℓ) and the underloaded group with largest index (denoted j). Due to the invariant we know that $j < \ell$, i.e., there is an overloaded machine with larger index than any underloaded machine. We decrease x_ℓ and increase x_j by the same amount until one of the groups becomes saturated. This transformation preserves the invariant. The procedure eventually terminates, since we saturate at least one group in each iteration.

We determine the change in cost in one iteration as follows. Consider the initial situation, i.e., the Nash equilibrium with machine-loads $w_1 \geq \cdots \geq w_m$ and group-loads $c_1 \geq \cdots \geq c_k$. Let $w_i^{\min}$, respectively $w_i^{\max}$ denote the minimum, respectively maximum load of any machine in group M_i. Let $j < \ell$, note that $w_j^{\min} > w_\ell^{\max}$, and observe that

$$\frac{c_j}{m_j} \geq \frac{m_j w_j^{\min}}{m_j} = w_j^{\min} > w_\ell^{\max} = \frac{m_\ell w_\ell^{\max}}{m_\ell} \geq \frac{c_\ell}{m_\ell}.$$

Thus, initially, not only the group-loads c_i are in decreasing order, but also the relative loads c_i/m_i. Now consider a transformation in which load is moved from group M_ℓ to M_j. As every iteration increases x_j over c_j and decreases x_ℓ under c_ℓ, the inequality continues to hold for the values of x during the execution of our algorithm: $\frac{x_j}{m_j} \geq \frac{c_j}{m_j} > \frac{c_\ell}{m_\ell} \geq \frac{x_\ell}{m_\ell}$. Now suppose that M_j receives load $\delta > 0$ from M_ℓ. The change of the cost is $\delta^2(\frac{1}{m_j} + \frac{1}{m_\ell}) + 2\delta(\frac{x_j}{m_j} - \frac{x_\ell}{m_\ell})$. Since $\frac{x_j}{m_j} > \frac{x_\ell}{m_\ell}$, in every iteration our algorithm increases the cost. Hence, it transforms the Nash profile $c_1, \ldots, c_k$ into the optimum profile $c_1^*, \ldots, c_k^*$ without decreasing the cost. This yields $\text{cost}(s^*) \geq \sum_i \frac{c_i^2}{m_i}$ and the proof is complete. □

For the proof of Theorem 2 we assemble the lower bound for s^* and a simple upper bound for any Nash equilibrium s. Then with a similar Azuma-Hoeffding argument as in the proof of Theorem 1 the result follows.

2.1 Average-Case Analysis of an Optimization Algorithm

In this short section, we point out that Theorem 2 also has an algorithmic perspective. By proving upper bounds on the expected price of anarchy of restricted selfish scheduling, we obtain an average-case analysis for an algorithm for the non-economical latency optimization problem (e.g. the standard scheduling variant of the game) as a byproduct. We consider the algorithm, which we call NASHIFY, due to Gairing et al. [7] introduced for maximum latency social cost. The algorithm begins with an arbitrary assignment and uses the idea of blocking flows to compute a Nash equilibrium. It has running time $O\left(nmA(\log t + m^2)\right)$ with $A = \sum_i |A_i|$. It is remarkable that the algorithm also performs well for total latency minimization for restricted scheduling, see Theorem 3 below. In the scheduling problem, the objective is to minimize $\sum_j \ell_j$, regardless if it is a Nash equilibrium or not. Let $\text{cost}(s)$ and $\text{cost}(s^*)$ denote the objective values of a schedule obtained by NASHIFY and by an (potentially exponential time) optimum algorithm OPT. While $\text{cost}(s)/\text{cost}(s^*)$ is called the *performance ratio*, for random task-lengths T_j the expectation $\mathbb{E}\left[\text{cost}(S)/\text{cost}(S^*)\right]$ is called the *expected performance ratio* of the algorithm NASHIFY. Here S and S^* are the associated random variables of s and s^*. The result below follows directly from [7] and Theorem 2.

Theorem 3. *Under the assumptions of Theorem 2, the bounds stated therein are upper bounds for the expected performance ratio of the algorithm* NASHIFY.

3 Maximum Latency Cost

Here we consider the social cost function $\text{cost}(s) = \ell_{\max} = \max_j \ell_j$. Define the parameter $r = \text{cost}(s^*)/t_{\max}$ of the game. Awerbuch et al. [2] showed a bound of $\Theta\left(\log m/(r \log(1 + \frac{\log m}{r}))\right)$ on the price of anarchy. This gives $1 + \epsilon$ price of anarchy for $r = \Omega(\log m/\epsilon^2)$. We contribute the following alternative bound, where the total traffic t is the parameter.

Theorem 4. *For our game with general restrictions and maximum latency social cost, we have the following:*

(1) For every $t = \sum_j t_j > 0$*, it holds that* $\text{PoA}(\Sigma) \leq 1 + \frac{m^2}{t}$.
(2) For $T = \sum_j T_j$ *with* $\mathbb{E}[T] = \omega\left(\sqrt{n \log n}\right)$ *and independent* T_j *we have* $\text{EPoA}(\Sigma) \leq 1 + \frac{m^2}{\mathbb{E}[T]}(1 + o(1))$.

Both bounds also hold for the (expected) performance of the algorithm NASHIFY.

References

1. Awerbuch, B., Azar, Y., Epstein, A.: The price of routing unsplittable flow. In: Proc. 37th STOC, pp. 57–66 (2005)
2. Awerbuch, B., Azar, Y., Richter, Y., Tsur, D.: Tradeoffs in worst-case equilibria. In: Proc. 1st WAOA, pp. 41–52 (2003)
3. Caragiannis, I., Flammini, M., Kaklamanis, C., Kanellopoulos, P., Moscardelli, L.: Tight bounds for selfish and greedy load balancing. In: Bugliesi, M., Preneel, B., Sassone, V., Wegener, I. (eds.) ICALP 2006. LNCS, vol. 4051, pp. 311–322. Springer, Heidelberg (2006)
4. Christodoulou, G., Koutsoupias, E.: The price of anarchy of finite congestion games. In: Proc. 37th STOC, pp. 67–73 (2005)
5. Czumaj, A., Krysta, P., Vöcking, B.: Selfish traffic allocation for server farms. In: Proc. 34th STOC, pp. 287–296 (2002)
6. Czumaj, A., Vöcking, B.: Tight bounds for worst-case equilibria. In: Proc. 13th SODA, pp. 413–420 (2002)
7. Gairing, M., Lücking, T., Mavronicolas, M., Monien, B.: Computing nash equilibria for scheduling on restricted parallel links. In: Proc. 36th STOC, pp. 613–622 (2004)
8. Gairing, M., Lücking, T., Mavronicolas, M., Monien, B., Rode, M.: Nash equilibria in discrete routing games with convex latency functions. In: Díaz, J., Karhumäki, J., Lepistö, A., Sannella, D. (eds.) ICALP 2004. LNCS, vol. 3142, pp. 645–657. Springer, Heidelberg (2004)
9. Gairing, M., Lücking, T., Mavronicolas, M., Monien, B., Spirakis, P.: Structure and complexity of extreme Nash equilibria. Theoretical Computer Science 343(1-2), 133–157 (2005)

10. Hoefer, M., Souza, A.: Tradeoffs and average-case equilibria in selfish routing. In: Arge, L., Hoffmann, M., Welzl, E. (eds.) ESA 2007. LNCS, vol. 4698, pp. 63–74. Springer, Heidelberg (2007)
11. Kolliopoulos, S., Stein, C.: Approximation algorithms for single-source unsplittable flow. SIAM Journal on Computing 31, 919–946 (2002)
12. Kontogiannis, S., Spirakis, P.: Atomic selfish routing in networks: A survey. In: Deng, X., Ye, Y. (eds.) WINE 2005. LNCS, vol. 3828, pp. 989–1002. Springer, Heidelberg (2005)
13. Koutsoupias, E., Papadimitriou, C.: Worst-case equilibria. In: Meinel, C., Tison, S. (eds.) STACS 1999. LNCS, vol. 1563, pp. 404–413. Springer, Heidelberg (1999)
14. Lenstra, J.K., Shmoys, D., Tardos, E.: Approximation algorithms for scheduling unrelated parallel machines. Mathematical Programming 46, 259–271 (1990)
15. Mavronicolas, M., Panagopoulou, P., Spirakis, P.: A cost mechanism for fair pricing of resource usage. In: Deng, X., Ye, Y. (eds.) WINE 2005. LNCS, vol. 3828, pp. 210–224. Springer, Heidelberg (2005)
16. Scharbrodt, M., Schickinger, T., Steger, A.: A new average case analysis for completion time scheduling. In: Proc. 34th STOC, pp. 170–178 (2002)
17. Souza, A., Steger, A.: The expected competitive ratio for weighted completion time scheduling. In: Diekert, V., Habib, M. (eds.) STACS 2004. LNCS, vol. 2996, pp. 620–631. Springer, Heidelberg (2004)
18. Suri, S., Toth, C., Zhou, Y.: Selfish load balancing and atomic congestion games. Algorithmica 47(1), 79–96 (2007)

Congestion Games with Linearly Independent Paths: Convergence Time and Price of Anarchy

Dimitris Fotakis

Dept. of Information and Communication Systems Engineering
University of the Aegean, 83200 Samos, Greece
fotakis@aegean.gr

Abstract. We investigate the effect of linear independence in the strategies of congestion games on the convergence time of best response dynamics and on the pure Price of Anarchy. In particular, we consider symmetric congestion games on extension-parallel networks, an interesting class of networks with linearly independent paths, and establish two remarkable properties previously known only for parallel-link games. More precisely, we show that for arbitrary non-negative and non-decreasing latency functions, any best improvement sequence converges to a pure Nash equilibrium in at most n steps, and that for latency functions in class $\mathcal{D}$, the pure Price of Anarchy is at most $\rho(\mathcal{D})$.

1 Introduction

Congestion games provide a natural model for non-cooperative resource allocation in large-scale communication networks and have been the subject of intensive research in algorithmic game theory. In a *congestion game*, a finite set of non-cooperative players, each controlling an unsplittable unit of load, compete over a finite set of resources. All players using a resource experience a latency (or cost) given by a non-negative and non-decreasing function of the resource's load (or congestion). Among a given set of resource subsets (or strategies), each player selects one selfishly trying to minimize her *individual cost*, that is the sum of the latencies on the resources in the chosen strategy. A natural solution concept is that of a *pure Nash equilibrium*, a configuration where no player can decrease her individual cost by unilaterally changing her strategy.

The prevailing questions in recent work on congestion games have to do with quantifying the inefficiency due to the players' selfish behaviour (see e.g. [19,20,14,5,7,4,6]), and bounding the convergence time to pure Nash equilibria if the players select their strategies in a selfish and decentralized fashion (see e.g. [11,18,1]). In this work, we investigate the effect of linear independence in the strategies of congestion games on the convergence time of best improvement sequences and on the inefficiency of pure Nash equilibria. In particular, we consider symmetric congestion games on extension-parallel networks, an interesting class of networks whose paths are linearly independent, in the sense that every path contains an edge not included in any other path. For this class of congestion games, which comprises a natural and non-trivial generalization of the extensively studied class of parallel-link games (see e.g. [19,20,14,11,18,6]), we provide best possible answers to both research questions above.

B. Monien and U.-P. Schroeder (Eds.): SAGT 2008, LNCS 4997, pp. 33–45, 2008.

Convergence Time to Pure Nash Equilibria. Rosenthal [23] proved that the pure Nash equilibria of congestion games correspond to the local optima of a natural potential function. Hence Rosenthal established that every congestion game admits at least one pure Nash equilibrium (PNE) reached in a natural way when players iteratively select strategies that minimize their individual cost given the strategies of other players. Nevertheless, this may take an exponential number of steps, since computing a PNE is PLS-complete even for asymmetric network congestion games as shown by Fabricant *et al.* [12]. In fact, the proof of Fabricant *et al.* establishes the existence of instances where any best improvement sequence is exponentially long. Even for symmetric network congestion games, where a PNE can be found efficiently by a min-cost flow computation [12], Ackermann *et al.* [1] presented instances where any best improvement sequence is exponentially long.

A natural approach to circumvent the negative results of [12,1] is to identify large classes of congestion games for which best improvement sequences reach a PNE in a polynomial number of steps. For instance, it is well known that for symmetric singleton congestion games (aka parallel-link games), any best improvement sequence converges to a PNE in at most n steps, where n denotes the number of players. Ieong *et al.* [18] proved that even for asymmetric singleton games with non-monotonic latencies, best improvement sequences reach a PNE in polynomial time. Subsequently, Ackermann *et al.* [1] generalized this result to *matroid* congestion games, where the strategy space of each player consists of the bases of a matroid over the set of resources. Furthermore, Ackermann *et al.* proved that the matroid property on the players' strategy spaces is necessary for guaranteeing polynomial-time convergence of best improvement sequences if one does not take into account the global structure of the game.

Contribution. The negative results of [12,1] leave open the possibility that some particular classes of symmetric network congestion games can guarantee fast convergence of best improvement sequences. We prove that for symmetric congestion games on extension-parallel networks with arbitrary non-negative and non-decreasing latency functions, any best improvement sequence converges to a PNE in at most n steps[1]. In particular, we show that in a best improvement sequence, every player moves at most once. This result is best possible, since there are instances where reaching a PNE requires that every player moves at least once.

Price of Anarchy. Having reached a PNE, selfish players enjoy a minimum individual cost given the strategies of other players. However, the public benefit is usually measured by the *total cost* incurred by all players. Since a PNE does not need to minimize the total cost, one seeks to quantify the inefficiency due to the players' non-cooperative and selfish behaviour. The *Price of Anarchy* was introduced by Koutsoupias and Papadimitriou [19] and has become a widely accepted measure of the performance degra-

[1] We highlight that matroid games and games on extension-parallel networks have a different combinatorial structure and may have quite different properties. For example, a network consisting of two pairs of parallel links connected in series is not extension-parallel, but the corresponding network congestion game is a symmetric matroid game. For another example, Milchtaich [22, Example 4] proved that weighted congestion games on extension-parallel networks may not admit a PNE. On the other hand, Ackermann *et al.* [2, Theorem 2] proved that every weighted matroid congestion game admits a PNE.

dation due to the players' selfish behaviour. The (pure) Price of Anarchy is the worst-case ratio of the total cost of a (pure) Nash equilibrium to the optimal total cost. Many recent contributions have provided strong upper and lower bounds on the pure Price of Anarchy (PoA) for several classes of congestion games, mostly congestion games with affine and polynomial latency functions and congestion games on parallel links[2].

For the special case of parallel links with linear latency functions, Lücking *et al.* [20] proved that the PoA is $4/3$. For parallel links with polynomial latency functions of degree d, Gairing *et al.* [14] proved the PoA is at most $d+1$. Awerbuch *et al.* [5] and Christodoulou and Koutsoupias [7] proved independently that the PoA of congestion games is $5/2$ for affine latency functions and $d^{\Theta(d)}$ for polynomial latency functions of degree d. Subsequently, Aland *et al.* [4] obtained exact bounds on the PoA of congestion games with polynomial latency functions. In the non-atomic setting, where the number of players is infinite and each player controls an infinitesimal amount of load, Roughgarden [24] proved that the PoA is independent of the strategy space and equal to $\rho(\mathcal{D})$, where ρ depends on the class of latency functions $\mathcal{D}$ only (e.g. ρ is equal to $4/3$ for affine and 1.626 for quadratic functions). Subsequently, Correa *et al.* [8] introduced $\beta(\mathcal{D}) = 1 - \frac{1}{\rho(\mathcal{D})}$ and gave a simple proof of the same bound. Recently Fotakis [13] and independently Caragiannis *et al.* [6, Theorem 23] proved that the PoA of (atomic) congestion games on parallel links with latency functions in class $\mathcal{D}$ is also $\rho(\mathcal{D})$.

Contribution. Despite the considerable interest in the PoA of congestion games, it remains open whether some better upper bounds close to ρ are possible for symmetric congestion games on simple networks other than parallel links (e.g. extension-parallel networks, series-parallel networks), or strong lower bounds similar to the lower bounds of [5,7,4] also apply to them. As a first step in this direction, we prove that the PoA of symmetric congestion games on extension-parallel networks with latency functions in class $\mathcal{D}$ is at most $\rho(\mathcal{D})$. On the negative side, we show that this result cannot be further generalized to series-parallel networks.

Related Work on Congestion Games with Linearly Independent Strategies. There has been a significant volume of previous work investigating the impact of linear independent strategies on properties of congestion games. Holzman and Law-Yone [16] proved that a symmetric strategy space admits a strong equilibrium[3] for any selection of non-negative and non-decreasing latency functions iff it consists of linearly independent strategies. Furthermore, Holzman and Law-Yone showed that for symmetric congestion games with linearly independent strategies, every PNE is a strong equilibrium and also a minimizer of Rosenthal's potential function. Subsequently, Holzman and Law-Yone [17] proved that the class of congestion games on extension-parallel networks is the network equivalent of congestion games with linearly independent strategies.

Milchtaich [21] was the first to consider networks with linearly independent paths (under this name). Milchtaich proved that an undirected network has linearly independent paths iff it is extension-parallel. Furthermore, Milchtaich showed that extension-parallel networks is the only class of networks where for any selection of non-negative

[2] Here we cite only the most relevant results on the pure PoA for the objective of total cost. For a survey on the PoA of congestion games , see e.g. [15].

[3] A configuration is a *strong equilibrium* if no coalition of players can deviate in a way profitable for all its members.

and increasing (resp. non-decreasing) latency functions, all equilibria in the non-atomic setting are (resp. weakly) Pareto efficient.

Recently Epstein *et al.* [10,9] considered fair connection games and congestion games on extension-parallel networks. In [10], they proved that fair connection games on extension-parallel networks admit a strong equilibrium. In [9], they showed that extension-parallel networks is the only class of networks where for all non-negative and non-decreasing latencies, any PNE minimize the maximum players' cost.

2 Model and Preliminaries

For any integer $k \geq 1$, we let $[k] \equiv \{1, \dots, k\}$. For a vector $x = (x_1, \dots, x_n)$, we let $x_{-i} \equiv (x_1, \dots, x_{i-1}, x_{i+1}, \dots, x_n)$ and $(x_{-i}, x'_i) \equiv (x_1, \dots, x_{i-1}, x'_i, x_{i+1}, \dots, x_n)$.

Congestion Games. A *congestion game* is a tuple $\Gamma(N, E, (\Sigma_i)_{i \in N}, (d_e)_{e \in E})$, where N denotes the set of players, E denotes the set of resources, $\Sigma_i \subseteq 2^E \setminus \{\emptyset\}$ denotes the strategy space of each player i, and $d_e : \mathbb{N} \mapsto \mathbb{R}_{\geq 0}$ is a non-negative and non-decreasing latency function associated with each resource e. A congestion game is *symmetric* if all players have a common strategy space.

A *configuration* is a vector $\sigma = (\sigma_1, \dots, \sigma_n)$ consisting of a strategy $\sigma_i \in \Sigma_i$ for each player i. For every resource e, we let $\sigma_e = |\{i \in N : e \in \sigma_i\}|$ denote the congestion induced on e by σ. The individual cost of player i in the configuration σ is $c_i(\sigma) = \sum_{e \in \sigma_i} d_e(\sigma_e)$. A configuration σ is a *pure Nash equilibrium* (PNE) if no player can improve her individual cost by unilaterally changing her strategy. Formally, σ is a PNE if for every player i and every strategy $s_i \in \Sigma_i$, $c_i(\sigma) \leq c_i(\sigma_{-i}, s_i)$.

In the following, we let n denote the number of players. We focus on *symmetric network* congestion games, where the players' strategies are determined by a directed network $G(V, E)$ with a distinguished source s and sink t (aka $s - t$ network). The network edges play the role of resources and the common strategy space of the players is the set of (simple) $s - t$ paths in G, denoted $\mathcal{P}$. For any $s - t$ path p and any pair of vertices v_1, v_2 appearing in p, we let $p[v_1, v_2]$ denote the segment of p between v_1 and v_2 ($p[v_1, v_2]$ is empty if v_1 appears after v_2 in p). For consistency with the definition of strategies as resource subsets, we usually regard paths as sets of edges.

Flows and Configurations. Let $G(V, E)$ be a $s - t$ network. A $s - t$ *flow* f is a vector $(f_e)_{e \in E} \in \mathbb{R}^m_{\geq 0}$ that satisfies the flow conservation at all vertices other than s and t. The *volume* of f is the total flow leaving s. A flow is *acyclic* if there is no directed cycle in G with positive flow on all its edges. For a flow f and a path p, we let $f_p^{\min} = \min_{e \in p}\{f_e\}$.

Given a configuration σ for a symmetric network congestion game Γ, we refer to the congestion vector $(\sigma_e)_{e \in E}$ as the *flow* induced by σ. We say that a flow σ is *feasible* if there is a configuration inducing congestion σ_e on every edge e. Hence any configuration of Γ corresponds to a feasible flow. We always let the same symbol denote both a configuration and the feasible flow induced by it.

Best Improvement Sequences. A strategy $s_i \in \Sigma_i$ is a *best response* of player i to a configuration σ (or equivalently to σ_{-i}) if for every strategy $s'_i \in \Sigma_i$, $c_i(\sigma_{-i}, s_i) \leq c_i(\sigma_{-i}, s'_i)$. If i's current strategy σ_i is not a best response to the current configuration σ, a best response of i to σ is a *best improvement* of i. We consider best improvement sequences, where in each step, a player i whose strategy σ_i is not a best response to the current configuration σ switches to her best improvement. Using a potential function, Rosenthal [23] proved that any such sequence reaches a PNE in a finite number of steps.

Social Cost and the Price of Anarchy. To quantify the inefficiency of PNE, we evaluate configurations using the objective of *total cost*. The total cost $C(\sigma)$ of a configuration σ is the sum of players' costs in σ : $C(\sigma) = \sum_{i=1}^{n} c_i(\sigma) = \sum_{e \in E} \sigma_e d_e(\sigma_e)$. The *optimal configuration*, usually denoted o, minimizes the total cost among all configurations in $\mathcal{P}^n$. The pure *Price of Anarchy* (PoA) of a congestion game Γ is the maximum ratio $C(\sigma)/C(o)$ over all PNE σ of Γ.

Extension-Parallel Networks. Let $G_1(V_1, E_1)$ and $G_2(V_2, E_2)$ be two networks with sources s_1 and s_2 and sinks t_1 and t_2 respectively, and let $G'(V_1 \cup V_2, E_1 \cup E_2)$ be the union network of G_1 and G_2. The *parallel composition* of G_1 and G_2 results in a $s - t$ network obtained from G' by identifying s_1 and s_2 to the source s and t_1 and t_2 to the sink t. The *series composition* of G_1 and G_2 results in a $s - t$ network obtained from G' by letting s_1 be the source s, letting t_2 be the sink t, and identifying t_1 with s_2.

A directed $s-t$ network is *series-parallel* if it consists of either a single edge (s, t) or two series-parallel networks composed either in series or in parallel. A directed $s-t$ network is *extension-parallel* if it consists of either: (i) a single edge (s, t), (ii) a single edge and an extension-parallel network composed in series, or (iii) two extension-parallel networks composed in parallel. Every extension-parallel network is series-parallel, but the converse is true only if in every series composition, at least one component is a single edge.

A $s-t$ network has *linearly independent paths* if every $s-t$ path contains at least one edge not belonging to any other $s - t$ path[4]. Milchtaich [21, Proposition 5] proved that an undirected $s - t$ network has linearly independent paths iff it is extension-parallel. Therefore, every (directed) extension-parallel network has linearly independent paths (see also [17, Theorem 1]). Furthermore, [21, Propositions 3, 5] imply that a (directed) series-parallel network has linearly independent paths iff it is extension-parallel.

An interesting property of extension-parallel networks is that for any two $s - t$ paths p, p', the segments $p \setminus p'$ and $p' \setminus p$ where p and p' deviate from each other form two internally disjoint paths with common endpoints (see also [21, Proposition 4]). In addition, every $s-t$ path having an edge in common with $p \setminus p'$ does not intersect $p' \setminus p$ at any vertex other than its endpoints. The following proposition gives another interesting property of networks with linearly independent paths (and thus of extension-parallel networks).

Proposition 1. *Let Γ be a symmetric congestion game on a $s - t$ network G with linearly independent paths, let f be any configuration of Γ, and let π be any (simple) path with $f_\pi^{\min} > 0$. Then there exists a player i whose strategy in f includes π.*

Every configuration of a symmetric congestion game on a series-parallel (and thus on an extension-parallel) network corresponds to a feasible acyclic flow of volume n. Proposition 1 implies that for any congestion game Γ on an extension-parallel network, every

[4] The name is motivated by the fact that in such a network, it is not possible to express any path as the symmetric difference of some other paths [21, Proposition 6].

feasible acyclic $s - t$ flow corresponds to a unique Γ's configuration (uniqueness is up to players' permutation, see also [16, Section 6]). Therefore, for symmetric congestion games on extension-parallel networks, there is a correspondence between configurations and feasible acyclic flows.

3 Convergence Time to Pure Nash Equilibria

Next we show that for symmetric congestion games on extension-parallel networks, any best improvement sequence reaches a PNE after each player moves at most once.

Lemma 1. *Let Γ be a congestion game on an extension-parallel network, let σ be the current configuration, and let i be a player switching from her current strategy σ_i to her best improvement σ'_i. Then for every player j whose current strategy σ_j is a best response to σ, σ_j remains a best response of j to the new configuration $\sigma' = (\sigma_{-i}, \sigma'_i)$.*

Proof. For sake of contradiction, we assume that there is a player j whose current strategy σ_j is a best response to σ but not to σ'. Let σ'_j be a best response of j to σ', and let $p = \sigma_j \setminus \sigma'_j$ and $p' = \sigma'_j \setminus \sigma_j$ be the segments where σ_j and σ'_j deviate from each other. Due to the extension-parallel structure of the network, p and p' are internally disjoint paths with common endpoints, denoted u and w. Since p and p' are edge-disjoint and player j improves her individual cost in σ' by switching from p to p',

$$\sum_{e\in p} d_e(\sigma'_e) > \sum_{e\in p'} d_e(\sigma'_e + 1) \tag{1}$$

Using (1) and the fact that σ'_i is a best improvement of player i to σ, and exploiting the extension-parallel structure of the network, we establish that if player j prefers σ'_j to σ_j in the new configuration σ', then σ_j is not a best response of j to σ. In particular, we show that player j can also improve her individual cost in σ by switching from an appropriate segment of σ_j to the corresponding segment of σ'_i. Clearly, this contradicts the hypothesis that σ_j is a best response of j to σ and implies the lemma. The technical part of the proof proceeds by case analysis.

Case I, $u, w \in \sigma'_i$**:** We first consider the case where σ'_i contains u and w and thus $\sigma'_i[u, w]$ can serve as an alternative to p. We further distinguish between two subcases: *Case I.a,* $p \cap \sigma'_i = \emptyset$: We start with the case where σ'_i and p are edge-disjoint. We first consider the case where $\sigma'_i[u, w] \setminus p'$ does not contain any edges of σ_i (Fig. 1.a). Then,

$$\begin{aligned}
\sum_{e\in p'} d_e(\sigma'_e + 1) &\geq \sum_{e\in p'\cap\sigma'_i} d_e(\sigma_e + 1) + \sum_{e\in(p'\cap\sigma_i)\setminus\sigma'_i} d_e(\sigma_e) + \sum_{e\in(p'\setminus\sigma_i)\setminus\sigma'_i} d_e(\sigma_e + 1) \\
&\geq \sum_{e\in p'\cap\sigma'_i} d_e(\sigma_e + 1) + \sum_{e\in\sigma'_i[u,w]\setminus p'} d_e(\sigma_e + 1)
\end{aligned} \tag{2}$$

For the first inequality, we use that when player i switches from σ_i to σ'_i: (i) the congestion of any edge e in σ'_i does not decrease (i.e. $\sigma'_e \geq \sigma_e$), (ii) the congestion of any edge e decreases by at most 1 (i.e. $\sigma'_e \geq \sigma_e - 1$), and (iii) the congestion of any edge

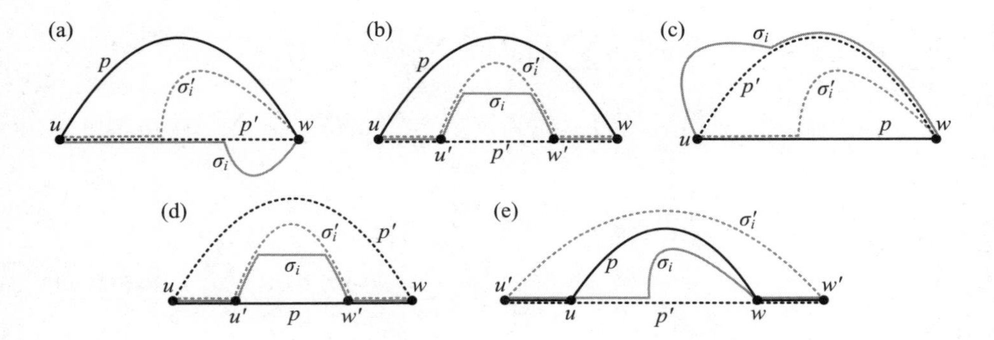

Fig. 1. The different cases considered in the proof of Lemma 1. In each case, the solid black path labeled p represents the best response of player j to σ between vertices u and v, the solid grey path labeled σ_i represents the strategy of player i in σ, and the dotted grey path labeled σ_i' represents the best improvement of player i. We assume that the best response of player j changes from p to the dotted black path labeled p' when player i switches from σ_i to σ_i' and establish a contradiction in all cases.

e not in $\sigma_i \cup \sigma_i'$ does not change (i.e. $\sigma_e' = \sigma_e$). For the second inequality, we observe that $\sum_{e \in (p' \cap \sigma_i) \setminus \sigma_i'} d_e(\sigma_e) + \sum_{e \in (p' \setminus \sigma_i) \setminus \sigma_i'} d_e(\sigma_e + 1)$ is the individual cost of player i on $p' \setminus \sigma_i'$ in σ (i.e. when the configuration of the remaining players is σ_{-i}) and that $\sum_{e \in \sigma_i'[u,w] \setminus p'} d_e(\sigma_e + 1)$ is the individual cost of i on $\sigma_i'[u, w] \setminus p'$ in σ (recall that $\sigma_i'[u, w] \setminus p'$ does not contain any edges of σ_i). Since σ_i' is a best response of i to σ_{-i}, the former cost is no less than the latter.

Using (2), we conclude that player j can improve her individual cost in σ by changing her path between u and w from p to $\sigma_i'[u, w]$, which contradicts the hypothesis that σ_j is a best response of player j to σ. Formally,

$$\sum_{e \in p} d_e(\sigma_e) \geq \sum_{e \in p} d_e(\sigma_e') > \sum_{e \in p'} d_e(\sigma_e' + 1) \geq \sum_{e \in \sigma_i'[u,w]} d_e(\sigma_e + 1)$$

The first inequality holds because $p \cap \sigma_i' = \emptyset$, and the congestion of the edges in p does not increase when player i switches from σ_i to σ_i'. The second inequality is (1) and the third inequality follows from (2).

If $\sigma_i'[u, w] \setminus p'$ contains some edges of σ_i, we can show that due to the extension-parallel structure of the network, the congestion of the edges in $p \cup p'$ does not change when player i switches from σ_i to σ_i' (see Fig. 1.b). This contradicts the hypothesis that the best response of player j changes from σ_j to σ_j' when player i moves from σ_i to σ_i'.

Case I.b, $p \cap \sigma_i' \neq \emptyset$: We proceed with the case where σ_i' and p are not edge-disjoint. Then, due to the extension-parallel structure of the network, σ_i' does not have any edges in common with p' and does not intersect p' at any vertex other than u and w. We first consider the case where $\sigma_i'[u, w] \setminus p$ does not contain any edges of σ_i (Fig. 1.c). Then,

$$
\begin{aligned}
\sum_{e\in p\cap\sigma'_i} d_e(\sigma'_e) + \sum_{e\in p\setminus\sigma'_i} d_e(\sigma_e) \geq \sum_{e\in p} d_e(\sigma'_e) > \sum_{e\in p'} d_e(\sigma'_e+1) \\
\geq \sum_{e\in p'\cap\sigma_i} d_e(\sigma_e) + \sum_{e\in p'\setminus\sigma_i} d_e(\sigma_e+1) \\
\geq \sum_{e\in\sigma'_i[u,w]} d_e(\sigma'_e) \\
= \sum_{e\in p\cap\sigma'_i} d_e(\sigma'_e) + \sum_{e\in\sigma'_i[u,w]\setminus p} d_e(\sigma_e+1)
\end{aligned}
$$

The first inequality holds because the congestion of any edge e not in σ'_i does not increase when player i switches from σ_i to σ'_i (i.e. $\sigma_e \geq \sigma'_e$). The second inequality is (1). The third inequality holds because when player i switches from σ_i to σ'_i: (i) the congestion of any edge e decreases by at most 1 (i.e. $\sigma'_e \geq \sigma_e - 1$), and (ii) the congestion of any edge e not in σ_i does not decrease (i.e. $\sigma'_e \geq \sigma_e$). For the fourth inequality, we observe that the left-hand side is equal to the individual cost of player i on p' in σ, and that the right-hand side is equal to the cost of player i on $\sigma'_i[u,w]$ in σ. Since σ'_i is a best response of player i to σ_{-i}, the former cost is not less than the latter. The equality holds because $\sigma'_i[u,w] \setminus p$ does not contain any edges of σ_i and thus the congestion of every edge $e \in \sigma'_i[u,w] \setminus p$ increases by 1 when player i switches from σ_i to σ'_i.

Therefore, $\sum_{e\in p\setminus\sigma'_i} d_e(\sigma_e) > \sum_{e\in\sigma'_i[u,w]\setminus p} d_e(\sigma_e+1)$, and player j can improve her individual cost in σ by switching from $p \setminus \sigma'_i$ to $\sigma'_i[u,w] \setminus p$. This contradicts the hypothesis that σ_j is a best response of player j to σ.

If $\sigma'_i[u,w] \setminus p$ contains some edges of σ_i, we can show that due to the extension-parallel structure of the network, the congestion of the edges in $p \cup p'$ does not change when player i switches from σ_i to σ'_i (see Fig. 1.d). This contradicts the hypothesis that the best response of player j changes from σ_j to σ'_j when player i moves from σ_i to σ'_i.

Case II, either $u \notin \sigma'_i$ or $w \notin \sigma'_i$: We proceed with the case where σ'_i does not contain either u or w. Then, σ'_i does not have any edges in common with p and p'.

If σ_i too does not contain either u or w, then σ_i does not have any edges in common with p and p'. Since $(\sigma_i \cup \sigma'_i) \cap (p \cup p') = \emptyset$, the congestion of the edges in $p \cup p'$ does not change when player i switches from σ_i to σ'_i. This contradicts the hypothesis that the best response of player j changes from σ_j to σ'_j when player i moves from σ_i to σ'_i.

Therefore, we can restrict our attention to the case where σ_i contains both u and w. Let $\sigma'_i \setminus \sigma_i$ and $\sigma_i \setminus \sigma'_i$ be the segments where σ_i and σ'_i deviate from each other. Due to the extension-parallel structure of the network, and since σ'_i does not contain either u or w and σ_i contains both u and w, $\sigma'_i \setminus \sigma_i$ and $\sigma_i \setminus \sigma'_i$ are (non-empty) internally disjoint paths with common endpoints, denoted u' and w'. Their first endpoint u' appears no later than u and their last endpoint w' appears no sooner than w in σ_i. Furthermore, either u is different from u' or w is different from w' (or both). Due to the extension-parallel structure of the network, and since σ_i deviates from at least one of p and p' between u and w, there is a unique path $\sigma_i[u',u]$ between u and u' and a unique path $\sigma_i[w,w']$ between w and w' (see Fig. 1.e). Let $z = \sigma_i[u',u] \cup \sigma_i[w,w']$. We highlight that both $\sigma_i[u', u]$ and $\sigma_i[w, w']$ are included in σ_j and σ'_j. In particular, $\sigma_j[u', w'] = z \cup p$. Using the previous observations, we obtain that:

$$\begin{aligned}\sum_{e \in \sigma_j[u',w']} d_e(\sigma_e) &\geq \sum_{e \in z} d_e(\sigma_e) + \sum_{e \in p} d_e(\sigma'_e) \\ &> \sum_{e \in z} d_e(\sigma_e) + \sum_{e \in p'} d_e(\sigma'_e + 1) \\ &\geq \sum_{e \in z} d_e(\sigma_e) + \sum_{e \in p' \cap \sigma_i} d_e(\sigma_e) + \sum_{e \in p' \setminus \sigma_i} d_e(\sigma_e + 1) \\ &\geq \sum_{e \in \sigma'_i[u',w']} d_e(\sigma_e + 1)\end{aligned}$$

The first inequality holds because the edges of p do not belong to σ'_i and the congestion of any edge $e \notin \sigma'_i$ does not increase when player i moves from σ_i to σ'_i (i.e. $\sigma_e \geq \sigma'_e$). The second inequality follows from (1). The third inequality holds because when player i switches from σ_i to σ'_i: (i) the congestion of any edge e decreases by at most 1 (i.e. $\sigma'_e \geq \sigma_e - 1$), and (ii) the congestion of any edge e not in σ_i does not decrease (i.e. $\sigma'_e \geq \sigma_e$). For the fourth inequality, we observe that the left-hand side is equal to the individual cost of player i on $\sigma_i[u', u] \cup p' \cup \sigma_i[w, w']$ in σ, and that the right-hand side is equal to the individual cost of player i on $\sigma'_i[u', w']$ in σ (recall that $\sigma'_i[u', w']$ and $\sigma_i[u', w']$ are edge disjoint). Since σ'_i is a best response of player i to σ_{-i}, the former cost is not less than the latter.

Therefore, player j can decrease her individual cost in σ by switching from $\sigma_j[u', w']$ to $\sigma'_i[u', w']$. This contradicts the hypothesis that σ_j is a best response of player j to σ. Since we have reached a contradiction in all different cases, this concludes the proof of the lemma. □

By Lemma 1, once a player moves to her best improvement strategy, she will not have an incentive to deviate as long as the subsequent players switch to their best improvement strategies. Hence we obtain the main result of this section:

Theorem 1. *For any n-player symmetric congestion game on an extension-parallel network, every best improvement sequence reaches a PNE in at most n steps.*

4 Bounding the Price of Anarchy

For a latency function $d(x)$, let $\rho(d) = \sup_{x \geq y \geq 0} \frac{xd(x)}{yd(y)+(x-y)d(x)}$, and let $\beta(d) = \sup_{x \geq y \geq 0} \frac{y(d(x)-d(y))}{xd(x)}$. For a class of latency functions $\mathcal{D}$, let $\rho(\mathcal{D}) = \sup_{d \in \mathcal{D}} \rho(d)$ and $\beta(\mathcal{D}) = \sup_{d \in \mathcal{D}} \beta(d)$. We note that $(1 - \beta(\mathcal{D}))^{-1} = \rho(\mathcal{D})$. In [24,8], it was shown that the PoA of non-atomic congestion games with latencies in class $\mathcal{D}$ is $\rho(\mathcal{D})$. Next we establish the same upper bound on the PoA of symmetric congestion games on extension-parallel networks. The proof is based on the following lemma.

Lemma 2. *Let Γ be a symmetric congestion game on an extension-parallel network $G(V, E)$, and let f be a PNE and g be any configuration of Γ. Then,*

$$\Delta(f, g) \equiv \sum_{e: f_e > g_e} (f_e - g_e) d_e(f_e) - \sum_{e: f_e < g_e} (g_e - f_e) d_e(f_e + 1) \leq 0$$

Proof. We assume wlog. that the configurations f and g are not identical and consider the corresponding feasible flows f and g. Let $\hat{G}(V, \hat{E})$ be the graph of the flow $f - g$. In particular, for each edge $(u, w) \in E$, $\hat{E}$ contains a *forward* edge (u, w) with flow $f_{(u,w)} - g_{(u,w)}$ if $f_{(u,w)} > g_{(u,w)}$, a *backward* edge (w, u) with flow $g_{(u,w)} - f_{(u,w)}$ if $f_{(u,w)} < g_{(u,w)}$, and no edge between u and w if $f_{(u,w)} = g_{(u,w)}$. For every cycle C of $\hat{G}$, let $C^+ = \{(u, w) \in E : (u, w) \in C \text{ and } f_{(u,w)} > g_{(u,w)}\}$ be the set of forward edges in C, and let $C^- = \{(u, w) \in E : (w, u) \in C \text{ and } f_{(u,w)} < g_{(u,w)}\}$ be the set of backward edges in C with their directions reversed (i.e. their directions are as in E).

Since f and g are feasible acyclic $s - t$ flows of the same volume, a flow decomposition of $f - g$ yields only cycles and no paths of $\hat{G}$. Let $\{C_1, \ldots, C_k\}$ be the set of (simple) cycles of $\hat{G}$ produced by the standard flow decomposition of $f - g$ (see e.g. the algorithm described in [3, Theorem 3.5]), and let s_i denote the amount of flow carried by each cycle C_i in that decomposition of $f - g$. Since f and g are feasible acyclic $s - t$ flows, every cycle C_i contains at least one forward and at least one backward edge.

By the properties of the standard flow decomposition algorithm, $\cup_{i \in [k]} C_i^+$ is equal to $\{e \in E : f_e > g_e\}$, and $\cup_{i \in [k]} C_i^-$ is equal to $\{e \in E : f_e < g_e\}$. Moreover, for every forward edge $(u, w) \in \hat{E}$, $\sum_{i:(u,w) \in C_i^+} s_i = f_{(u,w)} - g_{(u,w)}$, and for every backward edge $(w, u) \in \hat{E}$, $\sum_{i:(u,w) \in C_i^-} s_i = g_{(u,w)} - f_{(u,w)}$. Therefore,

$$\Delta(f, g) = \sum_{i=1}^{k} s_i \left(\sum_{e \in C_i^+} d_e(f_e) - \sum_{e \in C_i^-} d_e(f_e + 1) \right) \tag{3}$$

The following proposition shows that for every cycle C_i in the decomposition of $f - g$ (in fact, for every simple cycle of $\hat{G}$), $\sum_{e \in C_i^+} d_e(f_e) - \sum_{e \in C_i^-} d_e(f_e + 1) \leq 0$.

Proposition 2. *Let Γ be a symmetric congestion game on an extension-parallel network G, let f be a PNE and g be any configuration of Γ, and let $\hat{G}$ be the graph of the flow $f - g$. For every simple cycle C of $\hat{G}$,*

$$\sum_{e \in C^+} d_e(f_e) - \sum_{e \in C^-} d_e(f_e + 1) \leq 0$$

Proof sketch. Using induction on the extension-parallel structure of G, we prove that for every simple cycle C of $\hat{G}$, there are vertices u, w on C such that C^+ and C^- are two internally disjoint $u - w$ paths in G. Since C^+ consists of forward edges only, for every $e \in C^+$, $f_e > 0$. Hence by Proposition 1, there is a player i whose strategy in f includes C^+. Therefore, $\sum_{e \in C^+} d_e(f_e) \leq \sum_{e \in C^-} d_e(f_e + 1)$, since otherwise player i could switch from C^+ to C^- between u and w and improve her individual cost, which contradicts the hypothesis that f is a PNE. □

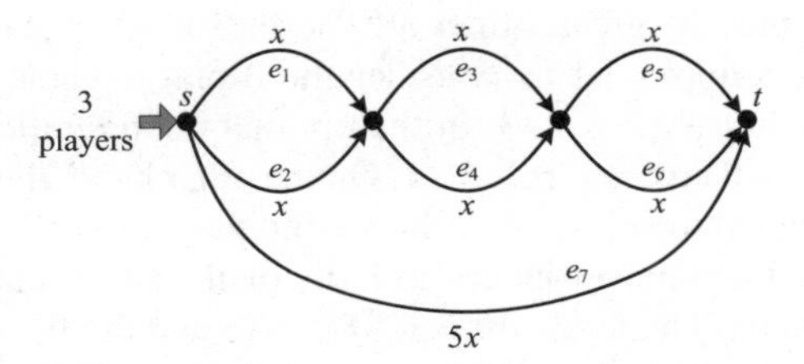

Fig. 2. A symmetric congestion game on a series-parallel network with linear latencies and PoA greater than $4/3$

Combining (3) and Proposition 2, we obtain that $\Delta(f, g) \leq 0$. □

Now we are ready to establish the main result of this section. The following theorem follows easily from Lemma 2 and the definition of $\rho(\mathcal{D})$.

Theorem 2. *For any symmetric congestion game on an extension-parallel network with latency functions in class $\mathcal{D}$, the PoA is at most $\rho(\mathcal{D})$.*

Proof. We consider a symmetric congestion game Γ on an extension-parallel network $G(V, E)$. The latency functions of Γ are such that $\{d_e(x)\}_{e \in E} \subseteq \mathcal{D}$. Let o be the optimal configuration, and let f be Γ's PNE of maximum total cost.

For every edge e with $f_e > o_e$,

$$\begin{aligned} f_e d_e(f_e) &= o_e d_e(f_e) + (f_e - o_e) d_e(f_e) \\ &\leq o_e d_e(o_e) + \beta(\mathcal{D}) f_e d_e(f_e) + (f_e - o_e) d_e(f_e) , \end{aligned} \tag{4}$$

where the inequality follows by applying the definition of $\beta(\mathcal{D})$ to the term $o_e d_e(f_e)$.

On the other hand, for every edge e with $f_e < o_e$,

$$\begin{aligned} f_e d_e(f_e) &= o_e d_e(o_e) - o_e d_e(o_e) + f_e d_e(f_e) \\ &\leq o_e d_e(o_e) - (o_e - f_e) d_e(f_e + 1) \end{aligned} \tag{5}$$

The inequality follows from $d_e(f_e) \leq d_e(f_e + 1)$ and $d_e(f_e + 1) \leq d_e(o_e)$, because the latency functions are non-decreasing and $f_e + 1 \leq o_e$ (recall that o_e and f_e are integral).

Using (4), (5), and Lemma 2, we obtain that:

$$\begin{aligned} C(f) &\leq \sum_{e \in E} o_e d_e(o_e) + \beta(\mathcal{D}) \sum_{e : f_e > o_e} f_e d_e(f_e) + \Delta(f, o) \\ &\leq C(o) + \beta(\mathcal{D}) C(f) , \end{aligned}$$

which implies that $C(f) \leq (1 - \beta(\mathcal{D}))^{-1} C(o) = \rho(\mathcal{D}) C(o)$. For the first inequality, we apply (4) to every edge e with $f_e > o_e$ and (5) to every edge e with $f_e < o_e$. The last inequality follows from Lemma 2, which implies that $\Delta(f, o) \leq 0$. □

Remark 1. The PoA may be greater than $\rho(\mathcal{D})$ even for series-parallel networks with linear latencies. For example, let us consider the 3-player game in Fig 2. Since the latency functions are linear, $\rho = 4/3$. In the optimal configuration o, every edge has congestion 1 and the total cost is $C(o) = 11$. On the other hand, there is a PNE f where the first player is assigned to (e_1, e_3, e_6), the second player to (e_1, e_4, e_5), and the third player to (e_2, e_3, e_5). Each player incurs an individual cost of 5 and does not have an incentive to deviate to e_7. The total cost is $C(f) = 15$ and the PoA is $15/11 > 4/3$. In this example, Lemma 2 fails because Proposition 1 does not hold.

References

1. Ackerman, H., Röglin, H., Vöcking, B.: On the Impact of Combinatorial Structre on Congestion Games. In: FOCS 2006, pp. 613–622 (2006)
2. Ackerman, H., Röglin, H., Vöcking, B.: Pure Nash Equilibria in Player-Specific and Weighted Congestion Games. In: Spirakis, P.G., Mavronicolas, M., Kontogiannis, S.C. (eds.) WINE 2006. LNCS, vol. 4286, pp. 50–61. Springer, Heidelberg (2006)
3. Ahuja, R.K., Magnanti, T.L., Orlin, J.B.: Network Flows: Theory, Algorithms, and Applications. Prentice-Hall, Englewood Cliffs (1993)
4. Aland, S., Dumrauf, D., Gairing, M., Monien, B., Schoppmann, F.: Exact Price of Anarchy for Polynomial Congestion Games. In: Durand, B., Thomas, W. (eds.) STACS 2006. LNCS, vol. 3884, pp. 218–229. Springer, Heidelberg (2006)
5. Awerbuch, B., Azar, Y., Epstein, A.: The Price of Routing Unsplittable Flow. In: STOC 2005, pp. 57–66 (2005)
6. Caragiannis, I., Flammini, M., Kaklamanis, C., Kanellopoulos, P., Moscardelli, L.: Tight Bounds for Selfish and Greedy Load Balancing. In: Bugliesi, M., Preneel, B., Sassone, V., Wegener, I. (eds.) ICALP 2006. LNCS, vol. 4051, pp. 311–322. Springer, Heidelberg (2006)
7. Christodoulou, G., Koutsoupias, E.: The Price of Anarchy of Finite Congestion Games. In: STOC 2005, pp. 67–73 (2005)
8. Correa, J.R., Schulz, A.S., Stier Moses, N.E.: Selfish Routing in Capacitated Networks. Mathematics of Operations Research 29(4), 961–976 (2004)
9. Epstein, A., Feldman, M., Mansour, Y.: Efficient Graph Topologies in Network Routing Games. In: NetEcon+IBC 2007 (2007)
10. Epstein, A., Feldman, M., Mansour, Y.: Strong Equilibrium in Cost Sharing Connection Games. In: EC 2007, pp. 84–92 (2007)
11. Even-Dar, E., Kesselman, A., Mansour, Y.: Convergence Time to Nash Equilibria in Load Balancing. ACM Transactions on Algorithms 3(3) (2007)
12. Fabrikant, A., Papadimitriou, C., Talwar, K.: The Complexity of Pure Nash Equilibria. In: STOC 2004, pp. 604–612 (2004)
13. Fotakis, D.: Stackelberg Strategies for Atomic Congestion Games. In: Arge, L., Hoffmann, M., Welzl, E. (eds.) ESA 2007. LNCS, vol. 4698, pp. 299–310. Springer, Heidelberg (2007)
14. Gairing, M., Lücking, T., Mavronicolas, M., Monien, B., Rode, M.: Nash Equilibria in Discrete Routing Games with Convex Latency Functions. In: Díaz, J., Karhumäki, J., Lepistö, A., Sannella, D. (eds.) ICALP 2004. LNCS, vol. 3142, pp. 645–657. Springer, Heidelberg (2004)
15. Gairing, M., Lücking, T., Monien, B., Tiemann, K.: Nash Equilibria, the Price of Anarchy and the Fully Mixed Nash Equilibrium Conjecture. In: Caires, L., Italiano, G.F., Monteiro, L., Palamidessi, C., Yung, M. (eds.) ICALP 2005. LNCS, vol. 3580, pp. 51–65. Springer, Heidelberg (2005)

16. Holzman, R., Law-Yone, N.: Strong Equilibrium in Congestion Games. Games and Economic Behaviour 21, 85–101 (1997)
17. Holzman, R., Law-Yone, N.: Network Structure and Strong Equilibrium in Route Selection Games. Mathematical Social Sciences 46, 193–205 (2003)
18. Ieong, S., McGrew, R., Nudelman, E., Shoham, Y., Sun, Q.: Fast and compact: A simple class of congestion games. In: AAAI 2005, pp. 489–494 (2005)
19. Papadimitriou, C.H., Koutsoupias, E.: Worst-Case Equilibria. In: Meinel, C., Tison, S. (eds.) STACS 1999. LNCS, vol. 1563, pp. 404–413. Springer, Heidelberg (1999)
20. Monien, B., Mavronicolas, M., Lücking, T., Rode, M.: A New Model for Selfish Routing. In: Diekert, V., Habib, M. (eds.) STACS 2004. LNCS, vol. 2996, pp. 547–558. Springer, Heidelberg (2004)
21. Milchtaich, I.: Network Topology and the Efficiency of Equilibrium. Games and Economic Behaviour 57, 321–346 (2006)
22. Milchtaich, I.: The Equilibrium Existence Problem in Finite Network Congestion Games. In: Spirakis, P.G., Mavronicolas, M., Kontogiannis, S.C. (eds.) WINE 2006. LNCS, vol. 4286, pp. 87–98. Springer, Heidelberg (2006)
23. Rosenthal, R.W.: A Class of Games Possessing Pure-Strategy Nash Equilibria. International Journal of Game Theory 2, 65–67 (1973)
24. Roughgarden, T.: The Price of Anarchy is Independent of the Network Topology. Journal of Computer and System Sciences 67(2), 341–364 (2003)

The Price of Anarchy on Uniformly Related Machines Revisited

Leah Epstein[1] and Rob van Stee[2,⋆]

[1] Department of Mathematics, University of Haifa, 31905 Haifa, Israel
lea@math.haifa.ac.il
[2] Max-Planck-Institut für Informatik, Saarbrücken, Germany
vanstee@mpi-inf.mpg.de

Abstract. Recent interest in Nash equilibria led to a study of the *price of anarchy* (POA) and the *strong price of anarchy* (SPOA) for scheduling problems. The two measures express the worst case ratio between the cost of an equilibrium (a pure Nash equilibrium, and a strong equilibrium, respectively) to the cost of a social optimum.

We consider scheduling on uniformly related machines. Here the atomic players are the jobs, and the delay of a job is the completion time of the machine running it, also called the load of this machine. The social goal is to minimize the maximum delay of any job, while the selfish goal of each job is to minimize its own delay, that is, the delay of the machine running it.

While previous studies either consider identical speed machines or an arbitrary number of speeds, focusing on the number of machines as a parameter, we consider the situation in which the number of different speeds is small. We reveal a linear dependence between the number of speeds and the POA. For a set of machines of at most p speeds, the POA turns out to be exactly $p+1$. The growth of the POA for large numbers of related machines is therefore a direct result of the large number of potential speeds. We further consider a well known structure of processors, where all machines are of the same speed except for one possibly faster machine. We investigate the POA as a function of both the speed of the fastest machine and the number of slow machines, and give tight bounds for nearly all cases.

1 Introduction

Scheduling on uniformly related machines is a basic assignment problem. In such problems, a set of jobs $J = \{j_1, j_2, \ldots, j_n\}$ is to be assigned to a set of m machines $M = \{M_1, \ldots, M_m\}$, where machine M_i has a speed s_i. The size of job j_k is denoted by p_k and it is equal to its running time on a unit speed machine. Moreover, the running time of this job on a machine of speed s is $\frac{p_k}{s}$. An assignment or schedule is a function $S : J \to M$. The completion time of

⋆ Research supported by the German Research Foundation (DFG). Work performed while the author was at the University of Karlsruhe, Germany.

B. Monien and U.-P. Schroeder (Eds.): SAGT 2008, LNCS 4997, pp. 46–57, 2008.

machine M_i, which is also called the delay of this machine, is $\sum_{k:S(j_k)=M_i} \frac{p_k}{s_k}$. The cost, or the *social cost* of a schedule is the maximum delay of any machine, i.e., the makespan.

Following recent interest of computer scientists in game theory [19,14,22], we study pure Nash equilibria and strong equilibria for this scheduling problem, and in particular the cost (makespan) of such equilibria compared to the cost of an optimal schedule. However, in contrast to previous work, we focus on the number of *speeds* of the machines as a parameter, and show tight bounds depending on this parameter. Our results show that the number of available speeds is an important parameter which is by itself enough to give tight bounds. In many practical settings, only few different speeds will be available. For instance, each speed could represent a different available technology.

We next define pure equilibria for scheduling problems. We see jobs as atomic players, thus we use terms such as choice and benefit for these players. A schedule is a *Nash equilibrium* if there exists no job that can decrease its delay by migrating to a different machine unilaterally. More precisely, consider an assignment $S : J \to \{M_1, \ldots, M_m\}$. The class of schedules $\mathcal{S}$ contains all schedules S' that differ from S only in the assignment of a single job. That is $S' \in \mathcal{S}$ if there exists a job k such that $S'(j_\ell) = S(j_\ell)$ for all $\ell \neq k$ and $S'(j_k) \neq S(j_k)$. We say that S is a (pure) Nash equilibrium if for any job j_k, the delay of j_k in any schedule $S' \in \mathcal{S}$, for which $S'(j_k) \neq S(j_k)$, is no smaller than its delay in S. Pure Nash equilibria do not necessary exist for all games (as opposed to mixed Nash equilibria). It is known that for scheduling games of this type, a pure Nash equilibrium always exists [10,6].

A schedule is a *strong equilibrium* if there exists no (non-empty) subset of jobs, such that if all jobs in this set migrate to different machines of their choice simultaneously, this results in a smaller delay for each and every one of them. More precisely, given a schedule S, we can define a class of schedules $\mathcal{S}$ which contains all sets of schedules $\mathcal{S}_K$, where $K \subseteq J$, $K \neq \emptyset$. For any $S' \in \mathcal{S}_K$, and $\ell \notin K$, we have $S'(j_\ell) = S(j_\ell)$ whereas for $\ell \in K$, we have $S'(j_\ell) \neq S(j_\ell)$. S is a strong equilibrium if for any $K \neq \emptyset$, and any $S' \in \mathcal{S}_K$, there exists at least of job $j_k \in K$ whose delay in S_K is no smaller than its delay in S. A strong equilibrium is always a pure Nash equilibrium (by definition). Strong equilibria do not necessarily exist. Andelman, Feldman and Mansour [1] were the first to study strong equilibria in the context of scheduling and proved that scheduling games (of a more general form) admit strong equilibria. More general studies of the classes of congestion games which admit strong equilibria were studied in [12,24].

In this paper, we study the price of anarchy (POA) and the strong price of anarchy (SPOA) for scheduling on uniformly related machines.

In our scheduling model, the *coordination ratio*, or *price of anarchy (*POA*)* (see [21]) is the worst case ratio between the cost of a pure Nash equilibrium and the cost (i.e., maximum delay or makespan) of an optimal schedule. Such an optimal schedule as well as its cost are denoted by OPT. The *strong price of anarchy (*SPOA*)* is defined similarly, but only strong equilibria are considered.

Therefore we refer to the pure price of anarchy by POA and when we discuss the mixed price of anarchy we call it the mixed POA. Note that a pure equilibrium is a special case of mixed equilibria.

It is noted an a series of papers (e.g., [14,18,20,4,3]) the model we study is a simplification of problems arising in real networks, that seems appropriate for describing basic problems in networks.

A number of papers studied equilibria for scheduling on uniformly related machines [14,18,4,7,8]. Chumaj and Vöcking [4] showed that the POA is $\Theta(\frac{\log m}{\log\log m})$ (and $\Theta(\frac{\log m}{\log\log\log m})$ for mixed strategies). Feldmann et al. [7] proved that the POA for $m = 2$ and $m = 3$ is $\frac{\sqrt{4m-3}+1}{2}$ which equals $\phi = \frac{\sqrt{5}+1}{2}$ for two machines and 2 for three machines. In [5], the exact POA and SPOA for two machines is found as a function of the machine speeds. The two measures given different results for the interval $(1.618, 2.247)$ of speeds ratios between the two machines, and identical results otherwise. As for the mixed POA, it was shown in [14] that it is at least $1 + \frac{s}{s+1}$ for $s \leq \phi$. Recently, Fiat et al. [8] showed that the SPOA for this model is $\Theta(\frac{\log m}{(\log\log m)^2})$.

For m identical machines, the POA is $\frac{2m}{m+1}$ which can be deduced from the results of [9] (the upper bound) and [23] (the lower bound). It was shown in [1] that the SPOA has the same value as the POA for every m. Note, however, that the mixed POA is non-constant already in this case, and equals $\Theta(\frac{\log m}{\log\log m})$, where the lower bound was shown by Koutsoupias and Papadimitriou [14] and the upper bound by Chumaj and Vöcking [4] and independently by Koutsoupias, Mavronicolas and Spirakis [13]. Tight bounds of $\frac{3}{2}$ on the mixed POA for two identical machines were shown by [14].

It can be seen that the POA and SPOA were studied mainly as a function of the number of machines. Another relevant parameter for uniformly related machines is the number of different speeds. A natural question is whether the POA and SPOA grow as the number of machines increases even if the number of different speeds is constant, or whether it is actually the number of speeds that needs to increase. Previous results, and in particular, the POA for identical machines already hints that the second option is the right one. We prove this property formally, specifically, we show that the POA for inputs with at most p different speeds is exactly $p + 1$. We note that it can be deduced from [8] that the SPOA for inputs with at most p different speeds is $\Omega(\frac{p}{\log p})$ (and $O(p)$ by our result), therefore the SPOA is quite close to the POA and is influenced by the number of different speeds as well. By the results mentioned above [4,14,13], the mixed POA can not be bounded as a function of p, since it can be arbitrarily large even for $p = 1$.

We further focus on a well known configuration of machines, which consists of a single "fast machine" of speed $s \geq 1$ together with $m - 1$ unit speeds machines. Such a structure, where one processor is fast and all others are identical, is natural and was studied in [17,11,2,16,15]. We give a nearly complete analysis of the exact POA as a function of the speed of the faster machines s and the number of identical machines $m' = m - 1$. We believe that our analysis contributes to a deeper understanding of the POA as a function of several parameters, rather than as a function of the number of machines as a single parameter. Our results imply that the worst case POA (the supremum POA over all values of s and m) for this special case of two different speeds is already 3. We conclude the paper by showing that the worst case SPOA for this variant is strictly smaller than the POA, already in this special case, but it is still strictly larger than the SPOA for m identical machines.

2 A Tight Bound on the Poa for p Speeds

In this section, we consider the general case of a machine set with a fixed number of different speeds, and show that the POA is linearly dependent on the number of speeds, namely, it is $p+1$ if there are p different speeds. We use ingredients of the proofs in [4], focusing on the load in different groups of machines. We assume $p > 1$ otherwise we get the case of identical machines, for which a tight bound is known [9,23,1].

Theorem 1. *The price of anarchy on m related machines that have at most p different speeds is exactly $p+1$.*

Proof. We first show the upper bound. Consider a job assignment to machines, denoted by S, that satisfies the conditions of a Nash equilibrium. Let $\sigma_1 \geq \ldots \geq \sigma_p$ be a sorted list of the speeds. We define the speed class ℓ as the subset of machines with speed σ_ℓ. We assume that machines are numbered by $1, \ldots, m$, and their speeds $s_1, \ldots, s_m$ are sorted by non-increasing speed (i.e., $s_1 \geq s_2 \geq \ldots \geq s_m$). Moreover, we assume that the machines of each speed class are sorted by non-increasing load in S. Let T be the maximum load over all machines and scale the instance so that OPT $= 1$. Assume $T > 1$, otherwise we are done. Note that since some machine has load that exceeds 1, there must exist at least one machine whose load is strictly smaller than 1.

Let C be the load of the least loaded machine of speed class 1, by the order defined above, that is, a machine r of speed $s_r = \sigma_1$ such that $s_{r+1} = \sigma_2$. We claim that $C \geq T - 1$. If the maximum load is achieved on this machine, then we have $C = T$ and we are done. Otherwise, let k be a machine of load T. For every job j of the instance, an optimal solution (which has makepsan 1) runs j on one of the machines, which we denote by i. Therefore we have that its size satisfies $w_j \leq s_i \leq \sigma_1$ and thus $\frac{w_j}{\sigma_1} \leq 1$. Since moving a job from machine k to machine r is not beneficial, for such a job we have $T \leq C + \frac{w_j}{\sigma_1} \leq C + 1$. This proves the claim. If $C \leq 1$ then $T \leq 2 < p+1$. Therefore we assume $C > 1$.

We introduce additional notations. Let $C' = \lceil C \rceil \geq 2$, and let $J_1, \ldots, J_{C'-1}$ and $I_1, \ldots, I_{C'-1}$ be indices of machines. We let I_i be the first machine with load strictly less than $C' - i$, and $J_i = I_i - 1$. We show that the values J_i are actual indices of machines (i.e., $J_i \geq 1$ for $i \geq 1$). Since machine r has load C and by definition $C' < C + 1$, we have that machine r has load $C > C' - 1$. By the ordering of machines, machines $1, \ldots, r-1$ have a load of at least $C' - 1$ as well. By the definition of the indices I_i, we have $I_1 \geq r+1$ and thus $J_1 \geq r \geq 1$.

Moreover, $I_i \geq I_{i-1}$ for all $2 \leq i \leq C' - 1$, thus we actually have $J_i \geq 1$ for all $i \geq 1$.

Thus the load of machines $1, \ldots, J_i$ is at least $C' - i$. Note that $I_{C'-1}$ is the first machine with load less than $C' + 1 - C' = 1$, so this last index must exist, since some machine must have load less than 1. However $I_{C'}$ cannot exist since this would imply a machine of load less than 0.

We now claim that the speed of I_i is no larger than σ_{i+1} for $i = 1, \ldots, C'$. We prove this by induction. For $i = 1$ we showed that $I_1 \geq r + 1$, so its speed is at most σ_2. For other values of i, we prove that the speed of I_i is strictly smaller than the speed of I_{i-1}. Let s' be the speed of I_{i-1}. All machines up to J_{i-1} have load of at least $C' - (i - 1) = C' + 1 - i > 1$ since $i \leq C' - 1$. Recall that $I_i \geq r + 1$ for $i \geq 1$. We showed that in S, machines $1, \ldots, J_{i-1}$ are loaded by more than 1. Thus in this schedule they must have a job that OPT schedules on one of the machines $I_{i-1}, \ldots, m$. Denote such a job and its size by a. The machine that runs it in S has load of at least $C' + 1 - i$. Let y be the machine to which a is assigned in OPT. We have $a \leq s_y \leq s'$ and $J_{i-1} < I_{i-1} \leq I_i$. If the speed of machine I_i is s' as well, moving job a to I_i will result in load of less than $(C' - i) + 1$, which would be a contradiction to S being a Nash equilibrium, since the load of the machine running a in S is larger.

From this claim it follows that the speed of $I_{C'-1}$ is at most $\sigma_{C'}$, i.e., $C' \leq p$ (since σ_p is the smallest speed). We conclude that $T \leq C + 1 \leq C' + 1 \leq p + 1$.

We now show a matching lower bound. Let $\varepsilon > 0$ be such that $1/\varepsilon \in \mathbb{N}$. We consider a set of machines with speeds in the set $\{2^{p-1}, 2^{p-2}, \ldots, 1\}$ for some integer $p \geq 2$. There are N_i machines of speed 2^i, where N_i will be determined later. In OPT, each machine of speed 2^i has a job of size $(1 - \varepsilon)2^i$, for $i \geq 1$. $4N_1$ of the machines of speed 1 have a single job of size $1 - \varepsilon$ and the rest have sand (jobs of size ε) of total size 1. We will define N_0 to be large enough to ensure $N_0 \geq 4N_1$. Therefore OPT = 1.

In the Nash equilibrium that we define, there is one machine of speed 2^{p-1} which contains $p + 1$ jobs of size $(1 - \varepsilon)2^{p-1}$. We let $N_{p-1} = p + 1$. Each one of the other machines of speed 2^{p-1} contains $2p$ jobs of size $(1 - \varepsilon)2^{p-2}$. We let $N_{p-2} = 2p(N_{p-1} - 1) = 2p^2$. For $1 \leq i \leq p - 2$, each machine of speed 2^i in the Nash equilibrium contains $2(i + 1)$ jobs of size $(1 - \varepsilon)2^{i-1}$. Therefore, for these values of i (except for $i = 1$), $N_{i-1} = 2(i + 1)N_i$. We let $N_0 = 4N_1/\varepsilon$. Thus if in the Nash equilibrium, each machine of speed 1 has a total of $1 - \varepsilon$ of sand, and in OPT, each machine except $4N_1$ machines have a total of 1 of sand, we get that the amount of sand is constant; $4N_1/\varepsilon - 4N_1 = (1 - \varepsilon)4N_1/\varepsilon$.

Moreover, the load of a machine of speed 2^i is $(1 - \varepsilon)(i + 1)$, except for one machine of speed 2^{p-1} which has a load of $(1 - \varepsilon)(p + 1)$.

To show that this is indeed a Nash equilibrium. We do not need to consider cases in which jobs move to faster machines, since they are more loaded. We first consider the case where a job of size $(1-\varepsilon)2^{p-1}$ moves from the machine of speed 2^{p-1} that contains all jobs of this size, to a machine of some speed 2^j $(j \leq p-1)$. It increases the load of the target machine by $(1 - \varepsilon)2^{p-1-j}$. The load of this machine was $(1-\varepsilon)(j+1)$, so we need to show $(1-\varepsilon)(j+1+2^{p-1-j}) \geq (1-\varepsilon)(p+1)$

or $2^{p-1-j} \geq p-j$. It is enough to show $2^{t-1} \geq t$ for $t \geq 1$. This is easily shown by induction.

We now consider a job of size $(1-\varepsilon)2^i$ moving from a machine of speed 2^{i+1} to a machine of speed 2^j, where $j \leq i$. The load of the target machine increases by $(1-\varepsilon)2^{i-j}$. The load there was $(1-\varepsilon)(j+1)$ so we need to show $2^{i-j}+j+1 \geq i+2$ for $i-j \geq 0$. Taking $t = i-j+1$, we again get $2^{t-1} \geq t$. □

Note that the SPOA increases rapidly as a function of the number of speeds as well. The lower bound construction of Fiat et al. [8] uses a parameter ℓ, such that the SPOA is $\Omega(\ell)$ and the number of speeds is $\Theta(\ell \log \ell)$. This implies a lower bound of $\Omega(\frac{p}{\log p})$ on the SPOA for instances with at most p different speeds.

3 One Fast Machine

Recall that the architecture of processors that we consider there consists of $m' = m-1$ identical slow machines of speed 1 (where $m' \geq 2$, since the case $m' = 1$ is fully covered in [5]) and one fast machine of speed s. We scale all sizes of jobs in the instances we consider so that OPT $= 1$. We can therefore assume that the sum of jobs sizes is at most $s + m'$. Moreover, all slow machines contain in an optimal schedule only jobs that are no larger than 1, and the largest job of any instance is no larger than s. Denote the load on the fast machine by x, and the number of jobs there by t. If $x > 1$ then the total size of jobs on the fast machine is $xs > s$ and therefore this machine must contain at least one job that is of size no larger than 1.

The price of anarchy is determined by the highest possible load of any machine. Obviously, if there is a machine with load above 1, there must also be a machine with load less than 1. To prove upper bounds we consider two basic cases; the price of anarchy is either determined by the fast machine, or by some other machine. We assume that we are given a specific schedule with is a pure Nash equilibrium and study its properties.

3.1 Tight Bounds for $1 \leq s \leq 2$ and All $m' \geq 2$

We define

$$\textsc{FastMax} = \frac{2m'+s}{m'+s} = 2 - \frac{s}{m'+s}$$

$$\textsc{SmallJobs}(t) = \frac{1+\frac{s}{m'}}{1+\frac{s}{m'}-\frac{s}{t}} = \frac{t(m'+s)}{t(m'+s)-m's}$$

We prove in the following lemma that $\textsc{SmallJobs}(t)$ is an upper bound for the load on the fast machine in case there are t jobs on the fast machine, and $t \geq xs$ (thus, the jobs have average size of at most 1).

Some of the lemmas hold not only for $s \leq 2$, and are used in other sections as well. When this is the case, we state it explicitly. Otherwise we may assume $s \leq 2$.

Lemma 1. *If $x > 1$, then $x \leq$ FASTMAX. If in addition $t \geq xs$, then $x \leq$ SMALLJOBS(t). This holds for any $s \geq 1$.*

Proof. The average load on the slow machines is at most

$$\frac{s + m' - xs}{m'} = 1 - (x-1)\frac{s}{m'} . \tag{1}$$

Since $x > 1$, and the optimal makespan is 1, there exists a job of size at most 1 on the fast machine. This job does not reduce its delay by moving to the least loaded slow machine. If it moves, the load on the machine that it moves to becomes at most $2 - (x-1)\frac{s}{m'}$. Therefore, this value must be at least x. This implies $x(1 + \frac{s}{m'}) \leq 2 + \frac{s}{m'}$, and therefore $x \leq$ FASTMAX.

If there are t jobs on the fast machine, the average size of jobs there is xs/t, so among these jobs there is at least one job of size at most xs/t. This constraint does not add new information unless $t > xs$, we therefore assume $t \geq xs$, and therefore $t \geq s$. Once again, since this job does not benefit from moving to the least loaded slow machine, using (1), we find $x \leq 1-(x-1)\frac{s}{m'} + \frac{xs}{t}$ which implies $x(1 + \frac{s}{m'} - \frac{s}{t}) \leq 1 + \frac{s}{m'}$, and therefore $x \leq$ SMALLJOBS(t) (since by $t \geq s$, we have $t(m'+s) - m's \geq ts > 0$). □

Lemma 2. *Assume that $t(m'+s) - m's > 0$. We have* SMALLJOBS$(t) \leq$ FASTMAX *if and only if* $\frac{s}{t}$FASTMAX ≤ 1.

Lemma 3. *If there are t jobs on the fast machine, we have $x \leq \min($*FASTMAX, SMALLJOBS$(t))$.

Proof. We assume $x > 1$, otherwise the claim holds trivially. The first term is an upper bound by Lemma 1. If SMALLJOBS$(t) \leq$ FASTMAX, then we have $\frac{s}{t}$FASTMAX ≤ 1 by Lemma 2. Since $x \leq$ FASTMAX, Lemma 1 implies that $x \leq$ SMALLJOBS(t). □

Definition 1. *Let y be the highest load of any slow machine. Let M_y be a slow machine with this load. Let z be the smallest load of any slow machine.*

Lemma 4. *If there is only one job on M_y, then $y \leq s$. If there are at least two jobs, then $y \leq 2z$ and $y \leq \frac{2(m'+s)}{m'+2s}$. This holds for any $s \geq 1$.*

Proof. The first bound follows as there cannot be a job larger than s if the optimal makespan is 1.

Suppose there are at least two jobs and $y > 2z$. The smallest job on M_y has size at most $y/2$ and (using $m' \geq 2$) it can improve by moving to a machine with load z where the load will be at most $z + y/2 < y$. Thus this is not an equilibrium, a contradiction.

Therefore $z \geq y/2$. Since none of the jobs on M_y can improve by moving to the fast machine, we find $y \leq x + y/(2s)$ or $x \geq \frac{2s-1}{2s}y$. Since the total size of jobs is at most $m' + s$, this implies $(m'-1)\frac{y}{2} + \frac{2s-1}{2}y + y \leq m' + s$, which gives

$$y\left(\frac{m'-1}{2} + \frac{2s-1}{2} + 1\right) = \frac{y}{2}(m' - 1 + 2s - 1 + 2)) = \frac{y}{2}(m' + 2s) \leq m' + s,$$

which implies the desired bound. □

Theorem 2. *For $s \le 2$ and $m' \ge 2$, we have*

$$\text{POA} = \max\left(\min\left(\textsc{SmallJobs}(2), \textsc{FastMax}, 1 + \frac{1}{s}\right),\right.$$
$$\left.\min\left(\textsc{SmallJobs}(3), \textsc{FastMax}\right), \frac{2(m'+s)}{m'+2s}, s\right).$$

Proof. The four terms represent the following situations in order: two jobs on the fast machine, at least three jobs on the fast machine, at least two jobs on M_y, one job of size s on M_y.

It is easy to see that this covers all the relevant possibilities: if there is only one job on the fast machine, then the POA is achieved on M_y since $x \le 1$. Therefore the upper bound will follow from showing the relevant upper bound in each one of the cases, according to the term which achieves the maximum.

In the examples for the lower bound, if the POA is achieved on the fast machine, all other machines will contain sand, that is, jobs of very small size. In such a case, each machine will receive the same amount of sand, which in all cases would be less than 1. This already ensures that none of these jobs can improve their delay moving to the fast machine (where the load will be more than 1). Thus we only need to check that the jobs on the fast machine cannot benefit from moving.

Due to space constraints, we only discuss one case here: the POA is achieved on the fast machine, where there are two jobs, and $\textsc{SmallJobs}(2) \le \min(1 + \frac{1}{s}, \textsc{FastMax})$. To prove the upper bound, we note that the first two terms in the minimum are implied by Lemma 3. The last term follows because the total size of any two jobs is at most $s+1$ if the optimal makespan is 1. We now show matching lower bounds using suitable instances for all three terms in the minimum.

We use $\textsc{SmallJobs}(2) \le \textsc{FastMax}$ to show that it is possible to enforce $x = \textsc{SmallJobs}(2)$. We have $\frac{s}{2}\textsc{SmallJobs}(2) \le \frac{s}{2}\textsc{FastMax} \le 1$ by Lemma 2. Consider the following instance. There are two jobs of size $\textsc{SmallJobs}(2) \cdot \frac{s}{2} \le 1$ which are running on the fast machine, i.e., $t = 2$. The total amount of sand is $m' + s - s \cdot \textsc{SmallJobs}(t)$. Each slow machine has sand, where the amount of sand on each slow machine is $1 - \frac{s^2}{t(m'+s) - m's}$. The optimal makespan is 1, by putting each large job on one machine, and adding sand to achieve an equal load on the machines. This schedule is an equilibrium since by moving a large job to a slow machine we get a delay of $1 - \frac{s^2}{t(m'+s)-m's} + \frac{t(m'+s)}{t(m'+s)-m's} \cdot \frac{s}{t} = \frac{t(m'+s)}{t(m'+s)-m's} = x$.

The other cases are treated similarly. □

Corollary 1. *For $s = 2$, POA $= 2$ for all $m' \ge 2$. For $1 \le s < 2$, POA < 2 for all $m' \ge 2$.*

Proof. All the upper bounds in Theorem 2 are at most 2 for $s = 2$ and any $m' \ge 2$, and the bound s is equal to 2. The second claim follows immediately from Lemma 3 and Lemma 4 (if $y > s$, then $y \le \frac{2m'+2s}{m'+2s} < 2$). □

3.2 Global Upper and Lower Bounds for the Poa for $s > 2$

Theorem 3. *For $2 \leq s < 3$ and large enough m', POA $= s$. For all $s \geq 2$ and $m' \geq 2$, POA $\leq s$.*

Proof. Fix $\varepsilon \in (0, 1]$. We will show a lower bound of s on the POA for $s = 3 - \varepsilon$ and m' large enough. Consider the following schedule.

There is one job of size s which is scheduled on a dedicated machine. There are six jobs of size $s(s-1)/6$ which are on the fast machine, so its load is $s - 1$. The remaining $m' - 1$ machines have sand, specifically, each machine has an amount of $(s-1) \cdot (1 - \frac{s}{6})$ which is less than 1 for $s < 3$. The amount of sand per machine ensures that none of the six jobs on the fast machine improves by moving to a slow machine: if such a job moves there, it adds $s(s-1)/6$ to the load, making the total load exactly $s - 1$. We need to make sure that the total size of all the jobs we use is not more than $m' + s$. This implies

$$m' + s \geq s + s(s-1) + (m'-1)(s-1)\left(1 - \frac{s}{6}\right)$$
$$\Rightarrow m' \geq \frac{7s^2 - 13s + 6}{s^2 - 7s + 12} = \frac{(7s-6)(s-1)}{(3-s)(4-s)} = \frac{(15-7\varepsilon)(2-\varepsilon)}{\varepsilon(1+\varepsilon)} .$$

For any $\varepsilon > 0$, this value is bounded from above. Since $x < 2$ by Lemma 1 and $y \leq s$ by Lemma 4 (the second bound there is at most $2 \leq s$), this proves the theorem. □

Lemma 5. *For any equilibrium instance, there exists an instance that is an equilibrium with the same loads on all machines, such that the fast machine has at most one job which is on the fast machine in the optimal solution. Specifically, it has at most one job larger than 1.*

Proof. If there are multiple such jobs, we can merge them into one job with size the total size of these jobs. This does not affect the optimal makespan, or the makespan of the schedule. Larger jobs can only benefit less from moving, thus the schedule is still an equilibrium if it was before. Regarding the second statement, clearly all jobs larger than 1 must be on the fast machine in an optimal solution with makespan 1. □

Lemma 6. *Any schedule that is in equilibrium satisfies*

$$y \leq \frac{xs}{s-1} . \tag{2}$$

Moreover, if M_y has a single job, this is a sufficient condition for this job not to benefit from moving.

Proof. Consider M_y. This machine has a job of size at most y, which does not benefit from moving to the fast machine. Therefore $y \leq x + \frac{y}{s}$, which implies the upper bound. If there is a single job of size y, then this is not only a necessary condition but also a sufficient condition. □

Lemma 7. *For $s \geq 2$ and $m' \geq 2$, if there exists an an equilibrium schedule where the* POA *is achieved on a slow machine, then the* POA *is achieved in an instance with $t \geq 2$.*

Note: this holds even after possibly merging jobs as in the proof of Lemma 5.

Proof. Suppose there is at most one job on the fast machine. The total size of the jobs on the fast machine and M_y (together) is then at most $s+1$. This means that $xs + y \leq s + 1$, or $xs \leq s + 1 - y$. But then Lemma 6 implies $y \leq \frac{s+1-y}{s-1}$, and therefore $y(1 + \frac{1}{s-1}) \leq \frac{s+1}{s-1}$, or $y \leq \frac{s+1}{s}$. But this value is smaller than $2 - s/(m' + s)$ for $s \geq 2$ and $m' \geq 2$. To prove this we note that $2 - s/(m' + s)$ is increasing in m'. For $m' = 2$, it is equal to $2 - s/(2 + s) = 1 + \frac{2}{2+s}$. However $\frac{2}{2+s} \geq \frac{1}{s}$ for $s \geq 2$. □

Definition 2. *Let*

$$\text{GLOBMAX} = \frac{s + 2m' - 1}{s + (m' - 1)(s - 1)/s} . \tag{3}$$

Lemma 8. *We have* FASTMAX < GLOBMAX *for all $s \geq 1$, $m' \geq 2$.*

Theorem 4. *For $s \geq 2$,* POA ≤ GLOBMAX.

Proof. By Lemma 8, the lemma holds if the POA is achieved on the fast machine. Therefore, suppose it is achieved on a slow machine M_y in some schedule. Denote the load there by $y > 1$. Then by Lemma 6, the load on the fast machine is at least $x = y \cdot \frac{s-1}{s}$, so the work there is $y(s - 1)$. By Lemma 5, the fast machine has at most one job larger than 1. By Lemma 7, the fast machine has at least two jobs, such at at least one of them is scheduled on a slow machine in an optimal schedule. Therefore, there is at least one job of size at most 1 on the fast machine. If this instance is in equilibrium, the load on each slow machine must then be at least $x - 1$. Finally, the total size of all the jobs must be at most $m' + s$. This implies

$$y\left(1 + (s-1) + (m'-1)\frac{s-1}{s}\right) - (m'-1) \leq s + m' \tag{4}$$

which holds if $y \leq \frac{s+2m'-1}{s+(m'-1)(s-1)/s} = \text{GLOBMAX}$. This proves the lemma. □

Let $m' \geq 2$ and $s \geq 2$. For this case, we show in the full paper that

$$(s - 1)\text{GLOBMAX} - \lceil s(\text{GLOBMAX} - 1)\rceil \geq 1 \tag{5}$$

implies that POA = GLOBMAX, and that this condition is satisfied as long as $s \geq \frac{5+\sqrt{17}}{2} \approx 4.562$. In the following table, for several values of m' the minimum value of s is given such that we can be certain that POA = GLOBMAX for all speeds of at least s, rounded to two decimal places.

m'	2	3	4	5	6	7	8
s	2.77	3.25	3.78	3.56	3.41	3.29	3.89

(6)

In all of these cases, (5) holds. Indeed, as m' grows large relative to s, GLOBMAX tends to $2/((s-1)/s) = 2s/(s-1)$. Then $s(\text{GLOBMAX}-1) = s(2s-s+1)/(s-1) = s(s+1)/(s-1)$ and $(s-1)\text{GLOBMAX} = 2s$. For these values, inequality (5) holds for $s \geq 4$, so for large m', the bound for s above which POA = GLOBMAX tends to 4.

Using a computer program, it can be found that in fact POA = GLOBMAX for $s \geq 4.365$ for all m', and that the value of m' for which the bound on s is maximized is 31.

3.3 SPOA for One Fast Machine

In the full paper, we demonstrate the fact that the SPOA is strictly smaller than the POA. We consider the overall bounds (i.e., the supremum bounds over all values of s and m') and compare them. The overall bound on the POA as implied by the previous sections is 3.

Theorem 5. *The* SPOA *is 2 for $m' \leq 5$. For any m',* SPOA $\leq \frac{3+\sqrt{5}}{2}$. *For $m' \geq 16$,* SPOA $\geq \frac{1+\sqrt{13}}{2} \approx 2.3027756$.

4 Conclusions

In this paper, we have shown the following results. The price of anarchy on uniformly related machines with at most p different speeds is $p+1$. For two speeds, this upper bound is approachable even with only one fast machine. However, the POA is only 2 (rather than 3) if the fast machine is twice as fast as the other machines, and it is less than 2 if the machine is slower than that. In the same setting, the SPOA is 2 if there are only few machines, but is between 2.3 and 2.6 for any fixed, sufficiently large m.

References

1. Andelman, N., Feldman, M., Mansour, Y.: Strong price of anarchy. In: Proc. of the 18th ACM-SIAM Symposium on Discrete Algorithms (SODA 2007), pp. 189–198 (2007)
2. Cho, Y., Sahni, S.: Bounds for List Schedules on Uniform Processors. SIAM Journal on Computing 9(1), 91–103 (1980)
3. Czumaj, A.: Selfish routing on the internet. In: Leung, J. (ed.) Handbook of Scheduling: Algorithms, Models, and Performance Analysis, vol. 42, CRC Press, Boca Raton (2004)
4. Czumaj, A., Vöcking, B.: Tight bounds for worst-case equilibria. ACM Transactions on Algorithms 3(1) (2007)
5. Epstein, L.: Equilibria for two parallel links: The strong price of anarchy versus the price of anarchy. manuscript (2007)
6. Even-Dar, E., Kesselman, A., Mansour, Y.: Convergence time to nash equilibria. In: Baeten, J.C.M., Lenstra, J.K., Parrow, J., Woeginger, G.J. (eds.) ICALP 2003. LNCS, vol. 2719, pp. 502–513. Springer, Heidelberg (2003)

7. Feldmann, R., Gairing, M., Lücking, T., Monien, B., Rode, M.: Nashification and the coordination ratio for a selfish routing game. In: Baeten, J.C.M., Lenstra, J.K., Parrow, J., Woeginger, G.J. (eds.) ICALP 2003. LNCS, vol. 2719, pp. 514–526. Springer, Heidelberg (2003)
8. Fiat, A., Kaplan, H., Levy, M., Olonetsky, S.: Strong price of anarchy for machine load balancing. In: Arge, L., Cachin, C., Jurdziński, T., Tarlecki, A. (eds.) ICALP 2007. LNCS, vol. 4596, pp. 583–594. Springer, Heidelberg (2007)
9. Finn, G., Horowitz, E.: A linear time approximation algorithm for multiprocessor scheduling. BIT Numerical Mathematics 19(3), 312–320 (1979)
10. Fotakis, D., Kontogiannis, S.C., Koutsoupias, E., Mavronicolas, M., Spirakis, P.G.: The structure and complexity of nash equilibria for a selfish routing game. In: Widmayer, P., Triguero, F., Morales, R., Hennessy, M., Eidenbenz, S., Conejo, R. (eds.) ICALP 2002. LNCS, vol. 2380, pp. 124–134. Springer, Heidelberg (2002)
11. Gonzalez, T., Ibarra, O.H., Sahni, S.: Bounds for LPT Schedules on Uniform Processors. SIAM Journal on Computing 6(1), 155–166 (1977)
12. Holzman, R., Law-Yone, N.: Strong equilibrium in congestion games. Games and Economic Behavior 21(1-2), 85–101 (1997)
13. Koutsoupias, E., Mavronicolas, M., Spirakis, P.G.: Approximate equilibria and ball fusion. Theory of Computing Systems 36(6), 683–693 (2003)
14. Koutsoupias, E., Papadimitriou, C.H.: Worst-Case Equilibria. In: Meinel, C., Tison, S. (eds.) STACS 1999. LNCS, vol. 1563, Springer, Heidelberg (1999)
15. Kovács, A.: Tighter Approximation Bounds for LPT Scheduling in Two Special Cases. In: Calamoneri, T., Finocchi, I., Italiano, G.F. (eds.) CIAC 2006. LNCS, vol. 3998, pp. 187–198. Springer, Heidelberg (2006)
16. Li, R., Shi, L.: An on-line algorithm for some uniform processor scheduling. SIAM Journal on Computing 27(2), 414–422 (1998)
17. Liu, J.W.S., Liu, C.L.: Bounds on scheduling algorithms for heterogeneous computing systems. In: Becvar, J. (ed.) MFCS 1979. LNCS, vol. 74, pp. 349–353. Springer, Heidelberg (1979)
18. Mavronicolas, M., Spirakis, P.G.: The price of selfish routing. In: Proc. of the 33rd Annual ACM Symposium on Theory of Computing (STOC 2001), pp. 510–519 (2001)
19. Nisan, N., Ronen, A.: Algorithmic mechanism design. Games and Economic Behavior 35, 166–196 (2001)
20. Papadimitriou, C.H.: Algorithms, games, and the internet. In: Proc. of the 33rd Annual ACM Symposium on Theory of Computing (STOC 2001), pp. 749–753 (2001)
21. Roughgarden, T.: Selfish routing and the price of anarchy. MIT Press, Cambridge (2005)
22. Roughgarden, T., Tardos, É.: How bad is selfish routing? Journal of the ACM 49(2), 236–259 (2002)
23. Schuurman, P., Vredeveld, T.: Performance guarantees of local search for multiprocessor scheduling. Informs Journal on Computing 19(1), 52–63 (2007)
24. Tennenholtz, M., Rozenfeld, O.: Strong and Correlated Strong Equilibria in Monotone Congestion Games. In: Spirakis, P.G., Mavronicolas, M., Kontogiannis, S.C. (eds.) WINE 2006. LNCS, vol. 4286, pp. 74–86. Springer, Heidelberg (2006)

Approximate Strong Equilibrium in Job Scheduling Games

Michal Feldman[1,*] and Tami Tamir[2]

[1] School of Business Administration and Center for the Study of Rationality, Hebrew University of Jerusalem
mfeldman@cs.huji.ac.il
[2] School of Computer Science, The Interdisciplinary Center, Herzliya, Israel
tami@idc.ac.il

Abstract. A Nash Equilibriun (NE) is a strategy profile that is resilient to unilateral deviations, and is predominantly used in analysis of competitive games. A downside of NE is that it is not necessarily stable against deviations by coalitions. Yet, as we show in this paper, in some cases, NE does exhibit stability against coalitional deviations, in that the benefits from a joint deviation are bounded. In this sense, NE approximates *strong equilibrium* (SE) [6].

We provide a framework for quantifying the stability and the performance of various assignment policies and solution concept in the face of coalitional deviations. Within this framework we evaluate a given configuration according to three measurements: (i) IR_{min}: the maximal number α, such that there exists a coalition in which the minimum improvement ratio among the coalition members is α (ii) IR_{max}: the maximum improvement ratio among the coalition's members. (iii) DR_{max}: the maximum possible damage ratio of an agent outside the coalition.

This framework can be used to study the proximity between different solution concepts, as well as to study the existence of approximate SE in settings that do not possess any such equilibrium. We analyze these measurements in job scheduling games on identical machines. In particular, we provide upper and lower bounds for the above three measurements for both NE and the well-known assignment rule *Longest Processing Time* (LPT) (which is known to yield a NE). Most of our bounds are tight for any number of machines, while some are tight only for three machines. We show that both NE and LPT configurations yield small constant bounds for IR_{min} and DR_{max}. As for IR_{max}, it can be arbitrarily large for NE configurations, while a small bound is guaranteed for LPT configurations. For all three measurements, LPT performs strictly better than NE.

With respect to computational complexity aspects, we show that given a NE on $m \geq 3$ identical machines and a coalition, it is NP-hard to determine whether the coalition can deviate such that every member decreases its cost. For the unrelated machines settings, the above hardness result holds already for $m \geq 2$ machines.

* Research partially supported by a grant of the Israel Science Foundation, BSF, Lady Davis Fellowship, and an IBM faculty award.

B. Monien and U.-P. Schroeder (Eds.): SAGT 2008, LNCS 4997, pp. 58–69, 2008.

1 Introduction

We consider job scheduling problems, in which n jobs are assigned to m identical machines and incur a cost which is equal to the total load on the machine they are assigned to[1]. These problems have been widely studied in recent years from a game theoretic perspective [21,3,10,11,15]. In contrast to the traditional setting, where a central designer determines the allocation of jobs into machines and all the participating entities are assumed to obey the protocol, in distributed settings, the situation may be different. Different machines and jobs may be owned by different *strategic* entities, who will typically attempt to optimize their own objective rather than the global objective. Game theoretic analysis provides us with the mathematical tools to study such situations, and indeed has been extensively used in recent years by computer scientists. This trend is motivated in part by the emergence of the Internet, which is composed of distributed computer networks managed by multiple administrative authorities and shared by users with competing interests [24].

Most game theoretic models applied to job scheduling problems, as well as other network games (e.g., [13,2,25,4]), use the solution concept of *Nash equilibrium* (NE), in which the strategy of each agent is a best response to the strategies of all other agents. While NE is a powerful tool for predicting outcomes in competitive environments, its notion of stability applies only to unilateral deviations. However, even when no single agent can profit by a unilateral deviation, NE might still not be stable against a group of agents *coordinating* a joint deviation, which is profitable to *all the members* of the group. This stronger notion of stability is exemplified in the *strong equilibrium* (SE) solution concept, coined by Aumann (1959). In a strong equilibrium, no coalition can deviate and improve the utility of *every* member of the coalition.

As an example, consider the configuration depicted in Figure 1(a). It is a NE since no job can reduce its cost through a unilateral deviation (recall that the cost of each job is defined to be the load on the machine it is assigned to, as assumed in many job scheduling models). One may think that a NE on identical machines is

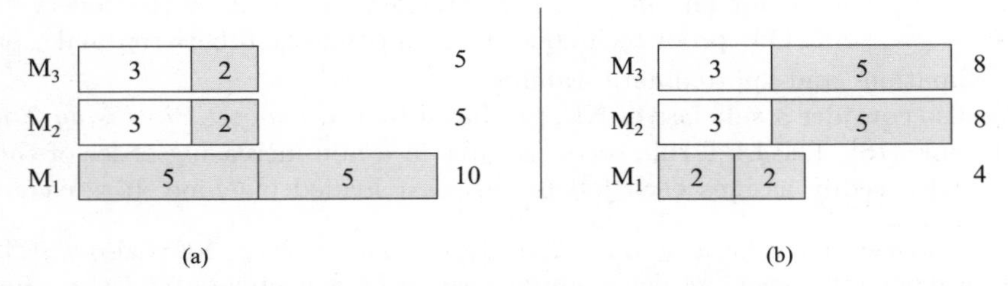

Fig. 1. An example of a configuration (a) that is a Nash equilibrium but is not resilient against coordinated deviations, since the jobs of load $\{5, 5, 2, 2\}$ all profit from the deviation demonstrated in (b)

[1] This cost function characterizes systems in which jobs are processed in parallel, or when all jobs on a particular machine have the same single pick-up time, or need to share some resource simultaneously.

also sustainable against joint deviations. Yet, as was already observed in [3], this may not be true[2]. For example, the configuration above is not resilient against a coordinated deviation of the coalition $\Gamma = \{5, 5, 2, 2\}$ deviating to configuration (b), where the jobs of load 5 decrease their costs from 10 to 8, and the jobs of load 2 improve from 5 to 4. Note that the cost of the two jobs of load 3 (which are not members of the coalition) increases.

In the example above, every member of the coalition improves its cost by a (multiplicative) factor of $\frac{5}{4}$. By how much more can a coalition improve? Is there a bound on the *improvement ratio*? As it will turn out, this example is in fact the most extreme one in a sense that will be clarified below. Thus, while NE is not completely stable against coordinated deviations, in some settings, it does provide us with some notion of approximate stability to coalitional deviations (or *approximate strong equilibrium*).

In this paper we provide a framework for studying the notion of approximate stability to coalitional deviations. In our analysis, we consider three different measurements. The first two measure the stability of a configuration, and the third measures the worst possible effect on the non-deviating jobs.

1. Minimum Improvement Ratio: This notion is discussed in Section 3, and refers to configurations from which no coalition of agents can deviate such that *every* member of the coalition improves by a large factor [3] . Formally, the improvement ratio of a job in the coalition is the ratio between its pre- and post-deviation cost. We say that a configuration s forms an α-SE if there is no coalition in which each agent can improve by a factor of more than α. This notion was also studied by [1] in the context of SE existence. There, the author showed that for a sufficiently large α, an α-SE always exists. The justification behind this concept is that agents may be willing to deviate only if they improve by a sufficiently high factor (due to, for example, some overhead associated with the migration).

For three machines, we show that every NE is a $\frac{5}{4}$-SE. That is, there is no coalition that can deviate such that every member improves by a factor larger than $\frac{5}{4}$. For this case, we also provide a matching lower bound (recall Figure 1 above), that holds for any $m \geq 3$. For arbitrary m, we show that every NE is a $(2 - \frac{2}{m+1})$-SE. Our proof technique draws a connection between makespan approximation[4] and approximate stability.

We also consider a subclass of NE, produced by the *Longest Processing Time* (LPT) rule [18]. The LPT rule sorts the jobs in a non-increasing order of their loads and greedily assigns each job to the least loaded machine. It is easy to

[2] This statement holds for $m \geq 3$. For 2 identical machines, every NE is also a SE [3].

[3] Throughout this paper, we define approximation by a *multiplicative* factor. Since the improvement and damage ratios for all the three measurements presented below are constants greater than one (as will be shown below), the *additive* ratios are unbounded. Formally, for any value a it is possible to construct instances (by scaling the instances we provide for the multiplicative ratio) in which the cost of all jobs is reduced, or the cost of some jobs is increased, by at least an additive factor of a.

[4] makespan is defined as the maximum load on any machine in the configuration.

verify that every configuration produced by LPT is a NE [17]. Is it also a SE? Note that for the instance depicted in Figure 1, LPT would have produced a SE. However, as we show, this is not always the case. Yet, for $m = 3$, every LPT-based configuration is a $\frac{2}{\sqrt{34}-4}$-SE (≈ 1.092), and we also provide a matching lower bound, that holds for any $m \geq 3$. For arbitrary m, we show an upper bound of $\frac{4}{3} - \frac{1}{3m}$. These results indicate that LPT is more stable than NE with respect to coalitional deviations.

2. Maximum Improvement Ratio: In Section 4 we study an alternative notion of approximate stability, in which there is no coalition such that *some* agent improves by a factor of more than α. This notion is similar in spirit to stability against a large *total* improvement. Interestingly, we find out that given a NE configuration, the improvement ratio of a single agent may not be bounded, for any $m \geq 3$. In contrast, for LPT-based configurations on three machines, no agent can improve by a factor of $\frac{5}{3}$ or more and this bound is tight. Thus, with respect to maximum IR, the relative stability of LPT compared to NE is significant. For arbitrary m, we provide a lower bound of $2 - \frac{1}{m}$, which we believe to be tight.

3. Maximum Damage Ratio: As is the case for the jobs of load 3 in Figure 1, some jobs might be hurt from a coalitional deviation. The third measurement that we consider is the worst possible effect of a deviation on these naive jobs. Formally, the *maximum damage ratio* is the maximal ratio between the pre- and post-deviation cost of a job. Note that it does not measure the stability of a configuration – we assume that an agent's motivation to deviate is not influenced by the potential damage it will cause others. However, this measurement is important since it guarantees a bound on the maximal damage that any agent can experience. In Section 5, we prove that the maximum damage ratio is less than 2 for any NE configuration, and less than $\frac{3}{2}$ for any LPT-based configuration. Both bounds hold for any $m \geq 3$, and for both we provide matching lower bounds. Note that the minimum damage ratio is of no practical interest.

In summary, our results in Sections 3-5 (see Table 1) indicate that NE-based configurations are approximately stable with respect to the IR_{min} measurement. Moreover, the performance of jobs outside the coalition would not be hurt by much as a result of a coalitional deviation. It would be interesting to study in what families of games NE are guaranteed to provide approximate SE. As for IR_{max}, our results provide an additional benefit of the LPT rule, which is already known to possess attractive properties (with respect to, e.g., makespan approximation and stability against unilateral deviations).

In Section 6, we study computational complexity aspects of coalitional deviations. We find that it is NP-hard to determine whether a NE configuration on $m \geq 3$ identical machines is a SE. Moreover, given a particular configuration and a set of jobs, it is NP-hard to determine whether this set of jobs can engage in a coalitional deviation. For unrelated machines (i.e., where each job incurs a different load on each machine), the above hardness results hold already for

Table 1. Our results for the three measurements. Unless specified otherwise, the results hold for arbitrary m.

	IR_{min}			IR_{max}		DR_{max}	
	upper bound		lower bound	upper bound	lower bound	upper bound	lower bound
	$m = 3$	$m \geq 3$					
NE	$\frac{5}{4}$	$2 - \frac{2}{m+1}$	$\frac{5}{4}$	unbounded		2	2
LPT	$\frac{2}{\sqrt{34}-4}$	$\frac{4}{3} - \frac{1}{3m}$	$\frac{2}{\sqrt{34}-4}$	$\frac{5}{3}$ $(m=3)$	$2 - \frac{1}{m}$	$\frac{3}{2}$	$\frac{3}{2}$

$m = 2$ machines. These results might have implications on coalitional deviations with computationally restricted agents.

Related work: NE is shown in this paper to provide approximate stability against coalitional deviations. A related body of work studies how well NE approximates the optimal outcome of competitive games. The Price of Anarchy was defined in [24,21] as the ratio between the worst-case NE and the optimum solution, and has been extensively studied in various settings, including job scheduling [21,10,11], network design [2,4,5,13], network routing [25,7,9], and more.

The notion of strong equilibrium (SE) [6] expresses stability against coordinated deviations. The downside of SE is that most games do not admit any SE, in contrast to NE which always exists (in mixed strategies). Various recent works have studied the existence of SE in particular families of games. [3] showed that in every job scheduling game and (almost) every network creation game, a SE exists. In addition, [12,19,20,26] provided a topological characterization for the existence of SE in different congestion games, including routing and cost-sharing connection games. The vast literature on SE [19,20,23,8] concentrate on pure strategies and pure deviations, as is the case in our paper. In job scheduling settings, [3] showed that if mixed deviations are allowed, it is often the case that no SE exists. When a SE exists, aside from its robustness, it has other appealing preoperties. For example, in many cases, the price of anarchy with respect to SE (denoted the *strong price of anarchy* in [3]) is significantly better than the price of anarchy with respect to NE [3,15,22].

2 Model and Preliminaries

In our job scheduling setting there is a set of m identical machines, $M = \{M_1, \ldots, M_m\}$, and n jobs, $N = \{1, \ldots, n\}$, where job j has load p_j, and is controlled by a single agent (in the remainder of the paper, we use agents and jobs interchangeably). A schedule $s \in S : N \rightarrow M$ (also denoted a configuration) is an assignment of jobs into machines. The load of a machine M_i in a configuration $s \in S$, denoted $C_i(s)$, is the sum of the loads of the jobs assigned to M_i, that is $C_i(s) = \sum_{\{j|s(j)=M_i\}} p_j$. In our model, the individual cost of player $j \in N$, denoted $c_j(s)$, is the total load on the machine job j is assigned to, i.e.,

$c_j(s) = C_i(s)$, where $s(j) = M_i$. Note that the internal order of the jobs on a particular machine does not affect the jobs' individual costs.

A configuration $s \in S$ is a pure **Nash Equilibrium** if no player $j \in N$ can benefit from unilaterally migrating to another machine. A configuration $s \in S$ is a pure **Strong Equilibrium** if no coalition $\Gamma \subseteq N$ can form a coordinated deviation in a way that *every* member of the coalition reduces its cost.

Recall that $C_i(s)$ denotes the load on machine i in configuration s. Let s' denote the post-deviation configuration. Then, $C_i(s')$ denotes the load on machine i after the deviation. When clear in the context, we abuse notation and denote the load on machine i before and after the deviation by C_i and C'_i, respectively. In addition, we let P_{i_1,i_2} be a binary indicator whose value is 1 if some job in the coalition migrates from M_{i_1} to M_{i_2}, and 0 otherwise. Since jobs in the coalition improve their cost by definition, $P_{i_1,i_2} = 1$ implies that $C'_{i_2} < C_{i_1}$. The *improvement ratio* of a job $j \in \Gamma$, migrating from machine M_{i_1} (with initial load C_{i_1}) to machine M_{i_2} (with post-deviation load C'_{i_2}), is $IR(j) = C_{i_1}/C'_{i_2}$. Clearly, for any job j in the coalition, $IR(j) > 1$. The *damage ratio* of a job $j \notin \Gamma$, assigned on machine M_i is $DR(j) = C'_i/C_i$. Clearly, for any job j not in the coalition, $IR(j) \leq 1$ (else j is part of the coalition). Finally, we refer to coalitions deviating from NE or LPT-based configurations as *NE-based* and *LPT-based coalitions*, respectively.

Definition 1. *A configuration s is an α-*strong equilibrium (α-SE) *if for any deviation and any coalition Γ, it holds that $min_{j \in \Gamma} IR(j) \leq \alpha$. We also say that for any Γ, $IR_{min}(s, \Gamma) \leq \alpha$.*

For the maximum improvement ratio, we say that $IR_{max}(s, \Gamma) \leq \alpha$ if for any deviation of a coalition Γ, it holds that $max_{j \in \Gamma} IR(j) \leq \alpha$.

For the maximum damage ratio, we say that $DR_{max}(s, \Gamma) \leq \alpha$ if for any deviation of a coalition Γ, it holds that $max_{j \notin \Gamma} DR(j) \leq \alpha$.

We next provide several useful observations and claims that prove useful in our analysis below. All missing proofs (from this section as well as other sections) are given in the full version of this paper [14].

Observation 1 *At least one job leaves any machine participating in an NE-based coalition.*

Proof: Suppose that there exists a machine to which a job migrates but no job leaves. Then, the job that migrates to it would also migrate alone, contradicting the original schedule is a NE. □

Definition 2. *Assume w.l.o.g that M_1 is the most loaded machine in a given configuration. We say that a coalition obeys the* flower structure *if for all $i > 1$, $P_{1,i} = P_{i,1} = 1$ and for all $i, j > 1$, $P_{i,j} = 0$.*

In particular, for $m = 3$, a coalition obeys the flower structure if $P_{1,2} = P_{2,1} = P_{1,3} = P_{3,1} = 1$ and $P_{2,3} = P_{3,2} = 0$.

Claim. Any NE-based coalition on three machines obeys the flower structure.

□

It is known [3] that any NE-schedule on two identical machines is also a SE. By the above claim, at least four jobs participate in any coalition on three machines. Clearly, at least four jobs participate in any coalition on $m > 3$ machines. Therefore,

Corollary 1. *For every NE-based coalition Γ, it holds that $|\Gamma| \geq 4$.*

3 α-Strong Equilibrium

In this section, the stability of configurations is measured by $min_{j \in \Gamma} IR(j)$. We first provide a complete analysis (i.e. matching upper and lower bounds) for $m = 3$ for both NE and LPT. For arbitrary m, we provide an upper bound for NE and LPT, and show that the lower bounds for $m = 3$ hold for any m.

Theorem 2. *Any NE schedule on three machines is a $\frac{5}{4}$-SE.* □

The above analysis is tight as shown in Figure 1. Moreover, this lower bound can be extended to any $m > 3$ by adding $m - 3$ machines and $m - 3$ heavy jobs assigned to these machines. Thus,

Theorem 3. *For $m \geq 3$, there exists a NE schedule s and a coalition Γ s.t. $IR_{min}(s, \Gamma) = \frac{5}{4}$.* □

For LPT-based configurations, the bound on the minimum improvement ratio is lower:

Theorem 4. *Any LPT-based schedule on three machines is a ($\frac{2}{\sqrt{34}-4} \approx 1.0924$)-SE.* □

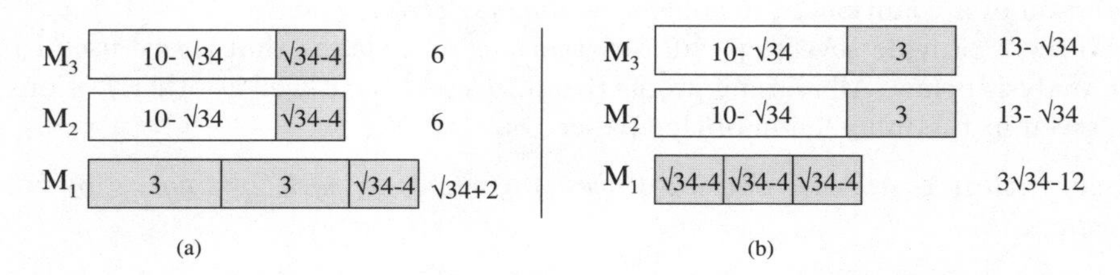

Fig. 2. An LPT-based coalition on 3 machines in which all migrating jobs improve by $\frac{2}{\sqrt{34}-4}$

The above analysis is tight as shown in Figure 2. Moreover, as for NE, this lower bound can be extended to any $m > 3$ by adding dummy jobs and machines. Thus,

Theorem 5. *For any $m \geq 3$, there exists an LPT schedule s and a coalition Γ s.t. $IR_{min}(s, \Gamma) = \frac{2}{\sqrt{34}-4}$.* □

We next provide upper bounds for arbitrary m.

Theorem 6. *Any schedule produced by LPT on m identical machines is a $(\frac{4}{3} - \frac{1}{3m}) - SE$.* □

Theorem 7. *Any NE schedule on m identical machines is a $(2 - \frac{2}{m+1}) - SE$.* □

4 Maximum Improvement Ratio

In this section, the stability of a configuration is measured by $max_{j \in \Gamma} IR(j)$. We provide a complete analysis for NE configurations and any $m \geq 3$, and for LPT configurations on three machines. The lower bound for LPT on three machines can be extended to arbitrary m. Our results show a significant difference between NE in general and LPT. While the improvement ratio of NE-based coalition can be arbitrarily high, for LPT-based coalition, the highest possible improvement ratio of any participating job is less than $\frac{5}{3}$.

Theorem 8. *For any $m \geq 3$ machines, the maximum improvement ratio of a NE-based coalition on m machines is not bounded.* □

Proof. Given r, consider the NE-schedule on 3 machines given in 3(a). The coalition consists of $\{1, 1, 2r, 2r\}$. Their improved schedule is given in Figure 3(b). The improvement ratio of the jobs of load 1 is $2r/2 = r$. For $m > 3$, dummy machines and jobs can be added. □

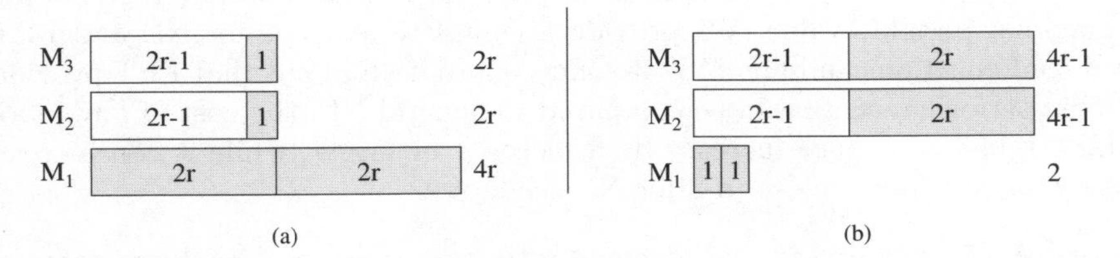

Fig. 3. An NE-based coalition in which the jobs of load 1 have improvement ratio r

In contrast to NE-based deviations, for LPT-based deviations we are able to bound the maximum improvement ratio by a small constant:

Theorem 9. *For any LPT schedule on three machines, the maximum improvement ratio of any coalition is less than $\frac{5}{3}$.* □

The above analysis is tight, as demonstrated in Figure 4 for $m = 3$ (where the improvement ratio is $2 - \frac{1}{m} = \frac{5}{3}$). Moreover, this figure shows that this lower bound can be generalized for any $m \geq 3$. The job of load $1 + \varepsilon$ that remains on M_1 improves its cost from $2m - 1 + \varepsilon$ to $m(1 + \varepsilon)$, that is, for this job, j, $IR(j) = \frac{2m-1+\varepsilon}{m(1+\varepsilon)} = 2 - \frac{1}{m} - \delta$. Formally,

Theorem 10. *For any $m \geq 3$, there exists an LPT-based configuration s and a coalition Γ such that $IR_{max}(s, \Gamma) = 2 - \frac{1}{m} - \delta$ for an arbitrarily small $\delta > 0$.*

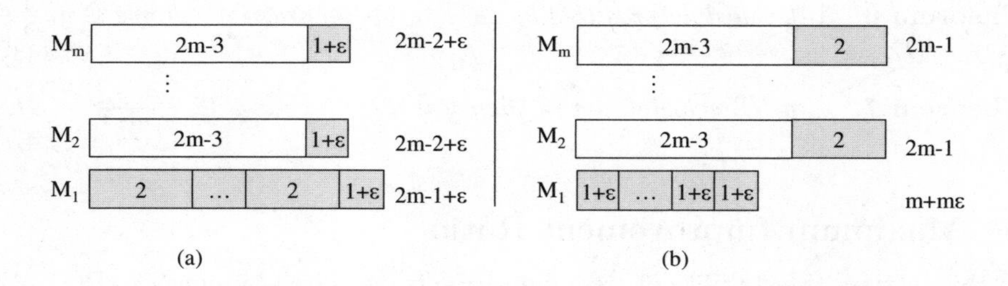

Fig. 4. An LPT-based coalition on m machines in which the job of load $1 + \varepsilon$ assigned to M_1 has improvement ratio arbitrarily close to $2 - \frac{1}{m}$

Note that the coalitional deviation in Figure 4 obeys the flower structure. We conjecture that the upper bound of $\frac{5}{3}$ for $m = 3$ can be generalized for any m, i.e., that for any LPT-based configuration s, and coalition Γ it holds that which $IR_{max}(s, \Gamma) < 2 - \frac{1}{m}$.

5 Maximum Damage Ratio

In this section, the quality of a configuration is measured by $max_{j \notin \Gamma} DR(j)$. Recall that $DR(j) = \frac{C'_i}{C_i}$, where i is the machine on which j is scheduled. For non-deviating jobs, this ratio might be larger than 1, and we would like to bound its maximal possible value. We provide a complete analysis for NE and LPT-based configurations and any $m \geq 3$. Once again, we find out that LPT provides a better performance guarantee compared to general NE: the cost of any job in an LPT schedule cannot increase by a factor $\frac{3}{2}$ or larger, while it can increase by a factor arbitrarily close to 2 for NE schedules.

Theorem 11. *For any m, the damage ratio caused by any NE-based coalition is less than* 2. □

The above analysis is tight as shown in Figure 3: The damage ratio of the jobs of load $2r - 1$ is $(4r - 1)/(2r)$, which can be arbitrarily close to 2. Formally,

Theorem 12. *For any $m \geq 3$, there exists a NE-based configuration s and a coalition Γ such that $DR_{max}(s, \Gamma) = 2 - \delta$ for an arbitrarily small $\delta > 0$.* □

For LPT-based coalitions we obtain a smaller bound:

Theorem 13. *For any m, the damage ratio caused by any LPT-based coalition is less than $\frac{3}{2}$.*

Proof. Let M_1 be the most loaded machine in the coalition. M_1 must have at least 2 jobs. Let x be the load of the last job assigned to M_1, and let $\ell = C_1 - x$. For every machine in the coalition, it must hold that $C_i \geq \ell$ (since else, x would not have been assigned to M_1), and $C'_i < \ell + x$ (since all jobs must improve).

case (a): $\ell \geq 2x$, and then for any machine M_i, $\frac{C'_i}{C_i} < \frac{\ell+x}{\ell} \leq \frac{3}{2}$.

case (b): $\ell < 2x$. We show that no coalition exists in this case. If $\ell < 2x$, then (by LPT) M_1 has exactly 2 jobs, of loads ℓ and x. By LPT, every other machine must have (i) one job of load at least ℓ (and possibly other small jobs), or (ii) two jobs of load at least x (and possible other small jobs). Let k and k' be the number of machines of type (i) and (ii), respectively (excluding M_1). Thus, there is a total of $k+1$ jobs of load ℓ and $2k'+1$ jobs of load x. After the deviation, no machine can have jobs of load ℓ and x together, nor can it have three jobs of load x. The $k+1$ machines assigned with the $k+1$ jobs of load ℓ after the deviation cannot be assigned any other job of load x. So, we end up with $2k'+1$ jobs of load x that should be assigned to k' machines. Thus, there must be a machine with at least three jobs of load x. Contradiction. □

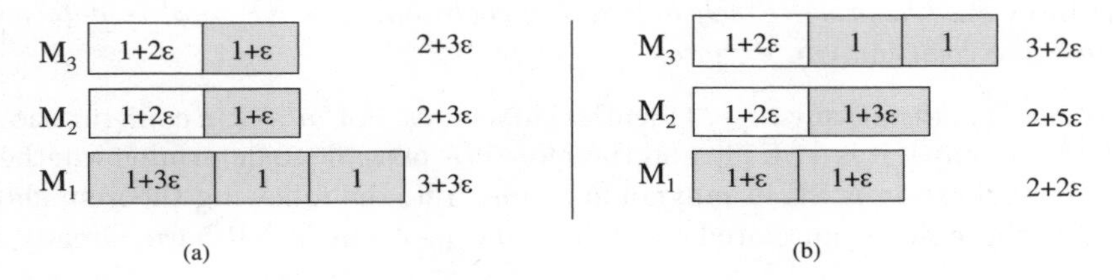

Fig. 5. An LPT-based coalition, in which the damage ratio of the job of load $1+2\varepsilon$ on M_3 is arbitrarily close to $\frac{3}{2}$

The above analysis is tight as shown in Figure 5. Moreover, by adding dummy machines and jobs it can be extended to any $m \geq 3$. Formally,

Theorem 14. *For any $m \geq 3$, there exists an LPT-based configuration s and a coalition Γ such that $DR_{max}(s, \Gamma) = \frac{3}{2} - \delta$ for an arbitrarily small $\delta > 0$.* □

6 Computational Complexity

It is easy to see that one can determine whether a given configuration is a NE in polynomial time. Yet, for SE, this task is more involved. In this section, we provide some hardness results about coalitional deviations.

Theorem 15. *Given a NE schedule on $m \geq 3$ identical machines, it is NP-hard to determine if it is a SE.*

Proof: We give a reduction from *Partition*. Given a set A of n integers $a_1, \ldots, a_n$ with total size $2B$, and the question whether there is a subset of total size B, construct the schedule in Figure 6(a). In this schedule on three machines there are $n+4$ jobs of loads $a_1, \ldots, a_n, B-2, B-2, B-1, B-1$. We assume w.l.o.g. that $min_i a_i \geq 3$, else the whole instance can be scaled. Thus, schedule 6(a) is a NE. For $m \geq 3$, add $m-3$ machines each with a single job of load $2B$.

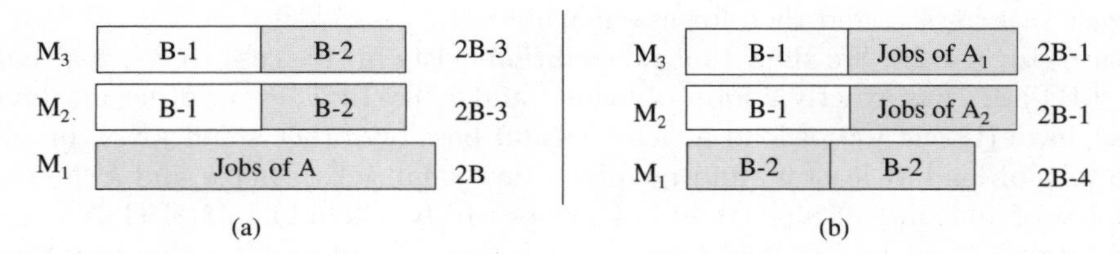

Fig. 6. Partition induces a coalition in a schedule on identical machines

Claim. The NE schedule in Figure 6(a) is a SE if and only if there is no partition. □

A direct corollary of the above proof is the following:

Corollary 2. *Given a NE schedule and a coalition, it is NP-hard to determine whether the coalition can deviate.*

Theorem 15 holds for any $m \geq 3$ identical machines. For $m \leq 2$, a configuration is a NE if and only if it is a SE [3], and therefore it is possible to determine whether a given configuration is SE in polynomial time. Yet, the following theorem shows that for the case of unrelated machines, the problem is NP-hard already for $m = 2$.

Theorem 16. *Given a NE schedule on $m \geq 2$ unrelated machines, it is NP-hard to determine if it is a SE.* □

A direct corollary of the above proof is the following:

Corollary 3. *Given an NE schedule on unrelated machines and a coalition, it is NP-hard to determine whether the coalition can deviate.*

It remains an open problem whether there exists a polynomial time approximation scheme that provides a $(1 + \varepsilon)$-SE.

Acknowledgments. We thank Oded Schwartz for helpful discussions.

References

1. Albers, S.: On the value of coordination in network design. In: SODA (2008)
2. Albers, S., Elits, S., Even-Dar, E., Mansour, Y., Roditty, L.: On Nash Equilibria for a Network Creation Game. In: SODA (2006)
3. Andelman, N., Feldman, M., Mansour, Y.: Strong Price of Anarchy. In: SODA (2007)
4. Anshelevich, E., Dasgupta, A., Kleinberg, J.M., Tardos, É., Wexler, T., Roughgarden, T.: The price of stability for network design with fair cost allocation. In: FOCS, pp. 295–304 (2004)
5. Anshelevich, E., Dasgupta, A., Tardos, E., Wexler, T.: Near-Optimal Network Design with Selfish Agents. In: STOC (2003)

6. Aumann, R.: Acceptable Points in General Cooperative n-Person Games. In: Conti, R., Ruberti, A. (eds.) Optimization Techniques 1973. LNCS, vol. 4, p. 1959. Springer, Heidelberg (1973)
7. Azar, Y., Tsur, D., Richter, Y., Awerbuch, B.: Tradeoffs in Worst-Case Equilibria. In: Solis-Oba, R., Jansen, K. (eds.) WAOA 2003. LNCS, vol. 2909, pp. 41–52. Springer, Heidelberg (2004)
8. Bernheim, D.B., Peleg, B., Whinston, M.D.: Coalition-proof nash equilibria: I concepts. Journal of Economic Theory 42, 1–12 (1987)
9. Christodoulou, G., Koutsoupias, E.: On the Price of Anarchy and Stability of Correlated Equilibria of Linear Congestion Games. In: Brodal, G.S., Leonardi, S. (eds.) ESA 2005. LNCS, vol. 3669, pp. 59–70. Springer, Heidelberg (2005)
10. Christodoulou, G., Koutsoupias, E., Nanavati, A.: Coordination Mechanisms. In: Díaz, J., Karhumäki, J., Lepistö, A., Sannella, D. (eds.) ICALP 2004. LNCS, vol. 3142, pp. 345–357. Springer, Heidelberg (2004)
11. Czumaj, A., Vöcking, B.: Tight bounds for worst-case equilibria. In: SODA, pp. 413–420 (2002)
12. Epstein, A., Feldman, M., Mansour, Y.: Strong Equilibrium in Cost Sharing Connection Games. In: ACMEC (2007)
13. Fabrikant, A., Luthra, A., Maneva, E., Papadimitriou, C., Shenker, S.: On a network creation game. In: PODC (2003)
14. Feldman, M., Tamir, T.: Approximate Strong Equilibrium in Job Scheduling Games. http://www.faculty.idc.ac.il/tami/Papers/approxSE.pdf
15. Fiat, A., Kaplan, H., Levi, M., Olonetsky, S.: Strong Price of Anarchy for Machine Load Balancing. In: Arge, L., Cachin, C., Jurdziński, T., Tarlecki, A. (eds.) ICALP 2007. LNCS, vol. 4596, pp. 583–594. Springer, Heidelberg (2007)
16. Finn, G., Horowitz, E.: A linear time approximation algorithm for multiprocessor scheduling. BIT Numerical Mathematics 19(3), 312–320 (1979)
17. Fotakis, D., Kontogiannis, S., Mavronicolas, M., Spiraklis, P.: The Structure and Complexity of Nash Equilibria for a Selfish Routing Game. In: Widmayer, P., Triguero, F., Morales, R., Hennessy, M., Eidenbenz, S., Conejo, R. (eds.) ICALP 2002. LNCS, vol. 2380, pp. 510–519. Springer, Heidelberg (2002)
18. Graham, R.: Bounds on multiprocessing timing anomalies. SIAM J. Appl. Math. 17, 263–269 (1969)
19. Holzman, R., Law-Yone, N.: Strong equilibrium in congestion games. Games and Economic Behavior 21, 85–101 (1997)
20. Holzman, R., Law-Yone, N.: Network structure and strong equilibrium in route selection games. Mathematical Social Sciences 46, 193–205 (2003)
21. Koutsoupias, E., Papadimitriou, C.H.: Worst-Case Equilibria. In: Meinel, C., Tison, S. (eds.) STACS 1999. LNCS, vol. 1563, pp. 404–413. Springer, Heidelberg (1999)
22. Leonardi, S., Sankowski, P.: Network Formation Games with Local Coalitions. In: PODC (2007)
23. Milchtaich, I.: Crowding games are sequentially solvable. International Journal of Game Theory 27, 501–509 (1998)
24. Papadimitriou, C.H.: Algorithms, games, and the internet. In: proceedings of the 33rd Annual ACM Symposium on Theory of Computing, pp. 749–753 (2001)
25. Roughgarden, T., Tardos, E.: How bad is selfish routing? Journal of the ACM 49(2), 236–259 (2002)
26. Rozenfeld, O., Tennenholtz, M.: Strong and correlated strong equilibria in monotone congestion games. In: working paper, Technion, Israel (2006)
27. Schuurman, P., Vredeveld, T.: Performance guarantees of local search for multiprocessor scheduling. INFORMS Journal on Computing (to appear)

Bertrand Competition in Networks

Shuchi Chawla[1,⋆] and Tim Roughgarden[2,⋆⋆]

[1] Computer Sciences Dept., University of Wisconsin - Madison
shuchi@cs.wisc.edu
[2] Department of Computer Science, Stanford University
tim@cs.stanford.edu

Abstract. We study price-of-anarchy type questions in two-sided markets with combinatorial consumers and limited supply sellers. Sellers own edges in a network and sell bandwidth at fixed prices subject to capacity constraints; consumers buy bandwidth between their sources and sinks so as to maximize their value from sending traffic minus the prices they pay to edges. We characterize the price of anarchy and price of stability in these "network pricing" games with respect to two objectives—the social value (social welfare) of the consumers, and the total profit obtained by all the sellers. In single-source single-sink networks we give tight bounds on these quantities based on the degree of competition, specifically the number of monopolistic edges, in the network. In multiple-source single-sink networks, we show that equilibria perform well only under additional assumptions on the network and demand structure.

1 Introduction

The Internet is a unique modern artifact given its sheer size, and the number of its users. Given its (continuing) distributed and ad-hoc evolution, as well as emerging applications, there have been growing concerns about the effectiveness of its current routing protocols in finding good routes and ensuring quality of service. Congestion and QoS based pricing has been suggested as a way of combating the ills of this distributed growth and selfish use of resources (see, e.g., [5,7,8,10,12]). Unfortunately, the effectiveness of such approaches relies on the cooperation of the multiple entities implementing them, namely the owners of resources on the Internet, or the ISPs. The ISPs' goals do not necessarily align with the social objectives of efficiency and quality of service; their primary objective is to maximize their own market share and profit.

In this paper we consider the following question: given a large combinatorial market such as the Internet, suppose that the owners of resources selfishly price their product so as to maximize their profit, and consumers selfishly purchase bundles of products to maximize their utility, how does this effect the functioning of the market as a whole?

⋆ Supported by NSF CAREER award CCF-0643763.

⋆⋆ Supported in part by NSF CAREER Award CCF-0448664, an ONR Young Investigator Award, and an Alfred P. Sloan Fellowship.

B. Monien and U.-P. Schroeder (Eds.): SAGT 2008, LNCS 4997, pp. 70–82, 2008.

We consider a simple model where each edge of the network is owned by a distinct selfish entity, and is subject to capacity constraints. Each consumer is interested in buying bandwidth along a path from its source to its destination, and obtains a fixed value per unit of flow that it can send along this path; consumers are therefore single-parameter agents. The game proceeds by the sellers first picking (per-unit-bandwidth) prices for their edges, and the consumers buying their most-desirable paths (or nothing if all the paths are too expensive). An outcome of the game (a collection of prices and the paths bought by consumers) is called a Nash equilibrium if no seller can improve her profit by changing her price single-handedly. Note that the consumers already play a best-response to the prices. We compare the performance of equilibria in this game to that of the best state achievable through coordination, under two metrics—the social value (efficiency) of the system, and the total profit earned by all the edges.

Economists have traditionally studied the properties of equilibria that emerge in pricing games with competing firms in single-item markets (see, e.g., [15,16] and references therein). It is well known [11], e.g., that in a single-good free market, oligopolies (two or a few competing firms) lead to a socially-optimal equilibrium[1]. On the other hand, a monopoly can cause an inefficient allocation by selfishly maximizing its own profit. Fortunately the extent of this inefficiency is bounded by a logarithmic factor in the (multiplicative) disparity between consumer values, as well as by a logarithmic factor in the number of consumers.

These classical economic models ignore the combinatorial aspects of network pricing, namely that consumers have different geographic sources and destinations for their traffic, and goods (i.e., edges) are not pure substitutes, but rather are a complex mix of substitutes and complements, as defined by the network topology. So a timely and basic research question is: which properties of standard price equilbrium models carry over to network/combinatorial settings? For example, are equilibria still guaranteed to exist? Are equilibria fully efficient? Does the answer depend in an interesting way on the network/demand structure? The network model captures the classical single-item setting in the form of a single-source single-sink network with a single edge (modeling a monopoly), or multiple parallel edges (modeling an oligopoly). In addition, we investigate these questions in general single-source single-sink networks, as well as multiple-source single-sink networks. Our work can we viewed as a non-trivial first step toward understanding price competition in general combinatorial markets.

Our results. We study the price of anarchy, or the ratio of the performance of the worst Nash equilibrium to that of an optimal state, for the network pricing game with respect to social value and profit. We give matching upper and lower bounds, as a function of the degree of competition in the network, and the ratio $\mathcal{L}$ of the maximum and minimum customer valuations. For instances with

[1] To be precise, there are two models of competition in an oligopolistic market—Bertrand competition, where the firms compete on prices, and Cournot competition, where they compete on quantity. The former always leads to a socially-optimal equilibrium; the latter may not. In this paper we will focus on the Bertrand model. See the full version [4] of this paper for a brief discussion of the Cournot model.

a high price of anarchy, a natural question is whether there exist any good equilibria for the instance. We provide a negative answer in most such cases, giving strong lower bounds on the price of stability, which quantifies the ratio of the performance of the *best* Nash equilibrium to that of an optimal solution.

For single-source single-sink networks, we provide tight upper and lower bounds on the prices of anarchy and stability (see Section 3). Although in a network with a single monopolistic edge, these quantities are $O(\log \mathcal{L})$ for social value, both become worse as the number of monopolies increases. The price of stability, for example, increases exponentially with the number k of monopolies, as $\Theta(\mathcal{L}^{k-1})$ for $k > 1$. The equilibrium prices in these instances are closely related to the min-cut structure of the instances.

With respect to profit, as is expected, networks that contain no monopolies display a large price of anarchy and stability because competition hurts the profits of all the firms, while networks with a single monopoly perform very well. One may suspect that as competition decreases further (the number of monopolies gets larger), collective profit improves. We show instead that the price of stability for profit also increases exponentially with the number of monopolies.

In multiple-source single-sink networks, the behavior of Nash equilibria changes considerably (see Section 4). In particular, equilibria do not always exist even in very simple directed acyclic networks. When they do exist, some instances display a high price of stability (polynomial in $\mathcal{L}$) despite strong competition in the network. In addition to the presence of monopolies, we identify other properties of instances that cause such poor behavior: (1) an uneven distribution of demand across different sources, and (2) congested subnetworks (congestion in one part of the network can get "carried over" to a different part of the network in the form of high prices due to the selfishness of the edges). We show that in a certain class of directed acyclic networks with no monopolies, in which equilibria are guaranteed to exist, the absence of the above two conditions leads to good equilibria. Specifically, the price of stability for social value in such networks is at most $1/\alpha$ where α is the sparsity of the network. Once again, we use the sparse-cut structure of the network to explicitly construct good equilibria.

Related work. The literature on quantifying the inefficiency of equilibria is too large to survey here; see [14] and the references therein for an introduction.

Recently, several researchers have studied the existence and inefficiency of equilibria in network pricing models where consumers face congestion costs from other traffic sharing the same bandwidth [9,1,2,13,17]. In these other works, the routing cost faced by each consumer has two components: the price charged by each edge on the path, and the latency faced by the consumer's flow owing to congestion on the path. In addition to selfish pricing, this congestion-based externality among consumers leads to highly inefficient outcomes even in very simple networks (such as single-source single-sink series-parallel networks [2]). The cost model considered by us is a special case of this latency-based cost function, in which the latency faced by a flow is 0 as long as all capacity constraints along the path are satisfied, and ∞ otherwise. Furthermore, in our model, latency (congestion) costs are paid by edges, rather than by consumers, and therefore force the edges to raise their prices just enough for the capacity constraints to be met. Owing to the generality of the latency functions they consider, these other papers study extremely simple network models. Acemoglu and Ozdaglar [1,2], for example, consider single-source single-sink networks with parallel links and assume that all consumers are identical and have unbounded values (i.e. they simply minimize their total routing cost). Hayrapetyan et al. [9] consider the same class of networks but in addition allow different values for different consumers. In contrast, we consider general single-source single-sink as well as multiple-source single-sink topologies with the simpler capacity-based cost model. In effect, our work isolates the impact of selfish pricing on the efficiency of the network in the absence of congestion effects. Although capacity constraints in our model mimic some congestion effects, we see interesting behavior even in the absence of capacity constraints when the market contains monopolies.

Another recent work closely related to ours is a network formation model introduced by Anshelevich et al. [3] in which neighboring agents form bilateral agreements to both buy and sell bandwidth simultaneously. The game studied in [3] can be thought of as a meta-level game played by agents when they first enter the network and install capacities based on anticipated demand. Furthermore, in their model there are no latencies or capacity constraints, instead there is a fixed cost for routing each additional unit of flow.

2 Model and Notation

A network pricing game (NPG) is characterized by a directed graph $G = (V, E)$ with edge capacities $\{c_e\}_{e \in E}$, and a set of users (traffic matrix) endowed with values. Each edge is owned by a distinct ISP. (Many of our results can be easily extended to the case where a single ISP owns multiple edges.) The value associated with each chunk of traffic represents the *per-unit monetary value* that the owner of that chunk obtains upon sending this traffic from its source to its destination. User values are represented in the form of *demand curves*[2], $\mathcal{D}_{(s,t)}$, for every source-destination pair (s,t), where for every ℓ, $\mathcal{D}_{(s,t)}(\ell)$ represents the amount of traffic with value at least ℓ. When the network has a single source-sink pair, we drop the subscript (s,t). We use $\mathcal{D}$ to denote the "demand suite", or the collection of these demand curves, one for each source-sink pair. Without loss of generality, the minimum value is 1, that is, $\mathcal{D}_{(s,t)}(1) = \mathbf{F}^{\text{tot}}_{s,t}$ for all pairs (s,t), and we use $\mathcal{L}$ to denote the maximum value—$\mathcal{L} = \sup\{\ell | \mathcal{D}_{(s,t)}(\ell) > 0\}$.

We extend the classic Bertrand model of competition to network pricing. The NPG has two stages. In the first stage, each ISP (edge) e picks a price π_e. In the second stage each user picks paths between its source and destination to send its traffic. We assume that users can split their traffic into infinitesimally small chunks, and spread it across multiple paths, or send fractional amounts of traffic. Each user picks paths to maximize her utility, $u = v - \min_P \sum_{e \in P} \pi_e$, where the minimum is over all paths P from the user's source to its destination, and v is its value (or sends no flow if the minimum total price is larger than its value v).

[2] We aggregate these curves over all users with the same source and destination pairs.

This selection of paths determines the amount of traffic f_e on each edge. ISP e's utility is given by $f_e \pi_e$ if $f_e \leq c_e$, and $-\infty$ otherwise. ISPs are selfish and set prices to maximize their utility.

A given state in a game (in this case consisting of a set of prices and flow) is called a Nash equilibrium if no agent wants to deviate from it unilaterally so as to improve its own utility. Note that in the NPG, users are price-takers, that is, they merely follow a best response to the prices set by ISPs, and the responses of different users are decoupled from each other. Therefore, given the first stage strategies, the second stage strategies always form a Nash equilibrium, and the dynamics of the system is determined primarily by the first stage game.

Note that by sending fractional flow, or splitting their traffic across multiple paths, users effectively mimick randomized strategies. ISPs, on the other hand, always pick a deterministic strategy (committing to a fixed price). Therefore, (pure strategy) equilibria do not always exist in these games (indeed in the full version of this paper [4] we present an example that admits no pure strategy equilibria). Nevertheless we identify some cases in which equilibria do exist, and characterize their performance in those cases.

Note also that if the flow f resulting from the users' strategies in the second stage is such that the capacity constraint on an edge e is violated, users using e still obtain their value from routing their flow, while e incurs a large penalty. Intuitively, the edge e is forced to compensate those users that are denied service due to capacity constraints, for not honoring its commitment to serve them at its declared price. This situation cannot arise at an equilibrium – any edge with a violated capacity can improve its profit by increasing the price charged by it.

We evaluate the Nash equilibria of these games with respect to two objectives—social value and profit. The social value of a state S of the network, $\mathbf{Val}(S)$, is defined to be the total utility of all the agents in the system, specifically, the total value obtained by all the users, minus the prices paid by the users, plus the profits (prices) earned by all the ISPs. Since prices are endogenous to the game, this is equivalent to the total value obtained by all the users, and we will use this latter expression to evaluate it throughout the paper. The worst such value over all Nash equilibria is captured by the price of anarchy: the price of anarchy of the NPG with respect to social value, $\mathbf{POA_{Val}}$, is defined to be the minimum over all Nash equilibria $S \in \mathcal{N}$ of the ratio of the social value of the equilibrium to the optimal achievable value $\mathbf{Val}^*$:

$$\mathbf{POA_{Val}}(G, \mathcal{D}) = \frac{\min_{S \in \mathcal{N}(G,\mathcal{D})} \mathbf{Val}(S)}{\mathbf{Val}^*}$$

Here, $\mathbf{Val}^*$ is the maximum total value achievable while satisfying all the capacity constraints in the network (this can be computed by a simple flow LP). Likewise, $\mathbf{POA_{Pro}}$ denotes the price of anarchy with respect to profit:

$$\mathbf{POA_{Pro}}(G, \mathcal{D}) = \frac{\min_{S \in \mathcal{N}(G,\mathcal{D})} \mathbf{Pro}(S)}{\mathbf{Pro}^*}$$

Here $\mathbf{Pro}(S)$ is the total utility of all the ISPs, or the total payment made by all users. The optimal profit $\mathbf{Pro}^*$ is defined to be the maximum profit over all states in which users are at equilibrium, and capacity constraints are satisfied.

In instances with a large price of anarchy, we also study the performance of the best Nash equilibria and provide lower bounds for it. The price of stability of a game is defined to be the *maximum* over all Nash equilbria in the game of the ratio of the value of the equilibrium to the optimal achievable value. We use $\mathbf{POS_{Val}}$ and $\mathbf{POS_{Pro}}$ to denote the price of stability with respect to social value and profit respectively.

3 The Network Pricing Game in Single-Source Single-Sink Networks

In this section we study the network pricing game in single commodity networks, that is, instances in which every customer has the same source and sink. As the single-item case suggests, the equilibrium behavior of the NPG depends on whether or not there is competition in the network. However, the extent of competition, specifically the number of monopolies, also plays an important role. In the context of a network (or a general combinatorial market), an edge monopolizes over a consumer if *all* the paths (bundles of items) desired by the customer contain the edge.

Definition 1. *An edge in a given network is called a monopoly if its removal causes the source of a commodity to be disconnected from its sink.*

No monopoly. In the absence of monopolies, the behavior of the network is analogous to competition in single-item markets. Specifically, competition drives down prices and enables higher usage of the network, thereby obtaining good social value but poor profit.

Theorem 1. *In a single commodity network with no monopolies,* $\mathbf{POA_{Val}} = 1$. *Furthermore, there exist instances with* $\mathbf{POS_{Pro}} = \Theta(\mathcal{L})$.

Proof. We first note that an equilibrium supporting the optimal flow (w.r.t. social value) always exists: consider an optimal flow of amount, say, f in the network; let $p = D^{-1}(f)$ if the flow saturates the network, and 0 otherwise; pick an arbitrary min-cut, and assign a price of p to every edge in the min-cut. These prices, along with the flow f form an equilibrium: edges cannot improve their profits by increasing prices unilaterally, because their customers can switch to a different cheaper path, and, edges with non-zero prices are saturated and cannot gain customers by lowering their price.

For a bound on the price of anarchy, consider any equilibrium in the given instance, and suppose that the network is not saturated. If all the traffic is admitted, then $\mathbf{POA_{Val}} = 1$. Otherwise, there exists an unsaturated edge, say e, with non-zero price that does not carry all of the admitted flow (if there exists a zero-price unsaturated path, then some users are playing suboptimally). Then there is a source-sink path P carrying flow with $e \notin P$. Edge e can then improve its profit by lowering its price infinitesimally and grabbing some of the flow on path P which is not among the cheapest paths any more. This contradicts the fact that the network is in equilibrium.

For the second part, we consider a network with unbounded capacity. Our argument above (that $\mathbf{POA_{Val}} = 1$) implies that in any equilibrium all the traffic is admitted. Therefore the price charged to each user is at most 1 (the minimum value), and the total profit of the network is $\mathbf{F}^{\mathrm{tot}}_{s,t}$. On the other hand, suppose that all but an infinitessimal fraction of the users have value $\mathcal{L}$, then a solution admitting only the high-value set of users (and charging a price of $\mathcal{L}$ to each user) has net profit almost $\mathcal{L}\mathbf{F}^{\mathrm{tot}}_{s,t}$.

Single monopoly. As we show below, the best-case and worst-case performance of single monopoly networks is identical to that of single-link networks.

Theorem 2. *In a single commodity network with* 1 *monopoly,* $\mathbf{POA_{Pro}} = 1$ *and* $\mathbf{POA_{Val}} = O(\log \mathcal{L})$*. Moreover, there exist instances with* $\mathbf{POS_{Val}} = \Theta(\log \mathcal{L})$*.*

Proof. The second part follows by considering the $1/x$ demand curve from 1 to $\mathcal{L}$ in a single link unbounded capacity network. The single link then behaves like a monopolist, and w.l.o.g. charges a price of $\mathcal{L}$, resulting in a social value of 1. Adding an infinitesimal point mass in the demand curve at $\mathcal{L}$ breaks ties among prices and ensures that this is the only equilibrium. The optimal social value, on the other hand, is the total value of all users $\int_1^{\mathcal{L}} 1/x dx = \log \mathcal{L}$.

For the first part of the theorem, we first note that in a single-link network (i.e. a single-item market), the above example is essentially the worst. Specifically, if at equilibrium an x amount of flow is admitted, and each user pays a price of p, then for each value $q < p$, $\mathcal{D}_{(s,t)}(q) \leq px/q$. Therefore, the total value foregone from not routing flow with value less than p is at most $\int_1^p (px/q - x) dq < px \log p < px \log \mathcal{L}$. With respect to profit, a single-link network is optimal by definition. We omit the straightforward extension to general single commodity networks (see [4]).

Multiple monopolies. The performance of the game with multiple monopolies degrades significantly – the price of anarchy can be unbounded even with 2 monopolies. As we show below, the best Nash equilibrium behaves slightly better but is still a polynomial factor worse than an optimal solution.

Theorem 3. *For every B, there exists a single-source single-sink instance of the NPG containing* 2 *monopolies, with* $\mathcal{L} = 2$*, and* $\mathbf{POA_{Val}}, \mathbf{POA_{Pro}} = \Omega(B)$*.*

Proof. Consider a network with a single source s, a single sink t, an intermediate node v, and two unit-capacity edges (s, v) and (v, t). $\mathbf{F}^{\mathrm{tot}}_{s,t} = 1$; all but a $1/B$ fraction of the traffic has a value of 1; the rest has a value of 2. We claim that $\pi_e = 1$ for each of the edges is an equilibrium: there is no incentive to increase price (and lose all customers), and, in order to get more customers, unilaterally any edge must decrease its price to 0. The social value and profit of this equilibrium are both $2/B$, whereas the optimal social value (with $\pi_e = 1/2$ for both the edges) is $1 + 1/B$ and the optimal profit is 1.

Theorem 4. *There exists a family of single-commodity instances with* $\mathbf{POS_{Val}}$*,* $\mathbf{POS_{Pro}} = \Omega(\mathcal{L}^{k-1})$*, where k is the number of monopolies. Moreover, in all single-commodity graphs with $k > 1$ monopolies,* $\mathbf{POS_{Val}}, \mathbf{POS_{Pro}} = O(\mathcal{L}^{k-1})$*.*

Proof. For the first part of the theorem, we consider a graph containing a single source-sink path with k edges and unbounded capacities. There are n users, each endowed with a unit flow. The ith user has value v_i with v_i recursively defined: $v_1 = 2$, $v_2 = (1 - \frac{1}{n})\frac{2k}{2k+1}$, $v_{i+1} = (1 - \frac{1}{n})\frac{ik}{ik+1}v_i$ for $i \in [3, n]$. (That is, $v_{i+1} = (1 - \frac{1}{n})^i \prod_{j \le i} \frac{kj}{kj+1}$ for $i > 1$.) This network contains a single equilibrium, one at which each edge charges a price of $v_1/k = 2/k$, and admits a single user.

Since the network has unbounded capacity, the optimal solution (for social value) admits the entire flow. Some algebra shows that $v_n = \Theta(n^{-1/k})$. So, the social value of the optimum is $\sum_i v_i = \Omega(n^{1-1/k}) = \Omega(\mathcal{L}^{k-1})$, as $\mathcal{L} = v_1/v_n = \Theta(n^{1/k})$. The total achievable profit is also at least $nv_n = \Omega(n^{1-1/k}) = \Omega(\mathcal{L}^{k-1})$. On the other hand, the social value of the equilibrium, as well as its profit, is $v_1 \cdot 1 = 2$. This concludes the proof of the first part of the theorem.

For the second part, let $\overline{D}$ denote the inverse-demand curve for the network, i.e., for every x, an x amount of flow has value at least $\overline{D}(x)$. Without loss of generality, $\overline{D}(0) = \mathcal{L}$, $\overline{D}(F) = 1$, where $F = \mathbf{F}_{s,t}^{\text{tot}}$ is the total optimal amount of flow. Let $x^* = \text{argmax}_{x \le F}\{x^{1/k}\overline{D}(x)\}$. We claim that the following is an equilibrium: each monopoly charges a price of $p^* = \overline{D}(x^*)/k$, and each non-monopoly charges 0. It is obvious that the non-monopolies have no incentive to increase their price. So, for the rest of the proof, we focus on the monopolies.

Suppose that a monopoly wants to deviate and change its price to $p' = p^* - \overline{D}(x^*) + \overline{D}(x') \ge 0$, for some $x' \in [0, F]$. Then, the total price of any source-sink path is $\overline{D}(x')$, and the total amount of flow admitted is no more than x'. The profit of the monopoly goes from p^*x^* to at most $p'x'$, which can be simplified as follows:

$$p'x' = \left(\frac{\overline{D}(x^*)}{k} - \overline{D}(x^*) + \overline{D}(x')\right) x' \le \frac{\overline{D}(x^*)x^*}{k} \left(\frac{x'}{x^*}(1-k) + k\left(\frac{x'}{x^*}\right)^{1-1/k}\right)$$
$$< \frac{\overline{D}(x^*)x^*}{k} \left(\frac{x'}{x^*}(1-k) + k + (k-1)\frac{x'}{x^*} - (k-1)\right) = p^*x^*$$

Here we used $(1+\epsilon)^\alpha < 1 + \alpha\epsilon$ for all $\epsilon > -1$ and for all $\alpha \in (0, 1)$. This proves that the agent has no incentive to deviate. It remains to show that this equilibrium achieves good social welfare. First note that $\overline{D}(F)F^{1/k} \le \overline{D}(x^*)(x^*)^{1/k}$. Therefore, $F \le x^*(\overline{D}(x^*))^k$. Likewise, $\forall y \in [0, F]$, $\overline{D}(y) \le \overline{D}(x^*)(x^*/y)^{1/k}$. So the total value of flow not admitted by the equilibrium is

$$\int_{y=x^*}^{y=F} \overline{D}(y)dy \le \int_{y=x^*}^{y=F} \overline{D}(x^*)(x^*/y)^{1/k}dy = \frac{\overline{D}(x^*)(x^*)^{1/k}}{(1-1/k)}(F^{1-1/k} - (x^*)^{1-1/k})$$
$$\le (1-1/k)^{-1}(\overline{D}(x^*)^k x^* - \overline{D}(x^*)x^*) < 2(\overline{D}(x^*))^k x^*$$

So, the maximum social welfare achievable is strictly less than $2(\overline{D}(x^*))^k x^*$ plus the social value of the above equilibrium, while the equilibrium achieves at least $\overline{D}(x^*)x^*$. The price of stability is therefore no more than $2(\overline{D}(x^*))^{k-1} + 1 \le 3\mathcal{L}^{k-1}$. It is easy to see that the same bound holds for profit as well.

4 Networks with Multiple Sources

Next we study the NPG in graphs with more general traffic matrix. Specifically different users have different sources, but a common sink. We assume that the network is a DAG with a single sink, and focus on instances that contain no monopolies[3]. Theorem 1 already shows that the price of stability with respect to profit can be quite large in this case. The main question we address here is whether competition drives down prices and enables a near socially optimal equilibrium just as in the single-commodity case.

The results are surprisingly pessimistic. We find that there are networks with no pure equilibria. (See [4] for proofs of the next two theorems.)

Theorem 5. *There exists a multi-source single-sink instance of the NPG with no monopolies that does not admit any pure Nash equilibria.*

In networks that admit pure equilibria, the price of stability for social value can be polynomial in $\mathcal{L}$. This can happen (Theorem 6 below) even when the network in question satisfies a certain strong-competition condition, specifically, (1) there is sufficient path-choice – from every node in the graph, there are at least two edge-disjoint paths to the sink, and (2) no edge dominates over a specific user in terms of the capacity available to that user – removing any single edge reduces the amount of traffic that any user or group of users can route by only a constant fraction. We therefore attempt to isolate conditions that lead to a high price of stability, and find two culprits:

1. Variations in demand curves across users—a very high value low traffic user can pre-empt a low value high traffic user.
2. Congestion in the network—congestion in one part of the network (owing to low capacity), can get "carried over" to a different part of the network (in the form of high prices) due to the ISPs' selfishness.

Each condition alone can cause the network to have a high price of stability.

Theorem 6. *There exists a family of multiple-source single-sink instances satisfying strong competition and containing uniform demand such that* $\mathbf{POS_{Val}} = \Omega(poly\ \mathcal{L},\ poly\ N)$*, where* N *is the size of the network. There exists a family of multiple-source single-sink instances satisfying strong competition and with sparsity* 1 *such that* $\mathbf{POS_{Val}} = \Omega(poly\ \mathcal{L}, poly\ N)$.

Here uniformity of demand and sparsity defined as follows.

Definition 2. *An instance of the NPG,* $(G, \mathcal{D})$*, with multiple commodities and a single sink* t *is said to contain* uniform demand *if there exists a demand curve* D *such that for all* s*,* $\mathcal{D}_{(s,t)}$ *is either zero, or equal to a scalar* $F_{s,t}$ *times* D.

Definition 3. *Given a capacitated graph and a demand matrix, the sparsity of a cut in the graph with respect to the demand is the ratio of the total capacity of the cut to the total demand between all pairs* (s,t) *separated by the cut. The sparsity of the graph is the minimum of these sparsities over all cuts in the graph.*

[3] We mainly give strong lower bounds on the price of stability. Naturally, the same bounds hold for instances containing monopolies.

Fortunately, in the absence of the two conditions above, the network behaves well. In particular, we consider a certain class of DAGs called traffic-spreaders in which equilibria are guaranteed to exist, and show that when demand is uniform, the price of stability with respect to the social value is at most $1/\alpha$, where α is the sparsity of the network. We conjecture that this bound on the price of stability holds for all DAGs that admit pure equilibria.

Definition 4. *A DAG with sink t is said to be a* traffic spreader *if for every node v in the graph, and every two distinct paths P_1 and P_2 from v to t, any maximal common subpath of P_1 and P_2 is a prefix of both the paths.*

Theorem 7. *Let $(G, \mathcal{D})$ be a uniform-demand instance of the NPG where G is a traffic spreader and contains no monopolies, and all sources in the graph are leaves, that is, their in-degree is 0. Then $(G, \mathcal{D})$ always admits a pure Nash equilibrium, and $\mathbf{POS_{Val}} \le 1/\alpha$, where α is the sparsity of G with respect to $\mathcal{D}$.*

We remark that for Theorem 7, we do not require the instance to satisfy strong competition. This indicates that the amount of competition in the network has lesser influence on its performance compared to its traffic distribution.

Proof of Theorem 7. We begin with some notation. Given a graph G and a flow f in G satisfying capacity constraints, $G[f]$ is the residual graph with capacities $c'_e = c_e - f_e$. For a graph $G = (V, E)$, set S of nodes, and set E' of edges, we use $G \setminus S$ to denote $(V \setminus S, E[V \setminus S])$, and $G \setminus E'$ to denote $(V, E \setminus E')$.

Given an instance $(G, \mathcal{D})$, $G = (V, E)$, satisfying the conditions in the theorem, we construct an equilibrium using the algorithm below. Let F_v denote the total traffic at source v, and D be a demand curve defined such that $\mathcal{D}_{v,t} = F_v D$ for all v. The algorithm crucially exploits the sparse-cut structure of the network. In particular, we use as subroutine a procedure for computing the maximum concurrent flow in a graph with some "mandatory" demand. We call this procedure **MCFMD** (for Maximum Concurrent Flow with Mandatory Demand).

MCFMD takes as input a DAG G with single sink t, a set of sources A with demands F_v at $v \in A$, and a set of mandatory-demand sources B with demands M_v at $v \in B$. It returns a cut C and a flow f. Let V_C denote the set of nodes from which t is not reachable in $G \setminus C$. The cut C minimizes "sparsity with mandatory demand" defined as follows:

$$\alpha_M(C) = \frac{\sum_{e \in C} c_e - \sum_{v \in B \cap V_C} M_v}{\sum_{v \in A \cap V_C} F_v}$$

The flow f routes the entire demand M_v of sources $v \in B$ to t, and an $\alpha_M(C)$ fraction of demands F_v at sources $v \in A$ to t. The next lemma asserts the correctness of this procedure (see [4] for a proof): sparsity is equal to maximum concurrent flow in DAGs with a single sink, even with mandatory demands.

Lemma 1. *Let (G, A, B) be an instance for* **MCFMD**, *and $\alpha = \alpha_M(C)$ be the sparsity of the cut C produced by the procedure. Then, there exists a flow in G that satisfies all capacity constraints, routes an M_v amount of flow from every $v \in B$ to t, an αF_v amount of flow from every $v \in A$ to t, and saturates C.*

Armed with this procedure, our algorithm for constructing an equilibrium is as follows. (Note that we do not care about computational efficiency here.)

1. Set $G_1 = G$, $V_1 = V$, $C = \emptyset$, $B_1 = \emptyset$, $i = 1$. Let $A_1 = A$ be the set of all sources in the instance. Let f denote a partial flow in the graph at any instant; initialize f to 0 at each edge.
2. Repeat until A_i is empty:
   (a) Run the procedure **MCFMD** on G_i with demands A_i and mandatory demands B_i. Let C_i be the resulting cut and f_i' be the resulting flow. Let $\alpha_i = \alpha_M(C_i)$, $X_i = A_i \cap V_{C_i}$, $Y_i = B_i \cap V_{C_i}$, and $C = C \cup C_i$. Define V_{i+1} to be the set of nodes with paths to t in $G \setminus C$, and S_i to be the subset of $V \setminus V_{i+1}$ reachable from X_i or Y_i in G.
   (b) Construct a partial flow from f_i' as follows. Let $B' = \{v : \exists u \text{ with } (u \to v) \in C_i\}$, and for all $v \in B'$ let $M_v = \sum_{u:(u \to v) \in C_i} c_{(u,v)}$. Let f_i be a partial flow of amount $\alpha_i F_v$ from each $v \in X_i$, and amount M_v from each $v \in Y_i$ to B', given by the prefices of some of the flow paths in f_i'. Let $f = f + f_i$, $A_{i+1} = A_i \setminus X_i$, and $B_{i+1} = (B \setminus Y_i) \cup B'$. Set $\ell_i = D^{-1}(\alpha_i)$.
   (c) Let $G_{i+1} = G_i \setminus S_i$; repeat for $i = i + 1$.
3. Route all the flow from B_i to t in G_i satisfying capacity constraints. Call this flow f_i, and set $f = f + f_i$.
4. Assign a "height" to every node v in the graph as follows: if there exists an i such that $v \in S_i$, then $h(v) = \min_{i:v \in S_i}\{\ell_i\}$; if there is no such i, then $h(v) = 0$. Furthermore, $h(t) = 0$ for the sink t.
5. For every edge $e = (u \to v)$, let $\pi_e = \max\{h(u) - h(v), 0\}$.

Let I be the final value of the index i. Recall that V_I is the set of nodes that can reach t in G_I. We will show that (π, f) is a Nash equilibrium. This immediately implies the result, because as we argue below, f admits an $\alpha_i \geq \alpha$ fraction of the most valuable traffic from all sources in X_i. We first state some facts regarding the heights $h(v)$ and the flow f (see [4] for the proofs of these lemmas).

Lemma 2. *f is a valid flow and routes an α_i fraction of the traffic from all $v \in X_i$ to t. Furthermore, for every i, $1 < i < I$, in the above construction, $\alpha_i \geq \alpha_{i-1}$, and $\alpha_1 > \alpha$, where α is the sparsity of the graph G.*

Lemma 3. *$V(G_i) = V_i$ for all $i \leq I$, and $h(v) = 0$ if and only if $v \in V_I$. For any source v with $v \in X_i$, $h(v) = \ell_i$.*

Lemma 4. *For every pair of nodes u and v with $h(u), h(v) > 0$ such that there is a directed path from u to v in G, $h(u) \geq h(v)$. Furthermore, for every node v with $h(v) > 0$, every path from v to t is fully saturated under the flow f.*

Lemma 5. *For every source v with $v \in X_i$, every path from v to t has total price at least ℓ_i. Furthermore, there exist at least two edge-disjoint paths P_1 and P_2 from v to t such that $\sum_{e \in P_1} \pi_e = \sum_{e \in P_2} \pi_e = \ell_i$.*

Lemma 6. *Let P be a flow carrying path from $v \in X_i$ to t. Then $\sum_{e \in P} \pi_e = \ell_i$.*

Finally, we claim that (π, f) is an equilibrium. First observe that we route an $\alpha_i F_v$ amount of flow for every v in X_i. Each chunk of traffic originating at v that gets routed has value at least $D^{-1}(\alpha_i) = \ell_i$. Therefore, Lemmas 5 and 6 imply that users follow best response. Next, consider any edge $e = (u \to v)$. Note that e has no incentive to increase its price – Lemma 5 ensures that all the traffic on this edge has an alternate path of equal total price. Finally, if the edge has non-zero price, it can gain from lowering its price only if this increases the traffic through it. Let C' be the mincut between u and t. Note that $h(u) > 0$. Lemma 4 implies that the cut C' is saturated. Suppose that e has non-zero residual capacity (i.e. $e \notin C'$) and by lowering its price, the edge gains extra traffic without violating the capacity of the cut C'. This means that the extra traffic on e was previously getting routed along a path that crosses the cut C', and furthermore shares a source with the edge e. This contradicts the fact that the network is a traffic spreader. Therefore, no edge has an incentive to deviate.

5 Discussion and Open Questions

We consider a simplistic model for network pricing. A more realistic model should take into account quality of service requirements of the users, which may be manifested in the form of different values for different paths between the same source-destination pairs. In general combinatorial markets it would also be interesting to consider the effect of production costs on the pricing game, and this may change the behavior of the market considerably. Finally, an alternate model of competition in two-sided markets is for the sellers to commit to producing certain quantities of their product, and allowing market forces to determine the demand and prices. This two-stage game, known as "Cournot competition", may lead to better or worse equilibria compared to Bertrand competition. We include a brief discussion of these extensions in the full version of this paper [4].

Acknowledgements

We are grateful to Aditya Akella, Suman Banerjee, and Cristian Estan for bringing to our notice their recent work on a system for QoS-based pricing in the Internet [6], which motivated the network pricing game we study in this paper.

References

1. Acemoglu, D., Ozdaglar, A.: Competition and efficiency in congested markets. Mathematics of Operations Research 32(1), 1–31 (2007)
2. Acemoglu, D., Ozdaglar, A.: Competition in parallel-serial networks. IEEE Journal on Selected Areas in Communications, special issue on Non-cooperative Behavior in Networking 25(6), 1180–1192 (2007)

3. Anshelevich, E., Shepherd, B., Wilfong, G.: Strategic network formation through peering and service agreements. In: Foundations of Computer Science (2006)
4. Chawla, S., Roughgarden, T.: Bertrand competition in networks, `http://www.cs.wisc.edu/shuchi/papers/Bertrand-competition.pdf`
5. Cole, R., Dodis, Y., Roughgarden, T.: Pricing network edges for heterogeneous selfish users. In: Proc. 34th ACM Symp. Theory of Computing (2003)
6. Estan, C., Akella, A., Banerjee, S.: Achieving good end-to-end service using bill-pay. In: HOTNETS-V (2006)
7. Fleischer, L., Jain, K., Mahdian, M.: Tolls for heterogeneous selfish users in multicommodity networks and generalized congestion games. In: Proc. 45th IEEE Symp. Foundations of Computer Science, pp. 277–285 (2004)
8. Gibbens, R.J., Kelly, F.P.: Resource pricing and the evolution of congestion control. Automatica 35(12), 1969–1985 (1999)
9. Hayrapetyan, A., Tardos, É., Wexler, T.: A network pricing game for selfish traffic. In: Proc. Symp. Principles of distributed Computing, pp. 284–291 (2005)
10. Mackie-Mason, J., Varian, H.: Pricing congestible network resources. IEEE J. Selected Areas Commun. 13(7), 1141–1149 (1995)
11. Mas-Colell, Whinston, Green: Microeconomic Theory. Oxford (1995)
12. Odlyzko, A.: Paris metro pricing for the internet. In: Proc. 1st ACM Conf. Electronic Commerce, pp. 140–147 (1999)
13. Ozdaglar, A., Srikant, R.: Incentives and pricing in communication networks. In: Algorithmic Game Theory, Cambridge Press, Cambridge (2007)
14. Roughgarden, T., Tardos, É.: Introduction to the inefficiency of equilibria. In: Nisan, N., Roughgarden, T., Tardos, É., Vazirani, V.V. (eds.) Algorithmic Game Theory, vol. 17, pp. 443–459. Cambridge University Press, Cambridge (2007)
15. Spence, A.M.: Entry, capacity, investment and oligopolistic pricing. The Bell Journal of Economics 8(2), 534–544 (1977)
16. Srinivasan, K.: Multiple market entry, cost signalling and entry deterrence. Management Science 37(12), 1539–1555 (1991)
17. Weintraub, G.Y., Johari, R., Van Roy, B.: Investment and market structure in industries with congestion. (Unpublished manuscript, 2007)

On the Approximability of Combinatorial Exchange Problems

Moshe Babaioff[1,⋆], Patrick Briest[2,⋆⋆], and Piotr Krysta[2,⋆⋆]

[1] Microsoft Research, Mountain View, CA, USA
moshe@microsoft.com
[2] Dept. of Computer Science, University of Liverpool, UK
{patrick.briest,p.krysta}@liverpool.ac.uk

Abstract. In a combinatorial exchange the goal is to find a feasible trade between potential buyers and sellers requesting and offering bundles of indivisible goods. We investigate the approximability of several optimization objectives in this setting and show that the problems of surplus and trade volume maximization are inapproximable even with free disposal and even if each agent's bundle is of size at most 3. In light of the negative results for surplus maximization we consider the complementary goal of social cost minimization and present tight approximation results for this scenario. Considering the more general supply chain problem, in which each agent can be a seller and buyer simultaneously, we prove that social cost minimization remains inapproximable even with bundles of size 3, yet becomes polynomial time solvable for agents trading bundles of size 1 or 2. This yields a complete characterization of the approximability of supply chain and combinatorial exchange problems based on the size of traded bundles. We finally briefly address the problem of exchanges in strategic settings.

1 Introduction

Following the emergence of the Internet as the world's foremost market place, much interest has been paid to problems naturally arising in a context where large scale economic problems need to be solved efficiently by computers. Many of these problems' essential difficulties can be captured by the class of combinatorial auction problems, which have in turn received a lot of attention from both practitioners and theoreticians in computer science. A major drawback of this model of abstraction, however, is the fact that it implicitly assumes a monopolistic market structure. While practitioners have therefore turned to more general (and more complex) supply chain models, these have not been subject to a rigorous theoretical investigation.

In this paper we consider computational aspects of combinatorial exchanges (CE's) and their extension to general supply chain problems. The CE model is

⋆ Partially supported by NSF ITR Award ANI-0331659.
⋆⋆ Supported by DFG grant Kr 2332/1-2 within Emmy Noether program.

B. Monien and U.-P. Schroeder (Eds.): SAGT 2008, LNCS 4997, pp. 83–94, 2008.

a generalization of combinatorial auctions, which departs from the assumption that a monopolist seller holds a set of products which are of no actual value to him, as the welfare of the outcome depends only on buyers' valuations. Instead, it is assumed that apart from the set of buyers interested in purchasing bundles of items from the auctioneer, he also has access to a number of sellers offering to supply different bundles of products at a certain price. Thus, the problem we are faced with is to simultaneously run both a forward and reverse auction, which should return a feasible trade leaving a reasonable profit margin for the auctioneer. In the more general supply chain model, we drop the assumption that each trader is either a seller or a buyer, but may in fact offer some bundle of products under the condition that he is supplied with a different bundle and some appropriate side payment in return.

Combinatorial auctions have drawn much recent research attention (see for example the book by Cramton, Shoham and Steinberg [5]), and their computational and communication hardness are well understood. Lehmann, O'Callaghan and Shoham [7] and Sandholm [15] have shown that combinatorial auctions are hard to approximate within $min(m^{1/2-\varepsilon}, n^{1-\varepsilon})$ for any $\varepsilon > 0$, unless P=NP, where m and n denote the numbers of products and bidders, respectively. Nisan and Segal [10] present communication lower bounds for combinatorial auctions. They show that if 2 buyers have general valuations exponential communication (in m) is required to find an allocation with maximum surplus, or even a $2^{1-\epsilon}$-approximation. Additionally, Nisan [9] shows that for n buyers distinguishing the case that the surplus is 1 and the case that the surplus is n requires exponential communication in m, assuming that $m^{1/2-\epsilon} > n$. Clearly, all these hardness results carry to the CE case, as has been observed before. It turns out, however, that exchanges are even essentially more difficult than combinatorial auctions even in quite restricted cases. On the other hand, they nevertheless allow for a number of positive results when the problem formulation is chosen carefully.

1.1 Preliminaries

We first describe the CE scenario, which most of the paper will be focused on. After that, we briefly explain the more general supply chain (SC) scenario. Assume that we are given a set of *agents* $\mathcal{A} = \mathcal{S} \cup \mathcal{B}$, where the collections $\mathcal{S}$, $\mathcal{B}$ are disjoint sets of *sellers* and *buyers*, respectively, with $|\mathcal{A}| = n$. We are interested in trades that include *indivisible* products $\mathcal{U}$, where $|\mathcal{U}| = m$. Each seller i is offering a bundle $q_i = (q_i^1, \ldots, q_i^m)$ of products at some price $v_i \in \mathbb{R}_+$. Buyer j is requesting to buy the bundle $q_j \in \mathbb{N}^m$ at price $v_j \in \mathbb{R}_+$. By $q_i^e, q_j^e \in \mathbb{N}$ we refer to the number of copies of product $e \in \mathcal{U}$ offered by seller i or requested by buyer j. For agent $k \in \mathcal{A}$, her bundle is a *set* if for all $e \in \mathcal{U}$, $q_k^e \in \{0, 1\}$.

A *feasible trade* $T = (S, B)$, $S \subseteq \mathcal{S}$ and $B \subseteq \mathcal{B}$, is a selection of sellers and buyers, such that

$$\sum_{i \in S} q_i^e \geq \sum_{j \in B} q_j^e \text{ for all } e \in \mathcal{U},$$

i.e., for every product $e \in \mathcal{U}$ the supply provided by sellers in S is sufficient to satisfy the requests of all buyers in B. Note, that we assume free disposal here, i.e., supply and demand do not have to match exactly. We say that trade $T = (S, B)$ has *surplus* [*volume*]

$$\mathrm{sur}(T) = \sum_{j \in B} v_j - \sum_{i \in S} v_i \qquad\qquad \mathrm{vol}(T) = \sum_{e \in \mathcal{U}} \sum_{j \in B} q_j^e.$$

The *CE surplus problem* is the problem of finding a trade which maximizes the surplus. The *CE volume problem* is the problem of finding a trade which maximizes the volume, subject to positive surplus. By *CE positive surplus problem* we refer to the problem variation in which we simply want to find any feasible trade with strictly positive surplus.

Following [14], as another objective we define the *social cost* of trade T as

$$\mathrm{cost}(T) = \sum_{j \notin B} v_j + \sum_{i \in S} v_i,$$

i.e., the cost is the sum of valuations of the trading sellers and non-trading buyers. Note that for every instance, the sum of the social cost and the surplus is constant, $\mathrm{cost}(T) + \mathrm{sur}(T) = \sum_{j \in \mathcal{B}} v_j$. Thus, a trade maximizes the surplus if and only if it minimizes the social cost. We consider the problem of minimizing the social cost subject to non-negative surplus, which we call the *CE social cost problem*. From a computational perspective, the social cost objective is preferable, because it allows us to derive approximation algorithms and express their approximation ratio in terms of a multiplicative factor, which is generally difficult if we are faced with any mixed sign objective and in fact turns out to be impossible in the case of CE's.

A natural generalization of the CE problem is obtained if we allow agents that are both sellers and buyers simultaneously and confront our algorithm with offers of the form "given bundle A I will supply bundle B for an additional payment of x". Formally, in the *supply chain (SC) problem* we are given a set $\mathcal{A}$ of n agents. Agent k is represented by $(\{\delta_k^e\}_{e \in \mathcal{U}}, v_k)$, where $\delta_k^e \in \mathbb{Z}$ denotes the number of copies of product e requested or supplied by agent k (modelled as $\delta_k^e \geq 0$ or $\delta_k^e < 0$) and $v_k \in \mathbb{R}$ is the additional payment offered or requested (modelled as $v_k \geq 0$ or $v_k < 0$). The objectives of surplus or volume maximization generalize naturally to the supply chain scenario. However, we need to adapt the notion of social cost to fit our generalized type of agents. We let $\mathcal{A}^+ = \{k \,|\, v_k > 0\}$ and $\mathcal{A}^- = \mathcal{A} \backslash \mathcal{A}^+$. Thus, $\mathcal{A}^+$ is the set of agents that have a positive utility for being included in the trade, agents in $\mathcal{A}^-$ incur a cost when included for which they need to be repaid. We can then naturally define the social cost of a trade $T \subseteq \mathcal{A}$ as

$$\mathrm{cost}(T) = \sum_{k \in \mathcal{A}^+ \backslash T} v_k - \sum_{k \in T \cap \mathcal{A}^-} v_k.$$

Finally, let us introduce some more notation that will come in handy. For a given instance $I = (\mathcal{S}, \mathcal{B}, v)$ of the CE problem we let $T^\star_{\mathrm{sur}}(I)$ refer to the surplus maximizing trade. The trade computed by some algorithm A on the same instance is denoted as $A(I)$. Analogous notation is used in the context of volume maximization or cost minimization, respectively.

1.2 Our Results

To our knowledge, the only previous results regarding approximability of CE's are the observations that exchanges are as hard to approximate as combinatorial auctions, of which they are a generalization, and are inapproximable[1] in the special case of *no free disposal* [17].

We remove the restrictive assumption of no free disposal and show that the CE surplus and volume maximization problems are nevertheless inapproximable, unless $P = NP$. They remain inapproximable even on instances in which each agent's bundle is a set of size at most 3, and when restricting to *multi-unit exchanges*, in which only a single type of product is traded. Our inapproximability result for bounded size sets is based on a reduction to a family of CE instances with sets of size at most 3 that has an interesting property. Once the quantity in which each product is bought is decided upon, the packing and covering problems defined by these quantities (finding the trading buyers and sellers, respectively) are polynomial time solvable. This implies that CE has inherent hardness that does not stem from the hardness of packing and covering alone. Finally, we prove that the problem remains inapproximable even when we restrict ourselves to instances with large *packing to covering factor*, i.e., in cases where there is a large gain from the optimal trade, and derive complementary inapproximability results for exchanges with sub-exponential communication based on the communication lower bounds of [9,10].

Sandholm and Suri [16] present an anytime algorithm for combinatorial auctions which also extends to CE's. Parkes et al. [11] have recently presented ICE, an iterative CE. Our hardness results have implications for the worst case performance of these algorithms and emphasize the need for algorithms with provable performance guarantees. We show that, focusing on the social cost objective rather than surplus or trade volume maximization, such algorithmic results can be obtained. More formally, we show that the social cost can be approximated within a factor of H_k, where k is the maximal size of any bundle and that this is essentially tight when agents are bidding for sets, as it is hard to achieve approximation guarantee $(1 - o(1))H_m$, unless NP $\subseteq$ DTIME$(n^{\mathcal{O}(\log \log n)})$. Similarly, for the relevant case of a single type of product (multi-unit exchange) we show that social cost minimization is NP-hard yet allows for an FPTAS.

In light of these positive results, we ask whether the more general supply chain social cost problem allows approximate social cost minimization, as well. It turns out that this is not the case and, in fact, social cost minimization in the supply chain scenario is inapproximable, even with bundles of size at most 3. This is interesting, as it is the first formal result separating CE's from the more general supply chain scenario. We then consider the special case in which agents are restricted to sets of size 1 or 2. While this case is polynomially solvable for combinatorial auctions, it was unknown whether this is true for CE's. We prove here that it holds, in fact, even for the supply chain scenario. Our algorithm works

[1] Inapproximable within any factor that is a polynomially computable function of n and m, and even if we allow for an additive term that is a polynomially computable function of n and m.

by reducing the problem to a weighted b-matching problem through a number of transformations and reformulation in terms of the social cost objective. This also shows that our hardness result for supply chains with sets of size 3 is tight.

Finally, although the computational problem of social cost minimization is much easier than surplus maximization, we show that there is another problem that arises in strategic settings, even when computation is not a problem at all. We move to consider the case that agents have privately known values for their bundles and we need to elicit this information from them via a truthful mechanism. A classical result by Myerson and Satterthwaite [8] shows that for bilateral trade, any mechanism that is truthful, individually rational and budget-balanced must sometimes be inefficient. This implies that a truthful, individually rational and budget-balanced mechanism cannot always minimize the social cost. We extend this result and show that for any constant $\alpha \geq 1$, α-approximation for social cost is impossible. Circumventing the Myerson and Satterthwaite impossibility result [8] for exchanges has been the subject of several papers. Parkes et al. [12] enforce budget-balance as a hard constraint, and explore payment rules that are fairly efficient and fairly truthful. On the other hand, Babaioff and Walsh [2] consider the problem of supply chains and present a truthful and budget-balanced mechanism with efficiency which depends on the size of the efficient trade.

The rest of the paper is organized as follows. Section 2 presents our inapproximability results for surplus and volume maximization. Section 3 presents approximation algorithms as well as inapproximability results for social cost minimization. Section 4 deals with supply chain problems. Finally, Section 5 shows that economic considerations prevent mechanisms that achieve social cost approximation. Most of the proofs are omitted from this extended abstract due to space limitations.

2 Surplus and Volume Inapproximability

We start by showing a number of strong inapproximability results for both surplus and volume maximization in CE's. Section 2.1 derives results for the single-minded setting under standard complexity theoretic assumptions. Section 2.2 presents results for general agents in terms of communication complexity.

2.1 Computational Hardness

We show that both the CE surplus and volume problems do not allow polynomial time approximation algorithms with any reasonable approximation guarantee. This is formalized in the following definition.

Definition 1. *Let $\alpha, \beta : \mathbb{N} \times \mathbb{N} \to \mathbb{N}$ be any polynomial-time computable functions. The CE surplus problem is* approximable *(for family of instances $\mathcal{F}$) if for some α, β there exists a poly-time approximation algorithm A, such that $\mathrm{sur}(T^\star_{\mathrm{sur}}(I)) \leq \alpha(n,m) \cdot \mathrm{sur}(A(I)) + \beta(n,m)$ holds on every problem instance $I \in \mathcal{F}$. The CE surplus problem is* inapproximable *if it is not approximable.*

Approximability in the volume maximization and social cost minimization case is defined analogously. We are interested in a very natural restricted class of families of instances.

Definition 2. *The family of instances $\mathcal{F}$ is* rational, *if all values v_i are rational numbers. $\mathcal{F}$ is* scalable *if it is closed under scaling of agents' valuations, i.e., if instance $I = (\mathcal{S}, \mathcal{B}, v) \in \mathcal{F}$ then $I' = (\mathcal{S}, \mathcal{B}, \gamma \cdot v) \in \mathcal{F}$ for any rational $\gamma > 0$.*

It turns out that for the families of instances defined above inapproximability can be derived from the fact that detecting any trade with positive surplus or volume is hard in itself.

Lemma 1. *If the CE positive surplus problem for a rational and scalable family $\mathcal{F}$ is NP-hard then the CE surplus (volume) problem for the family $\mathcal{F}$ is inapproximable, unless $P = NP$.*

Theorem 1. *The CE surplus (volume) problem is inapproximable, unless $P = NP$. It remains inapproximable under this assumption even for families $\mathcal{F}$ of instances with (1) only sets ($q_k^e \in \{0,1\}$) of size at most 3 ($\sum_{e \in \mathcal{U}} q_k^e \leq 3$ for every agent k) or (2) only one type of product (multi-unit exchange, $|\mathcal{U}| = 1$).*

Lemma 2. *The CE positive surplus problem for the family $\mathcal{F}$ of instances with only sets ($q_k^e \in \{0,1\}$ $\forall k \in \mathcal{A}, e \in \mathcal{U}$) of size at most 3 ($\sum_{e \in \mathcal{U}} q_k^e \leq 3$ $\forall k \in \mathcal{A}$) is NP-hard.*

Sketch of Proof. We show a reduction from the set packing problem, which is known to be NP-hard [6] even for sets of size at most 3. Let an unweighted set packing instance $S_1, \ldots, S_n \subseteq \mathcal{U}$, $|\mathcal{U}| = m$, be given and assume that $|S_j| \leq 3$ for all j. We ask whether a collection $P \subseteq \{1, \ldots, n\}$ of sets with $S_i \cap S_j = \emptyset$ for all $i, j \in P$ and $\bigcup_{i \in P} S_i = \mathcal{U}$ exists.

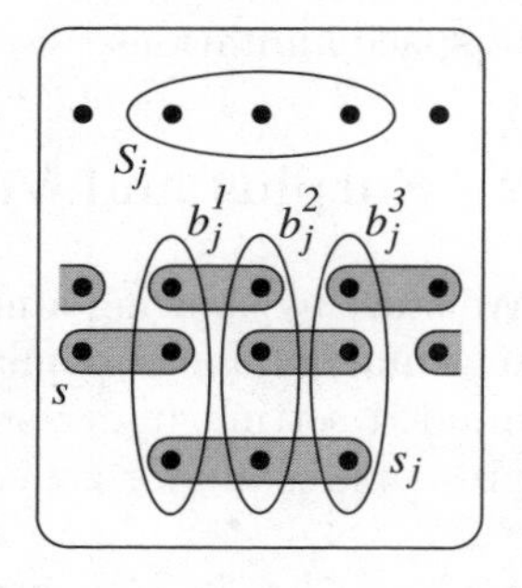

Fig. 1. Reduction from the proof of Lemma 2

Figure 1 illustrates the construction of the resulting CE instance. Points represent goods, ellipses represent sets of sellers and buyers. Sellers' sets are shaded. The ground set $\mathcal{U}'$ consists of two independent sets of goods each of which corresponds to the set packing ground set $\mathcal{U}$ and is supplied by sellers of type s (see Fig. 1), plus some additional goods corresponding to each set in the set packing instance. For a set S_j of size 3 we define 3 special goods supplied by a designated seller s_j. Three buyers b_j^1, b_j^2 and b_j^3 request one of these special products plus both products corresponding to one of the elements in set S_j each. For sets of size 1 or 2 the number of special goods is reduced accordingly.

Let now $|\mathcal{U}'| = \tau$ denote the number of products in the CE instance and define all valuations, such that every seller charges a *price per item* of 1, while buyers offer $\tau/(\tau-1)$ per product they request. The result follows from the following two observations. First, by the choice of valuations it holds that in every trade with positive surplus supply and demand for each product match exactly. Second, by the way buyers and sellers are entangled in the construction (see Fig. 1), any such trade must involve all products corresponding to the ground set of the set packing instance. □

Lemma 3. *The CE positive surplus problem for the family $\mathcal{F}$ of instances with only a single type of good (multi-unit exchange problem) is NP-hard.*

Finally, we consider cases where there are relatively large gains from the trade. Formally, we prove inapproximability for cases with a *packing to covering factor* as large as the known set packing lower bound.

Definition 3. *For instance I we define the* packing to covering factor $f(I)$ *as the maximum value of $\sum_{i\in B^*} v_i / \sum_{j\in S^*} v_j$ over all surplus maximizing trades $T^\star_{sur}(I) = (S^*, B^*)$.*

Theorem 2. *There exists a function $\gamma(n, m) = \Omega(\min\{n^{1-\varepsilon}, m^{1/2-\varepsilon}\})$ such that the CE surplus (volume) problem is inapproximable for the family $\mathcal{F}$ of instances which satisfy $f(I) \geq \gamma(n, m)$ $\forall I \in \mathcal{F}$, unless P=NP.*

2.2 Communication Lower Bounds

We next consider the problem of achieving approximation to the CE surplus (volume) problem when agents have general valuations (buyers are not single-minded). We show that the two goals cannot be approximated unless exponential communication in m is used. The inapproximability results hold even in the case of a single seller holding a set, and buyers that have general monotone valuations over sets (not multi-sets).

Let $G = (1, \ldots, 1)$ be the bundle with one item of each product. Assume that buyer i has a monotone valuation function $v_i : 2^m \rightarrow \mathbb{R}_+$. As we assume that a single seller offers G, the goal of a communication protocol P is to find a partition of the items to the buyers such that the surplus (volume) is maximized. We define inapproximability of a communication protocol similar to Definition 1 (with α, β using only m as their argument). Based on a result from Nisan and Segal [10] we show the following.

Theorem 3. *The CE surplus (volume) problem is not approximable in less than $\binom{m}{m/2}$ bits. This holds even with only a single seller and two buyers which have valuations over sets.*

Moreover, even if there is a large gain from the surplus maximzing trade the problem remains inapproximable. Based on a lower bound of Nisan [9] we prove that when $f(I) \geq n$ and $n < m^{1/2-\epsilon}$, any approximation requires exponential communication in m.

Theorem 4. *The CE surplus (volume) problem for instances with packing to covering factor at least n is not approximable with less than $e^{m/(2n^2)-5log\ n}$ bits. The lower bound holds for randomized and nondeterministic protocols.*

3 Approximating Social Cost

In the following section we present an algorithm that achieves a logarithmic approximation ratio for the objective of minimizing social cost of the trade. We additionally present a matching lower bound for the case without multi-sets. Our algorithm is based on the well known greedy approximation algorithm for the multi-set multi-cover problem [13]. For the remainder of this section, let $k_i = \sum_{e \in \mathcal{U}} q_i^e$ for all $i \in \mathcal{S}, \mathcal{B}$ and define $k = \max_i k_i$.

It is known that the greedy approximation algorithm approximates the covering integer program (CIP) in Figure 2 within H_k, where H_k denotes the k'th harmonic number. Essentially, Theorem 5 follows from the observation that this CIP is an exact formulation of the problem of finding a trade of minimal social cost. We briefly mention that the greedy approximation algorithm we apply is inherently monotone and, thus, yields a truthful exchange mechanism if combined with an appropriate (critical value based) payment scheme. However, we point out in Section 5 that there are other reasons that prevent us from obtaining reasonable truthful mechanisms.

Theorem 5. *Algorithm* Cover *is a H_k-approximation algorithm for the CE social cost problem.*

Proof. We can write the social cost minimization problem as the following integer linear program, where variables $x_i, x_j \in \{0,1\}$ indicate selected sellers and buyers and constraints (2) ensure feasibility of the trade:

$$\text{min.} \sum_{j \in \mathcal{B}} v_j(1 - x_j) + \sum_{i \in \mathcal{S}} v_i x_i \tag{1}$$

$$\text{s.t.} \sum_{i \in \mathcal{S}} q_i^e x_i \geq \sum_{j \in \mathcal{B}} q_j^e x_j \quad \forall e \in \mathcal{U} \tag{2}$$

Defining $\Delta_e = \sum_{j \in \mathcal{B}} q_j^e$ as in the algorithm we can rewrite $\sum_{j \in \mathcal{B}} q_j^e x_j$ as $\Delta_e - \sum_{j \in \mathcal{B}} q_j^e(1-x_j)$. Thus, constraints (2) become $\sum_{i \in \mathcal{S}} q_i^e x_i \geq \Delta_e - \sum_{j \in \mathcal{B}} q_j^e(1-x_j)$. Substituting a new variable y_j for $1 - x_j$ for all buyers $j \in \mathcal{B}$ we obtain exactly the covering integer program defined in algorithm Cover. By the fact that the greedy algorithm for multi-set multi-covering [13] has approximation ratio H_k we immediately obtain that $\text{cost}(T) \leq H_k \cdot \text{cost}(T^\star_{\text{cost}}(I))$. If trade T has negative surplus, we return the empty trade, which has even smaller cost in this case, instead. □

The following theorem states that the approximation ratio of algorithm Cover is essentially best possible, as parameter k is trivially upper bounded by the number m of distinct goods whenever we do not allow multi-sets.

Theorem 6. *The CE social cost problem cannot be approximated in polynomial time better than within $(1 - o(1)) \ln m$, unless $NP \subseteq DTIME(n^{\mathcal{O}(\log \log n)})$.* [2]

[2] Note that we can replace this assumption by $P \neq NP$ if we relax the lower bound to $\Omega(\ln m)$ by [1].

We finally mention that algorithm COVER can in fact be viewed as a generic reduction of social cost minimization in CE's to social cost minimization in reverse combinatorial auctions. Loosely speaking, we first allocate to all buyers in the exchange scenario their desired bundles at their offered price and then run a reverse auction algorithm considering all exchange participants as sellers. If one of the original sellers is selected, we buy her offered bundle. If one of the original buyers is selected, we buy the previously allocated bundle back from her. Setting demand for every product as done above, we ensure that the auction algorithm achieves sufficient supply for all buyers not returning their bundles.

1. Let $\Delta_e = \sum_{j \in \mathcal{B}} q_j^e$ for all products $e \in \mathcal{U}$.
2. Apply the greedy approximation algorithm to the following multi-set multi-cover problem:

$$\begin{aligned} \text{min.} \quad & \sum_{i \in \mathcal{S}} v_i x_i + \sum_{j \in \mathcal{B}} v_j y_j \\ \text{s.t.} \quad & \sum_{i \in \mathcal{S}} q_i^e x_i + \sum_{j \in \mathcal{B}} q_j^e y_j \geq \Delta_e \quad \forall e \in \mathcal{U} \\ & x_i, y_j \in \{0, 1\} \end{aligned}$$

3. Let $S = \{i \mid x_i = 1\}$, $B = \{j \mid y_j = 0\}$. If $\sum_{j \in B} v_j \geq \sum_{i \in S} v_i$ return trade $T = (S, B)$, else return $T = (\emptyset, \emptyset)$.

Fig. 2. Approximating optimal social cost by algorithm COVER

In the case of multi-unit exchanges (MUs) with only a single type of product, social cost minimization reduces to solving the min-knapsack (or reverse multi-unit auction) problem. Similarly to algorithm COVER, we can define algorithm MINKNAPSACK by applying the known monotone FPTAS [3] for min-knapsack to the covering formulation of multi-unit CE's. Similar to the proof of Theorem 6 a simple reduction from the partition problem (which is known to be NP-hard [6]) yields optimality of our algorithm's approximation guarantee.

Theorem 7. *Algorithm* MINKNAPSACK *is an FPTAS for the MU social cost problem. Furthermore, the MU social cost problem is NP-hard.*

4 The Supply Chain Problem

The objectives of surplus or volume maximization generalize naturally to the supply chain (SC) scenario. Thus, all the hardness results presented in Section 2.1 hold for this more general model, as well. However, we proceed by showing that the situation is even worse and that for the supply chain scenario even (approximate) social cost minimization is out of reach.

Theorem 8. *The SC social cost problem is inapproximable, unless P=NP.*

Sketch of Proof. We show a reduction from the decision version of the set packing problem, which is known to be NP-hard [6]. Let sets $S_1, \ldots, S_n$ over ground set $\mathcal{U}$, $|\mathcal{U}| = m$, and integer r be given. We want to decide whether there exist r non-intersecting sets. We define our supply chain social cost instance over ground set $\mathcal{U} \cup \{e^*\}$ as follows. For each set S_j we create a corresponding agent j who requests set S_j, supplies one copy of e^* and has valuation $v_j = -1$. Additionally, we define agent $n+1$ who supplies $\mathcal{U}$, requests r copies of e^* and offers to pay $v_{n+1} = \alpha(n,m)r + \beta(n,m) + 1$. It is then not difficult to check that a non-empty trade of cost r exists iff the set packing instance has r non-intersecting sets. Furthermore, the empty trade has cost $v_{n+1} = \alpha(n,m)r + \beta(n,m) + 1$, which yields the claim. □

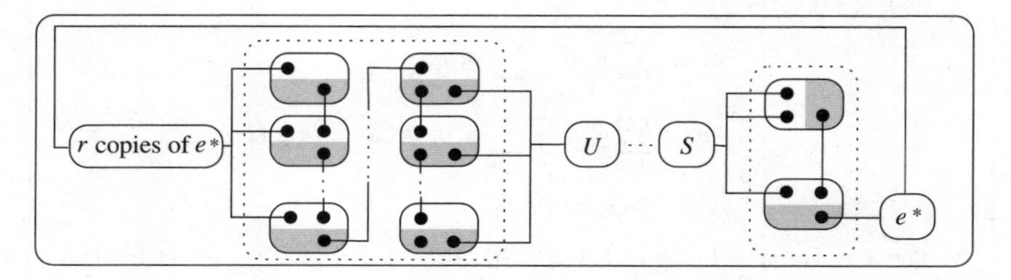

Fig. 3. Gadgets for the proof of Theorem 9. Agents are depicted as rectangles. Shaded areas indicate supplied products, lines connect products of identical type. Again, every feasible non-empty trade must contain agents corresponding to at least r non-intersecting sets.

A natural question to ask is whether similar hardness can be shown for supply chain instances with bounded size bundles, where we define the size of agent j's bundle as $\sum_{e \in \mathcal{U}} |\delta_j^e|$.

Theorem 9. *The SC social cost problem with bundles of size at most* 3 *is inapproximable, unless P=NP.*

The proof of Theorem 9 is based on the idea of simulating the construction from the proof of Theorem 8 by agents with bundles of bounded size, as illustrated in Fig. 3. This hardness result is tight, as we will see below that bundles of size 2 allow exact polynomial time algorithms.

It is well known that combinatorial auctions become solvable in polynomial time when the participating bidders are restricted to bid on sets of size at most 2, since in this case the problem can be formulated in terms of a weighted matching problem. For CE's or even supply chains, on the other hand, it is not immediate whether the problem reduces to matching in this case. As we shall see, the social cost objective turns out to be the key in obtaining such a problem formulation.

Theorem 10. *The SC surplus problem with sets of size* 1 *and* 2 *can be solved in polynomial time.*

Theorem 10 is a direct consequence of Lemmas 4, 5 and 6, which describe the reduction of the supply chain surplus problem to weighted b-matching.

Lemma 4. *The SC surplus problem with bundles of size at most k reduces to the CE surplus problem with bundles of size at most k.*

Sketch of Proof. Let an instance of the supply chain surplus problem with agents $\mathcal{A}$ be given and let $\alpha = \sum_{k \in \mathcal{A}^+} v_k + 1$, thus, $\mathrm{sur}(T^\star) < \alpha$. An agent j that both offers and requests products at price v_j is simulated by seller s_j supplying the same products as j and some new product e_j at price α and a buyer b_j who requests e_j and the products requested by j at price $\alpha + v_j$. □

Lemma 5. *The CE surplus problem with sets of size* 1 *and* 2 *reduces to the CE surplus problem with sets of size* 2.

Lemma 5 follows easily by adding some dummy products and traders. Lemma 6 follows by reformulating the social cost minimization problem similar to Theorem 5. Since all sets are of size exactly 2, the resulting ILP formulation is equivalent to a weighted matching problem, which can be solved in polynomial time [4].

Lemma 6. *The CE surplus problem with sets of size* 2 *can be solved in polynomial time.*

5 Non-existence of Mechanisms for Social Cost Approximation

In this section we discuss the existence of truthful mechanisms for the CE problem. A mechanism consists of some algorithm A that outputs a trade $T = (S, B)$ and additional payments $(p_i^s)_{i \in \mathcal{S}}$, $(p_j^b)_{j \in \mathcal{B}}$ determining the payments given to sellers and collected from buyers, respectively. A mechanism is *normalized* and satisfies *voluntary participation* (VP), if selected buyers never pay more than their declared valuation, selected sellers are never paid less than their valuation and payments to and from non-selected agents are 0. Furthermore, a mechanism is *budget-balanced* (BB) if the sum of payments is always non-negative ($\sum_{i \in S} p_i^s \le \sum_{j \in B} p_j^b$), α-approximately *cost-efficient* if it computes trades that are α-approximate with respect to social cost and *truthful* if it is a dominant strategy for every agent to declare their true valuations. A classical result by Myerson and Satterthwaite [8] shows that no truthful 1-approximately cost-efficient mechanism can satisfy both (VP) and (BB). This result extends to approximately cost-efficient mechanisms.

Theorem 11. *Fix some $\alpha \ge 1$ and let $M = (A, p^s, p^b)$ be a truthful and α-approximately cost-efficient CE mechanism satisfying VP. Then M is not budget-balanced.*

References

1. Alon, N., Moshkovitz, D., Safra, M.: Algorithmic Construction of Sets for k-Restrictions. ACM Transactions on Algorithms 2, 153–177 (2006)
2. Babaioff, M., Walsh, W.E.: Incentive-Compatible, Budget-Balanced, yet Highly Efficient Auctions for Supply Chain Formation. Decision Support Systems 39, 123–149 (2005)
3. Briest, P., Krysta, P., Vöcking, B.: Approximation Techniques for Utilitarian Mechanism Design. In: Proc. of STOC (2005)
4. Cook, W.J., Cunningham, W.H., Pulleyblank, W.R., Schrijver, A.: Combinatorial Optimization. John Wiley, Chichester (1998)
5. Cramton, P., Shoham, Y., Steinberg, R.: Combinatorial Auctions. MIT Press, Cambridge (2006)
6. Garey, M.R., Johnson, D.S.: Computers and Intractability: A Guide to the Theory of NP-completeness. W.H. Freeman, New York (1979)
7. Lehmann, D., O'Callaghan, L.I., Shoham, Y.: Truth Revelation in Approximately Efficient Combinatorial Auctions. J. of the ACM 49(5), 1–26 (2002)
8. Myerson, R.B., Satterthwaite, M.A.: Efficient Mechanisms for Bilateral Trading. Journal of Economic Theory 29, 265–281 (1983)
9. Nisan, N.: The Communication Complexity of Approximate Set Packing and Covering. In: Widmayer, P., Triguero, F., Morales, R., Hennessy, M., Eidenbenz, S., Conejo, R. (eds.) ICALP 2002. LNCS, vol. 2380, pp. 868–875. Springer, Heidelberg (2002)
10. Nisan, N., Segal, I.: The Communication Requirements of Efficient Allocations and Supporting Prices. Journal of Economic Theory (2006)
11. Parkes, D.C., Cavallo, R., Elprin, N., Juda, A., Lahaie, S., Lubin, B., Michael, L., Shneidman, J., Sultan, H.: ICE: An Iterative Combinatorial Exchange. In: Proc. of EC, pp. 249–258 (2005)
12. Parkes, D.C., Kalagnanam, J., Eso, M.: Achieving Budget-Balance with Vickrey-Based Payment Schemes in Exchanges. In: Proc. of IJCAI, pp. 1161–1168 (2001)
13. Rajagopalan, S., Vazirani, V.V.: Primal-Dual RNC Approximation Algorithms for (multi)Set (multi)Cover and Covering Integer Programs. SIAM J. of Computing 28(2), 525–540 (1998)
14. Roughgarden, T., Sundararajan, M.: New Trade-Offs in Cost-Sharing Mechanisms. Proc. of STOC, 79–88 (2006)
15. Sandholm, T.: Algorithm for Optimal Winner Determination in Combinatorial Auctions. Artificial Intelligence 135(1-2), 1–54 (2002)
16. Sandholm, T., Suri, S.: Improved Algorithms for Optimal Winner Determination in Combinatorial Auctions and Generalizations. In: Proc. of AAAI/IAAI, pp. 90–97 (2000)
17. Sandholm, T., Suri, S., Gilpin, A., Levine, D.: Winner Determination in Combinatorial Auction Generalizations. In: Proc. of AAMAS, Bologna, Italy, pp. 69–76. ACM Press, New York (2002)

Window-Games between TCP Flows

Pavlos S. Efraimidis and Lazaros Tsavlidis

Department of Electrical and Computer Engineering,
Democritus University of Thrace,
Vas. Sophias 12, 67100 Xanthi, Greece
{pefraimi,ltsavlid}@ee.duth.gr

Abstract. We consider network congestion problems between TCP flows and define a new game, the *Window-game*, which models the problem of network congestion caused by the competing flows. Analytical and experimental results show the relevance of the Window-game to the real TCP game and provide interesting insight on Nash equilibria of the respective network games. Furthermore, we propose a new algorithmic queue mechanism, called Prince, which at congestion makes a scapegoat of the most greedy flow. Preliminary evidence shows that Prince achieves efficient Nash equilibria while requiring only limited computational resources.

1 Introduction

Algorithmic problems of networks can be studied from a game-theoretic point of view. In this context, the flows are considered independent players who seek to optimize personal utility functions, like the goodput. The mechanism of the game is determined by the network infrastructure and the policies implemented at regulating network nodes, like routers and switches. The above described game theoretic approach has been used for example in [24] and in several recent works like [3,21].

In this work, we consider congestion problems of competing TCP flows, a problem that has been addressed in [13,3]. The novelty of our approach lies in the fact that we focus on the congestion window, a parameter that is in the core of modern AIMD (Additive-Increase Multiplicative-Decrease) based network algorithms. The size of the congestion window, to a large degree, controls the speed of transmission [12]. We define the following game, which we call the *Window-game*, as an abstraction of the congestion problem. The game is played synchronously, in one or more rounds. Every flow is a player that selects in each round the size of its congestion window. The router (the mechanism of the game) receives the actions of all flows and decides how the capacity is allocated. Based on how much of the requested window has been satisfied, each flow decides the size of its congestion window for the next round. The utility of each flow is the capacity that it obtains from the router in each round.

The motivation for this work is the following question, posed in [13,21]: *Of which game or optimization problem is TCP/IP congestion control the Nash*

B. Monien and U.-P. Schroeder (Eds.): SAGT 2008, LNCS 4997, pp. 95–108, 2008.

equilibrium or optimal solution? The first contribution of this work is the definition of the Window-game, a natural model that is simple enough to be studied from an algorithmic and game-theoretic point of view, while at the same time it captures essential aspects of the real TCP game. In particular, the Window-game aims to capture the interaction of the window sizes of competing TCP flows. Compared to the model used in [3], the Window-game approach is simpler and more abstract, but still sufficiently realistic to model real TCP games. We use the Window-game to study characteristic network congestion games. Furthermore, the plain structure of the Window-game allows us to study also one-shot versions of the game.

The second contribution is a new queue policy, called Prince (of Machiavelli), which aggressively drops packets from the most greedy flow. Under normal conditions, Prince rewards flows that do not exceed their fair share, while it punishes exemplarily the most greedy flow. Consequently, it drives the network towards efficient Nash equilibria. It is noteworthy that Prince is simple and efficient enough to be deployable in the demanding environment of network routers. We provide preliminary theoretical evidence and experimental results to support the above claims.

Outline. The rest of the paper is organized as follows: The Window-game is described in Section 2. An overview of TCP congestion control concepts is given in Section 3. We consider Window-games where the players are AIMD flows in Section 4 and Window-games with general flows in Section 5. Finally, a discussion of the results is given in Section 6. Due to lack of space some proofs are omitted.

2 The Window Game

The main entities of a Window-game is a router with capacity C and a set of $N \leq C$ flows, as depicted in Figure 2. The router uses a queue policy to serve in each round up to C workload. The N flows are the independent players of the game. Unless otherwise specified, the number N is considered unknown to the players and to the router. The game consists of one or more rounds. In each round, every player selects a size $w \leq C$ for its congestion window and submits it to the router. The router collects all requests and applies the queue policy to allocate the capacity to the flows. The common resource is the router's capacity, an abstract concept that corresponds to how much load the router can handle in each round[1].

Each round of the game is executed independently; no work is pending at the start of a round and no work is inherited to a following round. An important restriction is that the entities (the router and the flows) may use only limited computational resources like memory and processing power. In particular, the queue policy of the router should be stateless or use as little state information as

[1] To keep the window game simple, we intentionally avoid using the concept of queueing delay, even though it is considered to be a critical parameter of TCP networking.

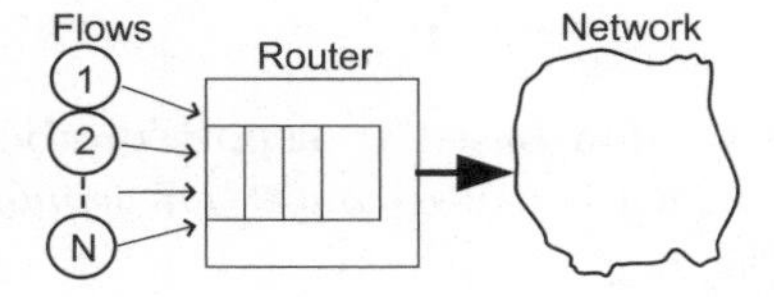

Fig. 1. The network model

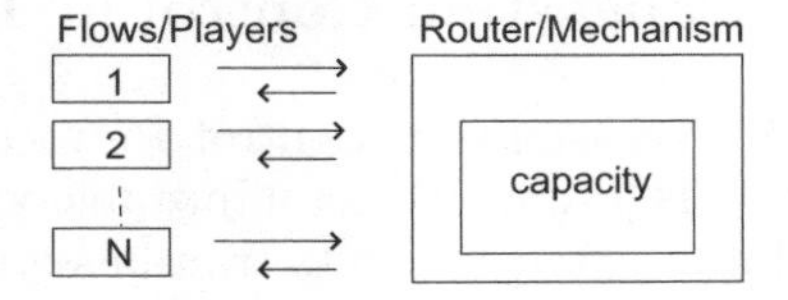

Fig. 2. One round of the window game

possible. This requirement is imposed by the real time conditions that a router[2] must work in. We consider several variations of the Window-game:

AIMD flows. First we consider Window-games where the players are AIMD flows. This class of Window-games is strongly related to the game(s) currently played by real TCP flows and exists implicitly in the analysis of [13,3]. Each AIMD flow selects the parameters (α, β) once, for the whole duration of the game. The utility for each flow is its average goodput (number of useful packets that are successfully delivered in each round) at steady state.

General flows. Then we consider Window-games where the flows can use arbitrary algorithms to choose their congestion window. We distinguish the following categories (in order of increasing complexity):
- One-shot game with complete information.
- One-shot game with incomplete information.
- Repeated game with incomplete information.

The action of a general flow is to choose the size of its congestion window for every round. The utility of the flow is the goodput in one-shot games and the average goodput at steady state for repeated games.

Assumptions. We make a set of simplifying assumptions similar to the assumptions of [3]. The window game is symmetric: All flows use the same network algorithms and parameters like Round Trip Time (RTT), loss recovery etc. All packets are of the same size and packet losses are caused only by congestion.

Solution concept. The solution concept for the Window-games is the Nash Equilibrium (NE) and in particular the Symmetric Nash Equilibrium (SNE). A good reason to start with SNE is that the analysis appears to be simpler compared to general NE. It is noteworthy, that the preliminary analytical results and the experiments show that in many cases there are only symmetric (or almost symmetric) NE. For each Window-game we study its SNE and discuss how efficient the network operates at it or them. In certain cases, we search experimentally for general, not necessarily symmetric, NE.

[2] In particular for the router: Stateful network architectures are designed to achieve fair bandwidth allocation and high utilization, but need to maintain state, manage buffers, and perform packet scheduling on a per flow basis. Hence, the algorithms used to support these mechanisms are less scalable and robust than those used in stateless routers. The stateless substance of nowadays IP networks allows Internet to scale with both the size of the network and heterogeneous applications and technologies [25,26].

3 Congestion Control in TCP

TCP (Transmission Control Protocol) is a window-based transport protocol. Every TCP-flow has an adjustable window, which is called *congestion window*, and uses it to control its transmission rate [12].

Congestion Window. The congestion window defines the maximum number of outstanding packets that the flow has not yet received acknowledgements for [14]. Essentially, the congestion window represents the sender's estimate of the amount of traffic that the network can absorb without becoming congested. The most common algorithm to increase or decrease the congestion window of a TCP-flow is AIMD.

AIMD. AIMD [6] can be considered as a probing algorithm designed to find the maximal rate at which flows can send packets under current conditions without incurring packet drops [13]. AIMD flows have two parameters α and β; upon success, the window is increased additively by α (in each round), and upon failure, the window is decreased multiplicatively by a factor β. AIMD is so far considered to be the optimal choice in the traditional setting of TCP Reno congestion control and FIFO drop-tail routers. If, however, we consider the developments like TCP SACK and active queue management, AIMD may no longer be superior [2].

Packet loss. When congestion occurs, packets are dropped. TCP variants use different loss recovery schemes to recover from packet losses. These schemes incur costs to the flow and can be thought of as a penalty on the flows which suspend their normal transmission rate until lost packets are retransmitted. We formed a penalty-based model, which is similar to the model of [13,3], to define a flow's behavior when losses occur.

Penalty Model. Assume that a flow with current window size w has lost $L \geq 1$ packets in the last round. Let γ be a small constant (eg. $\gamma = 1$). Then:

- Gentle penalty (resembles TCP SACK): The flow reduces its window to $\beta \cdot w - \gamma \cdot L$ in the next round.
- Severe penalty (TCP Tahoe): The flow timeouts for τ_s rounds and then continues with a window $w = \beta \cdot w$.
- Hybrid penalty (TCP Reno): If $L = 1$ the flow applies gentle penalty. If $L > 1$ a progressive severe penalty is applied. After a timeout of $\min\{2 + L, 15\}$ rounds, the flow restarts with a window equal to w/L. This penalty is justified by the experimental results of [7,19].

Router Policies. The router, and in particular the queue algorithm deployed by the router to allocate the capacity to the flows, defines the mechanism of the Window-game. We examine the common policies DropTail, RED, MaxMin, CHOKe and CHOKe+, as well as two variants of the proposed Prince policy and study their influence on the NE of the Window-game.

- **Drop-tail.** In each round, if the sum of the requested windows does not exceed the capacity C, all packets are served. Otherwise, the router drops a random sample of packets of size equal to the overflow.
- **RED (Random Early Detection) [8].** At overflow RED behaves like Drop-tail. However, for loads between min threshold $min_{th} = 70\%$ and a max threshold $max_{th} = 100\%$ of the capacity, packets are dropped with a probability p. In this case we simulate the behavior of the real RED: A random sample of packets is selected. The selected packets correspond to positions higher then min_{th} in a supposed queue. Each chosen packet is dropped with a probability proportional to the router's load.
- **MaxMin.** MaxMin is a stateful, fair queue policy [5] that conceptually corresponds to applying round-robin [11] for allocating the capacity.
- **CHOKe.** A stateless algorithm presented in [20]. We implement it in a way similar to RED. Every chosen packet that is above the min threshold, is compared to a packet chosen randomly from the packets below the threshold. The lower threshold, the upper threshold and the dropping probability are the same as in RED.
- **CHOKe+.** A variant of CHOKe presented in [3].
- **Prince.** Prince is an almost stateless queue policy. At congestion, packets are dropped from the flow with the largest congestion window. If the overflow is larger than the window of the most greedy flow, the extra packets are dropped from the remaining flows with drop-tail. For reasonable overflows, flows that do not exceed their fair share do not experience packet loss. Hence a greedy flow cannot plunder the goodput of other flows. We define a basic version of Prince that drops packets only in case of overflow and a RED-inspired version of Prince, called Prince-R, which applies its policy progressively starting from $min_{th} = 70\%$.

4 Window-Games with AIMD Flows

We discuss three characteristic Window-games between AIMD flows with gentle penalty, where the flows can choose the value of the α parameter[3]. We use in our analysis machinery from [3]. Assume N flows. Each flow i has parameters (α_i, β_i). At steady state, let N_i denote the window size after a packet loss, τ_i the number of rounds between two packet losses and L_i the number of packets lost at packet loss, for flow i. Then, as in [3],

$$N_i = \beta(N_i + (\alpha_i \cdot \tau_i - \gamma L_i)) \approx 1/2 \cdot (N_i + (\alpha_i \cdot \tau_i)) \tag{1}$$

[3] Experiments with parameter β show that in almost all cases selfish flows will use for β a value close to 1. The same conclusion can be made from the results in [3]. A simple argument for this behavior is that parameter β is very important when the flow has to quickly reduce its window size when the network conditions change. The TCP games examined in this work are rather static: Static bandwidth, static number of flows, static behavior of all flows during a game and hence there is no real reason for a flow to be adaptive.

and this gives

$$N_i = \alpha_i \cdot \tau_i. \tag{2}$$

Hence, the goodput of flow i is

$$G_i = 3/2 \cdot \alpha_i \cdot \tau_i. \tag{3}$$

Drop-tail router with synchronized packet losses and gentle penalty flows. All flows experience packet loss each time congestion occurs. Hence $\tau_1 = \tau_2 = \ldots = \tau_n$. Let $N = \sum_1^n N_i$ and $A = \sum_{i=1}^n \alpha_i$. Then $N = \beta(N + A \cdot \tau) \Rightarrow N = A \cdot \tau$. Since $N = \beta \cdot C = C/2$ we get $\tau = C/(2A)$. The goodput of flow i is

$$G_i = \frac{3 \cdot \alpha_i}{4 \cdot A \cdot C}. \tag{4}$$

G_i of flow i is an increasing function of α_i, regardless of the parameters α of the other flows. Hence, at Nash equilibrium all flows use the maximum possible value for their parameter α. This is an inefficient SNE, that resembles a "tragedy of the commons" situation where N players overuse a common resource, the network. The above claim is in agreement with the results in [3].

Drop-tail router with non-synchronized packet losses and gentle penalty flows. When congestion occurs, a random set of packets is dropped. A flow may or may not experience packet loss. The expected number of packets that it will lose is proportional to its window size. This case has not been studied analytically before. The fact that packet loss in not synchronized makes the analysis harder. Experimental results show that at SNE selfish flows will use large values for α (Figure 9). An explanation is that since $G_i = 3/2 \cdot \alpha_i \cdot \tau_i$, an increased value for α_i, for example $\alpha_i = 2$, increases the factor α_i of G_i. Even though flow i will experience packet loss more frequently (i.e., τ_i will decrease) the overall product $\alpha_i \cdot \tau_i$ will still increase. Intuitively, if the product would not increase then $\tau_2 = 2 \cdot \tau_1$ which cannot be true because in this case flow 2 must have on average a much larger window than flow 1. Consequently, they cannot have the same goodput.

We provide a proof for the case of 2 flows. Assume a router with capacity C and $N = 2$ flows with parameters $\alpha_1 = 1$, $\alpha_2 = z \cdot \alpha_1 = z$ and $\beta_1 = \beta_2 = 1/2$. We will show that flow 2 achieves a higher goodput by increasing its parameter α_2. At steady state, we know from Equations 2 and 3 that $N_i = \alpha_i \cdot \tau_i$ and $G_i = 3/2 \cdot \alpha_i \cdot \tau_i$, for $i = 1, 2$. A congestion round, is a round in which a packet loss occurs. A loss round for flow i is a congestion round in which flow i experiences packet loss. *Simplification:* We assume that at congestion only 1 packet is randomly selected and dropped. Hence only one flow will experience packet loss.

Assume x such that $G_2 = x \cdot G_1$. Let $w_{c,i}$ be the average window size of flow i at congestion rounds. Then $w_{c,1} + w_{c,2} \approx C$.

Claim. Assume y such that $w_{c,2} = y \cdot w_{c,1}$. Then $\tau_1 = y \cdot \tau_2$.

Proof. Let $w_{c,i}(k)$ be the window size of flow i at congestion round k. The probability that flow 1 experiences packet loss in congestion round k is $w_{c,1}(k)/C$. Hence we can assume a binary random variable $X_1(k)$ with mean value $w_{c,1}(k)/C$. The total number Ψ_1 of loss rounds of flow 1 is $\Psi_1 = \sum_k X_1(k)$ and the expected total number of loss rounds after K congestion rounds is $S_1 = E[\Psi_1] = w_1/C \cdot K$. Using an appropriate Hoeffding-Chernoff bound of [10, Page 200, Theorem 2.3] we can show (proof omitted) that with high probability $S_1 \in [(1-\epsilon)w_1/C \cdot K, (1+\epsilon)w_1/C \cdot K]$ for some positive constant ϵ. A sufficient number of rounds K can make the constant ϵ arbitrary small. Since we consider steady state, we can approximate $S_1 \approx w_1/C \cdot K$ for large K. We will assume $S_1 = w_1/C \cdot K$. Similarly $S_2 = w_2/C \cdot K$ and so $S_2 = y \cdot S_1$. This gives $\tau_1 = y \cdot \tau_2$.

Since $G_1 = 3/2 \cdot \alpha_1 \cdot \tau_1 = 3/2 \cdot \alpha_1 \cdot \frac{\tau_2}{y} = 3/2 \cdot \alpha_2 \tau_2 \cdot \frac{\alpha_1}{\alpha_2 y} = \frac{\alpha_1}{\alpha_2} y G_2 \Rightarrow$

$$x = \frac{\alpha_2}{\alpha_1} \cdot \frac{1}{y} = \frac{z}{y}. \tag{5}$$

Assuming that congestion rounds occur periodically, every $\tau+1$ rounds, gives that $G_2 = w_2 - 1/2 \cdot \alpha_2 \cdot \tau = y \cdot w_1 - 1/2 \cdot \alpha_2 \cdot \tau$. Also $G_1 = w_1 - 1/2 \cdot \alpha_1 \cdot \tau$ and $G_2 = x \cdot G_1 = x(w_1 - 1/2 \cdot \alpha_1 \cdot \tau)$. Combining the above relations gives

$$y \cdot w_1 - 1/2 \cdot \alpha_2 \tau = x \cdot w_1 - 1/2 \cdot x \alpha_1 \tau. \tag{6}$$

Substituting $w_1 = C/(y+1)$ and $\tau = \tau_1/(y+1)$ and solving for τ_1 we get:

$$\tau_1 = \frac{2zC/y - 2yC}{z\alpha_1/y - z\alpha_1}. \tag{7}$$

The goodput of flow 1 can also be calculated from (proof omitted):

$$G_1 = \frac{1}{\tau+1}\left((\tau+1) \cdot w_1 + \sum_{i=0}^{\tau} i \cdot \alpha_i - (\tau+1)\frac{w_1}{C}\frac{w_1}{2}\right), \tag{8}$$

where $(\tau+1) \cdot w_1$ is the number of packets in $\tau+1$ rounds starting at a congestion round, $\sum_{i=0}^{\tau} i \cdot \alpha_i$ is the number of packets due to the increment of the congestion window and $(\tau+1)\frac{w_1}{C}$ are the average packets lost due to a possible loss round. From this we get :

$$G_1 = w_1 - \frac{(w_1)^2}{2C} + \frac{\alpha_1 \tau}{2}. \tag{9}$$

Substituting w_1 and τ and solving for τ_1 gives:

$$\tau_1 = \frac{\frac{C}{y+1} - \frac{C}{2(y+1)^2}}{3/2 \cdot \alpha_1 - \frac{\alpha_1}{4(y+1)}}. \tag{10}$$

We combine equation 7 and 10 to eliminate τ_1 and solve for y. The outcome is the following polynomial:

$$12y^4 + 22y^3 + (10 - 16z)y^2 + (-20z)y + (-8z) = 0. \tag{11}$$

We use Mathematica [22] to solve the polynomial. Only one of the four solutions of y is a non-negative real number. The result is a complicated expression

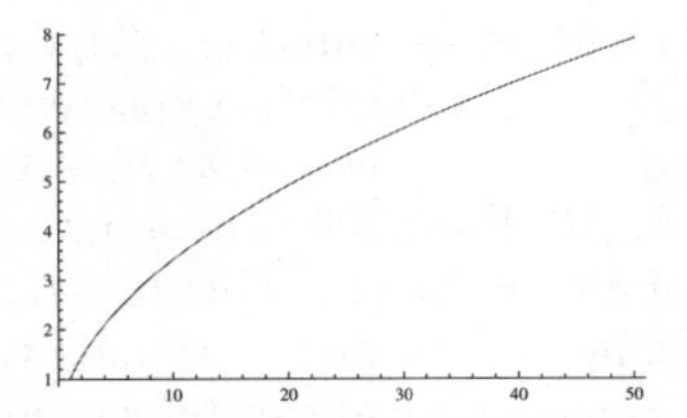

Fig. 3. y as a function of z

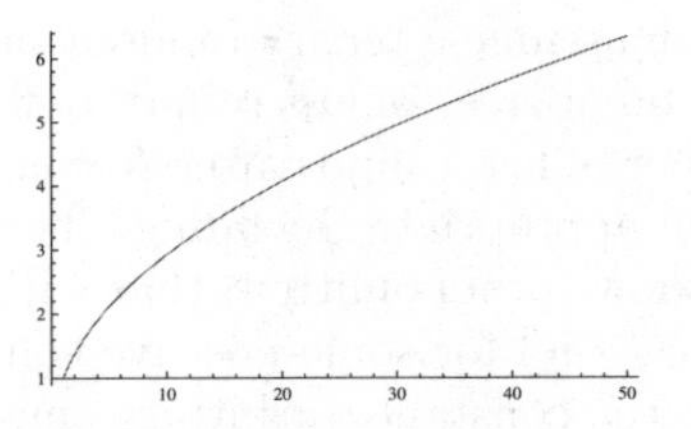

Fig. 4. x as a function of z

of z [17]. From y and Equation 5 we calculate x. The plots of y and x (Figures 3 and 4) show that both are increasing functions of z.

Since $G_2 = x \cdot G_1$, the previous results show that flow 2 will achieve a higher goodput then flow 1 if $z > 1$. The total goodput G is $G = G_1 + G_2$. Hence, $G_2 = (x/(x+1))G$, an increasing function of x. The total goodput G variates slowly as z increases. Overall, the goodput of flow 2 increases, when parameter α_2 is increased[4].

Prince router with gentle penalty flows. We present a preliminary argument for the effectiveness of Prince. Assume a router with capacity C and N flows $i = 1, \ldots, N$ with parameters $(\alpha_i, \beta_i = 1/2)$. At steady state, from Equation 2 we know that: $N_i = \alpha_i \cdot \tau_i$. Let w_{L_i} be the average window size of flow i at loss rounds (of flow i). Then $G_i = 3/4 \cdot w_{L_i}$. At congestion, $\sum_{i=1}^{N} w_i > C$. Clearly, $\max_{i=1,\ldots,N} w_i > C/N$. Consequently, $w_{L_i} \geq C/N$. Assume that flow i would have exclusive use of a router with capacity C/N. Then, by playing AIMD its average lost window w_L would be $w_L \leq C/N$ and its goodput $3/4 \cdot C/N$. Hence, with Prince the AIMD flow achieves a goodput at least as good as in the above (reasonably fair) case. A flow that does exceed its fair share does not loose any packet, unless a very large overflow ($> \max w_i$) occurs. *Interestingly, the fair share of flow i is ensured, regardless of the strategies of the competing flows.* This is strong evidence that with Prince, the network operates at an efficient state.

Experimental Results. We performed an extensive set of experiments with the network model of Figure 1, with $N = 10$ AIMD flows, a router with capacity $C = 100$, several queue policies and both the gentle and the hybrid penalty models. Parameter α takes values from the set $\{1, 2, .., 50\}$ and β from the set $\{0.5, 0.51, .., 0.99\}$. We focus on the results for varying parameter α when β is the fixed $\beta = 0.5$. First, we applied the iterative methodology of [3], we call it $M1$, to find SNE of the Window-game. Second, thanks to the simplicity of the

[4] The experiments showed that G started below $7/8 \cdot C$ for balanced flows ($z = 1$) and decreased slowly to above $6/8 \cdot C$ for completely unbalanced flows. An intuitive argument is that the more unbalanced the flows are, the larger (on average) the window of the flow that experiences packet loss is, and the larger the reduction on the overall goodput is. If w_L is the window of the flow that loses packet at a congestion round, then (extreme cases): If $w = 1/2 \cdot C$ at all loss rounds, then $G = 7/8 \cdot C$, and if $w = C$, then $G = 6/8 \cdot C$.

window game, we could perform a brute force search on all symmetric profiles to discover all possible SNE ($M2$). Finally, we used random non-symmetric starting points along with a generalization of the procedure of [3] to check if the network converges to non-symmetric NE (methodology $M3$). The experiments where performed with a simulator for the Window-game, which is called NetKnack[17]. NetKnack is implemented in Java and can perform from a simple experiment to massive series of experiments.

The methodology $M1$ is executed in iterations. In the first iteration, $\alpha^1 = 1$ for flows $F_1, \ldots, F_{n-1}$ and we search for the best response of flow F_n. Let $\alpha^{1,best}$ be the value α, with which F_n achieves the best goodput. By convention, a flow switches to a better value for α only if the average improvement of its utility is at least 2%. In the next iteration, flows $F_1, \ldots, F_{n-1}$ play with $\alpha^2 = \alpha^{1,best}$ and we search for the best α_n in this profile. If at iteration k, $\alpha^{k,best} = \alpha^k$ then this value, denoted by α_E, is the SNE of the game.

Every experiment consists of 2200 rounds. The first 200 rounds are used to allow the flows to reach steady state. To avoid synchronization of flows' windows the capacity C is variable and changes randomly, with plus 1 or minus 1 steps, in the region 100 ± 5 with an average of 100 packets. Finally, the measurements are averaged over 30 independent executions of each experiment.

We present graphs with the results of experiments where flows $i = 1, \ldots, 9$ use $\alpha = 1$ and flow 10 tries all possible values for $\alpha_{10} \in \{1, \ldots, 50\}$. Figures 5 and 6 show the results for gentle penalty flows and several queue policies. We separated the graphs into 2 figures for more clarity. From the above figures (gentle penalty) we see that Prince and MaxMin induce efficient Nash equilibria with small values for parameter α, while DropTail, RED, CHOKe and CHOKe+ actuate flow 10 to use large values for α. Figures 7 and 8 show that with Prince and MaxMin the deviator player 10 has clearly suboptimal performance for $\alpha_{10} > 1$. Hence, the profile α_i of the first iteration is a NE.

Figures 9 and 10 show the SNE that have been found with the methodology $M1$. We would like to note that depending on experiment parameters, like capacity C and number of players N, the value α for the NE may differ significantly. However, the NE for MaxMin and Prince are more efficient than the NE of the other policies. For flows with gentle penalty (Figure 9) the results show that the Nash Equilibria of Prince's variants are efficient but their per flow loss rate is not low under current game parameters, while all other policies result in

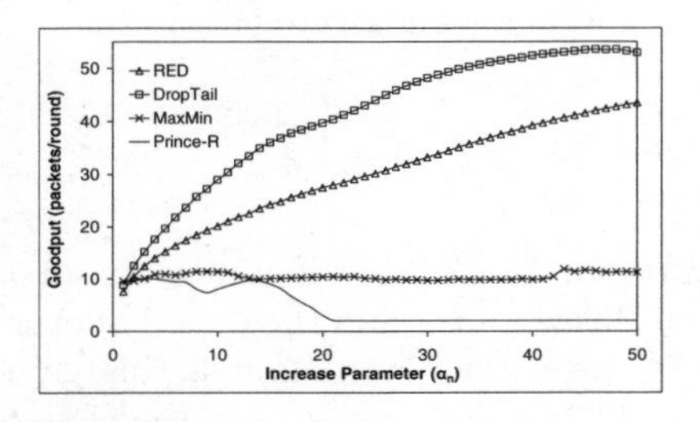

Fig. 5. Flows with gentle penalty (part 1)

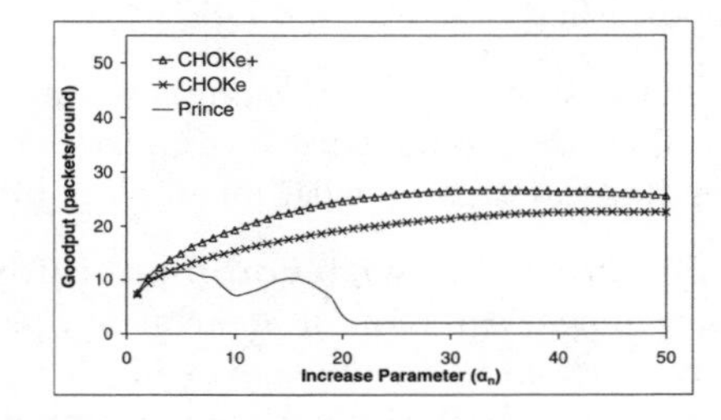

Fig. 6. Flows with gentle penalty (part 2)

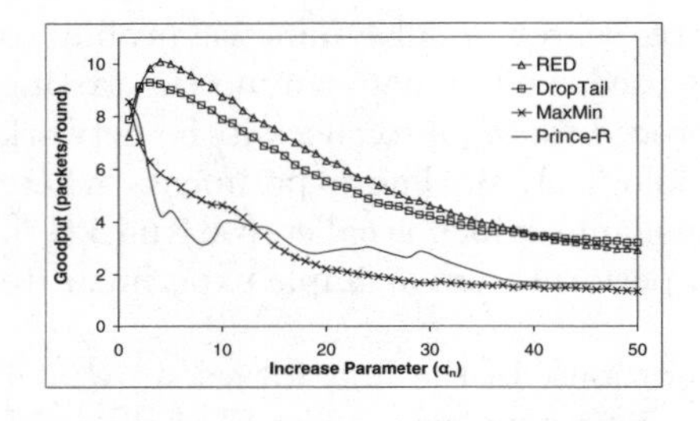

Fig. 7. Flows with hybrid penalty (part 1)

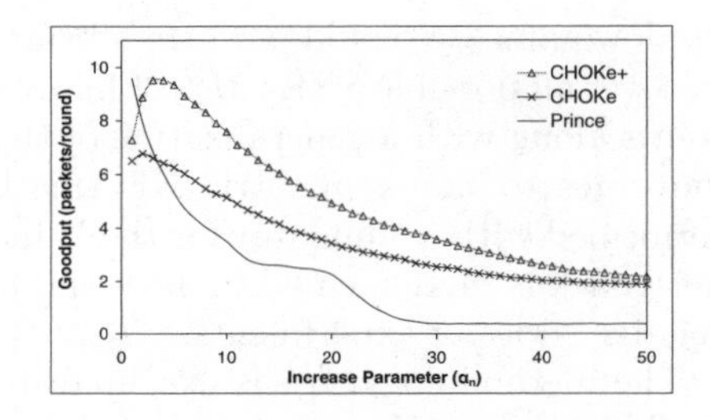

Fig. 8. Flows with hybrid penalty (part 2)

Queue Policy	parameter α			parameter β		
	α_E	goodput packets/round	loss rate (%)	β_E	goodput packets/round	loss rate (%)
DropTail	α_3=50	5,499	78,8	β_2=0,97	9,841	3,9
RED	α_2=49	5,485	78,5	β_2=0,97	8,309	4,4
CHOKe	α_3=49	3,363	86,8	β_2=0,94	7,761	5,4
CHOKe+	α_3=50	5,431	79,1	β_2=0,96	8,118	4,5
Prince-R	α_2=2	9,987	7,4	β_3=0,94	9,995	7,3
Prince	α_2=4	9,111	13,9	β_2=0,94	9,993	7,3
MaxMin	Osc. between a=9 and a=12			β_2=0,92	9,998	4,2

Fig. 9. SNE for gentle penalty

Queue Policy	parameter α			parameter β		
	α_E	goodput packets/round	loss rate (%)	β_E	goodput packets/round	loss rate (%)
DropTail	α_5=6	5,863	6,4	β_2=0,92	7,999	2,4
RED	α_2=4	6,427	4,4	β_2=0,96	7,591	3,0
CHOKe	α_2=2	6,182	3,6	β_4=0,78	6,674	2,6
CHOKe+	α_2=3	6,650	3,8	β_2=0,93	7,687	2,9
Prince-R	α_1=1	8,693	1,1	β_2=0,72	8,759	1,6
Prince	α_1=1	9,573	2,1	β_1=0,50	9,617	2,1
MaxMin	α_1=1	8,547	1,8	β_1=0,50	8,503	1,9

Fig. 10. SNE for hybrid penalty

an extremely undesirable NE. The results for hybrid penalty flows in Figure 10, show Prince and MaxMin to be preeminent over all other policies. The NE for DropTail, RED and CHOKe are inefficient and their loss rates are high.

Finally, the brute force search method $M2$ on Drop-tail and RED with gentle and hybrid flows revealed more SNE only for Drop-tail with hybrid penalty flows. The additional SNE use values of parameter α higher than the SNE found with the methodology $M1$ and are less efficient. The search with methodology $M3$ for non-symmetric NE for all queue policies and with both gentle and hybrid penalty flows, did not reveal any additional NE.

5 Window-Games with Non-AIMD Flows

We relax the restriction that the flows must be AIMD flows and consider the more general class of games where the flows can use an arbitrary strategy to choose their congestion window. We discuss the following strategic games:

- One-shot game with complete information.
- One-shot game with incomplete information.
- Repeated game with incomplete information.

One-shot game with complete information. There is one common resource, every player can request an arbitrary part of this resource and the payoff of every player depends on the moves of all players. Note that even though this game bears some similarity with congestion games [23], it does not fit into the class of (weighted) congestion games (See for example [23,16,15,9]). In the one-shot game the cost cannot be a timeout since there is only one round. Assume N flows with window sizes w_i, for $i = 1, \ldots, N$. Let $W = \sum_i w_i$. Let each successful packet give a profit of 1 and each packet loss cost $g \geq 0$.

DropTail. If $g = 0$ then

$$utility(N) = \begin{cases} w_N, & \text{if } W < C \\ \frac{C \cdot w_N}{W}, & \text{if } W \geq C \end{cases}$$

In both cases, the utility of flow N is an increasing function of w_N. There is a unique SNE, where all flows request the maximum possible value $w_i = C$. The SNE is very inefficient.

If $g > 0$, then a SNE can be calculated as follows: Assume the $N-1$ flows use $w_i = y$ and flow N uses $w_N = x$. The utility for flow N is the average number of successful packets minus g times the average number of lost packets (we assume $y \geq C/(N-1)$):

$$utility(y, x) = x \cdot \frac{C}{(N-1)y + x} - g \cdot x \cdot (1 - \frac{C}{(N-1)y + x}) \tag{12}$$

Solving the partial derivative of $utility(y, x)$ with respect to x we get one positive (and one negative) solution $x = y(1-N) + \sqrt{2Cy(N-1)}$. Using $x = y$ we get $y = \frac{2C(N-1)}{N^2}$. For example, if $C = 100$ and $N = 10$ then the SNE is at $w = 18$. Experimental results are presented in Figure 11.

MaxMin. In MaxMin there is a SNE where all flows play $w_i = C/N$. This SNE is the optimal solution for the Window-game problem. If $g > 0$, then this is the only NE of the game. As already discussed, the disadvantage of MaxMin is that it is a stateful policy.

Prince. If a flow i plays at most its fair share $w_i \leq C/N$ then it will experience no packet loss. Clearly, the profile where all flows play $w_i = C/N$ is a SNE. If the cost for packet loss is $g > 0$ then this is the only NE of the game.

One-shot game with incomplete information. If the players do not know the total number of players N then we get a one-shot game with incomplete information. We have to distinguish two different cases: If the players have no prior probabilities on the number of players, or else they have no distribution information for the unknown number N, then we get a game with no prior probability or a pre-Bayesian game [4,1].

Queue Policy	utility function			
	Passed packets	Passed-0.1*dropped	Passed-0.5*dropped	Passed-dropped
DropTail	100	⟩60	26-30	17,18
RED	100	⟩70	26-28	20
CHOKe	≈44	≈ 35	21-22	14-16
CHOKe+	100	⟩70	29	17-18
Prince-R	10,11	10,11	10	10
Prince	10,11	10,11	10	10
MaxMin	≥10	10	10	10

Fig. 11. Window sizes of NE for the one-shot game with complete information

In a pre-Bayesian game, we can apply ex-post equilibrium or safety-level equilibrium or robust equilibrium or other related equilibrium concepts. The common characteristic of these equilibrium concepts is that the player selects a very conservative action. In this case the players have to assume that the number of players N is the maximum possible number $N = C$. Hence, the players will play as in the game with complete information with $N = C$. For strategies like Prince and MaxMin, the choice of each player would be $w = 1$. Interestingly, common TCP implementations, like Reno or Tahoe, use a very conservative initial window $w = 1$ when starting a new flow.

However, in practice the flows are likely to have some prior information on the number N. Note that the TCP-game is a repeated game, which means that the flow would essentially in most cases[5] have an estimation on its fair share (except of the first round or a round after some serious network change). If the flow has prior information on the distribution of the unknown parameter N, we get a Bayesian game. We leave the analysis of this and the following case as future work.

Repeated game with incomplete information. The actual game that a TCP flow has to play is a repeated game with incomplete information. The problem has been addressed from an optimization and an on-line algorithm point of view in [13]. One other approach would be to consider this as a learning game: There are $N = C$ players and each player chooses either to participate in the game or to stay idle. The goal of each player is to "learn" the unknown number N of players who decide to participate. Each player may change its decision with a predetermined probability.

6 Discussion

We present a game-theoretic model for the interplay between the congestion windows of competing TCP flows. Preliminary theoretical and experimental results show that the model is relevant to the "real" TCP game. Furthermore we propose a simple queue policy, called Prince, with a sufficiently small state, and show that it achieves efficient SNE despite the presence of selfish flows.

Future work includes extending the analysis of the Window-game to the cases of the game with incomplete information. We also consider the proposed Prince policy of independent interest and intend to study further possible applications. We intend to investigate a realistic adaptation and implementation of Prince, possibly with streaming algorithms, on real TCP networking conditions or with the network simulator[18].

References

1. Aghassi, M., Bertsimas, D.: Robust game theory. Math. Program. 107(1), 231–273 (2006)
2. Akella, A., Seshan, S., Shenker, S., Stoica, I.: Exploring congestion control (2002)

[5] Unless we assume that the network conditions fluctuate a lot and therefore, the network load can change drastically and unpredictably from round to round.

3. Akella, A., Seshan, S., Karp, R., Shenker, S., Papadimitriou, C.: Selfish behavior and stability of the internet: a game-theoretic analysis of tcp. In: SIGCOMM 2002: Proceedings of the 2002 conference on Applications, technologies, architectures, and protocols for computer communications, pp. 117–130. ACM Press, New York (2002)
4. Ashlagi, I., Monderer, D., Tennenholtz, M.: Resource selection games with unknown number of players. In: AAMAS 2006: Proceedings of the fifth international joint conference on Autonomous agents and multiagent systems, pp. 819–825. ACM Press, New York (2006)
5. Bertsekas, D., Gallager, R. (eds.): Data networks. Publication New Delhi. Prentice-Hall of India Pvt. Ltd., Englewood Cliffs (1987)
6. Chiu, D.-M., Jain, R.: Analysis of the increase/decrease algorithms for congestion avoidance in computer networks. j-COMP-NET-ISDN 17(1), 1–14 (1989)
7. Fall, K., Floyd, S.: Simulation-based comparison of tahoe, reno, and sack tcp. Computer Communication Review 26, 5–21 (1996)
8. Floyd, S., Jacobson, V.: Random early detection gateways for congestion avoidance. IEEE/ACM Transactions on Networking 1(4), 397–413 (1993)
9. Fotakis, D., Kontogiannis, S., Spirakis, P.: Selfish unsplittable flows. Theor. Comput. Sci. 348(2), 226–239 (2005)
10. Habib, M., McDiarmid, C., Ramirez-Alfonsin, J., Reed, B.: Probabilistic Methods for Algorithmic Discrete Mathematics. Springer, Heidelberg (1998)
11. Hahne, E.L.: Round-robin scheduling for max-min fairness in data networks. IEEE Journal of Selected Areas in Communications 9(7), 1024–1039 (1991)
12. Jacobson, V.: Congestion avoidance and control. In: SIGCOMM 1988: Symposium proceedings on Communications architectures and protocols, pp. 314–329. ACM, New York (1988)
13. Karp, R., Koutsoupias, E., Papadimitriou, C., Shenker, S.: Optimization problems in congestion control. In: FOCS 2000: Proceedings of the 41st Annual Symposium on Foundations of Computer Science, Washington, DC, USA, p. 66. IEEE Computer Society, Los Alamitos (2000)
14. La, R.J., Anantharam, V.: Window-based congestion control with heterogeneous users. In: INFOCOM 2001. Twentieth Annual Joint Conference of the IEEE Computer and Communications Societies, vol. 3, pp. 1320–1329. IEEE, Los Alamitos (2001)
15. Milchtaich, I.: Congestion games with player-specific payoff functions. Games and Economic Behavior 13, 111–124 (1996)
16. Monderer, D., Shapley, L.: Potential games. Games and Economic Behavior 14, 124–143 (1996)
17. NetKnack: A simulator for the window-game, `http://utopia.duth.gr/~pefraimi/projects/NetKnack`
18. NS-2. The network simulator. `http://www.isi.edu/nsnam/ns/`
19. Padhye, J., Firoiu, V., Towsley, D.F., Kurose, J.F.: Modeling tcp reno performance: a simple model and its empirical validation. IEEE/ACM Trans. Netw. 8(2), 133–145 (2000)
20. Pan, R., Prabhakar, B., Psounis, K.: Choke, a stateless active queue management scheme for approximating fair bandwidth allocation. In: INFOCOM, pp. 942–951 (2000)
21. Papadimitriou, C.: Algorithms, games, and the internet. In: STOC 2001: Proceedings of the thirty-third annual ACM symposium on Theory of computing, pp. 749–753. ACM Press, New York (2001)

22. Wolfram Research. Mathematica (2007)
23. Rosenthal, R.W.: A class of games possessing pure-strategy nash equilibria. International Journal of Game Theory 2, 65–67 (1973)
24. Shenker, S.J.: Making greed work in networks: a game-theoretic analysis of switch service disciplines. IEEE/ACM Trans. Netw. 3(6), 819–831 (1995)
25. Stoica, I., Shenker, S., Zhang, H.: Core-stateless fair queueing: achieving approximately fair bandwidth allocations in high speed networks. SIGCOMM Comput. Commun. Rev. 28(4), 118–130 (1998)
26. Stoica, I., Zhang, H.: Providing guaranteed services without per flow management. SIGCOMM Comput. Commun. Rev. 29(4), 81–94 (1999)

Price Variation in a Bipartite Exchange Network

Ronen Gradwohl*

Department of Computer Science and Applied Mathematics
The Weizmann Institute of Science
Rehovot, 76100 Israel
ronen.gradwohl@weizmann.ac.il

Abstract. We analyze the variation of prices in a model of an exchange market introduced by Kakade et al. [11], in which buyers and sellers are represented by vertices of a bipartite graph and trade is allowed only between neighbors. In this model the graph is generated probabilistically, and each buyer is connected via preferential attachment to v sellers. We show that even though the tail of the degree distribution of the sellers gets heavier as v increases, the prices at equilibrium decrease exponentially with v. This strengthens the intuition that as the number of vendors available to buyers increases, the prices of goods decrease.

1 Introduction

This paper deals with an economic model of Kakade, Kearns, Ortiz, Pemantle, and Suri [11], in which goods are exchanged over a social network. As this model combines two parallel topics – economic models of exchange and social network theory – we begin by briefly describing each one separately, and then proceed to examine their union.

Mathematical models of exchange have a long history in economics, starting with the works of Walras [14] and Fisher [6]. One of the primary notions in these models is that of a market equilibrium, in which all agents maximize their utilities subject to budget constraints, and in addition the market clears (i.e. supply equals demand). In their landmark paper, Arrow and Debreu [1] proved that such an equilibrium exists under very general conditions.

More recently, research in economics has incorporated networks as models of interaction in exchange markets (see Jackson [9,8] for overviews of current literature). Of particular relevance to the current work's line of research are the papers of Kranton and Minehart [12] and Corominas-Bosch [4] – both analyze network effects on an economy in a setting in which agents bargain with neighboring agents. An additional related paper is that of Kakade et al. [10], in which the authors present a model that generalizes the frameworks of Walras and Fisher to include a network of interaction. In their model, each agent is represented by a vertex on some graph, and two agents are allowed to communicate and trade only if their respective vertices are connected in the graph.

* This research was supported by grant 1300/05 from the Israel Science Foundation.

B. Monien and U.-P. Schroeder (Eds.): SAGT 2008, LNCS 4997, pp. 109–120, 2008.

One particular type of graph that has received significant attention in recent literature was inspired by an attempt to analyze social networks. Empirical studies of various social networks, such as the Internet or the World Wide Web, have shown that such networks have various common properties, most notably a small diameter and a power-law degree distribution. There are numerous mathematical processes that model such networks – see the survey of Bollobás and Riordan [2] or the book of Chung and Lu [3]. Perhaps the simplest process that is commonly analyzed is preferential attachment, in which the graph is "grown", and each new vertex attaches to some random set of present vertices with probability proportional to their degree.

Kakade et al. [11] examine the marriage of a simple exchange market and a social network generated by preferential attachment. They superimpose a buyer-seller market on a bipartite social network, and examine the effects of the network's statistical properties on the distribution of prices in the market. They show, for example, that if the network contains a perfect matching, then there is no price variation in the network. Furthermore, they show that the degree of a seller is an upper bound on his price at equilibrium, and so the price distribution can be bounded above by the power-law distribution of the degrees.

Our Results. In the bipartite buyer-seller network of Kakade et al. [11], each new buyer enters the network with v neighbors (chosen by preferential attachment). As v increases, the tail of the degree distribution gets heavier. Thus, their upper bound on the price distribution increases. The intuition provided by proponents of a free market, however, suggests that when buyers have more options, the prices should go down, not up. In this work we prove that this intuition is correct in the model of [11] by showing that the price distribution decreases exponentially with v. More specifically, we show that the fraction of sellers with price greater than w is roughly $1/w^{v+1}$.

Organization. The rest of the paper is organized as follows. Sections 2 and 3 respectively describe the exchange market and the bipartite network in greater detail, and Section 4 contains our main theorems and most proofs.

2 The Market

The market we consider is the bipartite exchange economy formalized by Kakade et al. [11] (see also Even-Dar et al. [5] and Suri [13]). In this model there is a bipartite graph $G = (B, S, E)$ representing buyers and sellers. Each buyer has 1 infinitely divisible unit of cash and each seller has 1 infinitely divisible unit of wheat. Buyers have monotone increasing utility for wheat and sellers have monotone increasing utility for cash. Neither type of player has any utility for the commodity with which they are endowed.

For each seller j, let w_j^s be the price per unit wheat offered by seller j. Similarly, for each buyer b, let w_j^b be the price buyer j will pay per unit wheat. Finally, denote by x_{ij} the amount of wheat buyer i purchases from seller j.

Then exchange rates $\{w^s_j\}$ and $\{w^b_j\}$ and consumption plans $\{x_{ij}\}$ are a market equilibrium in G if the following conditions hold (see [11]):

- The market clears: For every seller j, $\sum_{i\in\Gamma(j)} x_{ij} = 1$, where $\Gamma(j) = \{i : (j,i) \in E\}$ is the set of neighbors of j in the graph G.
- Every buyer i maximizes his utility. Formally, this means that $x_{ij} > 0$ only if seller j has the cheapest price of all of i's neighbors:

$$w^b_i = \min_{k\in\Gamma(i)} w^s_k.$$

If for every seller j, $\Gamma(j) \neq \emptyset$, then there exists a market equilibrium. Furthermore, the prices $\{w^s_j\}$ of the equilibrium are unique, although the consumption plans are not [7]. For further discussion of the properties of this bipartite exchange economy, as well as a comparison with more general models, see [13].

3 The Network

The network on which our market takes place is generated as follows. We begin with one buyer and one seller with $v + 1$ edges between them. At step i, we add a new buyer and a new seller. The buyer chooses v sellers out of the previously added $i - 1$ at random, with probability proportional to their degrees, and connects to them. The seller chooses 1 buyer with probability proportional to his degree, and connects to him. This is repeated until there are n buyers and n sellers in the graph.

There are two differences between this model and the one considered by Kakade et al. [11]. First, our buyer chooses the sellers with replacement – that is, we allow the possibility that a buyer chooses the same seller more than once, in which case there will be multiple edges between the players. We sample with replacement solely for technical simplicity; this does not significantly affect the degree distribution of the resulting graph. The second difference is that the model of [11] is more general, as they have an additional parameter α. In their network, each buyer/seller samples a neighbor uniformly at random with probability α, and via preferential attachment with probability $1 - \alpha$. Here we only consider the special case $\alpha = 0$, as we are interested in the second parameter v.

The parameter v, which we think of as some arbitrary constant (independent of n), has a strong impact on the statistical properties of G. In what follows, we denote by G_v the graph sampled as above with parameter v, in which $|B| = |S| = n$. The following is essentially a theorem of [11] (except for the slight difference in the model mentioned above). The proof is analogous to that of Lemma 5.

Theorem 1. *Let $D(i)$ be a random variable that denotes the degree of the i'th seller in G_v, when G_v is randomly generated as above. Then*

$$\mathrm{E}\left[D(i)\right] = O\left(\left(\frac{n}{i}\right)^{\frac{v}{v+1}}\right).$$

Kakade et al. [11] show that the degrees of the sellers constitute an upper bound on the price distribution, yielding the following theorem.

Theorem 2 ([11]). *When the network is sampled according to G_v, the proportion of sellers j with price $w_j^s > w$ is at most $O(w^{-(v+1)/v})$.*

4 Price Distribution in the Network

Note that as v increases, $w^{-(v+1)/v}$ increases as well, since $(v+1)/v$ approaches 1 from above. Thus, the tail of the price distribution in the upper bound gets heavier with v. In this paper we prove the following theorem, which states that the opposite is in fact true.

Theorem 3 (Upper Bound). *Let $\alpha > 0$ and $\varepsilon > 0$ be arbitrarily small constants, let v be some positive integer constant, let q be an arbitrarily large constant, and fix $w = n^\alpha$. Let $P(w)$ be a random variable that denotes the proportion of sellers j with price $w_j^s > w$ in a random network G_v. Then*

$$\Pr\left[P(w) < \frac{1}{w^{v+1-\varepsilon}}\right] > 1 - \frac{1}{n^q}.$$

Theorem 3 states that when w is large enough (n^α), the fraction of sellers with price at least w is at most roughly $w^{-(v+1)}$. Note that this fraction decreases exponentially as v increases. We also prove a matching lower bound. Unfortunately, we are only able to prove the lower bound in expectation, and not with high probability.

Theorem 4 (Lower Bound). *Let α, ε, v, and w be as above. Then*

$$\mathrm{E}[P(w)] > \frac{1}{w^{v+1+\varepsilon}}.$$

The proofs of the theorems proceed in several stages. The lower bound on $P(w)$ follows from the upper bound, and so we first focus on the latter. Instead of directly analyzing the distribution of prices at equilibrium, we begin by analyzing the prices generated by a different process, which we call the most-recently added commitment (MRAC) process: Each buyer commits to purchasing from the most recently added seller to which he is connected (ignoring incoming edges from sellers), generating the following prices. If k buyers committed to purchasing from some seller, then his price will be k. The reason this seems reasonable (at least on an intuitive level) is that the most recently added sellers will probably have smaller degrees, and then hopefully they should have lower prices. In any case, the final step is to bound the prices at market equilibrium in G_v by those generated by the MRAC process. The following subsections contain the various parts of the proof.

4.1 Most-Recently Added Commitment Process

We begin by sampling a graph according to G_v, but for now restricting attention to edges chosen by buyers (and not by sellers connecting to buyers). Suppose sellers do not connect to buyers, but only vice versa. Buyers sample sellers with probability proportional to their degrees plus 1. When we sample such a graph, the resulting degrees and prices in the MRAC process are random variables that are determined by the graph that is eventually chosen. Since the graph G_v is sampled in stages, it seems reasonable to also analyze the degrees and prices in an iterative manner, as in [11].

Denote by $W(s,t)$ the weight (degree plus 1) of the s'th seller after stage t in the generation of the graph (i.e. after t buyer-seller pairs have been added). Note that $W(s,t) = 0$ for $t < s$, $W(s,s) = 1$, and $W(s,t)$ is nondecreasing with t. It will also be useful to denote by $W(\leq s,t) = \sum_{i=1}^{s} W(s,t)$.

Additionally, denote by $C(s,t)$ the price of the s'th seller after stage t in the MRAC process. Recall that this price is simply the number of buyers who have committed to the s'th seller, which occurs whenever the s'th seller is the most-recently added seller connected to those buyers. Note that $C(s,t) = 0$ for $t \leq s$ and that $C(s,t)$ is nondecreasing with t.

Our first lemma bounds the expected MRAC price of a seller in G_v.

Lemma 1. *Let v be some positive constant integer. Then in G_v, for all n and $s \in [n]$,*

$$\mathrm{E}[C(s,n)] \leq O\left(\left(\frac{n}{s}\right)^{\frac{1}{v+1}}\right).$$

Note that the expected MRAC prices given in Lemma 1 are "correct" in the following sense: If $w = (n/s)^{1/(v+1)}$, then $s = n/w^{v+1}$. Hence, the fraction of sellers with expected MRAC prices greater than w is $1/w^{v+1}$, which is close to what we wish to prove.

Proof: We will sample a graph G_v in stages, and denote by F_m the σ-field of information up to stage m. F_m includes information such as the current degrees of vertices and which edges were chosen. We have

$$\begin{aligned}
&\mathrm{E}[C(s,m+1)|F_m] = C(s,m) \\
&\qquad + \Pr\left[(m+1)\text{'th seller connects to buyer } s \text{ and other older buyers } |F_m\right] \\
&\leq C(s,m) + v \cdot \frac{W(s,m)}{(v+1)m} \cdot \left(\frac{W(\leq s,m)}{(v+1)m}\right)^{v-1} \\
&\leq C(s,m) + O\left(\frac{W(s,m) \cdot W(\leq s,m)^{v-1}}{m^v}\right),
\end{aligned}$$

where the second inequality holds since v is constant. We now take expectations on both sides, yielding

$$
\begin{aligned}
\mathrm{E}[C(s,m+1)] &\le \mathrm{E}[C(s,m)] + O\left(\frac{\mathrm{E}\left[W(s,m)\cdot W(\le s,m)^{v-1}\right]}{m^v}\right)\\
&\le \mathrm{E}[C(s,m)] + O\left(\frac{\mathrm{E}\left[W(s,m)\right]\cdot \mathrm{E}\left[W(\le s,m)\right]^{v-1}}{m^v}\right) \qquad (1)\\
&\le \mathrm{E}[C(s,m)] + O\left(\frac{1}{m^v}\cdot\left(\frac{m}{s}\right)^{\frac{v}{v+1}}\cdot\left(m^{\frac{v}{v+1}}s^{\frac{1}{v+1}}\right)^{v-1}\right)\\
&\le \mathrm{E}[C(s,m)] + O\left(s^{-\frac{1}{v+1}}\cdot m^{-\frac{v}{v+1}}\right),
\end{aligned}
$$

where (1) follows from Lemma 2 (see the end of this section). Hence,

$$
\begin{aligned}
\mathrm{E}[C(s,n)] &\le \sum_{m=s+1}^{n} O\left(s^{-\frac{1}{v+1}}\cdot m^{-\frac{v}{v+1}}\right)\\
&\le O\left(\frac{1}{s^{1/(v+1)}}\int_s^n m^{-v/(v+1)}dm\right)\\
&\le O\left(\left(\frac{n}{s}\right)^{\frac{1}{v+1}}\right). \qquad \blacksquare
\end{aligned}
$$

The proof of Lemma 1 utilizes the following lemma, which we will also use later to show some sort of measure concentration on the MRAC prices.

Lemma 2. *For all n, $s \in [n]$, and nonnegative integer constants i, j, and k, there exists a constant $c = c(i,j,k,v)$ such that in G_v,*

$$\mathrm{E}[W(s,n)^i \cdot W(\le s,n)^j \cdot C(s,n)^k] \le c\,\mathrm{E}[W(s,n)]^i \cdot \mathrm{E}[W(\le s,n)]^j \cdot \mathrm{E}[C(s,n)]^k.$$

In order to prove Lemma 2 we need the following lemma, whose proof is deferred to the full version of the paper du to space constraints.

Lemma 3. *If for constants b and d,*

$$x_n \le x_{n-1}\left(1+\frac{b}{n}+\frac{d}{n^2}\right)+c_n,$$

then for all $i \in [n]$,

$$x_n = O\left(x_i\left(\frac{n}{i}\right)^b + \sum_{k=i}^{n} c_k\left(\frac{n}{k}\right)^b\right).$$

We now sketch the proof of Lemma 2.

Proof Sketch: The proof is by induction on i, j, and k. The initial base cases are when $j = k = 0$ and when $i = k = 0$. Given these cases, we then prove the lemma with only one of i, j, k equal to 0. Due to space constraints, we defer the proof of the base cases to the full version of the paper.

We now use induction on k, where for each k, we assume the statement has been proven with $k-1$ and all i and j, as well as the same k but smaller i and j. In what follows, we have i, j and k fixed, and assume the statement is true for i', j', and k, where either $i' < i$ and $j' \leq j$ or $j' < j$ and $i' \leq i$. Furthermore, we assume the statement is true for i', j' and $k-1$, where i' and j' are arbitrarily large constant integers.

For ease of notation, let $C = C(s,n)$, $X = W(s,n)$, and $Z = W(\leq s,n)$, and let F_m be the σ-field of information up to stage m. We show that

$$\begin{aligned}
&\mathrm{E}[W(s,n+1)^i \cdot W(\leq s,n+1)^j \cdot C(s,n+1)^k | F_n] \\
&= X^i Z^j C^k \left(1 + \frac{v(i+j)}{(v+1)n}\right) + O\left(\frac{X^{i-1}Z^jC^k}{n} + \frac{X^{i+1}Z^{j+v-1}C^{k-1}}{n^v}\right)
\end{aligned}$$

(again the proof is deferred to the full version). Taking expectation on both sides and using the inductive hypothesis yields

$$\begin{aligned}
&\mathrm{E}[W(s,n+1)^i \cdot W(\leq s,n+1)^j \cdot C(s,n+1)^k] \\
&\quad = \mathrm{E}\left[X^i Z^j C^k\right]\left(1 + \frac{v(i+j)}{(v+1)n}\right) \\
&\qquad + O\left(\frac{\mathrm{E}[X]^{i-1}\,\mathrm{E}[Z]^j\,\mathrm{E}[C]^k}{n} + \frac{\mathrm{E}[X]^{i+1}\,\mathrm{E}[Z]^{j+v-1}\,\mathrm{E}[C]^{k-1}}{n^v}\right) \\
&\quad \leq \mathrm{E}\left[X^i Z^j C^k\right]\left(1 + \frac{v(i+j)}{(v+1)n}\right) + O\left(\left(\frac{n^{v(i+j)+k-v-1}}{s^{vi+k-j}}\right)^{\frac{1}{v+1}}\right).
\end{aligned}$$

Using Lemma 3, we get that

$$\begin{aligned}
&\mathrm{E}[W(s,n)^i \cdot W(\leq s,n)^j \cdot C(s,n)^k] \\
&\qquad \leq O\left(s^j \left(\frac{n}{s}\right)^{\frac{v(i+j)}{v+1}} + \sum_{x=s}^{n}\left(\frac{x^{v(i+j)+k-v-1}}{s^{vi+k-j}}\right)^{\frac{1}{v+1}}\left(\frac{n}{x}\right)^{\frac{v(i+j)}{v+1}}\right) \\
&\qquad \leq O\left(\frac{n^{\frac{v(i+j)}{v+1}}}{s^{\frac{vi-j}{v+1}}} + \frac{n^{\frac{v(i+j)}{v+1}}}{s^{\frac{vi-j+k}{v+1}}} \cdot \int_s^n x^{\frac{k-v-1}{v+1}}\,dx\right) \leq O\left(\frac{n^{\frac{v(i+j)+k}{v+1}}}{s^{\frac{vi-j+k}{v+1}}}\right) \\
&\qquad \leq O\left(\mathrm{E}[W(s,n)]^i \cdot \mathrm{E}[W(\leq s,n)]^j \cdot \mathrm{E}[C(s,n)]^k\right). \qquad \blacksquare
\end{aligned}$$

4.2 Upper Bound on Prices at Equilibrium

In this section we relate the MRAC prices obtained in the previous section to the prices at market equilibrium, and thus prove Theorem 3. We begin by discussing the bottleneck decomposition of a bipartite graph and its relation to equilibrium prices. The decomposition proceeds iteratively, where $G = G_1 = (B_1, S_1, E)$. In the i'th iteration:

- Let $U_i = \max_{U \subseteq B_i} |U|/|\Gamma_i(U)|$, where $\Gamma_i(U)$ denotes the set of all neighbors of vertices in U that are in the graph G_i.

– Fix $G_{i+1} = (B_i \setminus U_i, S_i \setminus \Gamma_i(U_i), E)$.
– Denote by $w_i = |U_i|/|\Gamma_i(U_i)|$.

Even-Dar et al. [5] and Wu and Zhang [15] relate the bottleneck decomposition of a graph to the equilibrium prices. In particular, they show that $w_1 \geq w_2 \geq \ldots$, and that the sellers that are in $\Gamma_i(U_i)$ have equilibrium price w_i. We prove the following related lemma.

Lemma 4. *Let $G = (B, S, E)$ be some fixed bipartite graph, and let $V = \{i \in S : w_i^s > w\}$, where w_i^s is the equilibrium price of seller i. Then the average* MRAC price *of sellers in V is greater than w.*

Proof: Consider the bottleneck decomposition of G, and let $a = \max\{i : w_i > w\}$. That is, the sellers in the first a stages of the decomposition are precisely the ones with equilibrium price greater than w. Note that $V = \bigcup_{i=1}^{a} S_i$, and let $U = \bigcup_{i=1}^{a} B_i$. By the properties of the bottleneck decomposition, $|U| > w \cdot |V|$.

Now, the average MRAC price of vertices in V is equal to the number of buyers whose most-recently added neighbor is in V, divided by $|V|$. But the number of such buyers is at least $|U|$, since buyers in U have **all** neighbors in V, and in particular the most-recently added one. Hence, the average MRAC price of vertices in V is at least $|U|/|V| > w$. ■

To use this lemma, we first need to show that the MRAC prices of a random G_v graph are close to their expectations. Fix some vertex s, and recall that $\mathrm{E}[C(s,n)] \leq O\left((n/s)^{1/(v+1)}\right)$. For any positive λ and constant positive integer k,

$$\begin{aligned}
\Pr\left[C(s,n) > \lambda \cdot \mathrm{E}[C(s,n)]\right] &= \Pr\left[C(s,n)^k > \lambda^k \cdot \mathrm{E}[C(s,n)]^k\right] \\
&\leq \frac{\mathrm{E}[C(s,n)^k]}{\lambda^k \cdot \mathrm{E}[C(s,n)^k]} \qquad (2) \\
&\leq O\left(\frac{\mathrm{E}[C(s,n)]^k}{\lambda^k \cdot \mathrm{E}[C(s,n)^k]}\right) \qquad (3) \\
&= O\left(\frac{1}{\lambda^k}\right),
\end{aligned}$$

where (2) is a Markov bound and (3) follows from Lemma 2, with $i = j = 0$.

If $\lambda = n^\beta$ for some positive constant β, then we can take a union bound:

$$\Pr\left[\forall s, C(s,n) \leq \lambda \cdot \mathrm{E}[C(s,n)]\right] \geq 1 - \frac{n}{\lambda^k} = 1 - n^{1-\beta k}.$$

For any constant β we can take k large enough so that $\beta k > q + 1$, where q is the constant from Theorem 3.

Now suppose we sample a graph according to G_v, and that $\forall s, C(s,n) \leq \lambda \cdot \mathrm{E}[C(s,n)]$. Fix some set $V \subseteq S$, $|V| = t$, and consider the average MRAC price of vertices in V. Since all sellers are within $\lambda = n^\beta$ of their expectations, this is maximal when V consists of the first t sellers, and their price is exactly λ times their expected price. In this case, their average MRAC price is at most

$$\frac{1}{t}\sum_{i=1}^{t}\lambda \mathrm{E}[C(i,n)] = \frac{n^\beta}{t}\sum_{i=1}^{t}\left(\frac{n}{i}\right)^{\frac{1}{v+1}} = O\left(n^\beta\left(\frac{n}{t}\right)^{\frac{1}{v+1}}\right).$$

Thus, by Lemma 4, if the number of sellers with equilibrium price greater than w is t, then

$$w \leq O\left(n^\beta\left(\frac{n}{t}\right)^{\frac{1}{v+1}}\right).$$

Turning things around, we get that (with probability $1-n^{-q}$) the number of sellers with equilibrium price greater than w is at most $O(n^{1+\beta(v+1)}/w^{v+1})$. When $w = n^\alpha$, the upper bound of Theorem 3 follows, since β can be made an arbitrarily small positive constant.

4.3 Lower Bound on Prices at Equilibrium

Throughout this section, we assume that $w = n^\alpha$ is fixed. Since we are interested in a lower bound on the number of sellers with price greater than w, we will show, roughly, that the first s sellers added to the graph all have such high prices. We choose $s = n/w^{v+1}$, which will yield the claimed bound. First, however, we bound from below the expected number players whose outgoing edges all fall in the first s sellers. This follows from the following two lemmas.

Lemma 5. *In G_v, for all n and $s \in [n]$,*

$$\mathrm{E}[W(\leq s,n)] = \Omega\left(n^{\frac{v}{v+1}}s^{\frac{1}{v+1}}\right).$$

Proof: For ease of notation, let $W(n) = W(\leq s,n)$ and $W = W(n)$. Then

$$\begin{aligned}
\mathrm{E}[W(n+1)|F_n] &= W + \sum_{i=1}^{v}\binom{v}{i}\left(\frac{W}{(v+1)n}\right)^i\left(1-\frac{W}{(v+1)n}\right)^{v-i}\cdot i \\
&= W + \sum_{i=1}^{v} v\cdot\binom{v-1}{i-1}\left(\frac{W}{(v+1)n}\right)^i\left(1-\frac{W}{(v+1)n}\right)^{v-i} \\
&= W + \frac{vW}{(v+1)n}\sum_{j=0}^{v-1}\binom{v-1}{j}\left(\frac{W}{(v+1)n}\right)^j\left(1-\frac{W}{(v+1)n}\right)^{v-j-1} \\
&= W\left(1+\frac{v}{(v+1)n}\right).
\end{aligned}$$

Taking expectations implies that $\mathrm{E}[W(n+1)] = \mathrm{E}[W(n)](1+v/(v+1)n)$.

Now, suppose that for all n,

$$\mathrm{E}[W(n)] = (v+1)\cdot n^{\frac{v}{v+1}}s^{\frac{1}{v+1}}$$

as desired. Then

$$\frac{\mathrm{E}[W(n+1)]}{\mathrm{E}[W(n)]} = \left(1+\frac{1}{n}\right)^{v/(v+1)} \leq \left(1+\frac{v}{(v+1)n}\right).$$

This implies that $\mathrm{E}[W(n+1)] = \mathrm{E}[W(n)](1 + v/(v+1)n) \geq \mathrm{E}[W(n)](1 + 1/n)^{v/(v+1)}$, which is the recurrence generated by the closed formula $\mathrm{E}[W(n)] = (v+1) \cdot n^{v/(v+1)} s^{1/(v+1)}$. Thus, this closed formula is a lower bound on our recurrence, completing the proof of the claim. ■

Lemma 6. *Let $s = n/w^{v+1}$, and let $Q(s)$ be a random variable that denotes the number of buyers b in G_v such that $\Gamma(b) \in [s]$. Then*

$$\mathrm{E}[Q(s)] = \Omega\left(\frac{n}{w^v}\right).$$

Proof: From the previous lemma we know that

$$\mathrm{E}[W(\leq s, n/2)] = \Omega\left(\frac{n}{w}\right).$$

For a buyer that entered after the $n/2$'th stage, what is the probability that all of his v outgoing edges lie in $[s]$? Fix $W = W(\leq s, n/2)$. Then for every such buyer b,

$$\Pr\left[b \text{ connects only to } [s] | F_{n/2}\right] \geq \left(\frac{W}{(v+1)n}\right)^v = P,$$

where P is the random variable that denotes the probability of hitting $[s]$. Now,

$$\begin{aligned}
\mathrm{E}[P] &= \mathrm{E}\left[\left(\frac{W}{(v+1)n}\right)^v\right] \\
&= \frac{\mathrm{E}\left[W^v\right]}{((v+1)n)^v} \\
&\geq \frac{\mathrm{E}\left[W\right]^v}{((v+1)n)^v} \\
&= \Omega\left(\frac{1}{w^v}\right),
\end{aligned}$$

where the inequality follows from Jensen's inequality. For each buyer $i \in \{n/2, \ldots, n\}$, let Y_i be an indicator random variable such $Y_i = 1$ if i's v outgoing edges hit $[s]$. Note that $\Pr[Y_i] \geq P$. Thus, we have that

$$\mathrm{E}[Q(s)] = \sum_{i=n/2+1}^{n} \mathrm{E}[Y_i] = \Omega(n\,\mathrm{E}[P]) = \Omega\left(\frac{n}{w^v}\right).$$

Recall that in addition to each buyer's v outgoing edges, some may also have incoming edges that the sellers chose. As is argued in [13], however, this only decreases the above expectation by a constant factor, completing the proof of the lemma. ■

Putting the above lemmas together and recalling that $s = n/w^{v+1}$, we get that the expected total price of the first s vertices is n/w^v, which means that the expected average price is w, as desired. This does not yet complete the proof of the lower bound, however, since it is still possible that few sellers have very high price, while the majority have low price. Recall that we are trying to show that many sellers have high price. This is where we use the stronger upper bound – we will show that high priced sellers do not contribute too much to the total price of the s sellers.

Lemma 7. *Let $\beta > 0$ and $\varepsilon > 0$ be arbitrarily small constants, and let r be an arbitrarily large constant. Let $C(i)$ be a random variable that denotes the price of seller i in G_v. Then, with probability $1 - 1/n^r$,*

$$\sum_{i:C(i)>w^{1+\beta}} C(i) < \frac{n}{w^{v+\varepsilon}}.$$

Proof: Consider the contribution of sellers i such that $C(i) \in [w^{1+k\beta}, w^{1+(k+1)\beta}]$, for a positive integer k. By Theorem 3, we know that with high probability, the number of such vertices is at most $n/w^{(1+k\beta)(v+1-\varepsilon)}$, for an arbitrarily small ε. The contribution of each such seller is at most $w^{1+(k+1)\beta}$, and so the total contribution of all these sellers is at most

$$\frac{nw^{1+(k+1)\beta}}{w^{(1+k\beta)(v+1-\varepsilon)}} = \frac{n}{w^{v+\beta(kv-k\varepsilon-1)-\varepsilon}}.$$

Note that this contribution decreases as k increases, and so the total contribution

$$\sum_{i:C(i)>w^{1+\beta}} C(i) < \frac{1}{\alpha\beta} \cdot \frac{n}{w^{v+\beta(v-\varepsilon-1)-\varepsilon}}$$

(with probability $1 - n^{-q}/\alpha\beta$), where α is the constant such that $w = n^{\alpha}$. Since $v > 1$ and using the fact that both α and β are constants, we can choose ε small enough so that $\beta(v - \varepsilon - 1) > 3\varepsilon$, implying the claim of the lemma. ∎

We are now ready to complete the proof of Theorem 4. With high probability,

$$\sum_{i:C(i)>w^{1+\beta}} C(i) < \frac{n}{w^{v+\varepsilon}},$$

so the contribution of sellers $i \in [s]$ with $C(i) < w^{1+\beta}$ is at least $n/w^v - n/w^{v+\varepsilon} = \Omega(n/w^v)$ (in expectation). Hence, the expected number of sellers in $[s]$ with $C(i) > w$ is at least

$$\Omega\left(\frac{n}{w^v} \cdot \frac{1}{w^{1+\beta}}\right) \geq \frac{n}{w^{v+1+2\beta}}.$$

Since $\beta > 0$ is an arbitrarily small constant, the lower bound of Theorem 4 follows.

Acknowledgements. I would like to thank Beni Gradwohl and Omer Reingold for very interesting and helpful conversations.

References

1. Arrow, K., Debreu, G.: Existence of an equilibrium for a competitive economy. Econometrica 22, 265–290 (1954)
2. Bollobás, B., Riordan, O.: Mathematical results on scale-free random graphs. In: Handbook of graphs and networks: from the genome to the internet, pp. 1–34. Wiley-VCH, Chichester (2002)
3. Chung, F., Lu, L.: Complex Graphs and Networks. In: Regional Conference Series in Mathematics, p. 107. American Mathematical Society, Providence, RI (2006)
4. Corominas-Bosch, M.: Bargaining in a network of buyers and sellers. Journal of Economic Theory 115, 35–77 (2004)
5. Even-Dar, E., Kearns, M., Suri, S.: A network formation game for bipartite exchange economies. In: ACM-SIAM Symposium on Discrete Algorithms (SODA) (2007)
6. I. Fisher. Ph.D. thesis, Yale University (1891)
7. Gale, D.: Theory of Linear Economic Models. McGraw-Hill, New York (1960)
8. Jackson, M.: The study of social networks in economics. In: Podolny, J., Rauch, J.E. (eds.) Forthcoming in The Missing Links: Formation and Decay of Economic Networks, Russell Sage Foundation, Thousand Oaks
9. Jackson, M.: The economics of social networks. Chapter 1 in Volume I of Advances in Economics and Econometrics. In: Blundell, R., Newey, W., Persson, T. (eds.) The economics of social networks., vol. 1, Cambridge University Press, Cambridge (2006)
10. Kakade, S., Kearns, M., Ortiz, L.: Graphical Economics. In: Shawe-Taylor, J., Singer, Y. (eds.) COLT 2004. LNCS (LNAI), vol. 3120, pp. 17–32. Springer, Heidelberg (2004)
11. Kakade, S., Kearns, M., Ortiz, L., Pemantle, R., Suri, S.: Economic properties of social networks. In: Saul, L.K., Weiss, Y., Bottou, L. (eds.) Advances in Neural Information Processing Systems, vol. 17, MIT Press, Cambridge (2005)
12. Kranton, R., Minehart, M.: A theory of buyer-seller networks. American Economic Review 91, 485–508 (2001)
13. Suri, S.: The effects of network topology on strategic behavior. Ph.D. thesis, University of Pennsylvania (2007)
14. Walras, L.: Éléments d'économie politique pure; ou, Théorie de la richesse sociale (Elements of Pure Economics, or the Theory of social wealth), Lausanne, Paris, vol. 4 (1874) (1899, 4th ed., 1954 Engl. transl.)
15. Wu, F., Zhang, L.: Proportional response dynamics leads to market equilibrium. In: Symposium on the Theory of Computing (STOC), pp. 354–363 (2007)

Atomic Congestion Games: Fast, Myopic and Concurrent*

Dimitris Fotakis[3], Alexis C. Kaporis[1,2], and Paul G. Spirakis[1,2]

[1] Dept. of Computer Eng. and Informatics, Univ of Patras, 26500 Patras, Greece
[2] Research Academic Comp. Tech. Inst., N. Kazantzaki Str, 26500 Patras, Greece
[3] Dept. of Information & Communication Systems Eng., Univ. of the Aegean, Samos, Greece
fotakis@aegean.gr, kaporis@ceid.upatras.gr, spirakis@cti.gr

Abstract. We study here the effect of concurrent greedy moves of players in atomic congestion games where n selfish agents (players) wish to select a resource each (out of m resources) so that her selfish delay there is not much. The problem of "maintaining" global progress while allowing concurrent play is exactly what is examined and answered here. We examine two orthogonal settings : (i) A game where the players decide their moves without global information, each acting "freely" by sampling resources randomly and locally deciding to migrate (if the new resource is better) via a random experiment. Here, the resources can have quite arbitrary latency that is load dependent. (ii) An "organised" setting where the players are pre-partitioned into selfish groups (coalitions) and where each coalition does an improving coalitional move. Our work considers concurrent selfish play for arbitrary latencies for the first time. Also, this is the first time where fast coalitional convergence to an approximate equilibrium is shown.

1 Introduction

Congestion games (CG) provide a natural model for non-cooperative resource allocation and have been the subject of intensive research in algorithmic game theory. A *congestion game* is a non-cooperative game where selfish players compete over a set of resources. The players' strategies are subsets of resources. The cost of each player from selecting a particular resource is given by a non-negative and non-decreasing latency function of the load (or congestion) of the resource. The individual cost of a player is equal to the total cost for the resources in her strategy. A natural solution concept is that of a pure Nash equilibrium (NE), a state where no player can decrease his individual cost by unilaterally changing his strategy. In a classical paper, Rosenthal [27] showed that pure Nash equilibria on atomic congestion games correspond to local minima of a natural potential function. Twenty years later, Monderer and Shapley [24] proved that congestion games are equivalent to potential games. Many recent contributions have provided considerable insight into the structure and efficiency (e.g. [14,3,7,17]) and tractability [12,1] of NE in congestion games. Given the non-cooperative nature

* The 2nd and 3rd author were partially supported by the IST Program of the European Union under contract number IST-015964 (AEOLUS). This work was partially supported by the Future and Emerging Technologies Unit of EC (IST priority – 6th FP), under contract no. FP6-021235-2 (project ARRIVAL).

B. Monien and U.-P. Schroeder (Eds.): SAGT 2008, LNCS 4997, pp. 121–132, 2008.

of congestion games, a natural question is whether the players trying to improve their cost converge to a pure NE in a reasonable number of steps. The potential function of Rosenthal [27] decreases every time a *single* player changes her strategy and improves her individual cost, while this is not true when concurrent selfish moves are performed. Hence every sequence of improving moves will eventually converge to a pure Nash equilibrium. However, this may require an exponential number of steps, since the problem is *PLS-complete* [12]. A pure Nash equilibrium of a *symmetric network* atomic congestion game can be found by a min-cost flow computation [12]. Even better, for *singleton* CG (aka CG on parallel links), for CG with *independent resources*, and for *matroid* CG, every sequence of improving moves reaches a pure Nash equilbrium in a polynomial number of steps [21,1]. An alternative approach to circumvent the PLS-completeness of computing a pure Nash equilibrium is to seek an *approximate* NE. [6] considers *symmetric* congestion games with a weak restriction on latency functions and proves that several natural families of ε-moves converge to an ε-NE in time polynomial in n and ε^{-1}. However, sequential moves take $\Omega(n)$ steps in the worst case to reach an (approximate) NE and requires central coordination. A natural question is whether concurrent and autonomous play can convergence to an approximate pure Nash equilibrium. In this work, we investigate the effect of concurrent moves on the rate of convergence to approximate pure Nash equilibria.

1.1 Singleton Games with Myopic Players

Related Work and Motivation. The Elementary Step System hypothesis, under which at most one user performs an improving move in each round, greatly facilitates the analysis of [8,11,17,18,22,23,26]. This is not an appealing scenario to modern networking, where simple decentralized distributed protocols can reflect better the essence of net's liberal nature on decision making. All the above manifest the importance of distributed protocols that allow an arbitrary number of users to reroute per round, on the basis of selfish migration criteria. This is an Evolutionary Game Theory [31] perspective, see also [28] with a current treatment of both nonatomic games and of evolutionary dynamics. In this setting, the main concern is on studying the *replicator-dynamics*, that is to model the way that users revise their strategies throughout the process.

Discrete setting. The work in [10] considers n players concurrently sample for a better link amongst m parallel links per round (singleton CG). Link j has linear latency $s_j x_j$, where x_j is the number of players and s_j is the constant speed of the link j. This migration protocol uses global info: only users with latency exceeding the overall average link latency $\overline{L}_t$ at round t are allowed with an appropriate probability to sample for a new link j. Also global info is used to amplify favorable links: link j is sampled *proportionally* to $d_t(j) = n_t(j) - s_j \overline{L}_t$, where $n_t(j)$ is the number of users on link j, and reaches in expectedly $O(\log \log n + \log m)$ rounds a NE. In [4] it was given the analysis of a concurrent protocol on identical links and players. On parallel during round t, each user b on resource i_b with load $X_{i_b}(t)$ selects a random resource j_b and if $X_{i_b}(t) > X_{j_b}(t)$ then b migrates to j_b with probability $1 - X_{j_b}(t)/X_{i_b}(t)$. It reaches an ε-NE in $O(\log \log n)$, or an exact NE in $O(\log \log n + m^4)$ rounds, in expectation. *Continuous setting.* The work in [29] gives a general definition of nonatomic potential games, and shows convergence to Nash equilibrium in these games, under a very broad class of evolutionary dynamics. A series of papers [5,13] on the *Wardrop* model give strong intuition on this subject. In [13] the significance of the *relative slope* parameter d is shown. A latency function ℓ has relative slope d if $x\ell'(x) \leq d\ell(x)$. Each user on path P in commodity i, either with probability β selects a uniformly random path Q in i, or with probability $1-\beta$ selects a path Q with probability proportional to its flow f_Q. If $\ell_Q < \ell_P$ user migrates to sampled Q with probability $\frac{\ell_P - \ell_Q}{d(\ell_P + \alpha)}$, where parameter α is arbitrary. In [5] it was shown that as along as all players concurrently employ arbitrary *no-regret* policies, they will eventually achieve convergence.

Contribution. We study a simple distributed protocol for congestion games on parallel links under very general assumptions on the latency functions. In parallel each player selects a link uniformly at random in each round and checks whether she can significantly decrease her latency by moving to the chosen link. If this is the case, the player becomes a potential migrant. The protocol selects at most one potential migrant to defect from each link. This is a local decision amongst users on the same link, allowing a realistic amount of parallelism amongst entities on different resources. Details on this, falling in the context of *dimension-exchange* protocols on load balancing, can be found in [2,9,16,20]. We prove that if the number of players is $\Theta(m)$, the protocol reaches an almost-NE in $O(\log(\Phi_0/\Phi^*))$ time, where Φ_0 is Rosental's potential value as the game starts and Φ^* is the corresponding value at a NE. The proof of convergence is technically involved and interesting and comprises the main technical merit of this work. Our notion of approximate pure Nash equilibrium, see Definition 2, is a bit different from similar approximate notions considered in previous work [6,10] in an atomic setting, while it is close in nature to the stable state defined in [13, Def. 4] for the Wardrop model. An *almost-Nash equilibrium* is a state where at most $o(m)$ links have latency either considerably larger or considerably smaller than the current average latency. This definition relaxes the notion of exact pure NE and introduces a meaningful notion of approximate (bicriteria) NE for our fully myopic model of migration described above. In particular, an almost-NE guarantees that unless a player uses an overloaded link (i.e. a link with latency considerably larger than the average latency), the probability that she finds (by uniform sampling) a link to migrate and significantly improve her latency is at most $o(1)$. Furthermore, it is unlikely that the almost-NE reached by our protocol assigns any number of players to overloaded an almost-NE). As it will become clear from the analysis, the reason that users do not accumulate on overloaded links, is that the number of players on such links is a strong super-martingale. In addition, by the fact that any bin initially has $O(\log n)$ load we get that in $O(\log n)$ rounds the overloaded bins will drain from users.

Our results extend the results in [4,10] in the sense that (i) we consider arbitrary and unknown latency functions subject only to the α-bounded jump condition [6, Section 2], (ii) it requires no other global info. Also, the strategy space of player i may be extended to all subsets of resources of cardinality k_i such that $\sum_i k_i = O(m)$, see also independent resource CG [21].

1.2 Congestion Games with Coalitions

In many practical situations however, the competition for resources takes place among coalitions of players instead of individuals. For a typical example, one may consider a telecommunication network where antagonistic service providers seek to minimize their operational costs while meeting their customers' demands. In this and many other natural examples, the number of coalitions (e.g. service providers) is rather small and essentially independent of the number of players (e.g. users). In addition, the coalitions can be regarded as having a quite accurate picture of the current state of the game and moving greedily and sequentially. In such settings, it is important to know how the competition among coalitions affects the rate of convergence to an (approximate) pure Nash equilibrium. Motivated by similar considerations, [19,15] proposed *congestion games with coalitions* as a natural model for investigating the effects of non-cooperative resource allocation among static coalitions. In congestion games with coalitions, the coalitions are static and the selfish cost of each coalition is the total delay of its players. [19] mostly considers congestion games on parallel links with identical users and convex delays. For this class of games, [19] establishes the existence and tractability of pure NE, presents examples where coalition formation deteriorates the efficiency of NE, and bounds the efficiency loss due to coalition formation. [15] presents a potential function for linear congestion games with coalitions.

Contribution. In this setting, we present an upper bound on the rate of convergence to approximate pure Nash equiliria in single-commodity linear congestion games with static coalitions. The restriction to linear latencies is necessary because this is the only class of latency functions for which congestion games with static coalitions is known to admit a potential function and a pure Nash equilibrium. We consider ε-moves, i.e. deviations that improve the coalition's total delay by a factor more than ε. Combining the approach of [6] with the potential function of [15, Theorem 6], we show that if the coalition with the largest improvement moves in every round, an approximate NE is reached in a small number of steps. More precisely, we prove that for any initial configuration s_0, every sequence of largest improvement ε-moves reaches an approximate NE in at most $\frac{kr(r+1)}{\varepsilon(1-\varepsilon)} \log \Phi(s_0)$ steps, where k is the number of coalitions, $r = \lceil \max_{j\in[k]}\{n_j\} / \min_{j\in[k]}\{n_j\} \rceil$ denotes the ratio between the size of the largest coalition and the size of the smallest coalition, and $\Phi(s_0)$ is the initial potential. This bound holds even for coalitions of different size, in which case the game is *not symmetric*. Since the recent results of [6] hold for symmetric games only, this is the first non-trivial upper bound on the convergence rate to approximate NE for a natural class of *asymmetric* congestion games. This bound implies that in *network* congestion games, where a coalition's best response can be computed in polynomial times by a min-cost flow computation [12, Theorem 2], an ε-Nash equilibrium can be computed in polynomial time. Moreover, in the special case that the number of coalitions is constant and the coalitions are almost equisized (i.e. $k = \Theta(1)$ and $r = \Theta(1)$), the number of ε-moves to reach an approximate NE is logarithmic in the initial potential.

2 Concurrent Atomic Congestion Games

Model. There is a finite set of players $\{1, \dots, n\}$ and a set of edges (or resources) $E = \{e_1, \dots, e_m\}$. The strategy space S_i of player i is E. It is assumed that $n = O(m)$. The game consists of a sequence of rounds $t = 0, \dots, t^*$. It starts at round $t = 0,$

where each player i selects myopically strategy $s_i(0) \in S_i$. In each subsequent round $t = 1, \ldots, t^*$, concurrently and independently, each player updates his current strategy $s_i(t)$ to $s_i(t+1)$ according to the simple, oblivious and distributed protocol `Greedy` presented in Section 2.1. That is, at round t the state $s(t) = \langle s_1(t), \ldots, s_n(t) \rangle \in S_1 \times \ldots \times S_n$ of the game is a combination of strategies over players. The number $f_e(t)$ of players on edge $e \in E$ is $f_e(t) = |\{j : e \in s_j(t)\}|$. Edge e has a latency $\ell_e(f_e(t))$ measuring the common delay of players on it at state $s(t)$. The cost $c_i(t)$ of player i equals the sum of latencies of all edges belonging in his current strategy $s_i(t)$, that is $c_i(t) = \sum_{e \in s_i(t)} \ell_e(f_e(t))$. Let the average delay of the resources be $\overline{\ell}(t) = \frac{1}{m} \sum_{e \in E} \ell_e(f_e(t))$. Consider the value of Rosenthal's potential $\Phi(t) = \sum_{e \in E} \sum_{x=1}^{f_e(t)} \ell_e(x)$. We assume no latency-info other than the α-bounded jump condition:

Definition 1. *[6] Consider a set of m resources E each $e \in E$ incurring latency $\ell_e(x)$ when x players use it, $x \in \{0, \ldots, n\}$. Let $\alpha = \min_a\{a | \forall x = 0, \ldots, n, \forall e \in E$ it holds $\ell_e(x+1) \leq a\ell_e(x)\}$. Then each $e \in E$ satisfies the* α-bounded jump condition.

This condition imposes a minor restriction on the increase-rate of the latency function $\ell_e()$ of any resource $e \in E$. For example $\ell_e(x) = \alpha^x$ is α-bounded, which is also true for polynomials of degree $d \leq \alpha$. Our bicriterial equilibria (see [13, Def. 4]) follow.

Definition 2. *An* almost-NE *is a state where $o(m)$ used edges have latency $> \alpha\overline{\ell}(t)$ and $\forall \epsilon > 0$, $\nexists S \subseteq E : |S| \geq \epsilon m$ with used edges in S of latency $< \frac{1}{\alpha_S}\overline{\ell}(t)$, where α_S is the jump-parameter with respect to edges in S.*

Target. We establish the following for protocol `Greedy` presented in Section 2.1.

Theorem 1. *The expected number of rounds until* `Greedy` *reaches an almost-NE is at most* $2\lceil p^{-1} \ln(2\Phi_{\max}/\Phi_{\min}) \rceil$.

Constant $p = \Theta(1)$ is defined in Theorem 2, intuitively it provides a bound on the expected potential's drop caused by `Greedy` within all consecutive rounds which are not on an almost-NE. Theorem 1 follows easily (see the proof in the full version of the paper in [30]) from Theorem 2, see in turn its proof plan in Section 2.2. Here $\Phi_{\max}$, ($\Phi_{\min}$) denote the initial (final) value of the potential (value of the potential at an exact NE).

Taking into account the very limited info that our protocol extracts per round, our analysis suggests that an almost-NE of this kind is a meaningful notion of a stable state that can be reached quickly. In particular, the almost-NE reached by our protocol is a relaxation of an exact NE where the probability that a significant number of players can find (by uniform sampling) links to migrate and significantly improve their cost is small.

More precisely, in an exact NE, no used link has latency greater than $\alpha\overline{\ell}(t)$ and no link with positive load has latency less than $\overline{\ell}(t)/\alpha$, while the definition of an almost-NE imposes the same requirements on all but $o(m)$ links. Hence the notion of an almost-NE is a relaxation of the notion of an exact NE. In addition, a player not assigned to an overloaded link (i.e. a link with latency greater than $\alpha\overline{\ell}(t)$) can significantly decrease her cost (i.e. by a factor greater than α^2) only if she samples an underloaded link (i.e. a link with latency less than $\overline{\ell}(t)/\alpha$). Therefore, in an almost-NE, the probability that a player not assigned to an overloaded link samples a link where she can migrate and significantly decrease her cost is $o(1)$. Furthermore, it is unlikely that the almost-NE reached by our protocol assigns a large number of players to overloaded links [1].

Theorem 2. *If round t is not an almost-NE then* $\mathbb{E}[\Phi(t+1)] \leq (1-p)\mathbb{E}[\Phi(t)]$, *with* p *bounded bellow by a positive constant.*

The proof plan of this theorem is presented in Section 2.2. Its proof will be given in Section 2.6 which combines results proved in Section 2.3, 2.4 and 2.5.

2.1 Concurrent Protocol `Greedy`

Initialization: $\forall i \in \{1, \ldots, n\}$ select a random $e \in \{1, \ldots, m\}$.

During round t, do in parallel $\forall e \in E$:

1. Select 1 player i from e at random.
2. Let player i sample for a destination edge e' u.a.r. over E.
3. If $\ell_{e'}(f_{e'}(t))(\alpha + \delta_\vartheta) < \ell_e(f_e(t))$ then allow player i migrate to e' with probability $\vartheta = \Omega(1)$.

For $\vartheta, \delta_\vartheta$ see Section 2.3 Lemma 2, Corollary 1, and Section 2.5 Case 1 and 2.

2.2 Convergence of `Greedy` - Overview

The idea behind main Theorem 1 is to show that, starting from $\Phi(0) = \Phi_{\max}$, per round t of `Greedy` not in an almost-NE, the expected $\mathbb{E}[\Delta\Phi(t)]$ potential drop is a positive portion of the potential $\Phi(t)$ at hand. Since the minimum potential $\Phi_{\min}$ is a positive value, the total number of round is at most logarithmic in $\frac{\Phi_{\max}}{\Phi_{\min}}$. We present below how Sections 2.3, 2.4 and 2.5 will be combined together towards showing that `Greedy` gives a large "bite" to the potential $\mathbb{E}[\Phi(t)]$ at hand, per round not in an almost-NE, and prove key Theorem 2. Section 2.3 shows that $\mathbb{E}[\Delta\Phi(t)]$ is at most the total expected cost-drop $\sum_i \mathbb{E}[\Delta c_i(t)]$ of users allowed by `Greedy` to migrate and proves that $\sum_i \mathbb{E}[\Delta c_i(t)] < 0$, i.e. *super-martingale* [25, Def. 4.7]. Hence, showing large potential drop per round not in an almost-NE reduces to showing $\sum_i \mathbb{E}[\Delta c_i(t)]$ equals a positive number times $-\mathbb{E}[\Phi(t)]$. This is achieved in Sections 2.4 and 2.5 which show that $|\sum_i \mathbb{E}[\Delta c_i(t)]|$ and $\mathbb{E}[\Phi(t)]$ are both closely related to $\mathbb{E}[\overline{\ell}(t)] \times m$, i.e. both are a corresponding positive number times $\mathbb{E}[\overline{\ell}(t)] \times m$. First, Section 2.4 shows that $\mathbb{E}[\Phi(t)]$ is a portion of $\mathbb{E}[\overline{\ell}(t)] \times m$. Having this, fast convergence reduces to showing $\sum_i \mathbb{E}[\Delta c_i(t)]$ equals a positive number times $-\mathbb{E}[\overline{\ell}(t)] \times m$ which is left to Section 2.5 & 2.6. At the end, Section 2.6 puts together Sections 2.3, 2.4 and 2.5 and completes the proof of our key Theorem 2.

[1] Due to the initial random allocation of the players to the links, the overloaded links (if any) receive $O(\log n)$ players with high probability. Lemma 3 and Corollary 2 show that the number of players on any overloaded link is a strong super-martingale during each round. Thus, such overloaded links will drain from users in expectedly $O(\log n)$ rounds.

2.3 Showing That $\sum_{i\in\mathcal{A}(t)} \mathbb{E}[\Delta c_i(t)]$ Upper Bounds $\mathbb{E}[\Delta\Phi(t)]$

Let $\mathcal{A}(t)$ the migrants allowed in step (3) of `Greedy` in Section 2.1. Linearity of expectation by Lemma 1 yields $\sum_{i\in\mathcal{A}(t)} \mathbb{E}[\Delta c_i(t)] \geq \mathbb{E}[\Delta\Phi(t)]$. $\sum_{i\in\mathcal{A}(t)} \mathbb{E}[\Delta c_i(t)] < 0$ follows by Lemma 2 and Corollary 1 below: user $i \in \mathcal{A}(t)$, by selfish criterion in step (3) of `Greedy`, decreases expectedly its cost if the latency on i's departure link is $> (\alpha + \delta_\vartheta)$ times the latency on its destination. Here ϑ is the migration probability in step (3) of `Greedy`.

Lemma 1. $\sum_{i\in\mathcal{A}(t)} \Delta[c_i(t)] \geq \Delta[\Phi(t)]$. *Equality holds if* $\Delta[f_e(t)] \leq 1, \forall e \in E$.

Proof. See the proof in the full version of the paper in [30]. □

Lemma 2. *For every positive constant* δ, *if migration probability* ϑ *of* `Greedy` *is at most* $\min\{\frac{\delta}{\alpha(\alpha-1)}, 1\}$, *the expected latency of a destination link e in the next round* $t+1$ *is:*

$$\mathbb{E}[\ell_e(f_e(t+1))] \leq (1+\delta/\alpha)\ell_e(f_e(t)+1) \leq (\alpha+\delta)\ell_e(f_e(t))$$

Proof. See the proof in the full version of the paper in [30]. □

Corollary 1. $\mathbb{E}[\Delta c_i(t)|\, c_i(t)] \leq \ell_{e'}(f_{e'}(t))(\alpha+\delta_\vartheta) - c_i(t) < 0, \forall i \in \mathcal{A}(t)$ *migrating* $e \to e'$.

Proof. See the proof in the full version of the paper in [30]. □

2.4 Showing That $\mathbb{E}[\Phi(t)]$ Is at Most a Portion of $\mathbb{E}[\overline{\ell}(t)] \times m$

By `Greedy`'s initialization the load is Binomially distributed, thus at round $t = 0$ we easily get (see the full version in [30]):

$$\mathbb{E}[\overline{\ell}(0)] \leq e^{\alpha\frac{n}{m-1}} e^{-\frac{n}{m}} = O(1), \text{ and } \mathbb{E}[\Phi(0)] = O(\mathbb{E}[\overline{\ell}(0)] \times m), \tag{1}$$

However, `Greedy` may affect badly the initial distribution of bins, thus making (1) invalid for each $t > 0$. We shall show that similar to round 0 strong tails will make (1) true for each round $t > 0$. To see this, consider the concurrent random process `Blind` (a simplification of `Greedy` in Section 2.1). At $t = 0$ throw randomly $n = O(m)$ balls to m bins (`Blind`'s and `Greedy`'s initializations are identical). Initially, the load distribution has Binomial tails from deviating from expectation $O(n/m) = O(1)$. During round $t > 0$, `Blind` draws exactly 1 random ball from each loaded bin (as Step 1 of `Greedy`). Let $n(t)$ the subset of drawn balls during round t. Round t ends by throwing at random these $|n(t)|$ drawn balls back into the m bins (then $|n(t)|$ allowed by `Blind` to migrate is at least the migrants allowed by `Greedy`, since no selfish criterion is required). Any bin is equally likely to receive any ball, thus, `Blind` preserves per round $t > 0$ strong Binomial tails from deviating from the constant expectation $O(n/m) = O(1)$ reminiscent to ones for $t = 0$. The above make true (1) for each round $t > 0$ of `Blind`.

Towards showing that `Greedy` also behaves, on a proper subset of bins, similarly to `Blind` it is useful the following lemma. Lemma 3 and Corollary 2 prove a supermartingale property on the load of bins with latency greater than a critical constant.

This will help us to identify this subset of critical bins that will preserve similar bounds to (1) for each round $t > 0$ of `Greedy`.

Lemma 3. *Let ν be any integer no less than $\lceil 2n/m \rceil + 1$. For any round $t \geq 0$, every link e with $\ell_e(f_e(t)) \geq \alpha^\nu$ has $\mathbb{E}[f_e(t+1)] \leq f_e(t)$.*

Proof. See the proof in the full version of the paper in [30]. □

Corollary 2. *Consider the corresponding numbers ν's defined in Lemma 3. We can find a constant $L^* : \forall t \geq 0$ on each edge with latency $\geq L^*$ the load is super-martingale.*

Let the constant L^* be as in Corollary 2 and define $\mathcal{A}_{L^*}(t) = \{e \in E : \ell_e(f_e(t)) < L^*\}$ and $\mathcal{B}_t^{L^*} = E \setminus \mathcal{A}_{L^*}(t)$. The target of Lemma 4 is to show that $\mathcal{B}_t^{L^*}$ is the subset of critical bins that will preserve similar bounds to (1) for each round $t > 0$ of `Greedy`.

Lemma 4. $\displaystyle\sum_{e \in \mathcal{B}_t^{L^*}} \frac{\mathbb{E}[\ell_e(f_e(t))]}{m} = O(1), \sum_{e \in \mathcal{B}_t^{L^*}} \frac{\mathbb{E}[f_e(t)\ell_e(f_e(t))]}{m} = O(\mathbb{E}[\overline{\ell}(t)])$

Proof. See the full version of the paper in [30]. □

Now, Fact 3 proves that $\mathbb{E}[\Phi(t)]$ is at most a portion of $\mathbb{E}[\overline{\ell}(t)] \times m$.

Fact 3. *If round t is not an almost-NE then $\mathbb{E}[\overline{\ell}(t)]m \geq \frac{\mathbb{E}[\Phi(t)]}{r(1+y_t)+1+x_t}$, $r = n/m$ and $r, y_t, x_t = \Theta(1)$.*

Proof. See the proof in the full version of the paper in [30]. □

2.5 Showing That $\sum_{i \in \mathcal{A}(t)} \mathbb{E}[\Delta c_i(t)]$ Is a Portion of $-\overline{\ell}(t) \times m$

Sketch of Case 1 and 2 below. According to Definition 2, a round is not at an almost-NE if $\geq \varepsilon m$ links are either *overloaded* (of latency $\geq \alpha \times \overline{\ell}(t)$) or *underloaded* (of latency $\leq \frac{1}{\alpha} \times \overline{\ell}(t)$) ones. We study separately each of these options in Cases 1 and 2 below. In both cases we relate $\sum_{i \in \mathcal{A}(t)} \mathbb{E}[\Delta c_i(t)]$ to $-\overline{\ell}(t) \times m$. The idea beyond both Case 1 and 2 is simple: each migrant from $O(t)$ to $\mathcal{U}(t)$ will contribute to $\sum_{i \in \mathcal{A}(t)} \mathbb{E}[\Delta c_i(t)]$ her little portion of $-\overline{\ell}(t)$ at hand (by the martingale property on the expected gain per user $i \in \mathcal{A}(t)$ proved in Corollary 1 Section 2.3). It remains to show that such migrations have as high impact as to boost the tiny atomic gain of order $\overline{\ell}(t)$, when considered in the overall population of migrants $\mathcal{A}(t)$, up to a portion of $\overline{\ell}(t) \times m$. Towards this, Fact 4 and 5 below show that, as long as the state is not an almost-NE, it induces imbalance amongst link-costs, which in turn influences a sufficient amount of migrations as to get cost-drop of order $-\overline{\ell}(t) \times m$.

Case 1. Here we define *underloaded* links in round t be $\mathcal{U}(t) = \{e \in E : \ell_e(f_e(t)) < (1-\delta)\overline{\ell}(t)\}$, while *overloaded* ones are $\mathcal{O}(t) = \{e \in E : \ell_e(f_e(t)) \geq \alpha\overline{\ell}(t)\}$. Let us assume that we are not at an almost-NE because $|\mathcal{O}(t)| \geq \varepsilon m$, with constant $\varepsilon \in (0,1)$.

Fact 4. *For every $\alpha > 1$ if $|\mathcal{O}(t)| \geq \varepsilon m$, then $|\mathcal{U}(t)| \geq \delta m$, with $\delta = \frac{\varepsilon}{2}(\alpha - 1)$.*

Proof. See the proof in the full version of the paper in [30]. □

Therefore, for every $e \in \mathcal{O}(t)$, a player migrates from e to a link in $\mathcal{U}(t)$ with probability at least $\vartheta\delta$ (see step (3) of Greedy, Section 2.1). Using Lemma 2 with $\vartheta = \varepsilon/4$, we obtain that the expected decrease in its cost is at least $\frac{\delta}{2}\alpha\bar{\ell}(t)$ (see the proof in the full version in [30]).

Given that k migrants switch from a link in $\mathcal{O}(t)$ to a link in $\mathcal{U}(t)$ we obtain that their expected cost-drop is at least $\frac{\delta}{2}\alpha\bar{\ell}(t)$ times their number k. Let $p_{O\to U}(k)$ the probability to have k such migrants. The expected number $\sum_k kp_{O\to U}(k)$ of such migrants is at least $\varepsilon\vartheta\delta m$, since for every $e \in \mathcal{O}(t)$ with $|\mathcal{O}(t)| \geq \epsilon m$, exactly 1 player migrates from e to a link in $\mathcal{U}(t)$ with probability at least $\vartheta\delta$ (see Fact 4 and step (3) of Greedy, Section 2.1). Now, the unconditional on k expected cost-drop due to migrants switching from links in $\mathcal{O}(t)$ to links in $\mathcal{U}(t)$ is at least

$$\sum_k \left(\tfrac{\delta}{2}\alpha\bar{\ell}(t)k \times p_{O\to U}(k)\right) \geq \tfrac{\delta}{2}\alpha\bar{\ell}(t) \times \varepsilon\vartheta\delta m = \varepsilon\vartheta\tfrac{\delta^2}{2}\alpha m\bar{\ell}(t) \tag{2}$$

By (2) we finally prove (for Case 1) the result of this section:

$$\sum_{i\in\mathcal{A}(t)} \mathbb{E}[\Delta c_i(t)] \leq -\varepsilon\vartheta\tfrac{\delta^2}{2}\alpha \times \bar{\ell}(t)m \tag{3}$$

Case 2. Here we define as *underloaded* links in round t be $\mathcal{U}(t) = \{e \in E : \ell_e(f_e(t)) < \frac{1}{\alpha}\bar{\ell}(t)\}$ and *overloaded* ones in $\mathcal{O}(t) = \{e \in E : \ell_e(f_e(t)) \geq (1+\delta)\bar{\ell}(t)\}$. Let us assume that we are not at an almost-NE because $|\mathcal{U}(t)| \geq \varepsilon m$.

Fact 5. *If* $|\mathcal{U}(t)| \geq \varepsilon m$, *then* $\sum_{e\in\mathcal{O}(t)} \ell_e(f_e(t)) > \delta\bar{\ell}(t)m$, *with* $\delta = \frac{\varepsilon(\alpha-1)}{2\alpha}$.

Proof. See the proof in the full version of the paper in [30]. □

Since $|\mathcal{U}(t)| \geq \varepsilon m$, a player migrates from each $e \in \mathcal{O}(t)$ to a link in $\mathcal{U}(t)$ with probability at least $\vartheta\varepsilon$ (see step (3) of Greedy, Section 2.1). Using Lemma 2 with $\vartheta = \frac{\varepsilon}{4\alpha}$, we obtain that the expected decrease in the cost of such a player is at least $\frac{\delta}{2(1+\delta)}\ell_e(f_e(t)) \geq \frac{\delta}{4}\ell_e(f_e(t))$ (see the proof in the full version in [30]). Using Fact 5, we obtain that the expected cost-drop due to migrants leaving overloaded links $\mathcal{O}(t)$ and entering $\mathcal{U}(t)$ in round t is at least:

$$\vartheta\varepsilon \times \frac{\delta}{4}\sum_{e\in\mathcal{O}(t)} \ell_e(f_e(t)) > \vartheta\varepsilon \times \frac{\delta}{4} \times \delta\,\bar{\ell}(t)m > \frac{\vartheta\varepsilon\delta^2}{4}\,\bar{\ell}(t)m \tag{4}$$

By (4) we finally prove (for Case 2) the result of this section:

$$\sum_{i\in\mathcal{A}(t)} \mathbb{E}[\Delta c_i(t)] \leq -\tfrac{\vartheta\varepsilon\delta^2}{4} \times m\bar{\ell}(t) \tag{5}$$

2.6 Proof of Key Theorem 2

Here we combine the results in Section 2.3, 2.4 and 2.5 and prove Theorem 2. From Section 2.3 we get $\mathbb{E}[\Delta\Phi(t)] \leq \sum_{i\in\mathcal{A}(t)} \mathbb{E}[\Delta c_i(t)] < 0$. As long as Greedy does not reach an almost-NE because: (i) The *overloaded* links, with respect to the realization $\overline{\ell}(t)$, are $|\mathcal{O}(t)| \geq \varepsilon m$. Then, we get from Expression (3) in Section 2.5 that $\mathbb{E}[\Delta\Phi(t)|\overline{\ell}(t)] \leq \sum_{i\in\mathcal{A}(t)} \mathbb{E}[\Delta c_i(t)|\overline{\ell}(t)] < -\varepsilon\vartheta\frac{\delta^2}{2}\alpha \times \overline{\ell}(t)m$. (ii) The *underloaded* links, with respect to the realization $\overline{\ell}(t)$, are $|\mathcal{U}(t)| \geq \varepsilon m$. Then, we get from Expression (5) in Section 2.5 that $\mathbb{E}[\Delta\Phi(t)|\overline{\ell}(t)] \leq \sum_{i\in\mathcal{A}(t)} \mathbb{E}[\Delta c_i(t)|\overline{\ell}(t)] < -\frac{\vartheta\varepsilon\delta^2}{4}\times\overline{\ell}(t)m$ In either Case 1 or 2 such that an almost-NE is not reached by realization $\overline{\ell}(t)$, we conclude from the above:

$$\mathbb{E}[\Delta\Phi(t)|\overline{\ell}(t)] \leq \sum_{i\in\mathcal{A}(t)} \mathbb{E}[\Delta c_i(t)|\overline{\ell}(t)] < -\frac{\vartheta\varepsilon\delta^2}{4} \times \overline{\ell}(t)m \tag{6}$$

Consider the space of all realizations $\overline{\ell}(t)$ not in an almost-NE due to $\geq \varepsilon m$ overloaded or underloaded links in round t. Let $p_{\overline{\ell}}(t)$ the probability to obtain a realization $\overline{\ell}(t)$ in this space. Removing the conditional on $\overline{\ell}(t)$, Expression (6) becomes:

$$\begin{aligned}\mathbb{E}[\Delta\Phi(t)] &= \sum_{\overline{\ell}(t)} \mathbb{E}[\Delta\Phi(t)|\overline{\ell}(t)]p_{\overline{\ell}}(t) \leq \sum_{\overline{\ell}(t)} \left[\sum_{i\in\mathcal{A}(t)} \mathbb{E}[\Delta c_i(t)|\overline{\ell}(t)]\right] p_{\overline{\ell}}(t)\\ &\leq \sum_{\overline{\ell}(t)} \left[-\frac{\vartheta\varepsilon\delta^2}{4} \times \overline{\ell}(t)m\right] p_{\overline{\ell}}(t) = -\frac{\vartheta\varepsilon\delta^2}{4} \times \mathbb{E}[\overline{\ell}(t)]m\end{aligned}$$

From Fact 3 the above becomes: $\mathbb{E}[\Delta\Phi(t)] \leq -\frac{\vartheta\varepsilon\delta^2}{4} \times \frac{\mathbb{E}[\Phi(t)]}{r(1+y_t)+1+x_t}$, $r = n/m$ and $r, x_t, y_t = \Theta(1)$.

3 Approximate Equilibria in Congestion Games with Coalitions

3.1 Model and Preliminaries

A *congestion game with coalitions* consists of a set of identical players $N = [n]$ ($[n] \equiv \{1,\ldots,n\}$) partitioned into k coalitions $\{C_1,\ldots,C_k\}$, a set of resources $E = \{e_1,\ldots,e_m\}$, a strategy space $\Sigma_i \subseteq 2^E$ for each player $i \in N$, and a non-negative and non-decreasing latency function $\ell_e : \mathbb{N} \mapsto \mathbb{N}$ associated with every resource e. In the following, we restrict our attention to games with linear latencies of the form $\ell_e(x) = a_e x + b_e$, $a_e, b_e \geq 0$, and symmetric strategies (or *single-commodity* congestion games), where all players share the same strategy space, denoted Σ. The congestion game is played among the coalitions instead of the individual players. We let n_j denote the number of players in coalition C_j. The strategy space of coalition C_j is Σ^{n_j} and the strategy space of the game is $\Sigma^{n_1} \times \cdots \times \Sigma^{n_k}$. A pure strategy $s_j \in \Sigma^{n_j}$ determines a (pure) strategy $s_j^i \in \Sigma$ for every player $i \in C_j$. We should highlight that if the coalitions have different sizes, the game is *not symmetric*. We let $r \equiv \lceil \max_{j\in[k]}\{|C_j|\}/\min_{j\in[k]}\{|C_j|\}\rceil$ denote the ratio between the size of the largest coalition to the size of the smallest coalition. Clearly, $1 \leq r < n$. For every resource $e \in E$, the load (or congestion) of e due to C_j in s_j is $f_e(s_j) = |\{i \in C_j : e \in s_j^i\}|$. A tuple $s = (s_1,\ldots,s_k)$ consisting of a pure strategy $s_j \in \Sigma^{n_j}$ for every coalition C_j is a *state* of the game. For every resource $e \in E$, the load of e in s is $f_e(s) = \sum_{j=1}^{k} f_e(s_j)$. The delay of a strategy $\alpha \in \Sigma$ in state s is $\ell_\alpha(s) = \sum_{e\in\alpha} \ell_e(f_e(s))$. The selfish cost of each coalition C_j in state s is given by the *total delay* of its players, denoted $\tau_j(s)$. Formally, $\tau_j(s) \equiv \sum_{i\in C_j} \ell_{s_j^i}(s) = \sum_{e\in E} f_e(s_j)\ell_e(f_e(s))$ Computing a coalition's best response in a network congestion game can be performed by first applying a transformation similar to that in [12, Theorem 2] and then computing a min-cost flow. A state s is a *Nash equilibrium* if for every coalition C_j and every strategy $s'_j \in \Sigma^{n_j}$, $\tau_j(s) \leq \tau_j(s_{-j}, s'_j)$, i.e. the total delay of coalition C_j cannot decrease by C_j's unilaterally changing its strategyFor every $\varepsilon \in (0,1)$, a state s is an *ε-Nash equilibrium* if for every coalition C_j and every strategy $s'_j \in \Sigma^{n_j}$, $(1-\varepsilon)\tau_j(s) \leq \tau_j(s_{-j}, s'_j)$. An *$\varepsilon$-move* of coalition C_j is a deviation from s_j to s'_j that decreases the total delay of C_j by more than $\varepsilon\tau_j(s)$. Clearly, a state s is an ε-Nash equilibrium iff no coalition has an ε-move available.

3.2 Convergence to Approximate Equilibria

To bound the convergence time to ε-Nash equilibria, we use the following potential function: $\Phi(s) = \frac{1}{2}\sum_{e\in E}[f_e(s)\ell_e(f_e(s)) + \sum_{j=1}^{k} f_e(s_j)\ell_e(f_e(s_j))]$, where [15, Theorem 6] proves that Φ is an exact potential function for (even multi-commodity) congestion games with static coalitions and *linear* latencies. We prove that for single-commodity linear congestion games with coalitions, the *largest improvement ε-Nash dynamics* converges to an ε-Nash equilibrium in a polynomial number of steps. Hence in network congestion games, where a coalition's best response can be computed in polynomial times by a min-cost flow computation, an ε-Nash equilibrium can be computed in polynomial time. If the current strategies profile is not an ε-Nash equilibrium, there may be many coalitions with ε-moves available. In the largest improvement ε-Nash dynamics, the coalition that moves is the one whose best response is an ε-move and results in the largest improvement in its total delay (and consequently in the potential). In the full version of the paper [30], the following theorem is proven.

Theorem 6. *In a single-commodity linear congestion game with n players divided into k coalitions, the largest improvement ε-Nash dynamics starting from an initial state s_0 reaches an ε-Nash equilibrium in at most $\frac{kr(r+1)}{\varepsilon(1-\varepsilon)} \log \Phi(s_0)$ steps, where $r = \lceil \max_{j\in[k]}\{n_j\} / \min_{j\in[k]}\{n_j\} \rceil$ denotes the ratio between the size of the largest coalition and the size of the smallest coalition.*

References

1. Ackermann, H., Roeglin, H., Voecking, B.: On the impact of combinatorial structure on congestion games. In: FOCS (2006)
2. Arndt, H.: Load balancing: dimension exchange on product graphs. In: 18th International Symposium on Parallel and Distributed Processing (2004)
3. Awerbuch, B., Azar, Y., Epstein, A.: The Price of Routing Unsplittable Flow. In: STOC (2005)
4. Berenbrink, P., Friedetzky, T., Goldberg, L.A., Goldberg, P., Hu, Z., Martin, R.: Distributed selfish load balancing. In: SODA (2006)
5. Blum, A., Even-Dar, E., Ligett, K.: Routing without regret: on convergence to nash equilibria of regret-minimizing algorithms in routing games. In: PODC (2006)
6. Chien, S., Sinclair, A.: Convergece to Approximate Nash Equilibria in Congestion Games. In: SODA (2007)

7. Christodoulou, G., Koutsoupias, E.: The Price of Anarchy of Finite Congestion Games. In: STOC (2005)
8. Christodoulou, G., Mirrokni, V.S., Sidiropoulos, A.: Convergence and approximation in potential games. In: Durand, B., Thomas, W. (eds.) STACS 2006. LNCS, vol. 3884, pp. 349–360. Springer, Heidelberg (2006)
9. Cybenko, G.: Load balancing for distributed memory multiprocessors. J. Parallel Distrib. Comput. 7, 279–301 (1989)
10. Even-Dar, E., Mansour, Y.: Fast convergence of selfish rerouting. In: SODA (2005)
11. Even-Dar, E., Kesselman, A., Mansour, Y.: Convergence Time to Nash Equilibria. In: Baeten, J.C.M., Lenstra, J.K., Parrow, J., Woeginger, G.J. (eds.) ICALP 2003. LNCS, vol. 2719, pp. 502–513. Springer, Heidelberg (2003)
12. Fabrikant, A., Papadimitriou, C., Talwar, K.: The Complexity of Pure Nash Equilibria. In: STOC (2004)
13. Fischer, S., Räcke, H., Vöcking, B.: Fast convergence to wardrop equilibria by adaptive sampling methods. In: STOC (2006)
14. Fotakis, D., Kontogiannis, S., Spirakis, P.: Selfish Unsplittable Flows. In: TCS, vol. 348, pp. 226–239 (2005)
15. Fotakis, D., Kontogiannis, S., Spirakis, P.: Atomic Congestion Games among Coalitions. In: Bugliesi, M., Preneel, B., Sassone, V., Wegener, I. (eds.) ICALP 2006. LNCS, vol. 4051, pp. 572–583. Springer, Heidelberg (2006)
16. Ghosh, B., Muthukrishnan, S.: Dynamic load balancing in parallel and distributed networks by random matchings. In: SPAA (1994)
17. Goemans, M.X., Mirrokni, V.S., Vetta, A.: Sink equilibria and convergence. In: FOCS 2005 (2005)
18. Goldberg, P.W.: Bounds for the convergence rate of randomized local search in a multiplayer load-balancing game. In: PODC (2004)
19. Hayrapetyan, A., Tardos, É., Wexler, T.: The Effect of Collusion in Congestion Games. In: STOC (2006)
20. Hosseini, S.H., Litow, B., Malkawi, M.I., McPherson, J., Vairavan, K.: Analysis of a graph coloring based distributed load balancing algorithm.. J. Par. Distr. Comp. 10, 160–166 (1990)
21. Ieong, S., McGrew, R., Nudelman, E., Shoham, Y., Sun, Q.: Fast and compact: A simple class of congestion games. In: AAAI (2005)
22. Libman, L., Orda, A.: Atomic resource sharing in noncooperative networks. Telecommunication Systems 17(4), 385–409 (2001)
23. Mirrokni, V., Vetta, A.: Convergence Issues in Competitive Games. In: Jansen, K., Khanna, S., Rolim, J.D.P., Ron, D. (eds.) RANDOM 2004 and APPROX 2004. LNCS, vol. 3122, pp. 183–194. Springer, Heidelberg (2004)
24. Monderer, D., Shapley, L.: Potential Games. Games& Econ. Behavior 14, 124–143 (1996)
25. Motwani, R., Raghavan, P.: Randomized Algorithms. Cambridge University Press, Cambridge (1995)
26. Orda, A., Rom, R., Shimkin, N.: Competitive routing in multiuser communication networks. IEEE/ACM Trans. on Net. 1(5), 510–521 (1993)
27. Rosenthal, R.W.: A Class of Games Possessing Pure-Strategy Nash Equilibria. International Journal of Game Theory 2, 65–67 (1973)
28. Sandholm, W.H.: Population Games and Evolutionary Dynamics. MIT Press (to be published), `http://www.ssc.wisc.edu/~whs/book/index.html`
29. Sandholm, W.H.: Potential Games with Continuous Player Sets. Journal of Economic Theory 97, 81–108 (2001)
30. Extended version. `http://students.ceid.upatras.gr/ kaporis/papers/fks-sagt-08.pdf`
31. Weibull, J.W.: Evolutionary Game Theory. MIT Press, Cambridge (1995)

Frugal Routing on Wireless Ad-Hoc Networks

Gunes Ercal, Rafit Izhak-Ratzin, Rupak Majumdar, and Adam Meyerson

University of California, Los Angeles

Abstract. We study game-theoretic mechanisms for routing in ad-hoc networks. Game-theoretic mechanisms capture the non-cooperative and selfish behavior of nodes in a resource-constrained environment. There have been some recent proposals to use incentive-based mechanisms (in particular, VCG) for routing in wireless ad-hoc networks, and some frugality bounds are known when the connectivity graph is essentially complete. We show frugality bounds for random geometric graphs, a well-known model for ad-hoc wireless connectivity. Our main result demonstrates that VCG-based routing in ad-hoc networks exhibits small frugality ratio (i.e., overpayment) with high probability. In addition, we study a more realistic generalization where sets of agents can form *communities* to maximize total profit. We also analyze the performance of VCG under such a community model and show similar bounds. While some recent truthful protocols for the traditional (individual) agent model have improved upon the frugality of VCG by selecting paths to minimize not only the cost but the overpayment, we show that extending such protocols to the community model requires solving NP-complete problems which are provably hard to approximate.

1 Introduction

We study the frugality ratio (FR), a measure of cost-efficiency, of the generalized VCG mechanism for reliable routing in the presence of non-cooperative behavior in ad-hoc networks. We model ad-hoc networks by random geometric graphs (RGG), and show that VCG-based routing exhibits small frugality ratio with high probability (w.h.p.). We generalize the standard model of agent behavior by allowing sets of nodes to form communities to maximize the total profit and demonstrate bounds on the frugality ratio for this model as well. Moreover, while some recent truthful protocols for the traditional (individual) agent model have improved upon the frugality of VCG by selecting paths to minimize not only the cost but the overpayment, we show that extending such protocols to the community model requires solving NP-complete problems which are provably hard to approximate.

Reliable and cost-efficient routing in ad-hoc networks is a well-studied problem, with numerous proposals for routing protocols. Many of these protocols assume that the nodes in the network behave co-operatively. In resource-scarce environments, such as ad-hoc networks, this co-operativeness assumption is suspect. Forwarding a packet incurs some cost and in the absence of other incentives, nodes belonging to one community may refuse to forward packets belonging to

B. Monien and U.-P. Schroeder (Eds.): SAGT 2008, LNCS 4997, pp. 133–144, 2008.

another community. Under these assumptions, it is more reasonable to model a network as a game played between independent selfish agents, and to apply game theoretic reasoning to develop incentive-based routing protocols [1,2].

In an incentive-based routing protocol, a node is paid monetary compensation in return for forwarding a packet. The compensation covers the cost incurred by the node in forwarding the packet. Specifically, in order to route a packet from node s to node t, each node in the graph demands some payment commensurate with the cost it incurs to handle the packet. The minimum cost path is chosen as the route, each node along the path getting the payment it demanded. Unfortunately, in most cases, the actual cost incurred is information private to the community owning the node and the protocol must assume that the community sets its own price. This can lead to cheating: communities will tend to inflate their operating costs to maximize the benefits received, leading to instability in the protocol. Thus, the protocol must be designed so that individual communities have no incentive to cheat. Such a *truthful mechanism* [1,3,4] will ensure that each community will demand a payment equal to its actual cost. The VCG mechanism [4,5,6,7] implements a truthful mechanism: the chosen route is the minimum cost according to the demanded payments, and each community gets paid the maximum amount it could have demanded to still be part of the chosen route, all other communities' demands remaining the same.

Since VCG is truthful, the chosen route is indeed the cheapest path with respect to the true cost. However, the payment made to the communities can be significantly greater than the solution cost. Hence, one has to analyze the amount by which the mechanism overpays, called the *frugality* of the mechanism [8,9,10]. This is measured by the *frugality ratio*, the maximum over all source-sink pairs of the ratio of the total payment made to the actual cost of the route.

The VCG mechanism and associated FR have been studied for shortest path routing on graphs, where each node or edge is considered an independent agent. We demonstrate in this work that the mechanism extends to the presence of *communities.* This captures the real-world nature of ad-hoc networks where nodes are organized into communities acting together, for example mobile users who group together following common social interests [11,12,13]. While this extension is simple for the standard VCG mechanism, we show that many natural extensions to VCG that remain computationally tractable in the usual case become intractable once communities are explicitly added to the model.

Random geometric graphs (RGG) [14] have been well-studied as theoretical models of ad-hoc networks [15,16,17,18]. Such graphs are constructed by placing nodes at random in the unit square, and adding an edge between two nodes if they are closer than the parameter r, which represents the broadcast radius. We consider various organizations of the nodes into k communities, including the traditional individual agent model in which each node is its own community (and $k = n$). We consider both the model where each node belongs to a uniformly at random selected community and the case where the node belongs to an arbitrary community (with no known underlying distribution). For any given community we assume that the per node cost is identical for all nodes of the community. We take this to be a reasonable simplifying assumption which reflects the cooperative nature of nodes within a community, including that they may agree amongst themselves upon a fixed per node price. It may also reflect other forms of commonality of a given community's nodes, such as being of the same provider, being of the same general type, or sharing some locality in the clustered cases.

For a random geometric graph with k communities populated uniformly at random, where the costs are chosen uniformly at random from the interval $[c, c+B]$, we prove that the FR is bounded by $2\sqrt{2}(1+\frac{2B(\log\log n)^2}{c\log n})$ w.h.p. For the individual node model (where each node is a different community), we show that the FR is bounded by $2(1+\frac{B}{c})$ (respectively, $2(1+\frac{B\log\log n}{c\log n})$) w.h.p. when costs are chosen arbitrarily (respectively, uniformly at random) from the interval $[c, c+B]$. Our proof techniques use the connectivity properties of RGG [15], together with iterated applications of the coupon collector's problem [19]. We also show a logarithmic bound in expectation when the number of communities in the network is small.

We also performed extensive network model simulations to see how VCG-based routing behaves in practice. The FR obtained in these simulations were always lower (better) than the theoretical upper bounds we provide. Our experiments also demonstrate that the FR goes up as the number of communities increase. This indicates that in the presence of many communities, a mechanism which minimizes the FR by weighting paths based on the number of communities may be desirable. In fact this is the intuition behind the result of [10] to improve over the FR of VCG. Unfortunately, we show that in the community model such weighting schemes become computationally intractable (NP-hard and even hard to approximate), implying that these improved mechanisms will be difficult to implement in practice. Due to space limitations, we have defered the simulation results and some proofs to an extended version of the paper [20].

2 Related Work

The theory of algorithmic mechanism design was initiated by Nisan and Ronen in [4,21], in which they considered the generalized Vickrey-Clarke-Groves (VCG) mechanism [5,6,7] for various computational problems, including shortest path auctions. Although [4] considers VCG for general set systems, most subsequent work on truthful mechanisms for path auctions and the frugality thereof is restricted to the case where every edge is owned by an independent agent. Du et.al. [22] discuss a model where communities can own multiple edges, however in their model the identity of the community owning an edge is private, and they show that for such a model no truthful mechanism exists. In our work, we extend VCG for path auctions in the presence of communities where ownership is public but costs remain private. With the observation that VCG overpayments can be quite excessive for path auctions in worst cases, work has been put forth towards finding more frugal truthful mechanisms [8,9]. Karlin [10] proposed the $\sqrt{n}$ mechanism, which is within a $\sqrt{2}$ factor of the frugality ratio for the best truthful mechanism on any given graph, and in some cases performs up to $O(\sqrt{n})$ more frugally than VCG. In Section 6 we show that it is NP-hard to generalize many classes of truthful mechanisms for path auctions in the standard model, including the $\sqrt{n}$ mechanism of [10], to the community model.

As we are interested in path auctions for ad-hoc networks, we study the performance of VCG for RGG [14], a model for the theoretical analysis of ad-hoc networks [15,16,17,18]. In particular, Gupta and Kumar [15] model ad-hoc networks as RGGs in their analysis of the critical radius required for asymptotic connectivity.

An alternative to the VCG is the first path auction where the agents on the winning path are paid their bid value. Immorlica et. al. [23] characterized all strong ϵ-Nash equilibria of a first path auction and showed that the total payment of this mechanism is often better than the VCG total payment. However, the drawback is that there is no guarantee that the bidders will reach an equilibrium, moreover, unlike the VCG, the preferred bid may depend on the communicating pair, which might not be known in advance.

VCG and variations thereof have been previously considered for routing in networks, fitting into a recent body of research tackling the problem of game-theoretic formalization of routing incentives for various networking domains [2,24,25,26]. Closest to our work in this regard is Anderegg and Eidenbenz [2] paper in which they propose VCG for routing in ad-hoc networks. Although our work is nominally similar, there are crucial differences. In particular, while both consider VCG on ad-hoc networks, in their mechanism they consider nodes to have unbounded maximum potential radius, paying selected nodes to set their actual radius as desired according to how many bits they forward for the source-sink, and take each node to be an independent agent. We, on the other hand, consider a fixed topology in which radii are already set, and pay nodes to transmit according to some cost function set by their community.

Finally, we focus on previously unconsidered theoretical aspects of the problem, leaving the concrete implementation to a large body of work on implementation of internet currency [27] and other work dealing with the game-theoretic multi-hop routing [2,24,25,26] implementation.

3 Mechanism Design and the Payment Model

We model an ad-hoc network with k communities as a connected undirected graph $G = (V, E)$ where the nodes in V are partitioned into k subsets (the communities). Each community is assumed to be independently profit maximizing. We assume that there is no monopoly community in the graph, so that by removing one community from the graph the graph will still remain connected.

Given a k-community ad-hoc network (V, E), and nodes $s, t \in V$, our goal is to design a protocol that will let s route a packet to t by a cheapest-cost path from s to t. A community i charges money for any packet that one of its node transmits. We assume all nodes in a community charge the same price, however, the exact determination of this cost is information private to the community.

While nodes can change location and connectivity over time, we assume that the network is static during the routing phase. We use tools from mechanism design [4] and define our protocol as follows.

1. We define a game on a k-community ad-hoc network (V, E) with k players, each corresponding to a community, and two states $s, t \in V$ (the source and the sink for routing). We define the *allowed outcomes* O of the game to be the finite set of simple paths between s and t.
2. For each path $o \in O$, each community i has a private cost $t^i(o)$ which is a function of the number of community nodes in path o and the cost of forwarding a packet by a node belonging to the community. We simplify the model by assuming that all the nodes belong to the same community have the same packet transmitting cost. Under this assumption $t^i(o) = C_i \cdot n_i(o)$, where C_i is the cost of transmitting one packet by a node of community i, and $n_i(o)$ is the number of i's nodes lying on path o.
3. Each community defines a valuation function $p^i(o)$, which is the price it charges to transmit a packet on path o.
4. If the path $\hat{o}$ is chosen as the route from s to t, then the utility function of community i will be $u^i(\hat{o}) = p^i(\hat{o}) - t^i(\hat{o})$ where $p^i(\hat{o}) \geq 0$ is the payment the community receives from the mechanism. The goal of community i is to maximize its utility $u^i(\hat{o})$.

The payment p^i to the communities is used to ensure a truthful implementation, i.e., an implementation where the dominant strategy of each community is to set its valuation p^i to be equal to t^i. We use the following payment in our mechanism. Let $d_{G|i=\infty}$ be the shortest path that does not contain any node belongs to community i and let $d_{G|i=0}$ be the cost of the shortest path where all nodes on the shortest path that belong to i have a zero cost. Then, the payment function $p^i(\hat{o}) = 0$ if i is not on the shortest path $\hat{o}$, and $p^i(\hat{o}) = d_{G|i=\infty} - d_{G|i=0}$ measures the maximum amount community i could have charged to still be part of the chosen route. This is a generalization of the shortest path payment scheme in [4]. Since shortest paths is a *monotone selection rule* (i.e., a losing community cannot become part of the shortest path by raising its valuation), standard techniques [4,8] show that this payment scheme implements a truthful mechanism. The *frugality ratio* is the "over payment" ratio of the mechanism: $FR = \frac{\sum_i p^i(\hat{o})}{\sum_i (t^i(\hat{o})}$.

4 Graph and Cost Model

A random geometric graph (RGG) with n nodes and radius r is constructed by picking n points (nodes) uniformly at random from the unit square, and putting an edge between nodes u and v if the distance between u and v is less than or equal to r.

Following previous theoretical work on ad-hoc networks [15], we represent ad-hoc networks as random geometric graphs. We choose the radius r at least on the order of asymptotic connectivity $r_{con} = \Omega(\sqrt{\frac{\log n}{n}})$ [15], i.e., the radius that ensures that the graph is connected almost surely. Our models have four parameters: the number of nodes (n), the radius of the RGG (r), the number and choice of communities (k), and choice of transmission costs (F). We shall assume henceforth that $r \geq r_{con}$.

We consider three types of cost distribution functions F. First, we study *arbitrary bounded* cost distributions $F_A(c_{min}, B)$, where community picks an arbitrary cost from the interval $[c_{\min}, c_{\min} + B]$. As a special case, we study the *unit cost distribution* $F_C = F_A(1, 0)$ where each community charges unit cost per edge. Second, we study *uniformly-at-random bounded* cost distributions $F_U(c_{\min}, B)$, where each community j picks a cost c_j uniformly at random from the interval $[c_{\min}, c_{\min} + B]$. Third, we study *uniformly-at-random unbounded* cost distributions $F_{A,U}(\epsilon)$, where $\epsilon > 0$, and each community j picks a cost c_j uniformly at random from the interval $[\epsilon, 1]$. As $\epsilon \to 0$, this model represents the case where the ratios of costs can be unbounded. Our worst case bounds depend on B, which becomes unbounded as $\epsilon \to 0$. While this is not a realistic case; it is interesting to see how bad the practical results can be. We study the following models:

Individual agent model. In the individual agent model (IAM), each node of the graph is its own community. This corresponds to the traditionally studied shortest path VCG mechanism on graphs where each node is an independent agent. We write $NC = (n, r, F)$ for an IAM network cost model with n nodes, radius r, and cost distribution F.

Random graph with communities. Given a number k of communities, each node in the random graph is assigned a community uniformly at random. We write $NC = (n, r, k, F)$ for the network cost model where there are n nodes, the radius is r, there are k communities (each node selecting its community uniformly at random), and the costs are determined according to the cost distribution F.

5 Theoretical Results

5.1 Frugality Ratio with High Probability

In many of the bounds, we use the following well known lemma on occupancy.

Lemma 1 (Balls in Bins [19,17]). *For a constant $c > 1$, if one throws $n \geq c\beta \log \beta$ balls into β bins, then w.h.p. both the minimum and the maximum number of balls in any bin is $\Theta(\frac{n}{\beta})$. Moreover, for $c < 1$ if one throws $n \leq c\beta \log \beta$ balls into β bins, then w.h.p. there will exist an empty bin.*

Due to the critical nature of the above threshold, we are able to give bounds w.h.p. for uniform distributions of costs and communities.

As mentioned previously, we consider random geometric graphs with radius chosen to guarantee connectivity w.h.p. Recall that we assume $r \geq r_{con}$. Although we shall state results for such general radii, we are primarily interested in small radii r such that $r = \Theta(r_{con})$. In particular, we will satisfy a slightly stronger guarantee of *geo-denseness* [17], namely that, for any fixed arbitrary partitioning of the unit square into simple convex Euclidean regions β_i of area $\frac{r}{2\sqrt{2}} \times \frac{r}{2\sqrt{2}}$ each, every β_i will have the same order of nodes w.h.p. It follows from Lemma 1 that radius $\hat{r} = (2\sqrt{2}+\epsilon)\sqrt{\frac{\log n}{n}} \leq 3(r_{con,n})$ satisfies the geo-denseness property while still being on the same order as the radius for asymptotic connectivity. Henceforth, we will state some results for both general r and for $\hat{r}$ as defined here. Note further that our following theoretical results hold for geo-dense geometric graphs in general, not only random geometric graphs. Due to space limitations, some proofs have been deferred to the full version of the paper

Our first theorem considers the case of arbitrary costs in the Individual Agents Model (IAM), the standard model for path auctions.

Theorem 1 (IAM with Arbitrary Costs). *Given an IAM,* $NC = (n, r, F_A(c_{\min}, B))$*, for any* $r \geq \hat{r}$*, the FR of VCG is at most* $2(1 + \frac{B}{c_{\min}})$ *w.h.p.*

In particular, for IAM $NC = (n, r, F_C)$ with unit cost distribution, for any $r \geq \hat{r}$, the FR of VCG is at most 2. While unit costs do not seem to be a realistic assumption, and do not require notions of truthfulness, it yields insight into how the connectivity properties of a graph affect the overpayment. After all, with arbitrary costs one may obtain arbitrarily bad overpayments for any graph, but even with unit costs, the graph properties alone may yield bad overpayments. Therefore, the frugality ratio of VCG in the unit cost model is worthwhile to consider, and one that has been considered for other random graph models, namely Bernoulli graphs and random scale-free graphs, as well. A notable difference between random geometric graphs and those other two well-known random graph models is that while the hop diameter of the latter models is short w.h.p. the hop diameter of random geometric graphs is long w.h.p.

In standard shortest path auctions [4], unlike our model, costs are assigned on edges rather than nodes. For an IAM, $NC = (n, r, F_A(c_{\min}, B))$ where edge costs, we can similarly show that the FR is bounded by $2(1 + \frac{B}{c_{\min}})$ w.h.p.

When costs are distributed uniformly at random (u.a.r.), we may obtain provably better bounds than in the arbitrary case.

Theorem 2 (IAM with Random Costs). *Given* $NC = (n, r, F_U(c_{\min}, B))$*, for any* $r \geq \hat{r}$*, the FR is at most* $2(1 + \frac{B}{bc_{min}})$ *where* $b = \frac{\frac{nr^2}{8}}{2\log(\frac{nr^2}{8})}$ *w.h.p. In particular, for* $r = \hat{r}$*, if* $B = O(c_{min}\frac{\log n}{\log\log n})$*, the FR of VCG for NC is a constant w.h.p.*

Now, we give our results for models with communities. The bounds of arbitrary costs are almost identical to that of the IAM.

Theorem 3 (Community Model with Arbitrary Costs). *Given* $NC_C = (n, r, k, F_A(c_{\min}, B))$*, for any* $r \geq \hat{r}$*, the FR is at most* $2\sqrt{2}(1 + \frac{B}{c_{\min}})$ *w.h.p.*

In particular, for $NC = (n, r, k, F_C)$, with unit costs, for any $r \geq \hat{r}$, the FR is at most $2\sqrt{2}$ w.h.p. Again, for the u.a.r. case, we obtain better guarantees.

Theorem 4 (Community Model with Random Costs). *Let* $NC = (n, r, k, F_U(c_{\min}, B))$ *with* $r \geq \hat{r}$ *and* $k \leq \frac{8}{r^2}$ *communities. For* $b = \min\{\frac{k}{2\log k}, \frac{\frac{nr^2}{8}}{2\log\frac{nr^2}{8}}\}$ *the FR of VCG is at most* $2\sqrt{2}(1 + \frac{2B}{bc_{min}})$ *w.h.p. In particular, for* $r = \hat{r}$ *and* $\log n \leq k \leq \frac{n}{\log n}$*, if* $B = O(c_{min}\frac{\log n}{(\log\log n)^2})$*, the FR is a constant w.h.p.*

Proof. Let s and t be an arbitrary source and sink pair and $SP = \langle v_0, v_1, \cdots, v_d\rangle$ denote the shortest path between s and t. Since overpayments are made to communities rather than merely to nodes, partition SP into blocks $\langle L_1, \cdots, L_q\rangle$ where each block belongs to a single community and consecutive blocks do not belong to the same community. For each community j, let $K_j = \langle L_{j_1}, \cdots, L_{j_x}\rangle$ denote the set of blocks owned by community j. For each community j and block L_{j_i} denote by $v_{j_i,0}$ and $v_{j_i,f}$ the nodes in SP immediately preceding and succeeding L_{j_i} respectively, and let l_{j_i} be the line between $s' = v_{j_i,0}$ and $t' = v_{j_i,f}$. Partition l_{j_i} into $\frac{r}{2\sqrt{2}}$ length intervals (with at most one partial interval at the end of negligible effect) $y \in \{1, 2, \cdots, \frac{d(s',t')}{\frac{r}{2\sqrt{2}}}\}$. Depending on how close l_{j_i} is to a boundary of the unit square, it is clear that there must exist a $\frac{r}{2\sqrt{2}} \times d(s', t')$ rectangular area A_{j_i} with l_{j_i} as one of the sides lying entirely inside the unit square. Depending on the orientation of this rectangular area, for each interval y, let S_y denote the $\frac{r}{2\sqrt{2}} \times \frac{r}{2\sqrt{2}}$ square in A_{j_i} with interval y as one of the sides.

By Lemma 1 and the choice of r, there are $\Theta(\frac{nr^2}{8})$ nodes in each S_y w.h.p. Each node chooses amongst the k communities u.a.r. Each of k communities chooses its cost u.a.r. from $[c_{\min}, \cdots, c_{\min} + B]$. By the choice of b, w.h.p. the number of communities in each cost interval of the form $[c_{\min} + (\alpha - 1)\frac{B}{b}, c_{\min} + \alpha\frac{B}{b}]$ (for α from 1 to b) is $\Theta(\frac{k}{b})$. Therefore, since the number of communities in each cost interval is on the same order, each node in S_y picks amongst the cost intervals as well up to constant factors. Again, by the choice of b, the number of cost intervals and re-application of Lemma 1, we have that for each cost interval α there are $\Theta(\frac{nr^2}{8b})$ nodes of S_y having cost in interval α. Then, recalling that consecutive bins form a clique, we may route along nodes in the first two cost intervals in each square bin, depending upon which cost interval the corresponding community in SP lies. Then, for each A_{j_i}, we obtain a path of cost at most $2\sqrt{2}\frac{d(v_{j_i,0}, v_{j_i,f})}{r}(c_{\min} + \frac{2B}{b})$ other than L_{j_i} which has cost at least $d(v_{j_i,0}, v_{j_i,f})rc_{\min}$. So, for L_{j_i}, the FR is at most $2\sqrt{2}\frac{2B + c_{min}}{bc_{min}}$. Summing over each L_{j_i}, we obtain the same ratio. This characterizes the payment to community j. Moreover, the argument is the same for any community since the scaling by distance is lost. Thus, the theorem follows. □

5.2 Frugality Ratio in Expectation

The bounds so far are all with high probability. However, in the case of fewer communities we may find significantly improved bounds of VCG with communities for RGGs *in expectation*. When the number of communities k is $O(\frac{\log n}{\log\log n})$

(or, for general r, when k is $O(\frac{nr^2}{\log(nr^2)})$) we may note once again that every community occurs in every bin (of $\frac{r}{2\sqrt{2}}$x$\frac{r}{2\sqrt{2}}$ size). So, due to the aforementioned bin properties for RGGs, we need only bound the expected ratio of the second cheapest community to the cheapest community.

Theorem 5. *Let $NC = (n, r, k, F_U(c_{\min}, B))$ with radius $r \geq \hat{r}$ and $k \leq \frac{nr^2}{\log(nr^2)}$ communities. The expected FR of VCG for NC is $O(\min\{\log \frac{B}{c_{min}}, \frac{B}{kc_{min}}\})$ w.h.p.*

Proof. Due to aforementioned geometric bin properties and normalization, it suffices to show that the expected ratio of the second cheapest to the cheapest of k costs chosen u.a.r from $[1, B]$ is $O(\log B)$. As such, note that the probability that the cheapest is in $[x, x+dx]$ is $k\frac{dx}{B-1}(\frac{B-x}{B-1})^{k-1}$, corresponding to the choices for the cheapest variable and the event that that variable is in $[x, x+dx]$ and all rest are in $(x, B]$. Moreover, the expected value of the second cheapest given that the cheapest is x is the expected value of the cheapest of the $k-1$ restricted to interval $(x, B]$, which is easy to check to be $\frac{B+x(k-1)}{xk}$. Thus,

$$\begin{aligned} E_k[\tfrac{Y}{X}] &= \int_1^B \tfrac{k}{B-1}(\tfrac{B-x}{B-1})^{k-1}\tfrac{1}{x}\tfrac{B+x(k-1)}{xk}dx = \tfrac{B}{B-1}((\int_1^B (\tfrac{B-x}{B-1})^{k-1}\tfrac{dx}{x}) + \tfrac{k-1}{k}) \\ &\leq \tfrac{B}{B-1}(\min\{\log B, \tfrac{B-1}{k}\} + \tfrac{k-1}{k}) \end{aligned}$$

□

We may generalize the expected ratio of the second cheapest to the cheapest of k i.i.d. random costs given cumulative distribution F and density function f as follows: The probability that the minimum is in $[x, x+dx]$ is, taking over the k choices of the minimum variable, $kf(x)(1-F(x))^{k-1}$. Similarly, the probability that the second cheapest is in $[y, y+dy]$ given that the cheapest is x is the probability that the minimum of the remaining $k-1$ is in $[y, y+dy]$ given that all $k-1$ have cost greater than x. Thus, the expectation in question is:

$$\begin{aligned} E_k[\tfrac{Y}{X}] &= \int_1^\infty \tfrac{kf(x)(1-F(x))^{k-1}}{x}dx \int_x^\infty y(k-1)\tfrac{f(y)(1-F(y))^{k-2}}{(1-F(x))^{k-1}}dy \\ &= k(k-1)\int_1^\infty \tfrac{f(x)}{x}dx \int_x^\infty yf(y)(1-F(y))^{k-2}dy \end{aligned}$$

Substituting, we obtain the following results for some

Corollary 1. *Let $NC_\lambda = (n, r, k, F_\lambda, B))$ with $r \geq \hat{r}$ and $k \leq \frac{nr^2}{\log(nr^2)}$ communities and F_λ the exponential distribution translated by $+1$ with parameter λ. The expected FR of VCG for NC_λ is at most $4\sqrt{2}$ w.h.p..*

For the distribution F_{recip} obtained by taking reciprocals of random variables chosen according to the uniform distribution on the unit interval $(0, 1]$, in the model $NC_{recip} = (n, r, k, F_{recip}, B))$ with radius $r \geq \hat{r}$ and $k \leq \frac{nr^2}{\log(nr^2)}$ communities, we similarly get that the expected FR of VCG for NC_{recip} is $2\sqrt{2}\frac{k-1}{k-2}$ w.h.p. In fact, we can say something much stronger for this distribution.

Lemma 2. *Let $NC_{recip} = (n, r, k, F_{recip}, B))$ with radius $r \geq \hat{r}$ and $k \geq nr^2$ communities. The FR of VCG for NC_{recip} is at most $2e^3\sqrt{2}$ w.h.p..*

This holds because, by the geometric bin properties, it suffices to show that *within each bin* the probability that the second cheapest in that bin is more than e^3 times the cheapest in that bin is $O(\frac{1}{nm})$, where $m = \frac{8}{r^2}$ is the number of bins. Let q denote the number of communities occuring w.h.p. in every bin. By choice of k and r, we have $q = \Theta(\frac{nr^2}{8})$ by coupon collection. The event that $\frac{1}{X} \geq e^3 \frac{1}{Y}$ implies that $q-1$ reciprocals chosen u.a.r. all lay in $(0, \frac{1}{e^3})$, the probability of which is $\frac{q}{e^{3(q-1)}}$. Thus, $Pr[\frac{Y}{X} \geq e^3] = Pr[\frac{1}{X} \geq e^3 \frac{1}{Y}] < \frac{q}{e^{3(q-1)}}$, where X is the cheapest and Y is the second cheapest. Moreover, $\frac{q}{e^{3(q-1)}} \leq \frac{q}{n^2} = \frac{r^2}{n}$ by choice of q, completing the proof.

By noting that, for F_λ, the exponential distribution translated by $+1$, the probability that $q-1$ costs are higher than A is at most $ke^{-\lambda(A-1)(q-1)}$, a very similar argument gives the following.

Lemma 3. *Let $NC_\lambda = (n, r, k, F_\lambda, B))$ with radius $r \geq \hat{r}$ and $k \geq nr^2$ communities. The FR of VCG for NC_λ is $O(1)$ w.h.p..*

6 Hardness of Extensions

NP-Hardness of Extensions. Both simulation results and related work on the traditional path auction model [8,9,28,10] suggest that a mechanism that minimizes some weighting of total path costs by the number of communities on the path may have a lower FR than VCG. For example, the mechanism proposed in [10] is known to be up to $\sqrt{n}$ times more frugal than VCG. Unfortunately, as we show next, in the presence of communities, the implementation of this mechanism requires solving intractable problems.

The first step of the $\sqrt{n}$ mechanism of [10] is to find the least cost edge-disjoint cycle through s and t. In the community model, this would correspond to finding at least some community disjoint cycle through s and t. Note that the existence of two community disjoint paths is not guaranteed by the no-monopoly condition. For example, consider $k = 3$ and a graph consisting of three length paths P_1, P_2, P_3 from s to t where each path P_i excludes only community i.

By representing each community with a unique color, we color the nodes (or, alternately, edges, the results apply to both cases) according to their communities. Finding a community disjoint cycle is the same as finding a color-disjoint cycle. This problem is NP-Complete by a reduction from 3-SAT. A similar problem is independently shown to be NP-Complete in [29].

Lemma 4. *Consider the problem $\mathfrak{C}$: Given a graph $G = (V, E)$ with nodes arbitrarily colored from k colors, and a designated source-sink pair (s, t), find two color disjoint paths through s and t. $\mathfrak{C}$ is NP-Complete. The same is true considering edge colorings instead of node colorings.*

APX-Hardness of Natural Extensions. As a second possible extension, we can study the VCG under other cost models. For example, we could try to minimize the number of communities along the shortest path in order to try to reduce the FR. Unfortunately, we have found that many approaches in these directions turn out to be NP-complete, some even strongly approximation hard.

Here we show that any natural truthful mechanism with a selection rule incorporating some kind of minimization of the number of communities on the path is strongly approximation-hard to compute. Our reduction is an approximation preserving reduction from the Minimum Monotone Satifying Assignment ($MMSA_3$) problem, which is known to be $2^{\log^{1-o(1)} n}$ hard to approximate [30,31]. While there are closely related approximation hardness results under various names [32,33], our result and reduction are both more general and more direct. First, notice that for all $0 < x < 1$, we have $2^{\log^{1-o(1)} n} > n^x$. Now, we define a natural class of truthful mechanisms for path auctions in the community model in the following way. A truthful mechanism for path auctions in the community model (with per unit costs) is a (f, g) *min-agent mechanism* if its monotonic selection rule is of the following form. Given source s and destination t, select the path P from s to t that minimizes the product $f(q)g(p)$, for some strictly increasing, efficiently invertible function f and non-decreasing function g, where q is the number of communities on P and p is the total cost of P. Now, we proceed to our hardness result.

Theorem 6. *For any $0 < x < 1$, for any increasing, efficiently invertible function f and non-decreasing function g, the selection rule of a (f, g) min-agent mechanism is $f({k_n}^x)$ hard to approximate, where k_n is the total number of communities and n is the number of nodes.*

The same proof also implies the approximation-hardness of even computing VCG for various other cost-functions involving the community model, such as fixed community-network entrance fees (i.e., a one-time fee C_i for using any number of community i's nodes, which may be a more natural model for some service providers). The following is obtained by taking g to be a constant function in Theorem 6.

Corollary 2. *VCG for the community model under fixed community subnetwork entrance fees is hard to approximate to within k^x, for any $0 < x < 1$, given k total communities.*

Acknowledgments. We thank Deborah Estrin for many helpful discussions about models for ad-hoc networks. We also thank Chen Avin, Ilya Shpitser, and Sandra Batista for helpful discussions and comments. This research was sponsored in part by the NSF grants CCF-0427202 and CCF-0546170.

References

1. Papadimitriou, C.: Algorithms, games, and the internet. In: STOC (2001)
2. Anderegg, L., Eidenbenz, S.: Ad hoc-VCG: A truthful and cost-efficient routing protocol for mobile ad hoc networks with selfish agents. In: MOBICOM (2003)
3. Kreps, D.M.: A Course in Microeconomic Theory. Princeton University Press, Princeton (1990)
4. Nisan, N., Ronen, A.: Algorithmic mechanism design. Games and Economic Behavior 35, 166–196 (2001)
5. Vickrey, W.: Counterspeculation, auctions, and competitive sealed tenders. Journal of Finance 16, 8–37 (1961)

6. Clarke, E.: Multipart pricing of public goods. Public Choice 11, 17–33 (1971)
7. Groves, T.: Incentives in teams. Econometrica 41, 617–631 (1973)
8. Archer, A., Tardos, E.: Frugal path mechanisms. In: SODA (2002)
9. Talwar, K.: The price of truth: Frugality in truthful mechanisms. In: Alt, H., Habib, M. (eds.) STACS 2003. LNCS, vol. 2607, pp. 608–619. Springer, Heidelberg (2003)
10. Karlin, A., Kempe, D., Tamir, T.: Beyond VCG: Frugality of truthful mechanisms. In: FOCS (2005)
11. Kortuem, G., Segall, Z.: Wearable communities: Augmenting social networks with wearable computers. IEEE Pervasive Computing Magazine 2(1), 71–78 (2003)
12. Schulz, S., Herrmann, K., Kalckloesch, R., Schwotzer, T.: Towards trust-based knowledge management in mobile communities. In: IAAA (2003)
13. Musolesi, M., Hailes, S., Mascolo, C.: An ad hoc mobility model founded on social network theory. In: MSWiM (2004)
14. Penrose, M.D.: Random Geometric Graphs. Oxford University Press, Oxford (2003)
15. Gupta, P., Kumar, P.: The capacity of wireless networks. IEEE TIT, Los Alamitos (2000)
16. Goel, A., Rai, S., Krishnamachari, B.: Monotone properties of random geometric graphs have sharp thresholds. Annals of Applied Probability 15 (2005)
17. Avin, C., Ercal, G.: On the Cover Time of Random Geometric Graphs. In: Caires, L., Italiano, G.F., Monteiro, L., Palamidessi, C., Yung, M. (eds.) ICALP 2005. LNCS, vol. 3580, pp. 677–689. Springer, Heidelberg (2005)
18. Diaz, J., Petit, J., Serna, M.J.: Faulty random geometric networks. PPL (2000)
19. Motwani, R., Raghavan, P.: Randomized algorithms, Cambridge (1995)
20. Ercal, G., Izhak-Ratzin, R., Majumdar, R., Meyerson, A.: Frugal routing on wireless ad-hoc networks. Technical report, University of California, Los Angeles (2008)
21. Nisan, N., Ronen, A.: Computationally feasible VCG mechanisms. In: EC (2000)
22. Du, Y., Sami, R., Shi, Y.: Path auction games when an agent can own multiple edges. In: NetEcon (2006)
23. Immorlica, N., Karger, D., Nikolova, E., Sami, R.: First-price path auctions. In: EC (2005)
24. Sun, H., Song, J.: Strategy proof trust management in wireless ad hoc network. In: Electrical and Computer Eng' (2004)
25. Wang, W., Li, X.-Y., Eidenbenz, S., Wang, Y.: Ours: optimal unicast routing systems in non-cooperative wireless networks. In: MobiCom (2006)
26. Zhong, S., Li, L.E., Liu, Y.G., Yang, Y(R.): On designing incentive-compatible routing and forwarding protocols in wireless ad-hoc networks: an integrated approach using game theoretical and cryptographic techniques. In: MobiCom (2005)
27. Buttyán, L., Hubaux, J.: A virtual currency to stimulate cooperation in self-organized ad hoc networks. Technical Report DSC (2001)
28. Elkind, E., Sahai, A., Steiglitz, K.: Frugality in path auctions. In: SODA (2004)
29. Yuan, S., Jue, J.P.: Dynamic lightpath protection in WDM mesh networks under wavelength-continuity and risk-disjoint constraints. Comput. Netw (2005)
30. Dinur, I., Safra, S.: On the hardness of approximating label-cover. In: IPL (2004)
31. Alekhnovich, M., Buss, S.R., Moran, S., Pitassi, T.: Minimum propositional proof length is NP-hard to linearly approximate. In: Brim, L., Gruska, J., Zlatuška, J. (eds.) MFCS 1998. LNCS, vol. 1450, pp. 176–184. Springer, Heidelberg (1998)
32. Carr, R.D., Doddi, S., Konjevod, G., Marathe, M.: On the red-blue set cover problem. In: SODA (2000)
33. Wirth, H.C.: Multicriteria Approximation of Network Design and Network Upgrade Problems. PhD thesis (2001)

Facets of the Fully Mixed Nash Equilibrium Conjecture*

Rainer Feldmann[1], Marios Mavronicolas[2], and Andreas Pieris[3]

[1] Faculty of Computer Science, Electrical Engineering and Mathematics, University of Paderborn, 33102 Paderborn, Germany
obelix@uni-paderborn.de
[2] Department of Computer Science, University of Cyprus, Nicosia CY-1678, Cyprus
Currently visiting Faculty of Computer Science, Electrical Engineering and Mathematics, University of Paderborn, 33102 Paderborn, Germany
mavronic@cs.ucy.ac.cy
[3] Computing Laboratory, University of Oxford, Oxford OX1 3QD, United Kingdom
andreas.pieris@keble.ox.ac.uk

Abstract. In this work, we continue the study of the many facets of the *Fully Mixed Nash Equilibrium Conjecture*, henceforth abbreviated as the FMNE *Conjecture*, in selfish routing for the special case of n identical *users* over two (identical) parallel *links*. We introduce a new measure of *Social Cost*, defined to be the expectation of the square of the maximum *congestion* on a link; we call it *Quadratic Maximum Social Cost*. A *Nash equilibrium* (NE) is a stable state where no user can improve her (expected) latency by switching her mixed strategy; a *worst-case* NE is one that maximizes Quadratic Maximum Social Cost. In the *fully mixed* NE, all *mixed strategies* achieve full support.

Formulated within this framework is yet another facet of the FMNE *Conjecture*, which states that the fully mixed Nash equilibrium is the worst-case NE. We present an extensive proof of the FMNE *Conjecture*; the proof employs a mixture of combinatorial arguments and analytical estimations. Some of these analytical estimations are derived through some new bounds on *generalized medians* of the binomial distribution [22] we obtain, which are of independent interest.

1 Introduction

Motivation and Framework. In this work, we continue the study of the (multi-faceted) *Fully Mixed Nash Equilibrium Conjecture* [7], henceforth abbreviated as the FMNE *Conjecture*, in selfish routing. Specifically, we look at a special case of the KP model for selfish routing due to Koutsoupias and Papadimitriou [15]; here, a collection of n (*unweighted*) *users* wish to each transmit one unit of traffic from *source* to *destination*, which are joined through *two* (identical) parallel *links*. The *congestion* on a link is the total number of users choosing it; each

* This work has been partially supported by the IST Program of the European Union under contract number 15964 (AEOLUS).

B. Monien and U.-P. Schroeder (Eds.): SAGT 2008, LNCS 4997, pp. 145–157, 2008.

user makes her choice using a *mixed strategy*, which is a probability distribution over links. In the special case case of the KP model we look at, the *latency* on a link is identified with the congestion on it.

In a *Nash equilibrium* (NE) [20,21], no user can improve the expected congestion on the link she chooses by switching to a different (mixed) strategy. Originally considered by Kaplansky back in 1945 [14], *fully mixed Nash equilibria* have all their involved probabilities strictly positive; they were recently coined into the context of selfish routing by Mavronicolas and Spirakis [19]. Clearly, the fully mixed NE maximizes the randomization used in the mixed strategies of the players; so, it is a natural candidate to become a vehicle for the study of the effects of randomization on the quality of NEs.

We introduce a new measure of *Social Cost* [15] for the evaluation of NEs. The new measure is taken to be the expectation of the square of the maximum congestion on a link; call it *Quadratic Maximum Social Cost.* (The expectation is taken over all random choices of the users.) Note that the Quadratic Maximum Social Cost simultaneously generalizes the *Maximum Social Cost* (expectation of maximum latency) proposed in the seminal work of Koutsoupias and Papadimitriou [15], and the *Quadratic Social Cost* (expectation of the sum of the squares of the latencies) proposed in [16].

The motivation to consider the square of the latency comes from the real application of scheduling transmissions among nodes positioned on the Euclidian plane. The received power at a receiver is proportional to the power $-\delta$ of the (generalized) Euclidian distance from the sender to the receiver; δ is the *path-loss exponent*, for which it has been empirically assumed that $\delta \geq 2$ (cf. [13]). In many natural cases, the latency is proportional to the (generalized) Euclidian distance, and the proportionality constant may have to do with external conditions of the medium and the transmission power; in those cases, the received power is proportional to the power $-\delta$ of the latency. So, investigating the expected maximum latency to the power δ for the initial case $\delta = 2$ is expected to give insights about the optimization of received power in selfish transmissions.

For any particular definition of Social Cost, the FMNE *Conjecture* states that the fully mixed NE maximizes the Social Cost among all NEs. The validity of the FMNE *Conjecture* implies that computing the worst-case NE (with respect to the fixed Social Cost) for a given instance is trivial; it may also allow an approximation to the *Price of Anarchy* [15] in case where there is a FPRAS for approximating the Social Cost of the fully mixed NE (cf. [6]).

Contribution. In this proposed framework, we formulate a corresponding facet of the FMNE *Conjecture*:

Conjecture 1. The fully mixed NE maximizes the Quadratic Maximum Social Cost.

We present an extensive proof of this FMNE *Conjecture* using a wealth of combinatorial and analytical tools. The proof amounts to a very sharp comparison of the Quadratic Maximum Social Cost of an *arbitrary* NE to that of the fully mixed NE.

The proof has required some very *sharp* analytical estimates of various combinatorial functions that entered the analysis; this provides some evidence that the proved inequality among the two compared Quadratic Maximum Social Costs is very *tight*. The employed analytical estimates may be applicable elsewhere; so, they are interesting on their own right. In more detail, we have provided some new estimations for some generalizations of the *median* of the binomial distribution [11,22], which may be of independent interest.

Related Work. The FMNE *Conjecture* was first stated in [7]; it was motivated there by some initial observations in [6]. The FMNE *Conjecture* has been proved for the Maximum Social Cost for the cases of *(i)* two (*unweighted*) users and non-identical but *related links*, and *(ii)* an arbitrary number of (unweighted) users and two (identical) links in [17]. In fact, our estimation techniques significantly extend those for the case *(ii)* above in [17]; due to the increased complexity of the Quadratic Maximum Social Cost function (over Maximum Social Cost), far more involved estimations have been required in the present proof. Counterexamples to the FMNE *Conjecture* appeared *(i)* for the case of unrelated links in [17], and *(ii)* for the case of weighted users in [5]. In the context of selfish routing, the fully mixed NE and the FMNE *Conjecture* have attracted a lot of interest and

Table 1. The status of the studied facets of the FMNE *Conjecture*. A symbol $\surd$ (resp., $\times$) in the third column indicates that the FMNE *Conjecture* has been proven (resp., refuted) for the corresponding case. A number ρ in the third column indicates that an *approximate* version of the FMNE *Conjecture* has been shown: the Social Cost of an arbitrary NE is at most ρ times the one of the fully mixed. The symbol h denotes the factor by which the largest weight deviates from the average weight (in the case of weighted users).

Model assumptions	Social Cost	FMNE *Conjecture*?	Reference
$n = 2$, weighted users & identical links	MSC	$\surd$	[6]
unweighted users & related links	MSC	49.02	[6]
weighted users & identical links	MSC	$2h(1+\varepsilon)$	[9]
$n = 2$, unweighted users & related links	MSC	$\surd$	[17]
$m = 2$, unweighted users & identical links	MSC	$\surd$	[17]
$m = 2, n = 2$ & unrelated links	MSC	$\surd$	[17]
$m = 2, n = 3$ & unrelated links	MSC	$\times$	[17]
unweighted users & identical links	QSC	$\surd$	[16]
unweighted users & links with (identical) non-constant and convex latency functions	Σ_{IC}SC	$\surd$	[9]
unweighted users & identical links	PSC	$\surd$	[10]
weighted users & player-specific links	Σ_{IC}SC	$\surd$	[12]
weighted users & player-specific links	M_{IC}SC	$\surd$	[12]
weighted users & identical links	MSC	$\times$	[5]
weighted users with types & identical links	Σ_{IC}SC	$\surd$	[10]
weighted users with types & identical links	M_{IC}SC	$\surd$	[10]

attention; they both have been studied extensively in the last few years for a wide variety of theoretical models of selfish routing and Social Cost measures - see, e.g., [2,4,9,10,12,16,18].

The status of the studied facets of the FMNE *Conjecture* is summarized in Table 1. In the case of related links, latency is a linear function of congestion on a link; in the (special) case of identical links, the linear function is identity, while in the (more general) case of *player-specific links*, the linear function is specific to each player. In the (even more general) case of *unrelated links*, there is an additive contribution to latency on a link, which is both player-specific and *link-specific*. The *Quadratic Social Cost* [16], denoted as QSC, is the (expectation of the) sum of the squares of the latencies; more generally, the *Polynomial Social Cost*, denoted as PSC, is the (expectation of the) sum of polynomial functions of the latencies. The *Player-Average Social Cost* (considered in [9,12] and denoted as $\Sigma_{\mathsf{IC}}\mathsf{SC}$) is the sum of Individual Costs of the players; the *Player-Maximum Social Cost* (considered in [9,10] and denoted as $\mathsf{M}_{\mathsf{IC}}\mathsf{SC}$) is the maximum Individual Cost of a player.

2 Mathematical Tools

Notation. For any integer $n \geq 2$, denote $[n] = \{1, 2, \ldots, n\}$, $[n]_0 = [n] \cup \{0\}$. Denote $\mathbb{N}$ the set of integers $n \geq 1$, e the base of the natural logarithm. For a random variable X following the distribution $\mathbb{P}$, denote as $\mathbb{E}_{\mathbb{P}}(X)$ the *expectation* of X; $X \sim \mathbb{P}$ denotes that X follows the distribution $\mathbb{P}$. For an integer n, the predicates $\mathsf{Even}(n)$ and $\mathsf{Odd}(n)$ will be 1 when n is even and odd, respectively, and 0 otherwise.

Two Combinatorial Facts. The first fact is an extension of *Stirling's* approximation $n! \approx \sqrt{2\pi} n^{n+\frac{1}{2}} e^{-n}$ to $n!$. The extension yields a double inequality for $n!$ (cf. [3, Chapter 2, Section 9]).

Lemma 1. $\sqrt{2\pi} n^{n+\frac{1}{2}}\, e^{-n+\frac{1}{12n+1}} \leq n! \leq \sqrt{2\pi} n^{n+\frac{1}{2}} e^{-n+\frac{1}{12n}}$ *for all* $n \in \mathbb{N}$.

Applying Lemma 1 twice in fractional expansions of binomial coefficients yields:

Lemma 2. $n\sqrt{\frac{n}{2\pi}} e^{\frac{1}{12n+1}-\frac{1}{3n}} \leq \frac{n^2}{2^{n+1}}\binom{n}{\frac{n}{2}} \leq n\sqrt{\frac{n}{6}}$ *for all* $n \in \mathbb{N}$.

Lemma 3. $\sqrt{\frac{n}{2\pi}} e^{\frac{1}{12n+1}-\frac{1}{3n-3}} \leq \frac{n!}{2^n((\frac{n-1}{2})!)^2} \leq \left(\frac{n}{n-1}\right)^n \sqrt{\frac{n}{6}}$ *for all* $n \in \mathbb{N}$.

The second fact is a maximization property of the *Bernstein basis polynomial of order* k *and degree* n $\mathsf{b}_{k,n}(x) = \binom{n}{k} x^k (1-x)^{n-k}$, which forms a basis of the vector space of polynomials of degree n [1].

Lemma 4. $\max_{x \in [0,1]} b_{k,n}(x) = \binom{n}{k} k^k n^{-n} (n-k)^{n-k}$, *occurring at* $x = \frac{k}{n}$ *for all* $k \in [n]_0$.

Generalized Medians of the Binomial Distribution. Consider a sequence of N *Bernoulli* trials, each succeeding with probability p. The number of successes out of these N trials follows the *binomial distribution*; that is, the probability of obtaining at most $k \leq N$ successes is $\Sigma_{\ell=0}^{k}\binom{N}{\ell}p^{\ell}(1-p)^{N-\ell}$. Define $\mathsf{B}_{N,k}(p) : [0,1] \rightarrow \mathbb{R}$ with $\mathsf{B}_{N,k}(p) = \Sigma_{\ell=0}^{k}\binom{N}{\ell}p^{\ell}(1-p)^{N-\ell}$ to be the *binomial function.* Clearly, $\mathsf{B}_{N,k}(p)$ is strictly decreasing in (and continuous with) p, with $\mathsf{B}_{N,k}(0) = 1$ and $\mathsf{B}_{N,k}(1) = 0$. By continuity, it follows that $\mathsf{B}_{N,k}$ attains all intermediate values between 0 and 1. For any $\alpha \in [0,1]$, define the ***α-median*** of the binomial distribution, denoted as $\mathsf{M}_{N,p}(\alpha)$ with $\mathsf{M}_{N,p}(\alpha) = \min\{k \in [0,N] \mid \mathsf{B}_{N,k}(p) \geq \alpha\}$; intuitively, the α-median of the binomial distribution is the *least* integer k such that the probability of obtaining at most k successes is at least α. Clearly, $\mathsf{B}_{N,k}(p) < \alpha$ for all indices $k < \mathsf{M}_{N,p}(\alpha)$. This definition of α-median generalizes the classical definition of median of the binomial distribution (which is the $\frac{1}{2}$-median). We will use one known fact about medians [11, Theorem 2.3]:

Lemma 5. $\mathsf{M}_{N,\frac{1}{2}}\left(\frac{1}{2}\right) = \left\lfloor \frac{N}{2} \right\rfloor$ *for* $p < \frac{1}{2}$, $\mathsf{M}_{N,p}\left(\frac{1}{2}\right) \geq (N+1)p - 1$.

Furthermore, we establish in this work some **new** bounds on generalized medians, which shall be employed in some later proofs:

Lemma 6 (Generalized Medians). *For any* $\epsilon > 0$, *the following bounds hold on generalized medians of the binomial distribution, where* $p = \frac{1}{2} - \frac{r}{2(n-r-1)}$:

(1) $\mathsf{M}_{n-r-2,p}\left(\frac{1}{2}+\epsilon\right) > \left\lceil \frac{n-3}{2} \right\rceil - r - 1$, *where* $1 \leq r \leq \left\lfloor \frac{n-3}{2} \right\rfloor - 4$.
(2) $\mathsf{M}_{n-r-2,p}\left(\frac{3}{7}+\epsilon\right) > \left\lceil \frac{n-3}{2} \right\rceil - r - 1$, *where* $n \geq 134$ *is even and* $r = \left\lfloor \frac{n-3}{2} \right\rfloor - 3$.
(3) $\mathsf{M}_{n-r-2,p}\left(\frac{2}{5}+\epsilon\right) > \left\lceil \frac{n-3}{2} \right\rceil - r - 1$, *where* $n \geq 134$ *is even and* $r = \left\lfloor \frac{n-3}{2} \right\rfloor - 2$.
(4) $\mathsf{M}_{n-r-2,p}\left(\frac{1}{3}+\epsilon\right) > \left\lceil \frac{n-3}{2} \right\rceil - r - 1$, *where* $n \geq 134$ *is even and* $r = \left\lfloor \frac{n-3}{2} \right\rfloor - 1$.
(5) $\mathsf{M}_{n-r-2,p}\left(\frac{1}{4}+\epsilon\right) > \left\lceil \frac{n-3}{2} \right\rceil - r - 1$, *where* $n \geq 134$ *is even and* $r = \left\lfloor \frac{n-3}{2} \right\rfloor$.
(6) $\mathsf{M}_{n-r-2,p}\left(\frac{3}{11}+\epsilon\right) > \left\lceil \frac{n-3}{2} \right\rceil - r - 1$, *where* $n \geq 135$ *is odd and* $r = \left\lfloor \frac{n-3}{2} \right\rfloor - 3$.
(7) $\mathsf{M}_{n-r-2,p}\left(\frac{2}{9}+\epsilon\right) > \left\lceil \frac{n-3}{2} \right\rceil - r - 1$, *where* $n \geq 135$ *is odd and* $r = \left\lfloor \frac{n-3}{2} \right\rfloor - 2$.
(8) $\mathsf{M}_{n-r-2,p}\left(\frac{1}{7}+\epsilon\right) > \left\lceil \frac{n-3}{2} \right\rceil - r - 1$, *where* $n \geq 135$ *is odd and* $r = \left\lfloor \frac{n-3}{2} \right\rfloor - 1$
(9) $\mathsf{M}_{n-r-2,p}(\epsilon) > \left\lceil \frac{n-3}{2} \right\rceil - r - 1$, *where* $n \geq 135$ *is odd and* $r = \left\lfloor \frac{n-3}{2} \right\rfloor$.

3 Framework and Preliminaries

Our definitions are based on (and depart from) the standard ones for the KP model; see, e.g., [17, Section 2].

General. We consider a ***network*** consisting of **two** parallel *links* $1, 2$ from a ***source*** to a ***destination*** node. Each of $n \geq 2$ *users* $1, 2, \ldots, n$ wishes to route one unit of traffic from source to destination.

A ***pure strategy*** s_i for user $i \in [n]$ is some specific link; a ***mixed strategy*** σ_i is a probability distribution over pure strategies— so, σ_i is a probability distribution over links. The ***support*** of user i in her mixed strategy σ_i, denoted as $\mathsf{support}(\sigma_i)$, is the set of pure strategies to which i assigns strictly positive probability. A ***pure profile*** is a vector $\mathbf{s} = \langle s_1, \ldots, s_n \rangle$ of pure strategies, one for each user; a ***mixed profile*** is a vector $\boldsymbol{\sigma} = \langle \sigma_1, \ldots, \sigma_n \rangle$ of mixed strategies, one for each user. The mixed profile $\boldsymbol{\sigma}$ is ***fully mixed*** if for each user $i \in [n]$ and link $j \in [2]$, $\sigma_i(j) > 0$. Note that a mixed profile $\boldsymbol{\sigma}$ induces a (product) probability measure $\mathbb{P}_{\boldsymbol{\sigma}}$ on the space of pure profiles. A user i is pure in the mixed profile $\boldsymbol{\sigma}$ if $|\mathsf{support}(\sigma_i)| = 1$; so, a pure profile is the degenerate of a mixed profile where all users are pure. A user i is fully mixed in the mixed profile $\boldsymbol{\sigma}$ if $|\mathsf{support}(\sigma_i)| = 2$; so, a fully mixed profile is the special case of a mixed profile where all users are fully mixed.

Cost measures and Nash equilibria. The ***congestion*** on the link ℓ in the pure profile $\mathbf{s}$, denoted as $\mathsf{c}(\ell, \mathbf{s})$, is the number of users choosing link ℓ in $\mathbf{s}$; so, $\mathsf{c}(\ell, \mathbf{s}) = |\{i \in [n] : s_i = \ell\}|$. The ***Individual Cost*** of user i in the profile $\mathbf{s}$, denoted as $\mathsf{IC}_i(\mathbf{s})$, is the congestion on her chosen link; so, $\mathsf{IC}_i(\mathbf{s}) = \mathsf{c}(s_i, \mathbf{s})$. The ***expected congestion*** on the link ℓ in the mixed profile $\boldsymbol{\sigma}$, denoted as $\mathsf{c}(\ell, \boldsymbol{\sigma})$, is the expectation (according to $\boldsymbol{\sigma}$) of the congestion on link ℓ; so, $\mathsf{c}(\ell, \boldsymbol{\sigma}) = \mathbb{E}_{\mathbf{s} \sim \mathbb{P}_{\boldsymbol{\sigma}}}(\mathsf{c}(\ell, \mathbf{s}))$. The ***Expected Individual Cost*** of user i in the mixed profile $\boldsymbol{\sigma}$, denoted as $\mathsf{IC}_i(\boldsymbol{\sigma})$, is the expectation (according to $\boldsymbol{\sigma}$) of her Individual Cost; so, $\mathsf{IC}_i(\boldsymbol{\sigma}) = \mathbb{E}_{\mathbf{s} \sim \mathbb{P}_{\boldsymbol{\sigma}}}(\mathsf{IC}_i(\mathbf{s}))$.

The ***Maximum Social Cost*** of the mixed profile $\boldsymbol{\sigma}$, denoted as $\mathsf{MSC}(\boldsymbol{\sigma})$, is the expectation of the maximum congestion: $\mathsf{MSC}(\boldsymbol{\sigma}) = \mathbb{E}_{\mathbf{s} \sim \mathbb{P}_{\boldsymbol{\sigma}}} \left(\max_{\ell \in [2]} \mathsf{c}(\ell, \mathbf{s})\right)$. The ***Quadratic Maximum Social Cost*** of the mixed profile $\boldsymbol{\sigma}$, denoted as $\mathsf{QMSC}(\boldsymbol{\sigma})$, is the expectation of the square of the maximum congestion; so,

$$\begin{aligned}\mathsf{QMSC}(\boldsymbol{\sigma}) &= \mathbb{E}_{\mathbf{s} \sim \mathbb{P}_{\boldsymbol{\sigma}}} \left(\left(\max_{\ell \in [2]} \mathsf{c}(\ell, \mathbf{s}) \right)^2 \right) = \sum_{\mathbf{s} \in \mathcal{S}} \mathbb{P}_{\boldsymbol{\sigma}}(\mathbf{s}). \left(\max_{\ell \in [2]} \mathsf{c}(\ell, \mathbf{s}) \right)^2 \\ &= \sum_{\mathbf{s} \in \mathcal{S}} \left(\prod_{k \in [n]} \sigma_k(s_k) \right) . \left(\max_{\ell \in [2]} \mathsf{c}(\ell, \mathbf{s}) \right)^2 .\end{aligned}$$

The mixed profile $\boldsymbol{\sigma}$ is a ***(mixed)*** NE [20,21] if for each user $i \in [n]$, for each mixed strategy σ'_i of player i, $\mathsf{IC}_i(\boldsymbol{\sigma}) \leq \mathsf{IC}_i(\boldsymbol{\sigma}_{-i} \diamond \sigma'_i)$; so, player i has no incentive to unilaterally change her mixed strategy. (Note that $\boldsymbol{\sigma}_{-i} \diamond \sigma'_i$ is the mixed profile obtained by substituting the mixed strategy σ_i of player i in $\boldsymbol{\sigma}$ with the mixed strategy $\sigma_{i'}$.)

The fully mixed Nash equilibrium. We are especially interested in the fully mixed NE $\boldsymbol{\phi}$ which is known to exist uniquely in the setting we consider [19]; it is also known that for each pair of user $i \in [n]$ and a link $\ell \in [2]$, $\phi_i(\ell) = \frac{1}{2}$, so that all 2^n pure profiles are equiprobable, each occurring with probability $\frac{1}{2^n}$ [19, Lemma 15]. The Maximum Social Cost of $\boldsymbol{\phi}$ is given by $\mathsf{MSC}(\boldsymbol{\phi}) = \frac{n}{2} + \frac{n}{2^n} \binom{n-1}{\lceil \frac{n}{2} \rceil - 1}$ [17]. We now calculate the Quadratic Maximum Social Cost of the fully mixed NE $\boldsymbol{\phi}$.

Lemma 7. $\mathsf{QMSC}(\boldsymbol{\phi}) = \frac{n}{4} + \frac{n^2}{4} + \frac{n^2}{2^n} \binom{n-1}{\lceil \frac{n}{2} \rceil - 1}$.

The arbitrary Nash equilibrium. Fix now an arbitrary NE $\boldsymbol{\sigma}$. It is known that $\mathsf{MSC}(\boldsymbol{\phi}) \geq \mathsf{MSC}(\boldsymbol{\sigma})$ [17] (for the particular case of unweighted users and two identical links). We consider three sets:

- The set $\mathcal{U}_1 = \{i : \mathsf{support}(\sigma_i) = \{1\}\}$ of (pure) users choosing link 1.
- The set $\mathcal{U}_2 = \{i : \mathsf{support}(\sigma_i) = \{2\}\}$ of (pure) users choosing link 2.

– The set $\mathcal{U}_{12} = \{i : \mathsf{support}(\sigma_i) = \{1,2\}\}$ of (fully) mixed users choosing either link 1 or link 2.

Denote $u = \min\{|\mathcal{U}_1|, |\mathcal{U}_2|\}$. So, there exist $2u$ (pure) users of which u choose link 1 and the other u choose link 2 with probability 1. Denote $\widehat{\boldsymbol{\sigma}}$ the mixed profile derived from $\boldsymbol{\sigma}$ by eliminating those $2u$ users; note that $\widehat{\boldsymbol{\sigma}}$ is a (mixed) NE. Also, denote as $\widehat{\boldsymbol{\phi}}$ the fully mixed NE with $n-2u$ users. Note that $\widehat{\boldsymbol{\sigma}}$ has simpler form that $\boldsymbol{\sigma}$. Hence, it would be more convenient to compare $\mathsf{QMSC}(\widehat{\boldsymbol{\phi}})$ and $\mathsf{QMSC}(\widehat{\boldsymbol{\sigma}})$ (instead of comparing $\mathsf{QMSC}(\boldsymbol{\phi})$ and $\mathsf{QMSC}(\boldsymbol{\sigma})$). To do so, we need to prove a relation between $\mathsf{QMSC}(\widehat{\boldsymbol{\sigma}})$ and $\mathsf{QMSC}(\boldsymbol{\sigma})$, and another relation between $\mathsf{QMSC}(\widehat{\boldsymbol{\phi}})$ and $\mathsf{QMSC}(\boldsymbol{\phi})$.We first prove a relation between the Quadratic Maximum Social Costs of $\boldsymbol{\sigma}$ and $\widehat{\boldsymbol{\sigma}}$. Note that

$$\begin{aligned}\mathsf{QMSC}(\widehat{\boldsymbol{\sigma}}) &= \mathbb{E}_{\mathbb{P}_{\boldsymbol{\sigma}}}\left((\max\{\mathsf{c}(1,\boldsymbol{\sigma}),\mathsf{c}(2,\boldsymbol{\sigma})\}-u)^2\right)\\ &= \mathbb{E}_{\mathbb{P}_{\boldsymbol{\sigma}}}\left((\max\{\mathsf{c}(1,\boldsymbol{\sigma}),\mathsf{c}(2,\boldsymbol{\sigma})\})^2 - 2u\max\{\mathsf{c}(1,\boldsymbol{\sigma}),\mathsf{c}(2,\boldsymbol{\sigma})\}+u^2\right)\\ &= \mathbb{E}_{\mathbb{P}_{\boldsymbol{\sigma}}}\left((\max\{\mathsf{c}(1,\boldsymbol{\sigma}),\mathsf{c}(2,\boldsymbol{\sigma})\})^2\right) - 2u\mathbb{E}_{\mathbb{P}_{\boldsymbol{\sigma}}}\left(\max\{\mathsf{c}(1,\boldsymbol{\sigma}),\mathsf{c}(2,\boldsymbol{\sigma})\}\right)+u^2\\ &= \mathsf{QMSC}(\boldsymbol{\sigma}) - 2u\mathsf{MSC}(\boldsymbol{\sigma}) + u^2, \text{ hence it follows:}\end{aligned}$$

Lemma 8. $\mathsf{QMSC}(\widehat{\boldsymbol{\sigma}}) = \mathsf{QMSC}(\boldsymbol{\sigma}) - 2u\mathsf{MSC}(\boldsymbol{\sigma}) + u^2$.

We continue to compare the Quadratic Maximum Social Costs of $\boldsymbol{\phi}$ and $\widehat{\boldsymbol{\phi}}$. Lemma 7 implies that

$$\begin{aligned}&\mathsf{QMSC}(\boldsymbol{\phi}) - \mathsf{QMSC}(\widehat{\boldsymbol{\phi}})\\ &= \tfrac{n}{4} + \tfrac{n^2}{4} + \tfrac{n^2}{2^n}\binom{n-1}{\lceil\frac{n}{2}\rceil-1} - \tfrac{n-2u}{4} - \tfrac{(n-2u)^2}{4} - \tfrac{(n-2u)^2}{2^{n-2u}}\binom{n-2u-1}{\lceil\frac{n-2u}{2}\rceil-1}\\ &= -u\left(u-n-\tfrac{1}{2}\right) + \tfrac{n^2}{2^n}\binom{n-1}{\lceil\frac{n}{2}\rceil-1} - \tfrac{(n-2u)^2}{2^{n-2u}}\binom{n-2u-1}{\lceil\frac{n-2u}{2}\rceil-1}\\ &= -\mathsf{QMSC}(\widehat{\boldsymbol{\sigma}}) + \mathsf{QMSC}(\boldsymbol{\sigma}) - 2u\mathsf{MSC}(\boldsymbol{\sigma})\\ &\quad +u\left(n+\tfrac{1}{2}\right) + \tfrac{n^2}{2^n}\binom{n-1}{\lceil\frac{n}{2}\rceil-1} - \tfrac{(n-2u)^2}{2^{n-2u}}\binom{n-2u-1}{\lceil\frac{n-2u}{2}\rceil-1}.\end{aligned}$$

It follows that

$$\begin{aligned}&\mathsf{QMSC}(\boldsymbol{\phi}) - \mathsf{QMSC}(\boldsymbol{\sigma}) - (\mathsf{QMSC}(\widehat{\boldsymbol{\phi}}) - \mathsf{QMSC}(\widehat{\boldsymbol{\sigma}}))\\ &= -2u\mathsf{MSC}(\boldsymbol{\sigma}) + u\left(n+\tfrac{1}{2}\right) + \tfrac{n^2}{2^n}\binom{n-1}{\lceil\frac{n}{2}\rceil-1} - \tfrac{(n-2u)^2}{2^{n-2u}}\binom{n-2u-1}{\lceil\frac{n-2u}{2}\rceil-1}\\ &\geq -2u\mathsf{MSC}(\boldsymbol{\phi}) + u\left(n+\tfrac{1}{2}\right) + \tfrac{n^2}{2^n}\binom{n-1}{\lceil\frac{n}{2}\rceil-1} - \tfrac{(n-2u)^2}{2^{n-2u}}\binom{n-2u-1}{\lceil\frac{n-2u}{2}\rceil-1}\\ &= -2u\left(\tfrac{n}{2} + \tfrac{n}{2^n}\binom{n-1}{\lceil\frac{n}{2}\rceil-1}\right) + u\left(n+\tfrac{1}{2}\right) + \tfrac{n^2}{2^n}\binom{n-1}{\lceil\frac{n}{2}\rceil-1} - \tfrac{(n-2u)^2}{2^{n-2u}}\binom{n-2u-1}{\lceil\frac{n-2u}{2}\rceil-1}\\ &= \tfrac{u}{2} - 2u\tfrac{n}{2^n}\binom{n-1}{\lceil\frac{n}{2}\rceil-1} + \tfrac{n^2}{2^n}\binom{n-1}{\lceil\frac{n}{2}\rceil-1} - \tfrac{(n-2u)^2}{2^{n-2u}}\binom{n-2u-1}{\lceil\frac{n-2u}{2}\rceil-1}.\end{aligned}$$

We now prove a technical claim:

Lemma 9. *For all pairs of integers n and u such that $n \geq 2u$,*

$$-2u\tfrac{n}{2^n}\binom{n-1}{\lceil\frac{n}{2}\rceil-1} + \tfrac{n^2}{2^n}\binom{n-1}{\lceil\frac{n}{2}\rceil-1} - \tfrac{(n-2u)^2}{2^{n-2u}}\binom{n-2u-1}{\lceil\frac{n-2u}{2}\rceil-1} \geq 0.$$

Lemma 9 implies that (to prove that $\mathsf{QMSC}(\boldsymbol{\phi}) \geq \mathsf{QMSC}(\boldsymbol{\sigma})$) it suffices to prove that $\mathsf{QMSC}(\widehat{\boldsymbol{\phi}}) \geq \mathsf{QMSC}(\widehat{\boldsymbol{\sigma}})$. The rest of the paper is devoted to proving this inequality. For notational convenience, rename now the variables so that both $\widehat{\boldsymbol{\sigma}}$ and $\widehat{\boldsymbol{\phi}}$ henceforth refer to an instance with n users. All n users are fully mixed in $\widehat{\boldsymbol{\phi}}$; assume that in $\widehat{\boldsymbol{\sigma}}$, $r \geq 1$ (pure) users choose link 1 with prob. 1 and $n-r$ (mixed) users choose both links with probability > 0. Lücking *et al.* [17] proved:

Lemma 10. *For the* NE $\widehat{\boldsymbol{\sigma}}$, *for each mixed user* $i \in [n]$, $\sigma_i(1) = \frac{1}{2} - \frac{r}{2(n-r-1)}$. *Furthermore,* $r \leq \left\lfloor \frac{n-3}{2} \right\rfloor$. *(Henceforth, we shall denote, for each user* $i \in [n]$, $p = \sigma_i(1)$ *and* $q = \sigma_i(2)$, *where* $p + q = 1$.*)*

We now calculate $\mathsf{QMSC}(\widehat{\boldsymbol{\sigma}})$:

Lemma 11. $\mathsf{QMSC}(\widehat{\boldsymbol{\sigma}}) = \mathsf{Even}(n) \cdot \frac{n^2}{4} \binom{n-r}{\frac{n}{2}-r} p^{\frac{n}{2}-r} q^{\frac{n}{2}}$
$+ \sum_{i=\lfloor \frac{n}{2} \rfloor+1}^{n} i^2 \binom{n-r}{i-r} p^{i-r} q^{n-i} + \sum_{i=\lfloor \frac{n}{2} \rfloor+1}^{n-r} i^2 \binom{n-r}{i} p^{n-r-i} q^i$.

The next technical claim expresses $\mathsf{QMSC}(\widehat{\boldsymbol{\sigma}})$ in a different way by adding and subtracting terms.

Lemma 12. $\mathsf{QMSC}(\widehat{\boldsymbol{\sigma}}) = A + B - C + \mathsf{Even}(n) \cdot \frac{n^2}{4} \binom{n-r}{\frac{n}{2}-r} p^{\frac{n}{2}-r} q^{\frac{n}{2}}$, where :

$$A = \sum_{i=\lceil \frac{n+1}{2} \rceil}^{n} (i-r) \binom{n-r}{i-r} p^{i-r-1} q^{n+1-i} + \sum_{i=\lceil \frac{n+1}{2} \rceil}^{n-r} (n-r) \binom{n-r-1}{i-1} p^{n-r-i} q^i$$

$$B = \sum_{i=\lceil \frac{n+1}{2} \rceil}^{n} (n-r)(i-r-1) \binom{n-r-1}{i-r-1} p^{i-r-2} q^{n+2-i}$$
$$+ \sum_{i=\lceil \frac{n+1}{2} \rceil}^{n-r} (n-r)(n-r-1) \binom{n-r-2}{i-2} p^{n-r-i} q^i$$

$$C = \sum_{i=\lceil \frac{n+1}{2} \rceil}^{n} \binom{n-r}{i-r} \Big((i-r) p^{i-r-1} q^{n+1-i} + (i-r)^2 p^{i-r-2} q^{n+2-i}$$
$$-(i-r) p^{i-r-2} q^{n+2-i} - i^2 p^{i-r} q^{n-i} \Big).$$

We calculate that $A = q(n-r) + \mathsf{Odd}(n) \cdot q(n-r) \binom{n-r-1}{\frac{n-1}{2}-r} p^{\frac{n-1}{2}-r} q^{\frac{n-1}{2}}$,

$$B = q^2(n-r)(n-r-1)$$
$$\cdot \left(1 + \binom{n-r-2}{\lceil \frac{n-2}{2} \rceil - r} p^{\lceil \frac{n-2}{2} \rceil - r} q^{\lfloor \frac{n-2}{2} \rfloor} + \mathsf{Odd}(n) \cdot \binom{n-r-2}{\frac{n-3}{2}-r} p^{\frac{n-3}{2}-r} q^{\frac{n-1}{2}} \right),$$

$$C = (n-r) \Bigg(\big((pq - p^2) + q(q-p) \big) \sum_{i=\lceil \frac{n+3}{2} \rceil - r}^{n-r} \binom{n-r-2}{i-2} p^{i-2} q^{n-r-i}$$
$$+ (q^2 - p^2)(n-r-1) \sum_{i=\lceil \frac{n+1}{2} \rceil - r}^{n-r} \binom{n-r-2}{i-2} p^{i-2} q^{n-r-i}$$
$$+ (pq - p^2) \binom{n-r-2}{\lceil \frac{n-3}{2} \rceil - r} p^{\lceil \frac{n-3}{2} \rceil - r} q^{\lfloor \frac{n-1}{2} \rfloor} \Bigg)$$
$$> (n-r) \Bigg((q^2 - p^2)(n-r-1) \sum_{i=\lceil \frac{n+1}{2} \rceil - r}^{n-r} \binom{n-r-2}{i-2} p^{i-2} q^{n-r-i} +$$
$$(pq - p^2) \binom{n-r-2}{\lceil \frac{n-3}{2} \rceil - r} p^{\lceil \frac{n-3}{2} \rceil - r} q^{\lfloor \frac{n-1}{2} \rfloor} \Bigg).$$

4 The FMNE Conjecture Is Valid

The proof will use some estimations and technical claims which have been deferred to Sections 5 and 6, respectively. We establish:

Theorem 1. *For the fully mixed* NE $\widehat{\boldsymbol{\phi}}$ *and the* NE$\widehat{\boldsymbol{\sigma}}$, $\mathsf{QMSC}(\widehat{\boldsymbol{\phi}}) \geq \mathsf{QMSC}(\widehat{\boldsymbol{\sigma}})$.

Proof. Assume that $n \geq 134$. (For smaller n, the claim is verified directly.) Lemmas 7 and 12 imply that

$$\mathsf{QMSC}(\widehat{\boldsymbol{\phi}}) - \mathsf{QMSC}(\widehat{\boldsymbol{\sigma}}) \geq \frac{n}{4} + \frac{n^2}{4} + \frac{n^2}{2^n} \binom{n-1}{\lceil \frac{n}{2} \rceil - 1} - q(n-r) - q^2(n-r)(n-r-1)$$
$$+ (q^2 - p^2)(n-r)(n-r-1)\mathsf{Q} + \mathsf{D}, \quad \text{where}$$

$$\begin{aligned}\mathsf{D} = & -q^2(n-r)(n-r-1)\binom{n-r-2}{\lceil\frac{n-2}{2}\rceil-r}p^{\lceil\frac{n-2}{2}\rceil-r}\, q^{\lfloor\frac{n-2}{2}\rfloor} \\ & +(pq-p^2)(n-r)\binom{n-r-2}{\lceil\frac{n-3}{2}\rceil-r}p^{\lceil\frac{n-3}{2}\rceil-r}\, q^{\lfloor\frac{n-1}{2}\rfloor} \\ & -\mathsf{Odd}(n)\cdot q(n-r)\left(\binom{n-r-1}{\frac{n-1}{2}-r}p^{\frac{n-1}{2}-r}q^{\frac{n-1}{2}} + q(n-r-1)\binom{n-r-2}{\frac{n-3}{2}-r}p^{\frac{n-3}{2}-r}q^{\frac{n-1}{2}}\right) \\ & -\mathsf{Even}(n)\cdot\frac{n^2}{4}\binom{n-r}{\frac{n}{2}-r}p^{\frac{n}{2}-r}q^{\frac{n}{2}} \quad \text{and}\end{aligned}$$

$$\begin{aligned}\mathsf{Q} & = \sum_{i=\lceil\frac{n+1}{2}\rceil-r}^{n-r}\binom{n-r-2}{i-2}p^{i-2}q^{n-r-i} = \sum_{i=\lceil\frac{n-3}{2}\rceil-r}^{n-r-2}\binom{n-r-2}{i}p^{i}q^{n-r-2-i} \\ & = 1-\mathsf{B}_{n-r-2,\lceil\frac{n-3}{2}\rceil-r-1}(p).\end{aligned}$$

Note that Lemma 6 implies *lower* bounds on Q for various values of r, which will be used in some later proofs. We proceed by case analysis. We consider separately the two cases where n is even or odd.

Case 1: n is even By substituting p and q from Lemma 10, we get that

$$\begin{aligned}\mathsf{D} = & -\frac{(n-1)^2(n-r)}{4(n-r-1)}\binom{n-r-2}{\frac{n-2}{2}-r}p^{\frac{n-2}{2}-r}q^{\frac{n-2}{2}} \\ & +\frac{r(n-r)(n-2r-1)}{2(n-r-1)^2}\binom{n-r-2}{\frac{n-2}{2}-r}p^{\frac{n-2}{2}-r}q^{\frac{n-2}{2}} - \frac{n^2}{4}\binom{n-r}{\frac{n}{2}-r}p^{\frac{n}{2}-r}q^{\frac{n}{2}} \\ = & \binom{n-r-2}{\frac{n-2}{2}-r}p^{\frac{n-2}{2}-r}q^{\frac{n-2}{2}}\left(-\frac{(n-1)^2(n-r)}{4(n-r-1)} + \frac{r(n-r)(n-2r-1)}{2(n-r-1)^2}pq\frac{n(n-r)(n-r-1)}{n-2r}\right) \\ \geq & \binom{n-r-2}{\frac{n-2}{2}-r}p^{\frac{n-2}{2}-r}q^{\frac{n-2}{2}}\left(-\frac{(n-1)^2(n-r)}{4(n-r-1)} - \frac{n(n-1)(n-2r-1)(n-r)}{4(n-r-1)(n-2r)}\right)\end{aligned}$$

It follows that

$$\begin{aligned} & \mathsf{QMSC}(\widehat{\boldsymbol{\phi}}) - \mathsf{QMSC}(\widehat{\boldsymbol{\sigma}}) \\ \geq & \frac{n}{4}+\frac{n^2}{4}+\frac{n^2}{2^{n+1}}\binom{n}{\frac{n}{2}} - \frac{(n-1)(n-r)}{2(n-r-1)} - \frac{(n-1)^2(n-r)}{4(n-r-1)} + r(n-r)\mathsf{Q} \\ & -\binom{n-r-2}{\frac{n-2}{2}-r}p^{\frac{n-2}{2}-r}\, q^{\frac{n-2}{2}}\left(\frac{(n-1)^2(n-r)}{4(n-r-1)} + \frac{n(n-1)(n-2r-1)(n-r)}{4(n-r-1)(n-2r)}\right) \\ = & \frac{n^2}{2^{n+1}}\binom{n}{\frac{n}{2}} + r(n-r)\mathsf{Q} \\ & -\binom{n-r-2}{\frac{n-2}{2}-r}p^{\frac{n-2}{2}-r}\, q^{\frac{n-2}{2}}\left(\frac{(n-1)^2(n-r)}{4(n-r-1)} + \frac{n(n-1)(n-2r-1)(n-r)}{4(n-r-1)(n-2r)}\right) - \frac{r(n+1)}{4(n-r-1)} \\ > & \frac{n^2}{2^{n+1}}\binom{n}{\frac{n}{2}} + r(n-r)\mathsf{Q} \\ & -\binom{n-r-2}{\frac{n-2}{2}-r}p^{\frac{n-2}{2}-r}q^{\frac{n-2}{2}}\left(\frac{(n-1)^2(n-r)}{4(n-r-1)} + \frac{n(n-1)(n-r)}{4(n-r-1)}\right) - \frac{r(n+1)}{4(n-r-1)} \\ > & \underbrace{\frac{n^2}{2^{n+1}}\binom{n}{\frac{n}{2}} - \binom{n-r-2}{\frac{n-2}{2}-r}p^{\frac{n-2}{2}-r}q^{\frac{n-2}{2}}\left(\frac{n^2(n-r)}{2(n-r-2)}\right) + r(n-r)\mathsf{Q}}_{\mathsf{G}} - \frac{r(n+1)}{4(n-r-1)}.\end{aligned}$$

We proceed by case analysis on the range of values of r. For each range, we shall use the corresponding case(s) of Lemma 6 to infer a lower bound on Q.

1. $1 \leq r \leq \lfloor (n-3)/2 \rfloor - 4$: Note that in this case, Lemma 6 (Case (1)) implies that $\mathsf{Q} \geq \frac{1}{2}$. Hence, substituting p and q from Lemma 10, we get that

$$\begin{aligned}\mathsf{G} \geq & \frac{n^2}{2^{n+1}}\binom{n}{\frac{n}{2}} - \frac{n^2(n-r)}{2(n-r-2)}\binom{n-r-2}{\frac{n-2}{2}-r}\left(\frac{n-2r-1}{2(n-r-1)}\right)^{\frac{n-2r-2}{2}}\left(\frac{n-1}{2(n-r-1)}\right)^{\frac{n-2}{2}} + \frac{r(n-r)}{2} \\ = & \frac{n^2}{2^{n+1}}\binom{n}{\frac{n}{2}}\left(1 - \frac{n}{n-r-2}\,\frac{\prod_{i=0}^{r}(n-2r+2i)}{\prod_{i=1}^{r}(n-r+i)}\,\frac{(n-1)(n-2r-1)^{1-r}}{(n-r-1)^{3-r}}\left(\frac{n-2r-1}{n-r-1}\right)^{\frac{n-4}{2}}\left(\frac{n-1}{n-r-1}\right)^{\frac{n-4}{2}}\right) \\ & + \frac{r(n-r)}{2}.\end{aligned}$$

We consider two different subcases:

$\boxed{1.1.\quad 1 \leq r \leq 4:}$ We use Lemma 2 and the estimation in Lemma 13. Clearly,

$$\begin{aligned}
\mathsf{G} &\overset{(2),(13)}{\geq} n\sqrt{\tfrac{n}{6}}\left(1 - \frac{n}{n-r-2}\frac{\prod_{i=0}^{r}(n-2r+2i)}{\prod_{i=1}^{r}(n-r+i)}\frac{(n-1)(n-2r-1)^{1-r}}{(n-r-1)^{3-r}}\left(\frac{n-2r-1}{n-r-1}\right)^{\frac{n-4}{2}}\left(\frac{n-1}{n-r-1}\right)^{\frac{n-4}{2}}\right) \\
&\quad + \frac{r(n-r)}{2} \\
&\overset{(13)}{\geq} \frac{r(n+1)}{4(n-r-1)},
\end{aligned}$$

and the claim follows.

$\boxed{1.2.\quad 5 \leq r \leq \lfloor (n-3)/2 \rfloor - 4:}$ We use Lemmas 14, 18, and 19. Clearly,

$$\begin{aligned}
\mathsf{G} &\overset{(2),(14)}{\geq} n\sqrt{\tfrac{n}{2\pi}}e^{\frac{1}{12n+1}-\frac{1}{3n}} - \frac{n^2}{2}\frac{1}{\sqrt{2\pi}}\frac{n-r}{\sqrt{(n-r-2)(\frac{n}{2}-1)(\frac{n}{2}-r-1)}} \\
&\quad e^{\frac{1}{12(n-r-2)}-\frac{1}{6n-11}-\frac{1}{6n-12r-11}} + \frac{r(n-r)}{2} \\
&= n\sqrt{\tfrac{n}{2\pi}}e^{\frac{1}{12n+1}-\frac{1}{3n}} \\
&\quad \cdot\left(1 - \sqrt{\frac{n(n-r)^2}{(n-2)(n-r-2)(n-2r-2)}}e^{\frac{1}{12(n-r-2)}-\frac{1}{6n-11}-\frac{1}{6n-12r-11}-\frac{1}{12n+1}+\frac{1}{3n}}\right) \\
&\quad + \frac{r(n-r)}{2} \\
&\overset{(18)}{\geq} n\sqrt{\tfrac{n}{2\pi}}e^{\frac{1}{12n+1}-\frac{1}{3n}}\left(1 - \sqrt{\frac{n(n-r)^2}{(n-2)(n-r-2)(n-2r-2)}}\right) + \frac{r(n-r)}{2} \\
&\geq n\sqrt{\tfrac{n}{6}}\left(1 - \sqrt{\frac{n(n-r)^2}{(n-2)(n-r-2)(n-2r-2)}}\right) + \frac{r(n-r)}{2} \\
&\geq n\sqrt{\tfrac{n}{6}}\left(1 - \sqrt{\frac{n(n-r)^2}{(n-2)(n-r-2)(n-2r-2)}}\right) + \frac{r(n+1)}{4} \qquad \left(\text{since } r \leq \left\lfloor\tfrac{n-3}{2}\right\rfloor - 4\right) \\
&\overset{(19)}{\geq} \frac{r(n+1)}{4(n-r-1)},
\end{aligned}$$

and the claim follows.

$\boxed{2.\quad \left\lfloor\frac{n-3}{2}\right\rfloor - 3 \leq r \leq \left\lfloor\frac{n-3}{2}\right\rfloor:}$ We shall use the estimation in Lemma 15 (Case (1)). (Note that this way, we are implicitly using corresponding cases of Lemma 6 to get lower bounds on Q since the proof of Lemma 15 uses such lower bounds from Lemma 6.) By substituting p and q from Lemma 10, we get that

$$\begin{aligned}
&\mathsf{QMSC}(\widehat{\boldsymbol{\phi}}) - \mathsf{QMSC}(\widehat{\boldsymbol{\sigma}}) \\
&\geq \frac{n}{4} + \frac{n^2}{4} - \frac{(n-1)(n-r)}{2(n-r-1)} - \frac{(n-1)^2(n-r)}{4(n-r-1)} + r(n-r)\mathsf{Q} \\
&\quad - \binom{n-r-2}{\frac{n-2}{2}-r}\left(\frac{n-2r-1}{2(n-r-1)}\right)^{\frac{n-2}{2}-r}\left(\frac{n-1}{2(n-r-1)}\right)^{\frac{n-2}{2}}\left(\frac{(n-1)^2(n-r)}{4(n-r-1)} + \frac{n(n-1)(n-2r-1)(n-r)}{4(n-r-1)(n-2r)}\right) \\
&= \frac{n}{4} + \frac{n^2}{4} - (n-r)\cdot \\
&\quad \left(\frac{(n-1)(n+1)}{4(n-r-1)} - r\mathsf{Q} + \frac{\prod_{i=0}^{\frac{n-2r-4}{2}}(\frac{n}{2}+i)}{(\frac{n-2}{2}-r)!}\left(\frac{n-2r-1}{2(n-r-1)}\right)^{\frac{n-2r-2}{2}}\left(\frac{n-1}{2(n-r-1)}\right)^{\frac{n}{2}}\left(\frac{n^2-n-2nr+r}{n-2r}\right)\right) \\
&\geq \frac{n^2+n}{4} - \frac{n-r}{2}\left(\frac{(n-1)(n+1)}{2(n-r-1)} - 2r\mathsf{Q} + \frac{n(n-2r-1)^{\frac{n-2}{2}-r}(n^2-n-2nr+r)}{2^{\frac{n-2r-4}{2}}(\frac{n-2}{2}-r)!2(n-r-1)(n-2r)}\left(\frac{n-1}{2(n-r-1)}\right)^{\frac{n}{2}}\right) \\
&\geq -\frac{n-r}{2}\left(\frac{n^2}{2(n-r-1)} - \frac{n^2}{2(n-r)} - 2r\mathsf{Q} + \frac{n(n-2r-1)^{\frac{n-2}{2}-r}(n^2-n-2nr+r)}{2^{\frac{n-2r-4}{2}}(\frac{n-2}{2}-r)!(n-2r)}\left(\frac{n-1}{2(n-r-1)}\right)^{\frac{n}{2}}\right) \\
&\overset{(15)}{\geq} 0,
\end{aligned}$$

and the claim follows. This completes the proof for even n.

Case 2: n is odd Due to space restrictions we skip the proof for odd n.

5 Estimations

In this section, we collect together all estimations which were used in the proof of Theorem 1. Some of these estimations refer to the probabilities p and q introduced in Lemma 10. Some other estimations refer to the quantity $\mathsf{Q} = \sum_{i=\lceil \frac{n+1}{2} \rceil - r}^{n-r} \binom{n-r-2}{i-2} p^{i-2} q^{n-r-i}$ introduced in the proof of Theorem 1.

Lemma 13. *For all integers n and r such that $1 \leq r \leq 4$,*

$$(1) \quad \frac{n}{n-r-2} \frac{\prod_{i=0}^{r}(n-2r+2i)}{\prod_{i=1}^{r}(n-r+i)} \frac{(n-2r-1)^{1-r}}{(n-r-1)^{2-r}} \left(\frac{n-2r-1}{n-r-1}\right)^{\frac{n-4}{2}} \left(\frac{n-1}{n-r-1}\right)^{\frac{n-2}{2}} \geq 1.$$

$$(2) \quad n\sqrt{\frac{n}{6}} \left(1 - \frac{n}{n-r-2} \frac{\prod_{i=0}^{r}(n-2r+2i)}{\prod_{i=1}^{r}(n-r+i)} \frac{(n-1)(n-2r-1)^{1-r}}{(n-r-1)^{3-r}} \left(\frac{n-2r-1}{n-r-1}\right)^{\frac{n-4}{2}} \left(\frac{n-1}{n-r-1}\right)^{\frac{n-4}{2}}\right)$$
$$\geq \frac{r(n+1)}{4(n-r-1)} - \frac{r(n-r)}{2}.$$

Lemma 14. *For all integers n and r such that $5 \leq r \leq \lfloor \frac{n-3}{2} \rfloor - 4$,*

$$\binom{n-r-2}{\frac{n-2}{2}-r} p^{\frac{n-2}{2}-r} q^{\frac{n-2}{2}} \leq \sqrt{\frac{n-r-2}{2\pi(\frac{n}{2}-1)(\frac{n}{2}-r-1)}} e^{\frac{1}{12(n-r-2)} - \frac{1}{6n-11} - \frac{1}{6n-12r-11}}.$$

Lemma 15. *For all integers r such that $\lfloor \frac{n-3}{2} \rfloor - 3 \leq r \leq \lfloor \frac{n-3}{2} \rfloor$:*
(1) *For all even integers $n \geq 134$,*

$$\frac{n^2}{2(n-r-1)} - \frac{n^2}{2(n-r)} - 2r\mathsf{Q} + \frac{n(n-2r-1)^{\frac{n-2}{2}-r}(n^2-n-2nr+r)}{2^{\frac{n-2r-4}{2}}(\frac{n-2}{2}-r)!(n-2r)} \left(\frac{n-1}{2(n-r-1)}\right)^{\frac{n}{2}} \leq 0.$$

(2) *For all odd integers $n \geq 135$,*

$$\frac{n}{4} + \frac{n^2}{4} - \frac{n-r}{2} \left(\frac{(n-1)(n+1)}{2(n-r-1)} - 2r\mathsf{Q} + 2(n+1)\binom{n-r-1}{\frac{n-1}{2}-r}\left(\frac{n-2r-1}{2(n-r-1)}\right)^{\frac{n-1}{2}-r}\left(\frac{n-1}{2(n-r-1)}\right)^{\frac{n+1}{2}}\right) \geq 0.$$

Lemma 16. *For all integers n and r such that $1 \leq r \leq 4$,*

$$(1) \quad \frac{\prod_{i=0}^{r}(n-2r+1+2i)}{\prod_{i=0}^{r}(n-r+i)} \frac{(n-2r-1)^{-r}}{(n-r-1)^{-r}} \left(\frac{n-2r-1}{n-r-1}\right)^{\frac{n-1}{2}} \left(\frac{n-1}{n-r-1}\right)^{\frac{n+1}{2}} \geq 1.$$

$$(2) \quad \left(\frac{n}{n-1}\right)^{n} \sqrt{\frac{n}{6}} \left(1 - \frac{\prod_{i=0}^{r}(n-2r+1+2i)}{\prod_{i=0}^{r}(n-r+i)} \frac{(n-1)(n-r-1)^{r-1}}{(n-2r-1)^{r}} \left(\frac{n-2r-1}{n-r-1}\right)^{\frac{n-1}{2}} \left(\frac{n-1}{n-r-1}\right)^{\frac{n-1}{2}}\right)$$
$$\geq \frac{r(n+1)}{4(n-r-1)(n-1)} - \frac{1}{2}.$$

Lemma 17. *For all integers n and r,*

$$\binom{n-r-1}{\frac{n-1}{2}-r} p^{\frac{n-1}{2}-r} q^{\frac{n+1}{2}} \leq \frac{n-1}{\sqrt{2\pi(n-1)(n-r-1)(n-2r-1)}} e^{\frac{1}{12(n-r-1)} - \frac{1}{6n-5} - \frac{1}{6n-12r-5}}$$

6 Technical Claims

In this section, we collect together some simple technical claims which were used in the proof of Theorem 1.

Lemma 18. *For all $n \geq 1$, $r > 0$:* $\frac{1}{12(n-r-2)} - \frac{1}{6n-11} - \frac{1}{6n-12r-11} - \frac{1}{12n+1} + \frac{1}{3n} < 0.$

Lemma 19. *For all even $n \in \mathbb{N}$, $n \geq 134$ and $r \in \mathbb{N}$ such that $5 \leq r \leq \lfloor \frac{n-3}{2} \rfloor - 4$:*

$$n\sqrt{\frac{n}{6}} \left(1 - \sqrt{\frac{n(n-r)^2}{(n-2)(n-r-2)(n-2r-2)}}\right) + \frac{r(n+1)}{4} \geq \frac{r(n+1)}{4(n-r-1)}.$$

Lemma 20. *For all* $n \geq 1$, $r > 3$: $\frac{1}{12(n-r-1)} - \frac{1}{6n-5} - \frac{1}{6n-12r-5} - \frac{1}{12n+1} + \frac{1}{3n-3} < 0$.

Lemma 21. *For all odd* $n \in \mathbb{N}$, $n \geq 135$ *and* $r \in \mathbb{N}$ *such that* $4 \leq r \leq \lfloor \frac{n-3}{2} \rfloor - 4$:
$$n\sqrt{\frac{n}{6}}\left(1 - \sqrt{\frac{(n+1)^2(n-r)^2}{n(n-1)(n-r-1)(n-2r-1)}}\right) + \frac{r(n+1)}{4} \geq \frac{r(n+1)}{4(n-r-1)}.$$

7 Conclusions

We have presented an extensive proof for the validity of the FMNE *Conjecture* for a special case of the selfish routing model of Koutsoupias and Papadimitriou [15] where users are unweighted and there are only two identical (related) links. We adopted a new, well-motivated kind of Social Cost, called Quadratic Maximum Social Cost. To carry out the proof, we developed some new estimations of (generalized) medians of the binomial distribution, which are of independent interest and value. In turn, those estimations were used as tools, together with a variety of combinatorial arguments and other analytical estimations, in the main proof.

We believe that our work contributes significantly, both conceptually and technically, to enriching our knowledge about the many facets of the FMNE *Conjecture*. Based on this improved understanding, we extend the FMNE *Conjecture* formulated and proven in this work to an *Extended* FMNE *Conjecture* for the more general case with an arbitrary number of unweighted users, an arbitrary number of identical (related) links and Social Cost as the expectation of a polynomial with non-negative coefficients of the maximum congestion on a link. Settling this Extended FMNE *Conjecture* remains a major challenge.

Acknowledgements. We would like to thank Chryssis Georgiou and Burkhard Monien for helpful discussions.

References

1. Bernstein, S.: Démonstration du Théoreme de Weierstrass Fondée sur le Calcul des Probabilities. Commun. Soc. Math. Kharkow. 2(13), 1–2 (1912)
2. Elsässer, R., Gairing, M., Lücking, T., Mavronicolas, M., Monien, B.: A Simple Graph-Theoretic Model for Selfish Restricted Scheduling. In: Deng, X., Ye, Y. (eds.) WINE 2005. LNCS, vol. 3828, pp. 195–209. Springer, Heidelberg (2005)
3. Feller, W.: An Introduction to Probability Theory and its Applications. Wiley, Chichester (1968)
4. Ferrante, A., Parente, M.: Existence of Nash Equilibria in Selfish Routing Problems. In: Kralovic, R., Sýkora, O. (eds.) SIROCCO 2004. LNCS, vol. 3104, pp. 149–160. Springer, Heidelberg (2004)
5. Fischer, S., Vöcking, B.: On the Structure and Complexity of Worst-Case Equilibria. Theoretical Computer Science 378(2), 165–174 (2007)
6. Fotakis, D., Kontogiannis, S., Koutsoupias, E., Mavronicolas, M., Spirakis, P.: The Structure and Complexity of Nash Equilibria for a Selfish Routing Game. In: Widmayer, P., Triguero, F., Morales, R., Hennessy, M., Eidenbenz, S., Conejo, R. (eds.) ICALP 2002. LNCS, vol. 2380, pp. 123–134. Springer, Heidelberg (2002)

7. Gairing, M., Lücking, T., Mavronicolas, M., Monien, B., Spirakis, P.: Structure and Complexity of Extreme Nash Equilibria. Theoretical Computer Science 343(1–2), 133–157 (2005)
8. Gairing, M., Lücking, T., Mavronicolas, M., Monien, B.: The Price of Anarchy for Polynomial Social Cost. Theoretical Computer Science 369(1–3), 116–135 (2006)
9. Gairing, M., Lücking, T., Mavronicolas, M., Monien, B., Rode, M.: Nash Equilibria in Discrete Routing Games with Convex Latency Functions. In: Díaz, J., Karhumäki, J., Lepistö, A., Sannella, D. (eds.) ICALP 2004. LNCS, vol. 3142, pp. 645–657. Springer, Heidelberg (2004)
10. Gairing, M., Monien, B., Tiemann, K.: Selfish Routing with Incomplete Information. In: Proc. of ACM-SPAA 2005, pp. 203–212 (2005)
11. Göb, R.: Bounds for Median and 50 Percentage Point of Binomial and Negative Binomial Distribution. Metrika 41(1), 43–54 (1994)
12. Georgiou, C., Pavlides, T., Philippou, A.: Uncertainty in Selfish Routing. In: CD-ROM Proc. of IEEE IPDPS 2006 (2006)
13. Goussevskala, O., Oswald, Y.A., Wattenhofer, R.: Complexity in Geometric SINR. In: Proc. of ACM-MobiHoc 2007, pp. 100–109 (2007)
14. Kaplansky, I.: A Contribution to von-Neumann's Theory of Games. Annals of Mathematics 46(3), 474–479 (1945)
15. Papadimitriou, C.H., Koutsoupias, E.: Worst-Case Equilibria. In: Meinel, C., Tison, S. (eds.) STACS 1999. LNCS, vol. 1563, pp. 404–413. Springer, Heidelberg (1999)
16. Monien, B., Mavronicolas, M., Lücking, T., Rode, M.: A New Model for Selfish Routing. In: Diekert, V., Habib, M. (eds.) STACS 2004. LNCS, vol. 2996, pp. 547–558. Springer, Heidelberg (2004)
17. Lücking, T., Mavronicolas, M., Monien, B., Rode, M., Spirakis, P., Vrto, I.: Which is the Worst-Case Nash Equilibrium? In: Rovan, B., Vojtáš, P. (eds.) MFCS 2003. LNCS, vol. 2747, pp. 551–561. Springer, Heidelberg (2003)
18. Mavronicolas, M., Panagopoulou, P., Spirakis, P.: A Cost Mechanism for Fair Pricing of Resource Usage. In: Deng, X., Ye, Y. (eds.) WINE 2005. LNCS, vol. 3828, pp. 210–224. Springer, Heidelberg (2005)
19. Mavronicolas, M., Spirakis, P.: The Price of Selfish Routing. Algorithmica 48, 91–126 (2007)
20. Nash, J.F.: Equilibrium Points in *N*-Person Games. In: Proc. of the National Academy of Sciences of the USA. vol. 36, pp. 48–49 (1950)
21. Nash, J.F.: Non-Cooperative Games. Annals of Mathematics 54(2), 286–295 (1951)
22. Uhlmann, W.: Vergleich der Hypergeometrischen mit der Binomial-Verteilung. Metrika 10, 145–158 (1966)

Sensitivity of Wardrop Equilibria*

Matthias Englert[1], Thomas Franke, and Lars Olbrich[1]

[1]Dept. of Computer Science, RWTH Aachen University, Germany
{englert,lars}@cs.rwth-aachen.de

Abstract. We study the sensitivity of equilibria in the well-known game theoretic traffic model due to Wardrop. We mostly consider single-commodity networks. Suppose, given a unit demand flow at Wardrop equilibrium, one increases the demand by ε or removes an edge carrying only an ε-fraction of flow. We study how the equilibrium responds to such an ε-change.

Our first surprising finding is that, even for linear latency functions, for every $\varepsilon > 0$, there are networks in which an ε-change causes every agent to change its path in order to recover equilibrium. Nevertheless, we can prove that, for general latency functions, the flow increase or decrease on every edge is at most ε.

Examining the latency at equilibrium, we concentrate on polynomial latency functions of degree at most p with nonnegative coefficients. We show that, even though the relative increase in the latency of an edge due to an ε-change in the demand can be unbounded, the path latency at equilibrium increases at most by a factor of $(1+\varepsilon)^p$. The increase of the *price of anarchy* is shown to be upper bounded by the same factor. Both bounds are shown to be tight.

Let us remark that all our bounds are tight. For the multi-commodity case, we present examples showing that neither the change in edge flows nor the change in the path latency can be bounded.

1 Introduction

We analyze equilibria in the *Wardrop model* [15]. In this model we are given a network with load-dependent latency functions on the edges and a set of commodities, which is defined by source-sink pairs. For each commodity some demand (traffic flow) needs to be routed from the commodity's source to its sink. A common interpretation of the Wardrop model is that flow is controlled by an infinite number of selfish agents each of which carries an infinitesimal amount of flow. Each agent aims at minimizing its path latency. An allocation, in which no agent can improve its situation by unilaterally deviating from its current path is called *Wardrop equilibrium*.

Whereas the notion of equilibrium captures stability in closed systems, traffic is typically subject to external influences. Thus, from both the practical and the theoretical perspective it is a natural question, how equilibria respond to slight modifications of either the network topology or the traffic flow.

* Supported by DFG grant WE 2842/1 and by the DFG GK/1298 "AlgoSyn".

B. Monien and U.-P. Schroeder (Eds.): SAGT 2008, LNCS 4997, pp. 158–169, 2008.

To analyze this issue, we suppose, we are given an equilibrium flow for unit demand and increase the demand by ε or remove an edge carrying only an ε-fraction of flow. How does the equilibrium responds to such an ε-change in terms of change in flow and latency?

Consider the classical network exhibiting Braess's Paradox [2]. Suppose a unit demand needs to be routed from node s to node t. At equilibrium all traffic follows the zig-zag-path. Increasing the demand by $0 < \varepsilon \leq 1$, the paths containing the dashed edges gain an ε-fraction of flow, whereas the zig-zag-path loses an ε-fraction.

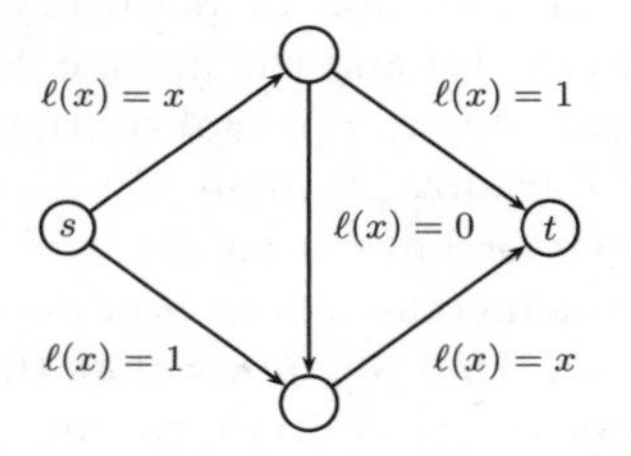

Latency functions ...

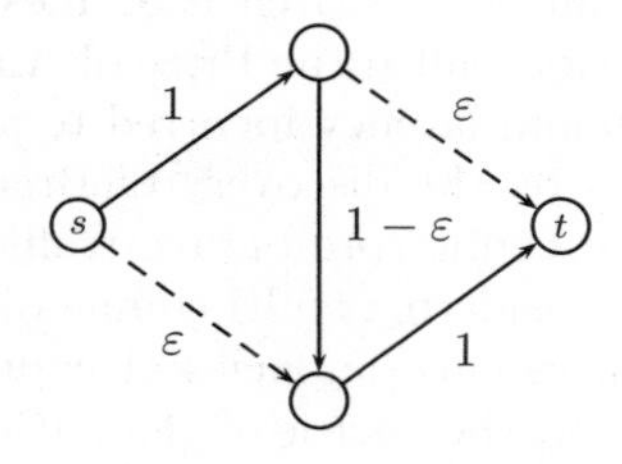

... and flow.

Thus, in general, neither path flows nor edge flows at equilibrium are monotone functions of the demand. This observation is one of the reasons why studying the effects of changes might be intriguing.

Our findings for single-commodity networks are as follows. Allowing non-decreasing, continuous latency functions, we show that for every $\varepsilon > 0$,

- there are networks, in which after an ε-change every agent is forced to change its path in order to recover equilibrium and
- the flow increase or decrease on every edge, however, is at most ε for every network.

Thus, in contrast to our surprising finding of global instability of equilibrium flow, we can prove that edge flows are locally stable.

Examining the latency at equilibrium, we concentrate on polynomial latency functions of degree at most p with nonnegative coefficients. We show that, due to an ε-change in the demand,

- the path latency at equilibrium increases at most by a factor of $(1+\varepsilon)^p$ (even though the relative increase in the latency of an edge can be unbounded).

This result yields the same bound on the increase in the *Price of Anarchy*, as well.

All presented bounds are best possible.

For the multi-commodity case, we present examples for every $\varepsilon > 0$, showing that neither the change in edge flows nor the increase in the path latency can be bounded. This holds already for networks equipped with linear latency functions.

1.1 Related Work

The game theoretic traffic model considered in this paper was introduced by Wardrop [15]. Beckmann, McGuire, and Winston [1] observe that such an equilibrium flow is an optimal solution to a related convex program. They give existence and uniqueness results for traffic equilibria (see also [4] and [11]). Dafermos and Sparrow [4] show that the equilibrium state can be computed efficiently under some assumptions on the latency functions and many subsequent papers gave increasingly efficient methods for computing equilibria.

Another line of research examines the degradation of performance due to selfish behavior, called the Price of Anarchy [8, 11] and the inverse, the increase of the maximum latency incurred to an agent due to optimal routing [12].

Motivated by the discovery of Braess's Paradox [2] many similarly counter-intuitive and counterproductive traffic behavior have been discovered. Fisk [5] shows that considering multi-commodities the increase of one flow demand might decrease others path latencies at equilibrium. Hall [6] shows that the vector of path flows and the vector of the path latencies are continuous functions of the input demand. Furthermore, he proves that for single-commodity networks the path latency at equilibrium is a monotone function of the input demand. Dafermos and Nagurney [3] show that equilibrium flow pattern depend continuously upon the demands and (even non-separable) latency functions. More recently, Patriksson [9] gave a characterization for the existence of a directional derivative of the equilibrium solution. In [7] Josefsson and Patriksson show that while equilibrium edge costs are directionally differentiable, this does not hold for edge flows itself.

1.2 Outline

In Section 2, we introduce Wardrop's traffic model. In Section 3, we establish global instability of equilibrium flows and local stability of edge flows at equilibrium for general latency functions. For polynomial latency functions with nonnegative coefficients, we give a tight upper bound on the increase of the path latency at equilibrium due to an ε-change of the demand (Section 4). Subsequently, the same bound on the increase of the Price of Anarchy is derived. In Section 5, we briefly present some negative results for the multi-commodity case.

2 Wardrop's Traffic Model

We consider Wardrop's traffic model originally introduced in [15]. We are given a directed graph $G = (V, E)$ with non-decreasing, continuous latency functions $\ell = (\ell_e)_e$ with $\ell_e : \mathbb{R}_{\geq 0} \rightarrow \mathbb{R}_{\geq 0}$. Furthermore, we are given a set of commodities $[k] = \{1, \ldots, k\}$ specified by source-sink pairs $(s_i, t_i) \in V \times V$ and flow demands d_i. The total demand is $d = \sum_{i \in [k]} d_i$. Let $\mathcal{P}_i$ denote the admissible paths of commodity i, i. e., all paths connecting s_i and t_i, and let $\mathcal{P} = \bigcup_{i \in [k]} \mathcal{P}_i$. Let $(G, (d_i), \ell)$ denote an instance of the routing problem.

A non-negative path flow vector $(f_P)_{P \in \mathcal{P}}$ is *feasible* if it satisfies the flow demands $\sum_{P \in \mathcal{P}_i} f_P = d_i$ for all $i \in [k]$. We denote the set of all feasible flow vectors by $\mathcal{F}$. A path flow vector $(f_P)_{P \in \mathcal{P}}$ induces an edge flow vector $f = (f_{e,i})_{e \in E, i \in [k]}$ with $f_{e,i} = \sum_{P \in \mathcal{P}_i : e \in P} f_P$. The total flow on edge e is $f_e = \sum_{i \in [k]} f_{e,i}$. The latency of an edge $e \in E$ is given by $\ell_e(f_e)$ and the latency of a path p is given by the sum of the edge latencies $\ell_P(f) = \sum_{e \in P} \ell_e(f_e)$. The weighted average latency of commodity $i \in [k]$ is given by $L_i(f) = \sum_{e \in E} \ell_e(f_e) \cdot f_{e,i}$. Finally, the total cost of a flow is defined as $C(f) = \sum_{P \in \mathcal{P}} \ell_P(f_P) f_P$ and can be expressed as $C(f) = \sum_{e \in E} \ell_e(f_e) f_e$. We drop the argument f whenever it is clear from the context. Whenever we consider a single-commodity network, we further drop the index i.

A flow vector is considered stable when no fraction of the flow can improve its sustained latency by moving unilaterally to another path. Such a stable state is generally known as *Nash equilibrium.* In our model a flow is stable if and only if all used paths have the same minimal latency, whereas unused paths may have larger latency. We call such a flow *Wardrop equilibrium.*

Definition 1. *A feasible flow vector f is at* Wardrop equilibrium *if for every commodity $i \in [k]$ and paths $P_1, P_2 \in \mathcal{P}_i$ with $f_{P_1} > 0$ it holds that $\ell_{P_1}(f) \leq \ell_{P_2}(f)$.*

It is well-known that Wardrop equilibria are exactly those allocations that minimize the following potential function introduced in [1]:

$$\Phi(f) = \sum_{e \in E} \int_0^{f_e} \ell_e(u) du \ .$$

The allocations in equilibrium do not only all have the same (optimal) potential but they also impose the same latencies on all edges. Thus, the path latencies $L_i = L_i(f)$ at equilibrium is uniquely determined. In this sense, the Wardrop equilibrium is essentially unique ([1], [4], [11]).

3 Sensitivity of Equilibrium Flows

For most of the paper we concentrate on the single-commodity case. First, for any given $\varepsilon > 0$, we present a network with linear latency functions, in which every agent needs to change its current path to recover equilibrium. Then we prove, that due to ε-changes the flow on every edge does not change by more than ε.

3.1 Instability of Equilibria: Every Agent Needs to Move

In [14] Roughgarden uses the *generalized Braess graphs* to show, that the path latency at equilibrium can arbitrarily decrease by removing several edges from a network. Our definition of B_k differs from the definition in [14] in the non-constant latency functions.

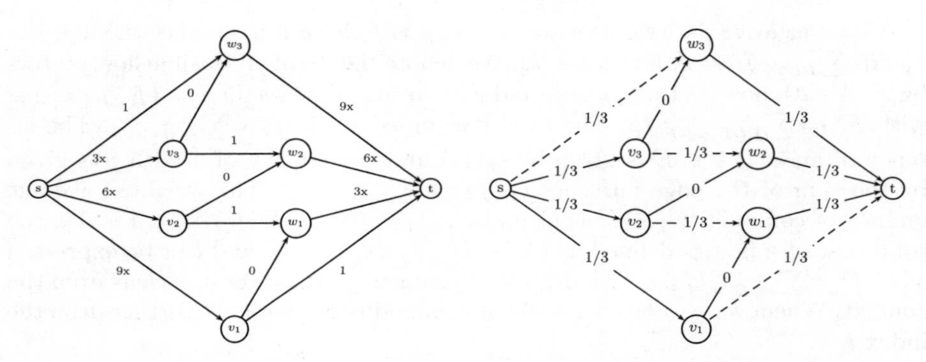

Fig. 1. Having unit demand, the solid paths in $B_{k=3}$ carry 1/3 of flow each and the dashed edges carry zero flow. After increasing the demand by $(1+\varepsilon) = (1+1/3)$, the solid paths lose all their flow and the paths containing the dashed edges gain flow of $(1+\varepsilon)/(k+1) = 1/3$ each.

Definition 2. *For every $k \in \mathbb{N}$, let $B_k = (V_k, E_k)$ be the graph with $V_k = \{s, v_1, \ldots, v_k, w_1, \ldots, w_k, t\}$ and $E_k = \{(s, v_i), (v_i, w_i), (w_i, t) : 1 \le i \le k\} \cup \{(v_i, w_{i-1}) : 2 \le i \le k\} \cup \{(s, w_k)\} \cup \{(v_1, t)\}$. Let B_k be equipped with the following latency functions.*

- $\ell^k_{v_i,w_i}(x) = 0$ *and* $\ell^k_{s,v_{k-i+1}}(x) = \ell^k_{w_i,t}(x) = i \cdot k \cdot x$ *for* $1 \le i \le k$,
- $\ell^k_{v_i,w_{i-1}}(x) = 1$ *for* $2 \le i \le k$ *and*
- $\ell^k_{s,w_k}(x) = \ell^k_{v_1,t}(x) = 1$.

Let B_k be called the kth Braess graph.

Let $\varepsilon > 0$ and consider the instance $(B_{\lceil 1/\varepsilon \rceil}, 1, \ell)$.
Let $(P_1, \ldots, P_{2k+1})^T = (P_{s,w_k,t}, P_{s,v_k,w_k,t}, P_{s,v_k,w_{k-1},t}, P_{s,v_{k-1},w_{k-1},t}, \ldots, P_{s,v_1,t})^T$ denote the corresponding path vector. The equilibrium flow is described by the vector (f_{P_j}) of path flows

$$f_{P_j} = \begin{cases} 0 & \text{for } j = 1, 3, \ldots, 2k+1 \\ 1/k & \text{for } j = 2, 4, \ldots, 2k \end{cases}$$

summing up to $\sum_P f_P = \sum_{j=1}^{2k+1} f_{P_j} = 1$.

All paths have path length $\ell_P(f) = k + 1$ and since any unilateral deviation strictly increases the sustained latency, the edge flows in equilibrium are unique (Figure 1).

Increasing the demand by $(1+\varepsilon)$, the equilibrium flow vector becomes (f'_{P_j}) with

$$f'_{P_j} = \begin{cases} (1+\varepsilon)/(k+1) & \text{for } j = 1, 3, \ldots, 2k+1 \\ 0 & \text{for } j = 2, 4, \ldots, 2k \end{cases}$$

summing up to $\sum_P f'_P = \sum_{j=1}^{2k+1} f'_{P_j} = 1 + \varepsilon$. The path latency can easily be computed to be $1 + \frac{k^2(1+\varepsilon)}{k+1}$.

Note that the path flow decomposition in equilibrium does not need to be unique. Nevertheless, we have uniqueness in B_k.

Definition 3. *An edge $e \in E$ carrying flow of at most ε is called* ε-edge.

Theorem 1. *Let $\varepsilon > 0$ and consider $(B_{\lceil \frac{1}{\varepsilon} \rceil}, 1, \ell)$. Then, increasing the flow by ε causes the entire demand to be redistributed to recover a Wardrop equilibrium, i.e., every agent is forced to change its path. Adding another edge to the network, one can achieve the same result for the removal of an ε-edge.*

Proof. For the path flow vector (f_{P_j}) and (f'_{P_j}) it holds, that, $f_{P_j} = 0 \Leftrightarrow f'_{P_j} > 0$ and $f_{P_j} > 0 \Leftrightarrow f'_{P_j} = 0$. For the second assertion, simply simulate a demand increase by directly connecting source s with sink t and choose the latency function, such that (s,t) carries an ε-fraction of flow. Then remove this edge. □

Let us remark that under mild conditions on the latency functions Theorem 1 can easily be transferred to optimal flows, i.e., flows minimizing the total cost. This is since optimal flows are Wardrop equilibria with respect to the so-called *marginal cost functions* $h_e(x) = (x \cdot \ell_e(x))' = \ell_e(x) + x \cdot \ell'_e(x)$, if $x \cdot \ell_e(x)$ are differentiable, convex functions for $e \in E$ (see [1]). Thus, it is sufficient to change the linear latency functions in $B_{\lceil \frac{1}{\varepsilon} \rceil}$.

3.2 Edge Flows Are Locally Stable

Let $f, f' \in \mathcal{F}$ be feasible flows for demands $d \leq d'$ and let $\Delta(f, f')$ denote the difference of f' and f,

$$(\Delta(f, f'))_e = f'_e - f_e \, , \forall e \in E \ .$$

An edge e is *positive* (with respect to f' and f), if $f'_e - f_e > 0$, and *negative* if $f'_e - f_e < 0$. A path is positive (or negative), if all its edges are positive (or negative). Observe that the flow conservation property holds for the difference of two network flows.

Definition 4. *A closed path consisting of flow carrying edges is called an* alternating flow cycle.

Lemma 1. *Let f denote an equilibrium flow for an instance $(G, 1, \ell)$ with non-decreasing, continuous latency functions. Then there is an equilibrium flow f' for $(G, 1+\varepsilon, \ell)$, such that $\Delta(f, f')$ does not contain an alternating flow cycle.*

Proof. Let f' denote an equilibrium flow for $(G, 1+\varepsilon, \ell)$. Assume there is an alternating flow cycle C in $\Delta(f, f')$. Since we can assume both equilibrium flows to be cycle free, we can assume that the alternating flow cycle C contains positive and negative edges. C can thus be divided into positive and negative path segments, $C = p_1 n_1 p_2 \dots n_k$, where p_i denotes a sequence of positive edges and n_i denotes a sequence of negative edges. Let u_i be the first node of p_i and denote the last node of n_i by v_i. Thus, there are two paths from u_1 to v_k in C (Figure 2).

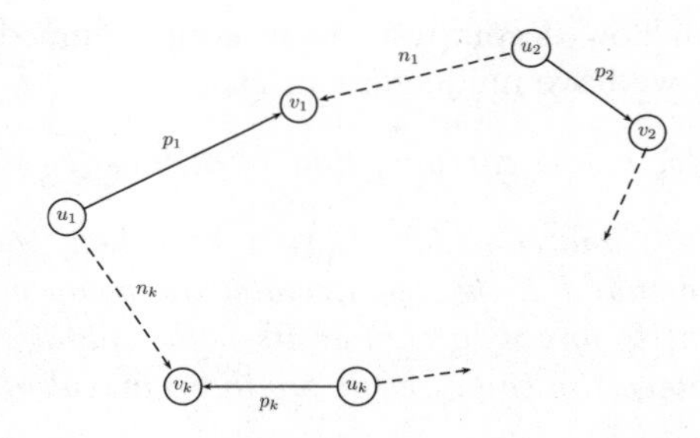

Fig. 2. An alternating flow cycle in $\Delta(f, f')$. Solid paths are positive, the dashed paths are negative. Thus, f certainly uses the dashed paths and possibly the solid paths and $f_e > f'_e$. For f' the converse holds.

For $u, v \in V$, let $\ell(u, v)$ denote the minimum path latency from u to v under f. For $u = s$, simply write $\ell(v)$. For f', write $\ell'(u, v)$ and $\ell'(v)$.

There are two facts we will make consistently use of. Since in equilibrium the flow routes only on shortest paths, we have

$$\ell(v) \leq \ell(u) + \ell(u, v) \text{ for any } u, v \in V \ , \tag{1}$$

and

$$\ell(v) = \ell(u) + \ell(u, v) \tag{2}$$

if there is a flow carrying path between s and v containing u. We show, that assuming f and f' being at equilibrium yields $\ell'(u_1, v_k) = \ell(u_1, v_k)$. On one hand, since n_k connects u_1 with v_k and there is more flow on every edge of n_k under f than under f' we have

$$\ell'(u_1, v_k) \leq \sum_{e \in n_k} \ell_e(f'_e) \leq \sum_{e \in n_k} \ell_e(f_e) = \ell(u_1, v_k) \ .$$

For the reverse direction, we show $\ell'(v_k) \geq \ell'(u_1) + \ell(u_1, v_k)$, since then $\ell(u_1, v_k) \leq \ell'(v_k) - \ell'(u_1) \leq \ell'(u_1, v_k)$.

In the following, we repeatedly make use of equations (1) and (2).

$$\begin{aligned}
\ell'(v_k) &= \ell'(u_k) + \ell'(u_k, v_k) \geq \ell'(v_{k-1}) - \ell'(u_k, v_{k-1}) + \ell'(u_k, v_k) \\
&= \ell'(u_{k-1}) + \ell'(u_{k-1}, v_{k-1}) - \ell'(u_k, v_{k-1}) + \ell'(u_k, v_k) \\
&\geq \ell'(u_1) + \sum_{i=1}^{k} \ell'(u_i, v_i) - \sum_{i=2}^{k} \ell'(u_i, v_{i-1}) \\
&\geq \ell'(u_1) + \sum_{i=1}^{k} \ell(u_i, v_i) - \sum_{i=2}^{k} \ell(u_i, v_{i-1}) \\
&\geq \ell'(u_1) + \sum_{i=1}^{k} (\ell(v_i) - \ell(u_i)) - \sum_{i=2}^{k} (\ell(v_{i-1}) - \ell(u_i)) \\
&= \ell'(u_1) - \ell(u_1) + \ell(v_k) = \ell'(u_1) + \ell(u_1, v_k) \ .
\end{aligned}$$

The third inequality is valid since f and f' route only on shortest paths. Explicitly, $\ell'(u_i, v_i) = \sum_{e \in p_i} \ell_e(f'_e) \geq \sum_{e \in p_i} \ell_e(f_e) \geq \ell(u_i, v_i)$ for each $i \in [k]$ and $\ell'(u_i, v_{i-1}) \leq \sum_{e \in n_i} \ell_e(f'_e) \leq \sum_{e \in n_i} \ell_e(f_e) = \ell(u_i, v_{i-1})$ for each $i \in \{2, \ldots, k\}$. Thus, $\ell'(u_1, v_k) = \ell(u_1, v_k)$. We deduce that the latency on every edge $e \in n_k$ does not change due to the flow change. Since the same analysis can be conducted for any path segment p_i and n_i, the latency of both paths on C connecting two arbitrary nodes remains unchanged. Therefore, by removing the bottleneck edge flow in C no edge latency is affected and the alternating flow cycle is eliminated. We may remove the set of alternating flow cycles in any order. Adding f to the altered difference, one gets the desired equilibrium flow for demand $1 + \varepsilon$. □

Thus, $(\Delta(f, f'))$ can be assumed a network flow of volume ε, when edges are allowed to be traversed in both directions. We can now state the following theorem.

Theorem 2. *Let f denote an equilibrium flow for an instance $(G, 1, \ell)$ with non-decreasing, continuous latency functions.*

- *Then there is an equilibrium flow f' for $(G, 1+\varepsilon, \ell)$, such that $|(\Delta(f, f'))_e| \leq \varepsilon$ for all $e \in E$.*
- *Consider an ε-edge (u, v) in G. There is an equilibrium flow f' for $(G' = (V, E - \{(u, v)\}), 1, \ell)$, such that $|(\Delta(f, f'))_e| \leq \varepsilon$ for all $e \in E$.*

Proof. Since the difference of f and f' can be assumed alternating flow cycle free, it constitutes a network flow of volume ε. To show the second assertion, let a single ε-edge (u, v) be removed. With the same argumentation as in Lemma 1, we can exclude alternating flow cycles in $(\Delta(f, f'))$ that do not include (u, v). Due to the flow conservation property for every node $u \neq w \neq v$, $(\Delta(f, f'))$ is a network flow from u to v of volume ε. □

Note, that since every edge gains or loses at most ε flow (Theorem 2), with respect to the number of paths $B_{\lceil \frac{1}{\varepsilon} \rceil}$ is a minimal example exhibiting global instability.

4 Stability of the Path Latency

The latency increase at equilibrium due to a demand increase clearly depends on the latency functions. Considering polynomials with nonnegative coefficients, the maximal degree is the critical parameter. Note, that the results in this section do not trivially result from Theorem 2, since the relative flow increase on an edge might be unbounded.

Theorem 3. *Let f and f' be equilibrium flows for instances $(G, 1, \ell)$ and $(G, 1+\varepsilon, \ell)$ with polynomial latency functions of degree at most p with nonnegative coefficients. Let L and L' denote the corresponding path latencies. Then $L' \leq (1 + \varepsilon)^p \cdot L$.*

Proof. Due to a scaling argument it is sufficient to consider monic monomials as latency functions. For equilibrium flows f and f' we have

$$L = \sum_{P \in \mathcal{P}} f_P \ell_P(f) = \sum_e f_e \ell_e(f_e) \text{ and } (1+\varepsilon) \cdot L' = \sum_e f'_e \ell_e(f'_e) \ ,$$

and we want to show that $\sum_e f'^{p_e+1}_e \leq (1+\varepsilon)^{p+1} \sum_e f^{p_e+1}_e$, where $\ell_e(x) = x^{p_e}$. Since equilibrium flows f and f' minimize the potential function

$$\Phi(x) = \sum_e \int_0^{x_e} \ell_e(u) du$$

over feasible flows x of volume 1 and $(1+\varepsilon)$, respectively, it holds that

$$(1+\varepsilon)^{p+1} \cdot \Phi(f) = (1+\varepsilon)^{p+1} \cdot \sum_e \frac{1}{p_e+1} f^{p_e+1}_e \leq \sum_e \frac{(1+\varepsilon)^{p-p_e}}{p_e+1} f'^{p_e+1}_e \ , \quad \text{(A)}$$

and similarly,

$$\Phi(f') = \sum_e \frac{1}{p_e+1} f'^{p_e+1}_e \leq \sum_e \frac{(1+\varepsilon)^{p_e+1}}{p_e+1} f^{p_e+1}_e \ . \quad \text{(B)}$$

For contradiction, assume

$$(1+\varepsilon)^{p+1} \sum_e f^{p_e+1}_e < \sum_e f'^{p_e+1}_e \ . \quad \text{(C)}$$

Calculating $p \cdot \text{(A)} + (p + (p+1)((1+\varepsilon)^p - 1)) \cdot \text{(B)} + ((1+\varepsilon)^p - 1) \cdot \text{(C)}$ yields

$$\sum_{k=0}^{p} c_k \sum_{p_e=k} f^{p_e+1}_e < \sum_{k=0}^{p} c'_k \sum_{p_e=k} f'^{p_e+1}_e \ , \quad (3)$$

with

$$c_k = p \cdot \frac{(1+\varepsilon)^{p+1}}{k+1} - ((p+1)(1+\varepsilon)^p - 1) \cdot \frac{(1+\varepsilon)^{k+1}}{k+1} + ((1+\varepsilon)^p - 1) \cdot (1+\varepsilon)^{p+1}$$

and

$$c'_k = p \cdot \frac{(1+\varepsilon)^{p-k}}{k+1} - ((p+1)(1+\varepsilon)^p - 1) \cdot \frac{1}{k+1} + ((1+\varepsilon)^p - 1) \ .$$

In the following we show that $c'_k \leq 0$ for $0 \leq k \leq p$. Analogous arguments can be used to show $c_k \geq 0$. Hence, we have a contradiction to equation (3).

For any $0 \leq k \leq p$ and $\varepsilon = 0$, we have $c'_k = 0$. We show that c'_k is monotonically decreasing in ε (for $\varepsilon \geq 0$). The derivative of c'_k with respect to $(1+\varepsilon)$ is

$$\frac{\partial c'_k}{\partial(1+\varepsilon)} = p \cdot (p-k) \cdot \frac{(1+\varepsilon)^{p-k-1}}{k+1} - p \cdot (p+1) \frac{(1+\varepsilon)^{p-1}}{k+1} + p \cdot (1+\varepsilon)^{p-1} \ .$$

Thus, it is sufficient to show that

$$\frac{1}{(1+\varepsilon)^{p-k-1}} \cdot \frac{\partial c'_k}{\partial(1+\varepsilon)} = p \cdot (p-k) \cdot \frac{1}{k+1} - p \cdot (p+1)\frac{(1+\varepsilon)^k}{k+1} + p \cdot (1+\varepsilon)^k \leq 0 \ .$$

For $\varepsilon = 0$, the left hand side equals 0. It remains to show that $\frac{1}{(1+\varepsilon)^{p-k-1}} \cdot \frac{\partial c'_k}{\partial(1+\varepsilon)}$ is monotonically decreasing in ε (for $\varepsilon \geq 0$). This is the case since

$$\frac{\partial(\frac{1}{(1+\varepsilon)^{p-k-1}} \cdot \frac{\partial c'_k}{\partial(1+\varepsilon)})}{\partial(1+\varepsilon)} = \frac{(k-p) \cdot p \cdot k}{k+1} \cdot (1+\varepsilon)^{k-1} \leq 0$$

and the proof is complete. □

The bound is tight, as shown by the network consisting of two nodes connected by an edge, equipped with the latency function $\ell(x) = x^p$. Allowing negative coefficients, the relative increase obviously can be unbounded.

4.1 Increase of the Price of Anarchy

The Price of Anarchy quantifies the degradation of performance due to selfish behavior.

Definition 5. *For an instance (G, d, ℓ) with equilibrium flow f and optimal flow f^* the* Price of Anarchy *is defined as* $\frac{C(f)}{C(f^*)}$.

In [13] the Price of Anarchy is shown to be asymptotically $\Theta(\frac{p}{\ln p})$ for polynomial latency functions of degree at most p with nonnegative coefficients.

Corollary 4. *Let ρ and ρ' denote the Price of Anarchy for instances $(G, 1, \ell)$ and $(G, 1+\varepsilon, \ell)$ with polynomial latency functions of degree at most p with non-negative coefficients. Then $\rho' \leq (1+\varepsilon)^p \cdot \rho$.*

Proof. Let $\bar{L}_d$ denote the average path latency for an optimal flow in (G, d, ℓ). Let $C_{\text{opt}}, C'_{\text{opt}}, C^*$ and C'^* denote the costs of an optimal flow and an equilibrium flow, respectively. Then $\rho = C^*/C_{\text{opt}}$ and $\rho' = C'^*/C'_{\text{opt}}$. Since $C_{\text{opt}} = 1 \cdot \bar{L}_1$ and $C'_{\text{opt}} = (1+\varepsilon) \cdot \bar{L}_{1+\varepsilon}$, we have

$$(1+\varepsilon) \cdot C_{\text{opt}} = (1+\varepsilon) \cdot \bar{L}_1 \leq (1+\varepsilon) \cdot \bar{L}_{1+\varepsilon} = C'_{\text{opt}} \ ,$$

since the average latency is clearly monotone in the demand. Thus, the increase of the Price of Anarchy can be bounded by

$$\frac{\rho'}{\rho} = \frac{C'^*/C'_{\text{opt}}}{C^*/C_{\text{opt}}} = \frac{L' \cdot (1+\varepsilon) \cdot C_{\text{opt}}}{L \cdot C'_{\text{opt}}} \leq \frac{L \cdot (1+\varepsilon)^p \cdot (1+\varepsilon) \cdot C_{\text{opt}}}{L \cdot C_{\text{opt}} \cdot (1+\varepsilon)} = (1+\varepsilon)^p \ ,$$

where the inequality is due to Theorem 3. □

This upper bound is tight in the following sense: There is a network family $(G, d, \ell(p))$, such that $\lim_p \frac{\rho'/\rho}{(1+\varepsilon)^p} = 1$ for every $\varepsilon > 0$. This holds for mildly modified instances of Pigou's example [10]. Assume two nodes to be connected via two edges equipped with latency functions $\ell_1(x) = x^p$ and $\ell_2(x) = (1+\varepsilon)^p$. We calculate $C^* = 1, C'^* = (1+\varepsilon)^{p+1}, C_{\text{opt}} = \frac{(1+\varepsilon)^{p+1}}{(p+1)^{(p+1)/p}} + (1 - \frac{1+\varepsilon}{(p+1)^{1/p}})(1+\varepsilon)^p$, and $C'_{\text{opt}} = \frac{(1+\varepsilon)^{p+1}}{(p+1)^{(p+1)/p}} + (1+\varepsilon - \frac{1+\varepsilon}{(p+1)^{1/p}})(1+\varepsilon)^p$. Thus, we have

$$\frac{\rho'}{\rho} = (1+\varepsilon)^p \cdot (1 - \frac{(p+1)^{1/p}\varepsilon\, p}{(p+1)^{(p+2)/p} - p(p+1)^{1/p}}) ,$$

and it holds that $\lim_p \frac{\rho'/\rho}{(1+\varepsilon)^p} = 1$ for every fixed $\varepsilon > 0$.

5 Instability in Multi-commodity Networks

There are no analogous results to Theorem 2 and 3 for the multi-commodity case. Figure 3 shows a network with two commodities, with both demands being 1, in which after increasing the demand of the second commodity or both demands by ε, the entire demand of the first commodity needs to be shifted to a single edge to recover an equilibrium state. If a single ε-edge is being removed, other edges might also lose an arbitrary fraction of the commodity's demand.

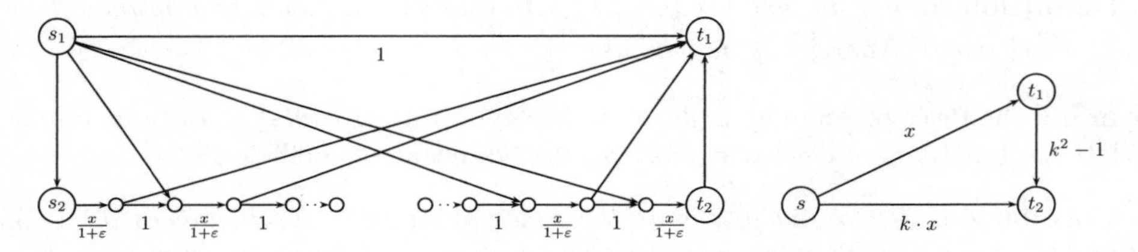

Fig. 3. *(left)* Unlabeled edges cause no latency. Assume there are $2 \cdot \lceil \frac{1}{\varepsilon} \rceil - 1$ many edges on the unique path connecting s_2 with t_2. For $d_1 = d_2 = 1$, the flow demand of commodity 1 is uniformly spread over all $\lceil 1/\varepsilon \rceil$ paths using one edge on the path connecting s_2 and t_2. After increasing d_2 by ε, we have $f_{(s_1,t_1)} = 1$. *(right)* For $d_1 = 1$ and $d_2 = k$, the path latency of the first commodity multiplicatively increases by $1+k\cdot\varepsilon$ if both demands are increased by a factor of $(1+\varepsilon)$.

Figure 3 also shows a network with 2 commodities. (Insisting on unit demands, one can split commodity 2 into k small commodities.) Increasing the demands by ε the path latency of commodity 1 increases by a factor of $1 + k \cdot \varepsilon$. Simple examples exhibit an even higher increase.

6 Open Problems

Suppose, given a unit demand flow at Wardrop equilibrium, one removes an edge carrying only an ε-fraction of flow. How does the path latency change after recovering equilibrium? Considering a network with two parallel edges, one gets a lower bound of $\frac{1}{(1-\varepsilon)^p}$. Is this bound tight?

Furthermore, we believe that our bound on the increase of the path latency holds for a broader class of latency functions, namely for latency functions with bounded *elasticity*.

What can be said about the sensitivity of equilibria in related models? For instance, are analogous results possible in atomic games, where every agents control some non-negligible amount of flow each?

Acknowledments

We wish to thank Alexander Skopalik for many insightful discussions.

References

[1] Beckmann, M., McGuire, C.B., Winston, C.B.: Studies in the Economics of Transportation. Yale University Press, New Haven (1956)
[2] Braess, D.: Über ein Paradoxon aus der Verkehrsplanung. Unternehmensforschung 12, 258–268 (1968)
[3] Dafermos, S., Nagurney, A.: Sensitivity Analysis for the Asymmetric Network Equilibrium Problem. Mathematical Programming 28, 174–184 (1984)
[4] Dafermos, S., Sparrow, F.T.: The Traffic Assignment Problem for a General Network. Journal of Research of the National Bureau of Standards 73(2), 91–118 (1969)
[5] Fisk, C.: More Paradoxes in the Equilibrium Assignment Problem. Transportation Research, Series B 13(4), 305–309 (1979)
[6] Hall, M.A.: Properties of the Equilibrium State in Transportation Networks. Transportation Science 12, 208–216 (1978)
[7] Josefsson, M., Patriksson, M.: Sensitivity Analysis of Separable Traffic Equilibria with Application to Bilevel Optimization in Network Design. Transportation Research Series B 41(1), 4–31 (2007)
[8] Koutsoupias, E., Papadimitriou, C.: Worst-Case Equilibria. In: Meinel, C., Tison, S. (eds.) STACS 1999. LNCS, vol. 1563, pp. 404–413. Springer, Heidelberg (1999)
[9] Patriksson, M.: Sensitivity Analysis of Traffic Equilibria. Transportation Research 38, 258–281 (2004)
[10] Pigou, A.C.: The Economics of Welfare. Macmillan, Basingstoke (1920)
[11] Roughgarden, T., Tardos, É.: How Bad is Selfish Routing. Journal of the ACM 49(2), 236–259 (2002)
[12] Roughgarden, T.: How Unfair is Selfish Routing. In: Roughgarden, T. (ed.) Proc. of th 13th Annual Symposium on Discrete Algorithms (SODA), pp. 203–204 (2002)
[13] Roughgarden, T.: The Price of Anarchy is Independent of the Network Topology. In: Proc. of th 34th Annual Symposium on Theory of Computing Discrete Algorithms (STOC), pp. 428–437 (2002)
[14] Roughgarden, T.: On the Severity of Braess's Paradox: Designing Networks for Selfish Users is Hard. Journal of Computer and System Sciences 72(5), 922–953 (2004)
[15] Wardrop, J.G.: Some Theoretical Aspects of Road Traffic Research. In: Proc. of the Institute of Civil Engineers Pt. II, pp. 325–378 (1952)

Prompt Mechanisms for Online Auctions

Richard Cole[1,⋆], Shahar Dobzinski[2,⋆⋆], and Lisa Fleischer[3,⋆⋆⋆]

[1] Computer Science Department, Courant Institute, New York University
cole@cs.nyu.edu
[2] The School of Computer Science and Engineering,
The Hebrew University of Jerusalem
shahard@cs.huji.ac.il
[3] Dartmouth College
lkf@cs.dartmouth.edu

Abstract. We study the following online problem: at each time unit, one of m identical items is offered for sale. Bidders arrive and depart dynamically, and each bidder is interested in winning one item between his arrival and departure. Our goal is to design truthful mechanisms that maximize the welfare, the sum of the utilities of winning bidders.

We first consider this problem under the assumption that the private information for each bidder is his value for getting an item. In this model constant-competitive mechanisms are known, but we observe that these mechanisms suffer from the following disadvantage: a bidder might learn his payment only when he departs. We argue that these mechanism are essentially unusable, because they impose several seemingly undesirable requirements on any implementation of the mechanisms.

To crystalize these issues, we define the notions of *prompt* and *tardy* mechanisms. We present two prompt mechanisms, one deterministic and the other randomized, that guarantee a constant competitive ratio. We show that our deterministic mechanism is optimal for this setting.

We then study a model in which both the value and the departure time are private information. While in the deterministic setting only a trivial competitive ratio can be guaranteed, we use randomization to obtain a prompt truthful $\Theta(\frac{1}{\log m})$-competitive mechanism. We then show that no truthful randomized mechanism can achieve a ratio better than $\frac{1}{2}$ in this model.

1 Introduction

1.1 Background

The field of algorithmic mechanism design attempts to handle the strategic behavior of selfish agents in a computationally efficient way. To date, most work

⋆ 251 Mercer Street, New York, NY 10012. This work was supported in part by NSF grants IIS0414763 and CCF0515127.

⋆⋆ Supported by the Adams Fellowship Program of the Israel Academy of Sciences and Humanities, by NSF grant CCF-0515127, and by grants from the Israel Science Foundation, the USA-Israel Bi-national Science Foundation.

⋆⋆⋆ 6211 Sudikoff, Dartmouth College, Hanover, NH 03755. Partially supported by NSF grants CCF-0515127 and CCF-0728869.

B. Monien and U.-P. Schroeder (Eds.): SAGT 2008, LNCS 4997, pp. 170–181, 2008.

in this field has sought to design truthful mechanisms for *static* settings, e.g., auctions. In reality, however, the setting of many problems is *online*, meaning that the mechanism has no prior information regarding the identity of the participating players, or that the goods that are for sale are unknown in advance. Examples include sponsored search auctions [12], single-good auctions [10], and even pricing WiFi at Starbucks [5].

This paper considers the following online auction problem: at each time unit exactly one of m identical items is offered for sale. The item at time t is called item t. There are n bidders, where bidder i arrives at time a_i and departs at time d_i, both unknown before bidder i's arrival. The interval $[a_i, d_i]$ will be called bidder i's *time window*, and the set of items offered in i's time window will be denoted by W_i. Each bidder is interested in winning at most one of the items within W_i. Let v_i denote the value to the ith bidder of getting an item in W_i. Our goal is to maximize the social welfare: the sum of the values of the bidders that get some item within their time window. As usual in online algorithms, our goal is to optimize the competitive ratio: the worst-case ratio between the welfare achieved by the algorithm and the optimal welfare.

In the full information setting, this problem is equivalent to the online scheduling of unit-length jobs on a single machine to maximize weighted throughput. This online problem and its variants have been widely studied (e.g., [1,8,3]). The best deterministic algorithm to date guarantees a competitive ratio of ≈ 0.547 [4,11], while it is known that no deterministic algorithm can obtain a ratio better than $\frac{2}{\sqrt{5}+1} \approx 0.618$ [2]. In the randomized setting, a competitive ratio of $1 - \frac{1}{e}$ is achieved by [1], and no algorithm can achieve a ratio better than 0.8 [2].

This problem provides an excellent example of the extra barriers we face when designing online mechanisms. The only general technique known for designing truthful mechanisms is the VCG payment scheme. In the offline setting we can obtain an optimal solution in polynomial time (with bipartite matching), and then we can apply VCG. In the online setting, however, it is impossible to find an optimal solution, and thus we cannot use VCG. Yet, truthful competitive mechanisms do exist. The competitive ratio of these mechanisms depends on the specific private-information model each mechanism was designed for. This paper considers two different natural models:

- **The Value-Only model:** Here, the private information of bidder i consists of just his value v_i, and the arrival time and the departure time are known to all (but both are unknown prior to the arrival of bidder i).
- **The Generalized Model:** The private information of bidder i consists of two numbers: his value v_i and his departure time d_i. The arrival time is public information (but unknown prior to the arrival of bidder i).

1.2 The Value-Only Model: Is Monotonicity Enough?

The only private information of a bidder in the value-only model is his value, and thus this model falls under the category of *single-parameter* environments – environments in which the private information of each bidder consists of only one number. Fortunately, designing truthful mechanisms for single-parameter environments is quite well understood: an algorithm is truthful if and only if it is monotone. That is, a winning bidder that raises his bid remains a winner.

Using the above characterization, it is possible to prove that the greedy algorithm is monotone [7] (see Section 2.4 for a description). Since [8] shows that greedy is 1/2 competitive, this gives a truthful mechanism that is $\frac{1}{2}$ competitive.

However, a closer look at this mechanism may make one wonder if it is indeed applicable. The notions of prompt and tardy mechanisms we define next highlight the issue.

Definition 1. *A mechanism for the online auction problem is* prompt *if a bidder that wins an item always learns his payment immediately after winning the item. A mechanism is* tardy *otherwise.*

As we show later in the paper, the tardiness in the greedy mechanism [7,8] is substantial: there are inputs for which a bidder learns his payment only when he departs. Tardy mechanisms seem very unintuitive for the bidders, and in addition they suffer from the following disadvantages:

- **Uncertainty:** A winning bidder does not know the cost of the item that he won, and thus does not know how much money he still has available. E.g., suppose the mechanism is used in a Las Vegas ticket office for selling tickets to a daily show. A tourist that wins a ticket is uncertain of the price of this privilege, and thus might not be able to determine how much money he has left to spend during his Las Vegas vacation.
- **Debt Collection:** A winning bidder might pay the mechanism long after he won the item. A bidder that is not honest may try to avoid this payment. Thus, the auctioneer must have some way of collecting the payment of a winning bidder.
- **Trusted Auctioneer:** A winning bidder essentially provides the auctioneer with a "blank check" in exchange for the item. Consequently, all bidders must trust the honesty of the auctioneer. Even if the bidders trust the auctioneer, they may still want to verify the exact calculation of the payment, to avoid over-payments that make winning the item less profitable, or even unprofitable. In order to verify this calculation, the bids of all bidders have to be revealed, leading to an undesirable loss of privacy.

Notice that all of these problems are due to the online nature of the setting, and do not arise in the offline setting. To overcome these problems, we present *prompt* mechanisms for the online auction problem. Prompt mechanisms are very intuitive to the bidders as they (implicitly) correspond to take-it-or-leave-it offers: a winning bidder is offered a price for one item exactly once *before* getting the item, and may reject the offer if it is not beneficial for him. We improve upon the greedy algorithm of [7,8] by showing a different mechanism that achieves the same competitive ratio, but is also prompt.

Theorem: There exists a $\frac{1}{2}$-competitive prompt and truthful mechanism for the online auction problem in the value-only model.

We show that this is the best possible by proving that no prompt deterministic mechanism can guarantee a competitive ratio better than $\frac{1}{2}$.

We also present a *randomized* mechanism that guarantees a constant competitive ratio. The achieved competitive ratio of the latter algorithm is worse than the competitive ratio of the deterministic algorithm. Yet, the core of the proof studies a balls-and-bins problem that might be of independent interest.

1.3 The Generalized Model

While truthful mechanisms for single-parameter settings are well characterized and thus relatively easy to construct, truthful mechanisms for multi-parameter settings, like the generalized model, are much harder to design. The online setting considered in this paper only makes the design of truthful mechanisms a more challenging task.

The online auction problem in the generalized model illustrates this challenge. Lavi and Nisan [9] introduced the online auction problem to the mechanism design community. They showed that no truthful deterministic mechanism for this multi-parameter problem can provide more than a trivial competitive ratio. As a result, Lavi and Nisan proposed a weaker solution concept, set-nash, and provided mechanisms with a constant competitive ratio under this notion. We stress that the set-nash solution concept is much weaker than the dominant-strategy truthfulness we consider.

By contrast with [9], instead of relaxing the solution concept, we use the well-known idea that randomization can help in mechanism-design settings [14]. We provide *randomized* upper and lower bounds in the generalized model for the online auction problem.

Theorem: There exists a prompt truthful randomized $\Theta(\frac{1}{\log m})$-competitive mechanism for the online auction problem in the generalized model.

The main idea of the mechanism is to extend the randomized mechanism for the value-only model to the generalized model. Specifically, we use the random-sampling method introduced in [6] to "guess" the departure time of each bidder, and then we use the above randomized mechanism with these guessed departures. This mechanism is also a prompt mechanism. We notice that it is quite easy to obtain mechanisms with a competitive guarantee of the logarithm of the ratio between the highest and lowest valuations. However, since this ratio might be exponential in the number of items or bidders, this guarantee is quite weak. By contrast, the competitive ratio our mechanism achieves is independent of the ratio between the highest and lowest valuations, and the mechanism is not required to know these valuations in advance.

Theorem: No truthful randomized mechanism for the online auction problem in the generalized model can obtain a competitive ratio better than $\frac{1}{2}$.

The proof of this bound is quite complicated. We start by defining a family of recursively-defined distributions on the input, and then show that no deterministic mechanism can obtain a competitive ratio better than $\frac{1}{2}$ on this family of distributions. We then use Yao's principle to derive the theorem.

The main open question left in the generalized model is to determine whether there is a truthful mechanism with a constant competitive ratio.

Paper Organization. In Section 2 we describe prompt mechanisms for the value-only case, and prove that no deterministic tardy algorithms can achieve a ratio better than $\frac{1}{2}$. Due to lack of space, lower and upper bounds for the generalized case are proved only in the full version (see `http://www.cs.huji.ac.il/~shahard`).

2 Prompt Mechanisms and the Value-Only Model

2.1 A Deterministic Prompt $\frac{1}{2}$-Competitive Mechanism

The mechanism maintains a *candidate* bidder c_j for each item j. To keep the presentation simple and without loss of generality, we assume an initialization of the mechanism in which each item j receives a candidate bidder c_j with a value of 0 for winning an item (i.e., $v_{c_j} = 0$).

The mechanism runs as follows: at each time t we look at all the bidders that arrived at time t. We consider these bidders one by one in some arbitrary order (independent of the bids): for each bidder i we look at all the candidates in i's time window, and let c_j be the candidate bidder with the smallest bid (if there are several such candidates, we select one arbitrarily). Formally, $c_j \in \arg\min_{k \in W_i} c_k$. We say that i *competes* on item j. Now, if $v_{c_j} < v_i$, we make i the candidate bidder for item j. After all the bidders that arrived at time t have been processed, we allocate item t to the candidate bidder c_t.

The next theorem proves that this algorithm is monotone, i.e., a bidder that raises his bid is still guaranteed to win. This is also a necessary and sufficient condition for truthfulness. We are still left with the issue of finding the payments themselves. First, observe that the payment of each winning bidder must equal his *critical value*: the minimum value he can declare and still win. Notice that this value is indeed well defined if the algorithm is monotone. For each bidder i this value can be found by using a binary search on the possible values of v_i. Clearly, this procedure takes a polynomial time. See, e.g., [13] for a more thorough discussion. By the discussion above, it is clear that a mechanism is prompt if and only if i's critical value can be found by the time i wins an item. In this case, the payment can also be calculated in polynomial time.

Theorem 1. *The mechanism is prompt and truthful. Its competitive ratio is* $\frac{1}{2}$.

Proof. To show that the mechanism is truthful we have to show that it is monotone: that is, a winning bidder i still wins an item by raising his value v_i to v'_i. First, observe that fixing the declarations of the other bidders, i competes on item j regardless of his value. We now compare two runs of the mechanism, with i declaring v_i and with i declaring v'_i, and show that at each time the candidate for any item j' is the same in both runs. In particular, it follows that the set of winners stays the same, and thus the mechanism is monotone.

First, observe that the two runs are identical until the arrival of i. Look at the next bidder e that arrives after i. For a contradiction, suppose that the candidate for some item changes after bidder e arrives. It follows that i declaring v_i' causes e to compete on an item different than the one that e competes on when i declares v_i. This is possible only if e is competing on j if i declares v_i, but if i declares v_i', e competes on $h \neq j$. It follows that if i declares, v_i' both i and e compete on j, and that i wins j. Thus, $v_i \geq v_e$. When i raises his bid e competes on h. Let c_h be the candidate for h at the time that e arrives. We have that $v_i' > v_{c_h} \geq v_i$, and thus $v_e < v_{c_h}$ so e does not become a candidate on h, and the set of candidates stays the same. To finish the monotonicity proof, look at the rest of the bidders one by one, and repeat the same arguments.

As for the promptness of the mechanism, observe that the identity of the item that i competes on is determined only by the information provided by bidders that had already arrived by the time of i's arrival. The winner of any item j is of course completely determined by the information provided by bidders that arrived by time j. Thus, we can calculate the payment of a winning bidder immediately after he wins an item.

We now analyze the competitive ratio of the mechanism. Let $OPT=(o_1, ..., o_m)$ be the optimal solution, and $ALG = (p_1, ..., p_m)$ be the solution constructed by the mechanism. That is, o_j is the bidder that wins item j in OPT and p_j is the bidder that wins item j in ALG. We will match each bidder i that wins an item in OPT to exactly one bidder l that wins an item in ALG. Furthermore, we will make sure that $v_i \leq v_l$, and that each bidder in ALG is associated with at most two bidders in OPT. This is enough to prove a competitive ratio of $\frac{1}{2}$.

The bidders are matched as follows: for each item j, let $o_{j_1}, \cdots, o_{j_{k_j}}$ be the bidders (ordered by their arrival time) that won an item in the optimal solution and are competing on j. Now match each o_{j_r} to $p_{j_{r+1}}$ for $r < k_j$. Match $o_{j_{k_j}}$ to p_j, the bidder that wins j in ALG (it is possible that $p_j = o_{j_{k_j}}$).

Observe that bidder p_j is associated with at most two bidders that win some item in OPT: bidder $o_{j_{k_j}}$, and at most one bidder, o_{j_i}, that is competing on an item j, where j is the item that o_j $(= o_{j_{i+1}})$ is competing on in ALG. To finish the proof, we only have to show that $v_{o_{j_{k_j}}} \leq v_{p_j}$ and $v_{o_{j_i}} \leq v_{p_j}$. Since $o_{j_{k_j}}$ and p_j both compete for slot j (possibly they are the same bidder) and p_j wins, $v_{o_{j_{k_j}}} \leq v_{p_j}$. Now we show the second claim. When $o_{j_{i+1}}$ arrives, o_{j_i} is already competing on slot j; as $o_{j_{i+1}}$ chooses to compete on slot j rather than slot j' which is also in its interval, thus the current candidate for slot j has value at least $v_{o_{j_i}}$. But the eventual winner of slot j, p_j, can only have a larger value; i.e. $v_{o_{j_i}} \leq v_{p_j}$. □

2.2 A Prompt Randomized Mechanism

We present a randomized prompt $O(1)$-competitive mechanism for the online auction problem in the value-only model. The analysis of the competitive ratio of the mechanism is related to a variant of the following balls-and-bins question:

Balls and Bins (intervals version): n balls are thrown to n bins, where the ith ball is thrown uniformly at random to bins in the interval $W_i = [a_i, d_i]$. We are given that the balls can be placed in a way such that all bins are filled, and each ball i is placed in exactly one bin in $[a_i, d_i]$. What is the expected number of full bins (bins with at least one ball)?

The theorem below proves that, for every valid selection of the a_i's and d_i's, in expectation at least $\frac{1}{10}$ of the bins will be full (notice that in the online auction problem the "balls" have weights). There is a gap between this ratio and the worst example we know: in Subsection 2.3 we present an example in which at most $\frac{11}{24}$ of the bins are full in expectation. Improving the analysis of the balls and bins question will almost immediately imply an improvement in the guaranteed competitive ratio of the mechanism.

The Mechanism

1. When bidder i arrives, assign it to exactly one item in W_i to compete on uniformly at random.
2. At time j conduct a second-price auction on item j among all the bidders that were selected to compete on item j in the first stage.

Theorem 2. *The mechanism is prompt and truthful, and guarantees a competitive ratio of* $\frac{1}{10}$.

Proof. To see that the mechanism is truthful, recall that in the value-only model the arrival time and the departure time of each bidder are public information. It follows that the identity of bidders competing on a certain item is determined only by the outcome of the random coin flips. It is well known that a second-price auction is truthful, and thus we conclude that the mechanism is truthful. Clearly, the mechanism is prompt since the price is determined by the second-price auction which is conducted before allocating the item to the winning bidder.

We now turn to analyzing the competitive ratio of the mechanism. Instead of analyzing this ratio directly, we analyze the competitive ratio of the following process. In addition to the input of the mechanism, the input of the process consists also of "forbidden" sets $S_1 \subseteq W_1, ..., S_n \subseteq W_n$. Later we will see how to construct these sets in a way that guarantees a constant competitive ratio.

1. For each bidder i that *won an item in the optimal solution*, select exactly one item j in W_i to compete on uniformly at random. If $j \in S_i$ then bidder i is not competing on any item at all.
2. At time j allocate item j to one bidder i, where bidder i is selected uniformly at random from the set of all bidders that are competing on item j.

We will compare runs of the mechanism and the process in which the same random coins are used in Step 1. We argue that the competitive ratio of the mechanism is at least as good as the competitive ratio of the process. To see this, observe that in the first step we are restricting ourselves only to bidders that won an item in the optimal solution. Furthermore, some of these bidders are eventually not competing on any item at all. Also, the bidder that is assigned item j is selected uniformly at random from the set of the bidders that are competing on item j, while in Step 2 of the mechanism the bidder with the highest valuation is assigned item j. Obviously, the mechanism does at least as well as the process. We will need the following technical lemma:

Lemma 1. *Let C_j be the random variable that denotes the number of bidders competing on item j (the* congestion *of item j). Let $U_{i,j}$ be the random variable that gets the value of the utility of bidder i from winning item j (that is, v_i if bidder i wins item j, and 0 otherwise). Then,*

$$E[U_{i,j}|i \textit{ is competing on item } j] \geq \frac{v_i}{E[C_j]+1}$$

Proof. We start by bounding from above $E[C_j|i \text{ is competing on item } j]$. That is, the expected congestion of item j given that bidder i is competing on j. Notice that the expected congestion produced by all other bidders apart from bidder i cannot exceed $E[C_j]$, since the item chosen for each bidder to compete on is selected independently. We are given that bidder i is already competing on item j, and thus we conclude that $E[C_j|i \text{ is competing on item } j] \leq E[C_j]+1$.

We now prove the main part of the lemma. Notice that $E[U_{i,j}|i \text{ is competing on item } j] = \Pr[i \text{ won item } j|i \text{ is competing on item } j] \cdot v_i$. Let E denote the set of all coin flips in which bidder i is competing on item j (observe that each event $e \in E$ occurs with equal probability). Let $n_j(e)$ be the congestion of item j in $e \in E$.

$$E[U_{i,j}|i \text{ is competing on item } j] = \Sigma_{e\in E}\frac{v_i}{|E| \cdot n_j(e)} \geq \frac{v_i}{E[C_j]+1}$$

where the first equality is by the definition of expectation, and the second inequality is by the convexity of the function $\frac{1}{x}$, and Jensen's inequality. □

As is evident from the lemma, if the expected congestion of all items that are in bidder i's time window is $O(1)$, then bidder i's expected utility is $\Theta(v_i)$. Unfortunately, it is quite easy to construct instances in which for every i, $S_i = \emptyset$ and some items face super-constant congestion. Instead, we will specify for each bidder i a set of items S_i, of size at most half of the size of his time window. We will see that by a proper choice of the S_i's the expected congestion of *every* item is bounded by 4.

Then, as each bidder i (that participates in the optimal solution) has a probability of at least one half of competing on some item, by Lemma 1 bidder i recovers in expectation at least $\frac{1}{2} \cdot \frac{1}{E[C_j]+1}$ of his value; by Lemma 2 this bidder receives in expectation at least $\frac{1}{10}$ of his value. Using the linearity of expectation, we conclude that the mechanism is $\frac{1}{10}$-competitive.

Lemma 2. *There exist sets $S_1, ..., S_n$ such that for each bidder i (that wins an item in the optimal solution), $S_i \subseteq W_i$, and $|S_i| \leq \frac{|W_i|}{2}$, and for each item j, $E[C_j] \leq 4$.*

Proof. The proof of the lemma consists of m stages. In each step we will consider bidders with time windows of length exactly t, where t will take values in descending order from m to 1. We will show for each bidder i with $|W_i| = t$ how to construct his set S_i. By the end of each step, we will be guaranteed that if $|W_i| \geq t$, then for each item in $W_i \setminus S_i$, the expected congestion is at most 4.

We start by handling the case where $t \geq \frac{m}{2}$. Fix some bidder i with $|W_i| \geq \frac{m}{2}$. We are considering only bidders that get an item in the optimal solution, and since there are m items, we need to take into account at most m bidders. Observe that since $W_i \geq \frac{m}{2}$, the average expected congestion of an item in W_i cannot exceed 2. We let S_i be the set of all items in W_i for which the expected congestion is at most 4. By simple Markov arguments, $|S_i| \leq \frac{|W_i|}{2}$. We now have that for every bidder i with $|W_i| \geq \frac{m}{2}$, and for each $j \in W_i \setminus S_i$, $E[C_j] \leq 4$.

Consider now Step t, where $t < \frac{m}{2}$. We first consider the congestion due to bidders with time windows of length at most t. Then we will see that our analysis remains almost the same when including bidders with larger time windows.

Fix some bidder i with $|W_i| = t$. We now bound from above the total congestion of the items in W_i. In the optimal solution, there are at most t bidders that won an item in W_i. Their contribution to the congestion of W_i is bounded from above by assuming that each one is competing on items in W_i time window with probability 1. Hence, the total contribution of these bidders is at most t.

Consider the bidders that won one item j, $a_i - t \leq j \leq a_i - 1$, in the optimal solution. (Our analysis will only improve if $a_i - t \leq 0$.) Clearly, if bidder b won item j in the optimal solution, then that item j is within b's time window. Since a bidder is selected to compete on an item uniformly, it is easy to verify that his contribution to the expected congestion of W_i is maximized when his arrival time is j and his departure time is $j+t-1$. (Recall that we are only considering bidders with time window of size at most t.) In this case, his contribution to the expected congestion of W_i is $\frac{j+t-a_i}{t}$. Summing over all bidders (with time windows of size at most t) that won one item j, $a_i - t \leq j \leq a_i - 1$, we get that the total contribution of these bidders is at most $\frac{t}{2}$.

Similarly, the total contribution of bidders with time windows of size at most t that won items $d_i + 1$ to $d_i + t$ in the optimal solution is at most $\frac{t}{2}$. It is easy to see that all other bidders with time windows of at most t contribute nothing to the expected congestion of items in W_i. In total, we get that the total expected congestion of items in W_i (due to bidders with time window of length at most t) is at most $\frac{t}{2} + \frac{t}{2} + t = 2t$, and thus the average expected congestion due to these items is at most 2.

As before, we let S_i be the set of all items in W_i for which the expected congestion is at most 4. Again, standard Markov arguments assure that $|S_i| \leq \frac{|W_i|}{2}$. We now have that for every bidder i with $|W_i| = t$, and for each $j \in W_i \setminus S_i$, the average expected congestion incurred by bidders with time windows of size at most t is at most 4. We still need to take into account the congestion incurred by bidders with time windows larger than t. Here we observe that by our construction of the S_i's, these bidders can only contribute to the congestion of items with an expected congestion of at most 4. Therefore, we claim that for each bidder i with $|W_i| \geq t$, and $j \in W_i \setminus S_i$, we have that $E[C_j] \leq 4$. We finish the proof of the lemma by considering smaller values of t, down to $t = 1$. □

2.3 A Bad Example

The following example shows that the mechanism presented has a competitive ratio strictly worse than $\frac{1}{2}$. The example is an instance of the balls and bins question presented earlier. For $1 \leq i \leq \frac{n}{3}$, we let $W_i = [i, \frac{2n}{3}]$. For $\frac{n}{3} < i \leq n$, we let $W_i = [\frac{n}{3}+1, i]$. The probability that bin i in $[1, \frac{n}{3}]$ will be empty is:

$$\Pr[\text{no ball falls in bin } i \in [1, \tfrac{n}{3}]\,] = \Pi_{t=1}^{i} \Pr[\text{ball } t \text{ does not fall to bin } i \in [1, \tfrac{n}{3}]\,]$$

$$= \Pi_{t=1}^{i}\left(1 - \frac{1}{\frac{2n}{3} - t + 1}\right) = \Pi_{t=1}^{i}\left(\frac{\frac{2n}{3} - t}{\frac{2n}{3} - t + 1}\right) = \frac{\frac{2n}{3} - i}{\frac{2n}{3}}$$

We now calculate the expected number of empty bins in the range $[1, \frac{n}{3}]$. Observe that the probability of bin $i \in [1, \frac{n}{3}]$ to be equal to the probability of bin $n-i+1$. Thus, the expected number of empty bins in $[1, \frac{n}{3}]$ is equal to the expected number of empty bins in $[\frac{2n}{3}, n]$:

$$\sum_{t=1}^{\frac{n}{3}} \frac{\frac{2n}{3} - t}{\frac{2n}{3}} = \frac{\frac{n}{3}(\frac{2n}{3} - 1 + \frac{n}{3})}{2 \cdot \frac{2n}{3}} = \frac{n-3}{4}$$

Next we handle bins in the range $[\frac{n}{3}, \frac{2n}{3}]$. By reasoning similar to the previous calculations, the probability that no ball i, $\frac{n}{3} \leq i \leq \frac{2n}{3}$, falls into bin t in this range is $\frac{t - \frac{n}{3} - 1}{\frac{n}{3}}$. The probability that no ball i, $1 \leq i \leq \frac{n}{3}$ falls in bin t is $\Pi_{j=1}^{\frac{n}{3}}(1 - \frac{1}{\frac{2n}{3} - i + 1}) = \frac{\frac{n}{3}}{\frac{2n}{3}} = \frac{1}{2}$. Similarly, the probability that no ball i, $\frac{2n}{3} \leq i \leq n$ falls in bin t is $\frac{1}{2}$. Thus, with probability $\frac{t - \frac{n}{3}}{\frac{n}{3}} \cdot \frac{1}{4}$ no ball falls into bin t, $\frac{n}{3} \leq t \leq \frac{2n}{3}$.

To conclude, the expected number of empty bins in the ranges $[1, \frac{n}{3}]$ and $[\frac{2n}{3}, n]$ together is $\approx \frac{n}{2}$. The expected number of empty bins in $[\frac{n}{3}, \frac{2n}{3}]$ is $\Sigma_{t=\frac{n}{3}}^{\frac{2n}{3}} \frac{t - \frac{n}{3}}{\frac{n}{3}} \cdot \frac{1}{4} \approx \frac{1}{8} \cdot \frac{n}{3}$. In total, about $\frac{13}{24}$ of the bins are empty in expectation. We note that this constant can be somewhat increased to $\frac{4}{7}$ by recursively applying this construction on balls in the middle third (and keeping the other balls' time windows the same). Details are omitted from this extended abstract.

2.4 Limitations of Deterministic Tardy Mechanisms

Here we show that the prompt mechanism of Section 2.1 is optimal. In order to develop some intuition about tardy mechanisms, we start by showing that the greedy mechanism of [7] is tardy.

Recall that the greedy mechanism allocates item t to the bidder with the highest valuation that is present at time t (and has not been assigned any item yet).

Consider the following example: two bidders, red and green, arrive at time 1. The red bidder has a value of 10 for winning an item, and his departure time is 5. The green bidder has a value of 6 and a departure time of 1. We consider two scenarios: in the first one, four bidders arrive at time 2, each of them with value 100 and a departure time of 5. In the second scenario, there are no more arrivals. Observe that the greedy mechanism assigns the red bidder the first item. To see that the red bidder cannot learn his payment immediately, recall the following characterization of the payment in single-parameter mechanisms: the payment of a winning bidder is equal to the minimum value he can bid and still win.

In order to win an item in the first scenario, the red bidder must declare a value of at least 6, and therefore this is his payment in this scenario. However, in the second scenario a declaration of 0 will make him win the second item. The mechanism cannot distinguish between the two scenarios when the red bidder wins at time 1, and thus cannot determine the payment at time 1. We conclude that the greedy mechanism is tardy.

The following proposition shows that every prompt deterministic mechanism for the online auction problem achieves a competitive ratio of no better than $\frac{1}{2}$.

Proposition 1. *Every prompt deterministic mechanism for online auctions (even in the value-only model) has a competitive ratio of no better than* $\frac{1}{2}$.

Proof. Consider the following setting: two bidders arrive at time 1, each having a value of 1, and a departure time of time 2. Suppose there are no more arrivals of other bidders. Any mechanism that achieves a competitive ratio better than 2 must assign one bidder the first item, and the other item to the second bidder. Let a be bidder that was assigned the first item, and b be the bidder that was assigned the second item.

Claim. Let M be a prompt mechanism with a finite competitive ratio. In the scenario described above, there is no declaration of a value v_b that makes bidder b win the first item.

Proof. Let P_b denote the payment of bidder b for winning the second item with a declaration of 1. Observe that $p_b < 1$[1]. We consider two cases, one in which b declares a value of $w > 1$, and one in which b declares a value of $w < 1$.

Suppose that bidder b raises his bid from 1 to w, and was assigned the first item. The mechanism is prompt, so the payment of bidder b is determined immediately. Suppose, for a contradiction, that this payment is higher than p_b. In this case, if bidder b's true value was w, he could improve his profit by declaring a value of 1, and be assigned the second item. Hence the payment must be at most p_b. Clearly, the payment can not be strictly less than p_b, since otherwise if b's true value is 1, he has an incentive to declare a value of w and increase his profit. Thus the payment must be equal to p_b, but now we will see that this cannot be the case. Consider the following setting: b's true value is 1, and therefore he does not win the first item. At time 2 a bidder c with value $w' >> w$ arrives. Bidder

[1] If p_i is equal to 1, we add some "noise" to the value to get a strict inequality.

c is going to depart immediately. In order to maintain a finite competitive ratio the mechanism must assign bidder c the second item. Thus, if bidder b's true value is 1, he has an incentive to declare a value of w (and therefore win the first item for a payment of p_b), and the mechanism is not truthful.

The other case is where b bids a value w, $w < 1$, and thereby wins the first item (with payment less than 1). As before, if a bidder c with a departure time of 2 and a very high value arrives at time 2, then the mechanism must assign c the second item in order to guarantee a finite competitive ratio. If bidder b's true value is 1, he has an incentive to declare w instead, and win the first item. □

Now alter the scenario described above, and let b's value be $w >> 1$. By the claim, bidder b will not be assigned the first item. However, if at time 2 bidder c with a departure time of 2 and a value of w arrives, the total welfare the mechanism achieves is at most $1 + w$, while the optimal welfare is $2 \cdot w$. □

References

1. Bartal, Y., Chin, F.Y.L., Chrobak, M., Fung, S.P.Y., Jawor, W., Lavi, R., Sgall, J., Tichý, T.: Online competitive algorithms for maximizing weighted throughput of unit jobs. In: Diekert, V., Habib, M. (eds.) STACS 2004. LNCS, vol. 2996, pp. 187–198. Springer, Heidelberg (2004)
2. Chin, F.Y.L., Fung, S.P.Y.: Online scheduling with partial job values: Does time-sharing or randomization help? Algorithmica 37(3), 149–164 (2003)
3. Chrobak, M., Jawor, W., Sgall, J., Tichý, T.: Improved online algorithms for buffer management in QoS switches. In: Albers, S., Radzik, T. (eds.) ESA 2004. LNCS, vol. 3221, pp. 204–215. Springer, Heidelberg (2004)
4. Englert, M., Westermann, M.: Considering suppressed packets improves buffer management. In: SODA 2007 (2007)
5. Friedman, E.J., Parkes, D.C.: Pricing wifi at starbucks: issues in online mechanism design. In: EC 2003 (2003)
6. Goldberg, A.V., Hartline, J.D., Karlin, A.R., Saks, M., Wright, A.: Competitive auctions. In: Games and Economic Behavior (2006)
7. Hajiaghayi, M.T., Kleinberg, R., Mahdian, M., Parkes, D.C.: Adaptive limited-supply online auctions. In: EC 2005 (2005)
8. Kesselman, A., Lotker, Z., Mansour, Y., Patt-Shamir, B., Schieber, B., Sviridenko, M.: Buffer overflow management in QoS switches. In: STOC, pp. 520–529 (2001)
9. Lavi, R., Nisan, N.: Online ascending auctions for gradually expiring items. In: SODA 2005 (2005)
10. Lavi, R., Nisan, N.: Competitive analysis of incentive compatible on-line auctions. In: ACM Conference on Electronic Commerce, pp. 233–241 (2000)
11. Li, F., Sethuraman, J., Stein, C.: Better online buffer management. In: SODA 2007 (2007)
12. Mahdian, M., Saberi, A.: Multi-unit auctions with unknown supply. In: EC 2006 (2006)
13. Mu'alem, A., Nisan, N.: Truthful approximation mechanisms for restricted combinatorial auctions. In: AAAI 2002 (2002)
14. Nisan, N., Ronen, A.: Algorithmic mechanism design. In: STOC (1999)

A Truthful Mechanism for Offline Ad Slot Scheduling

Jon Feldman[1], S. Muthukrishnan[1], Evdokia Nikolova[2], and Martin Pál[1]

[1] Google, Inc.
{jonfeld,muthu,mpal}@google.com
[2] Massachusetts Institute of Technology*
nikolova@mit.edu

Abstract. We consider the *Offline Ad Slot Scheduling* problem, where advertisers must be scheduled to *sponsored search* slots during a given period of time. Advertisers specify a budget constraint, as well as a maximum cost per click, and may not be assigned to more than one slot for a particular search. We give a truthful mechanism under the utility model where bidders try to maximize their clicks, subject to their personal constraints. In addition, we show that the revenue-maximizing mechanism is not truthful, but has a Nash equilibrium whose outcome is identical to our mechanism. Our mechanism employs a descending-price auction that maintains a solution to a certain machine scheduling problem whose job lengths depend on the price, and hence are variable over the auction.

1 Introduction

Sponsored search is an increasingly important advertising medium, attracting a wide variety of advertisers, large and small. When a user sends a query to a search engine, the advertisements are placed into *slots*, usually arranged linearly down the page. These slots have a varying degree of exposure, often measured in terms of the probability that the ad will be clicked; a common model is that the higher ads tend to attract more clicks. The problem of allocating these slots to bidders has been addressed in various ways. The most common method is to allocate ads to each search independently via a *generalized second price* (GSP) auction, where the ads are ranked by (some function of) their bid, and placed into the slots in rank order. (See [16] for a survey of this area.)

There are several important aspects of sponsored search not captured by the original models. Most advertisers are interested in getting many clicks throughout the day on a variety of searches, not just a specific slot on a particular search query. Also, many advertisers have daily budget constraints. Finally, search engines may have some knowledge of the distribution of queries that will occur, and so should be able use that knowledge to make more efficient allocations.

The *Offline Ad Slot Scheduling* problem is this: given a set of bidders with bids (per click) and budgets (per day), and a set of slots over the entire day where we know the expected number of clicks in each slot, find a schedule that places

* This work was done while the author was visiting Google, Inc., New York, NY.

B. Monien and U.-P. Schroeder (Eds.): SAGT 2008, LNCS 4997, pp. 182–193, 2008.

bidders into slots. The schedule must not place a bidder into two different slots at the same time. In addition, we must find a price for each bidder that does not exceed the bidder's budget constraint, nor their per-click bid. (See below for a formal statement of the problem.)

A good algorithm for this problem will have high revenue. Also, we would like the algorithm to be *truthful*; i.e., each bidder will be incented to report her true bid and budget. In order to prove something like this, we need a *utility function* for the bidder that captures the degree to which she is happy with her allocation. Natural models in this context (with clicks, bids and budgets) are *click-maximization*—where she wishes to maximize her number of clicks subject to her personal bid and budget constraints, or *profit-maximization*—where she wishes to maximize her profit (clicks $\times$ profit per click). In this paper we focus on click-maximization.[1]

We present an efficient mechanism for *Offline Ad Slot Scheduling* and prove that it is truthful. Interpreted another way, truthfulness under click-maximization says that clicks are monotonic in both declared bids and budgets, which is an important fact even under other utility functions. We also prove that the revenue-maximizing mechanism for *Offline Ad Slot Scheduling* is not truthful, but has a Nash equilibrium (under the same utility model) whose outcome is equivalent to our mechanism; this result is strong evidence that our mechanism has desirable revenue properties. Our results generalize to a model where each bidder has a personal *click-through-rate* that multiplies her click probability.

As far as we can tell, this is the first treatment of sponsored search that directly incorporates both multiple positions and budget constraints into an analysis of incentives (see below for a survey of related work). In its full generality, the problem of sponsored search is more complex than our model; e.g., since the query distribution is noisy, good allocation strategies need to be online and adaptive. Also, our mechanism is designed for a single query type, whereas advertisers are interested in enforcing their budget across multiple query types. However, the tools used in this paper may be valuable for deriving more general mechanisms in the future.

Methods and Results. A natural mechanism for *Offline Ad Slot Scheduling* is the following: find a feasible schedule and a set of prices that maximizes revenue, subject to the bidders' constraints. It is straightforward to derive a linear program for this optimization problem, but unfortunately this is not a truthful mechanism (see Example 1 in Section 2). However, there is a direct truthful mechanism—the *price-setting* mechanism we present in this paper—that results in the same outcome as an equilibrium of the revenue-maximizing mechanism.

[1] Our choice is motivated by the presence of budgets, which have a natural interpretation in this application: if an overall advertising campaign allocates a fixed portion of its budget to online media, then the agent responsible for that budget is incented to spend the entire budget to maximize exposure. Also, our choice of utility function is out of analytical necessity: Borgs et al. [4] show that under some reasonable assumptions, truthful mechanisms are impossible under budgets and a profit-maximizing utility.

We derive this mechanism by starting with the single-slot case in Section 2, where two extreme cases have natural, instructive interpretations. With only bids (and unlimited budgets), a winner-take-all mechanism works; with only budgets (and unlimited bids) the clicks are simply divided up in proportion to budgets. Combining these ideas in the right way results in a natural descending-price mechanism, where the price (per click) stops at the point where the bidders who can afford that price have enough budget to purchase all of the clicks.

Generalizing to multiple slots requires understanding the structure of feasible schedules, even in the special budgets-only case. In Section 3 we solve the budgets-only case by characterizing the allowable schedules in terms of the solution to a classical *machine scheduling problem* (to be precise, the problem $Q \mid pmtn \mid C_{\max}$ [11]). The difficulty that arises is that the lengths of the jobs in the scheduling problem actually depend on the price charged. Thus, we incorporate the scheduling algorithm into a descending-price mechanism, where the price stops at the point where the scheduling constraints are tight; at this point a block of slots is allocated at a fixed uniform price (dividing the clicks proportionately by budget) and the mechanism iterates. We extend this idea to the full mechanism by incorporating bids analogously to the single-slot case: the price descends until the set of bidders that can afford that price has enough budget to make the scheduling constraints tight. Finally we show that the revenue-optimal mechanism has a Nash equilibrium whose outcome is identical to our mechanism.

Related Work. There are some papers on sponsored search that analyze the *generalized second-price* (GSP) auction, which is the auction currently in use at Google and Yahoo. The equilibria of this auction are characterized and compared with VCG [7,15,2,21]. Here the utility function is the *profit-maximizing* utility where each bidder attempts to maximize her clicks $\times$ profit per click, and budget constraints are generally not treated.

Borgs et al. [4] consider the problem of budget-constrained bidders for multiple items of a single type, with a utility function that is profit-maximizing, subject to being under the budget (being over the budget gives an unbounded negative utility). Our work is different both because of the different utility function and the generalization to multiple slots with a scheduling constraint. Using related methods, Mahdian et al. [18,17] consider an online stochastic setting.

Our mechanism can be seen as a generalization of Kelly's fair sharing mechanism [14,13] to the case of multiple slots with a scheduling constraint. Nguyen and Tardos [20] give a generalization of [13] to general polyhedral constraints, and also discuss the application to sponsored search. Both their bidding language and utility function differ from ours, and in their words their mechanism "is not a natural auction mechanism for this case." It would be interesting to explore further the connection between their mechanism and ours.

There is some work on algorithms for allocating bidders with budgets to keywords that arrive online, where the bidders place (possibly different) bids on particular keywords [19,17]. The application of this work is similar to ours, but their concern is purely online optimization; they do not consider the game-theoretic aspects of the allocation. Abrams et al. [1] derive a linear program for the offline optimization problem of allocating bidders to queries, and handle multiple positions by using variables for "slates" of bidders. Their LP is related to ours, but they do not consider game-theoretic questions.

In our setting one is tempted to apply a *Fisher Market* model: here m divisible goods are available to n buyers with money B_i, and $u_{ij}(x)$ denotes i's utility of receiving x amount of good j. It is known [3,8,5] that under certain conditions a vector of prices for goods exists (and can be found efficiently [6]) such that the *market clears*, in that there is no surplus of goods, and all the money is spent. The natural way to apply a Fisher model to a slot auction is to regard the slots as commodities and have the utilities be in proportion to the number of clicks. However this becomes problematic because there does not seem to be a way to encode the scheduling constraints in the Fisher model; this constraint could make an apparently "market-clearing" equilibrium infeasible, and indeed plays a central role in our investigations.

Our Setting. We define the *Offline Ad Slot Scheduling* problem as follows. We have $n > 1$ bidders interested in clicks. Each bidder i has a budget B_i and a maximum cost-per-click (max-cpc) m_i. Given a number of clicks c_i, and a price per click p, the utility u_i of bidder i is c_i if both the true max-cpc and the true budget are satisfied, and $-\infty$ otherwise. In other words, $u_i = c_i$ if $p \leq m_i$ and $c_i p \leq B_i$; and $u_i = -\infty$ otherwise. We have n' advertising slots where slot i receives D_i clicks during the time interval $[0, 1]$. We assume $D_1 > \ldots > D_{n'}$.

In a *schedule*, each bidder is assigned to a set of (slot, time interval) pairs $(j, [\alpha, \beta))$, where $j \leq n'$ and $0 \leq \alpha < \beta \leq 1$. A *feasible schedule* is one where no more than one bidder is assigned to a slot at any given time, and no bidder is assigned to more than one slot at any given time. (Formally, the intervals for a particular slot do not overlap, and the intervals for a particular bidder do not overlap.) A feasible schedule can be applied as follows: when a user query comes at some time $\alpha \in [0, 1]$, the schedule for that time instant is used to populate the ad slots. If we assume that clicks come at a constant rate throughout the interval $[0, 1]$, the number of clicks a bidder is expected to receive from a schedule is the sum of $(\beta - \alpha)D_j$ over all pairs $(j, [\alpha, \beta))$ in her schedule.[2]

A *mechanism* for *Offline Ad Slot Scheduling* takes as input a declared budget B_i and declared max-cpc (the "bid") b_i, and returns a feasible schedule, as well as a price per click $p_i \leq b_i$ for each bidder. The schedule gives some number c_i of clicks to each bidder i that must respect the budget at the given price; i.e., we have $p_i c_i \leq B_i$. The *revenue* of a mechanism is $\sum_i p_i c_i$. A mechanism is *truthful* if it is a weakly dominant strategy to declare one's true budget and max-cpc; i.e., for any bidder i, given any set of bids and budgets declared by the other bidders, declaring her true budget B_i and max-cpc m_i maximizes u_i. A (pure strategy) *Nash equilibrium* is a set of declared bids and budgets such that no bidder wants to change her declaration of bid or budget, given that all other declarations stay fixed. An ϵ-*Nash equilibrium* is a set of bids and budgets where no bidder can increase her u_i by more than ϵ by changing her bid or budget.

[2] All our results generalize to the setting where each bidder i has a "click-through rate" γ_i and receives $(\beta - \alpha)\gamma_i D_j$ clicks (see Section 4). We leave this out for clarity.

Throughout the paper we assume some arbitrary lexicographic ordering on the bidders, that does not necessarily match the subscripts. When we compare two bids b_i and $b_{i'}$ we say that $b_i \succ b_{i'}$ iff either $b_i > b_{i'}$, or $b_i = b_{i'}$ but i occurs first lexicographically.

2 Special Case: One Slot

In this section we consider the case $k = 1$, where there is only one advertising slot, with some number $D := D_1$ of clicks. We will derive a truthful mechanism for this case by first considering the two extreme cases of infinite bids and infinite budgets. The proofs of all the theorems in this paper can be found in [9].

Suppose all budgets $B_i = \infty$. Then, our input amounts to bids $b_1 \succ b_2 \succ \ldots \succ b_n$. Our mechanism is simply to give all the clicks to the highest bidder. We charge bidder 1 her full price $p_1 = b_1$. We claim that reporting the truth is a weakly dominant strategy for this mechanism. Clearly all bidders will report $b_i \leq m_i$, since the price is set to b_i if they win. The losing bidders cannot gain from decreasing b_i. The winning bidder can lower her price by lowering b_i, but this will not gain her any more clicks, since she is already getting all D of them.

Now suppose all bids $b_i = \infty$; our input is just a set of budgets $B_1, \ldots, B_n$, and we need to allocate D clicks, with no ceiling on the per-click price. Here we apply a simple rule known as *proportional sharing* (see [14,13]): Let $\mathcal{B} = \sum_i B_i$. Now to each bidder i, allocate $(B_i/\mathcal{B})D$ clicks. Set all prices the same: $p_i = p = \mathcal{B}/D$. The mechanism guarantees that each bidder exactly spends her budget, thus no bidder will report $B'_i > B_i$. Now suppose some bidder reports $B'_i = B_i - \Delta$, for $\Delta > 0$. Then this bidder is allocated $D(B_i - \Delta)/(\mathcal{B} - \Delta)$ clicks, which is less than $D(B_i/\mathcal{B})$, since $n > 1$ and all $B_i > 0$.

Greedy First-Price Mechanism. A natural mechanism for the general single-slot case is to solve the associated "fractional knapsack" problem, and charge bidders their bid; i.e., starting with the highest bidder, greedily add bidders to the allocation, charging them their bid, until all the clicks are allocated. We refer to this as the *greedy first-price* (GFP) mechanism. Though natural (and revenue-maximizing as a function of bids) this is easily seen to be not truthful:

Example 1. Suppose there are two bidders and $D = 120$ clicks. Bidder 1 has ($m_1 = \$2$, $B_1 = \$100$) and bidder 2 has ($m_2 = \$1$, $B_2 = \$50$). In the GFP mechanism, if both bidders tell the truth, then bidder 1 gets 50 clicks for \$2 each, and 50 of the remaining 70 clicks go to bidder 2 for \$1 each. However, if bidder 1 instead declares $b_1 = \$1 + \epsilon$, then she gets (roughly) 100 clicks, and bidder 2 is left with (roughly) 20 clicks.

The problem here is that the high bidders can get away with bidding lower, thus getting a lower price. The difference between this and the unlimited-budget case above is that a lower price now results in more clicks. It turns out that in equilibrium, this mechanism will result in an allocation where a prefix of the top bidders are allocated, but their prices equalize to (roughly) the lowest bid in the prefix (as in the example above).

The Price-Setting Mechanism. An equilibrium allocation of GFP can be computed directly via the following mechanism, which we refer to as the *price-setting (PS) mechanism.* Essentially this is a descending price mechanism: the price stops descending when the bidders willing to pay at that price have enough budget to purchase all the clicks. We have to be careful at the moment a bidder is added to the pool of the willing bidders; if this new bidder has a large enough budget, then suddenly the willing bidders have *more* than enough budget to pay for all of the clicks. To compensate, the mechanism decreases this "threshold" bidder's effective budget until the clicks are paid for exactly.

Price-Setting (PS) Mechanism (Single Slot)
• Assume wlog that $b_1 \succ b_2 \succ \ldots \succ b_n \geq 0$.
• Let k be the first bidder such that $b_{k+1} \leq \sum_{i=1}^{k} B_i/D$. Compute price $p = \min\{\sum_{i=1}^{k} B_i/D, b_k\}$.
• Allocate B_i/p clicks to each $i \leq k-1$. Allocate $\hat{B}_k/p$ clicks to bidder k, where $\hat{B}_k = pD - \sum_{i=1}^{k-1} B_i$.

Example 2. Suppose there are three bidders with $b_1 = \$2$, $b_2 = \$1$, $b_3 = \$0.25$ and $B_1 = \$100$, $B_2 = \$50$, $B_3 = \$80$, and $D = 300$ clicks. Running the PS mechanism, we get $k = 2$ since $B_1/D = 1/3 < b_2 = \$1$, but $(B_1 + B_2)/D = \$0.50 \geq b_3 = \$0.25$. The price is set to $\min\{\$0.50, \$1\} = \$0.50$, and bidders 1 and 2 get 200 and 100 clicks at that price, respectively. There is no threshold bidder.

Example 3. Suppose now bidder 2 changes her bid to $b_2 = \$0.40$ (everything else remains the same as Example 2). We still get $k = 2$ since $B_1/D = 1/3 < b_2 = \$0.40$. But now the price is set to $\min\{\$0.50, \$0.40\} = \$0.40$, and bidders 1 and 2 get 250 and 50 clicks at that price, respectively. Note that bidder 2 is now a threshold bidder, does not use her entire budget, and gets fewer clicks.

Theorem 1. *The price-setting mechanism (single slot) is truthful.*

Price-Setting Mechanism Computes Nash Equilibrium of GFP. Consider the greedy first-price auction in which the highest bidder receives B_1/b_1 clicks, the second B_2/b_2 clicks and so on, until the supply of D clicks is exhausted. It is immediate that truthfully reporting *budgets* is a dominant strategy in this mechanism, since when a bidder is considered, her reported budget is exhausted as much as possible, at a fixed price. However, reporting $b_i = m_i$ is *not* a dominant strategy. Nevertheless, it turns out that GFP has an equilibrium whose outcome is (roughly) the same as the PS mechanism. One cannot show that there is a plain Nash equilibrium because of the way ties are resolved lexicographically; the following example illustrates why.

Example 4. Suppose we have the same instance as example 1: two bidders, $D = 120$ clicks, $(m_1 = \$2, B_1 = \$100)$ and $(m_2 = \$1, B_2 = \$50)$. But now suppose that bidder 2 occurs first lexicographically. In GFP, if bidder 2 tells the truth, and bidder 1 declares $b_1 = \$1$, then bidder 2 will get chosen first (since she is first lexicographically), and take 50 clicks. Bidder 2 will end up with the remaining 70 clicks. However, if bidder 1 instead declares $b_1 = \$1 + \epsilon$ for some $\epsilon > 0$, then she gets $100/(1 + \epsilon)$ clicks. But this is not a best response, since she could bid $1 + \epsilon/2$ and get slightly more clicks.

Thus, we prove instead that the bidders reach an ϵ-Nash equilibrium:

Theorem 2. *Suppose the PS mechanism is run on the truthful input, resulting in price p and clicks $c_1, \ldots, c_n$ for each bidder. Then, for any $\epsilon > 0$ there is a pure-strategy ϵ-Nash equilibrium of the GFP mechanism where each bidder receives $c_i \pm \epsilon$ clicks.*

3 Multiple Slots

Generalizing to multiple slots makes the scheduling constraint nontrivial. Now instead of splitting a pool of D clicks arbitrarily, we need to assign clicks that correspond to a feasible schedule of bidders to slots. The conditions under which this is possible add a complexity that we need to incorporate into our mechanism.

As in the single-slot case it will be instructive to consider first the cases of infinite bids or budgets. Suppose all $B_i = \infty$. In this case, the input consists of bids only $b_1 \succ b_2 \succ \ldots \succ b_n$. Naturally, what we do here is rank by bid, and allocate the slots to the bidders in that order. Since each budget is infinite, we can always set the prices p_i equal to the bids b_i. By the same logic as in the single-slot case, this is easily seen to be truthful. In the other case, when $b_i = \infty$, there is a lot more work to do.

Without loss of generality, we may assume the number of slots equals the number of bids (i.e., $n' = n$); if this is not the case, then we add dummy bidders with $B_i = b_i = 0$, or dummy slots with $D_i = 0$, as appropriate. We keep this assumption for the remainder of the paper.

Assigning Slots Using a Classical Scheduling Algorithm. First we give an important lemma that characterizes the conditions under which a set of bidders can be allocated to a set of slots, which turns out to be just a restatement of a classical result [12] from scheduling theory.

Lemma 1. *Suppose we would like to assign an arbitrary set $\{1, \ldots, k\}$ of bidders to a set of slots $\{1, \ldots, k\}$ with $D_1 > \ldots > D_k$. Then, a click allocation $c_1 \geq \ldots \geq c_k$ is feasible iff*

$$c_1 + \ldots + c_\ell \leq D_1 + \ldots + D_\ell \quad \text{for all } \ell = 1, \ldots, k. \tag{1}$$

Proof. In scheduling theory, we say a *job* with *service requirement* x is a task that needs x/s units of time to complete on a *machine* with *speed* s. The question of whether there is a feasible allocation is equivalent to the following scheduling problem: Given k jobs with service requirements $x_i = c_i$, and k machines with speeds $s_i = D_i$, is there a schedule of jobs to machines (with preemption allowed) that completes in one unit of time?

As shown in [12,10], the optimal schedule for this problem (a.k.a. $Q|pmtn|C_{\max}$) can be found efficiently by the *level algorithm*, and the schedule completes in time $\max_{\ell \leq k}\{\sum_{i=1}^{\ell} x_i / \sum_{i=1}^{\ell} s_i\}$. Thus, the conditions of the lemma are exactly the conditions under which the schedule completes in one unit of time. □

A Multiple-Slot Budgets-Only Mechanism. Our mechanism will roughly be a descending-price mechanism where we decrease the price until a prefix of budgets fits tightly into a prefix of positions at that price, whereupon we allocate that prefix, and continue to decrease the price for the remaining bidders.

The following subroutine, which will be used in our mechanism (and later in the general mechanism), takes a set of budgets and determines a prefix of positions that can be packed tightly with the largest budgets at a uniform price p. The routine ensures that all the clicks in those positions are sold at price p, and all the allocated bidders spend their budget exactly.

Routine "Find-Price-Block"

Input: Set of n bidders, set of n slots with $D_1 > D_2 > \ldots > D_n$.

- If all $D_i = 0$, assign bidders to slots arbitrarily and exit.
- Sort the bidders by budget and assume wlog that $B_1 \geq B_2 \geq ... \geq B_n$.
- Define $r_\ell = \sum_{i=1}^{\ell} B_i / \sum_{i=1}^{\ell} D_i$. Set price $p = \max_\ell r_\ell$.
- Let ℓ^* be the largest ℓ such that $r_\ell = p$. Allocate slots $\{1, \ldots \ell^*\}$ to bidders $\{1, \ldots, \ell^*\}$ at price p, using all of their budgets; i.e., $c_i := B_i / p$.

Note that in the last step the allocation is always possible since for all $\ell \leq \ell^*$, we have $p \geq r_\ell = \sum_{i=1}^{\ell} B_i / \sum_{i=1}^{\ell} D_i$, which rewritten is $\sum_{i=1}^{\ell} c_i \leq \sum_{i=1}^{\ell} D_i$, and so we can apply Lemma 1. Now we are ready to give the mechanism in terms of this subroutine; an example run is shown in Figure 1.

Price-Setting Mechanism (Multiple Slots, Budgets Only)

- Run Find-Price-Block on bidders $1, \ldots, n$, and slots $1, \ldots, n$. This gives an allocation of ℓ^* bidders to the first ℓ^* slots.
- Repeat on the remaining bidders and slots until all slots are allocated.

Let $p_1, p_2, \ldots$ be the prices used for each successive block assigned by the algorithm. We claim that $p_1 > p_2 > \ldots$; to see this, note then when p_1 is set, we have $p_1 = r_k$ and $p_1 > r_\ell$ for all $\ell > k$, where k is the last bidder in the block. Thus for all $\ell > k$, we have $p_1 \sum_{j \leq \ell} D_j > \sum_{i \leq \ell} B_j$, which gives $p_1 \sum_{k<j \leq \ell} D_j > \sum_{k<i \leq \ell} B_j$ using $p_1 = r_k$. This implies that when we apply Find-Price-Block the second time, we get $r'_\ell = \sum_{k<i \leq \ell} B_j / \sum_{k<j \leq \ell} D_j < p_1$, and so $p_2 < p_1$. This argument applies to successive blocks to give $p_1 > p_2 > \ldots$.

Theorem 3. *The price-setting mechanism (multi-slot, budgets only) is truthful.*

The Price-Setting Mechanism (General Case). The generalization of the PS mechanism combines the ideas from the bids-and-budgets version of the single

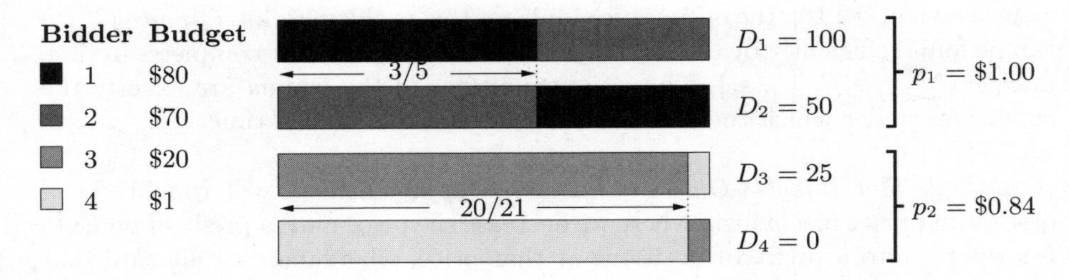

Fig. 1. An example of the PS mechanism (multiple slots, budgets only). The first application of Find-Price-Block computes $r_1 = B_1/D_1 = 80/100$, $r_2 = (B_1+B_2)/(D_1+D_2) = 150/150$, $r_3 = (B_1+B_2+B_3)/(D_1+D_2+D_3) = 170/175$, $r_4 = (B_1+B_2+B_3+B_4)/(D_1+D_2+D_3+D_4) = 171/175$. Since r_2 is largest, the top two slots make up the first price block with a price $p_1 = r_2 = \$1$; bidder 1 gets 80 clicks and bidder 2 gets 70 clicks, using the schedule as shown. In the second price block, we get $B_3/D_3 = 20/25$ and $(B_3 + B_4)/(D_3 + D_4) = 21/25$. Thus p_2 is set to $21/25 = \$0.84$, bidder 3 gets 500/21 clicks and bidder 4 gets 25/21 clicks, using the schedule as shown.

slot mechanism with the budgets-only version of the multiple-slot mechanism. As our price descends, we maintain a set of "active" bidders with bids at or above this price, as in the single-slot mechanism. These active bidders are kept ranked by *budget*, and when the price reaches the point where a prefix of bidders fits into a prefix of slots (as in the budgets-only mechanism) we allocate them and repeat. As in the single-slot case, we must be careful when a bidder enters the active set and suddenly causes an over-fit; in this case we again reduce the budget of this "threshold" bidder until it fits. We formalize this as follows:

Price-Setting Mechanism (General Case)
(i) Assume wlog that $b_1 \succ b_2 \succ \ldots \succ b_n = 0$.
(ii) Let k be the first bidder such that running Find-Price-Block on bidders $1, \ldots, k$ would result in a price $p \geq b_{k+1}$.
(iii) Reduce B_k until running Find-Price-Block on bidders $1, \ldots, k$ would result in a price $p \leq b_k$. Apply this allocation, which for some $\ell^* \leq k$ gives the first ℓ^* slots to the ℓ^* bidders among $1 \ldots k$ with the largest budgets.
(iv) Repeat on the remaining bidders and slots.

An example run of this mechanism is shown in Figure 2. Since the PS mechanism sets prices per slot, it is natural to ask if these prices constitute some sort of "market-clearing" equilibrium in the spirit of a Fisher market. The quick answer is no: since the price per click increases for higher slots, and each bidder values clicks at each slot equally, then bidders will always prefer the bottom slot. Note that by the same logic as the budgets-only mechanism, the prices $p_1, p_2, \ldots$ for each price block strictly decrease.

Theorem 4. *The price-setting mechanism (general case) is truthful.*

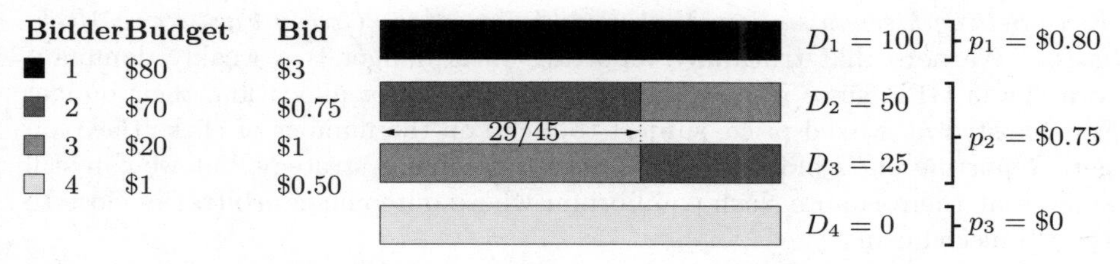

Fig. 2. Consider the same bidders and slots as in Figure 1, but now add bids as shown. Running Find-Price-Block on only bidder 1 gives a price of $r_1 = 80/100$, which is less than the next bid of \$1. So, we run Find-Price-Block on bidders 1 and 3 (the next-highest bid), giving $r_1 = 80/100$ and $r_2 = 100/150$. We still get a price of \$0.80, but now this is more than the next-highest bid of \$0.75, so we allocate the first bidder to the first slot at a price of \$0.80. We are left with bidders 2-4 and slots 2-4. With just bidder 3 (the highest bidder) and slot 2, we get a price $p = 20/50$ which is less than the next-highest bid of \$0.75, so we consider bidders 2 and 3 on slots 2 and 3. This gives a price of $\max\{70/50, 90/75\} = \$1.40$, which is more than \$0.50. Since this is also more than \$0.75, we must lower B_2 until the price is exactly \$0.75, which makes $B'_2 = \$36.25$. With this setting of B'_2, Find-Price-Block allocates bidders 2 and 3 to slots 2 and 3, giving 75(36.25/56.25) and 75(20/56.25) clicks respectively, at a price of \$0.75 per click. Bidder 4 is allocated to slot 4, receiving zero clicks.

Computational Efficiency. We give an $O(n^2)$ time algorithm for the PS mechanism, using the Gonzalez-Sahni algorithm [10] for scheduling related parallel machines as a subroutine (see [9] for details).

Greedy First-Price Mechanism for Multiple Slots. In the general case, as in the single-slot case, there is a natural *greedy first-price* (GFP) mechanism when the bidding language includes both bids and budgets: Order the bidders by bid $b_1 \succ b_2 \succ \ldots \succ b_n$. Starting from the highest bidder, for each bidder i compute the maximum possible number of clicks c_i that one could allocate to bidder i at price b_i, given the budget constraint B_i and the commitments to previous bidders $c_1, \ldots, c_{i-1}$. This reduces to the "fractional knapsack" problem in the single-slot case, and so one would hope that it maximizes revenue for the given bids and budgets, as in the single-slot case. This is not immediately clear, but does turn out to be true, as we will prove in this section.

As in the single-slot case, the GFP mechanism is not a truthful mechanism. However, we show that it does have a pure-strategy equilibrium, and that equilibrium has prices and allocation identical to the price setting mechanism.

Greedy is Revenue-Maximizing. Consider a revenue-maximizing schedule that respects both bids and budgets. We can assume wlog that each bidder i pays exactly b_i per click, since otherwise we can reduce the clicks c_i for bidder i and remain feasible with the same revenue. Thus, by Lemma 1, we can find a revenue-maximizing schedule $\mathbf{c}^* = (c_1^*, \ldots, c_n^*)$ by maximizing $\sum_i b_i c_i$ subject to $c_i \leq B_i/b_i$ and $c_1 + \ldots + c_\ell \leq D_1 + \ldots + D_\ell$ for all $\ell = 1, ..., n$.

Theorem 5. *The GFP auction gives a revenue-maximizing schedule.*

Price-Setting Mechanism is a Nash Equilibrium of the Greedy First Price Mechanism. We note that truthfully reporting one's budget is a weakly dominant strategy in GFP, since when a bidder is considered for allocation, their budget is exhausted at a fixed price, subject to a cap on the number of clicks they can get. Reporting one's bid truthfully is not a dominant strategy, but we can still show that there is an ϵ-Nash equilibrium whose outcome is arbitrarily close to the PS mechanism.

Theorem 6. *Suppose the PS mechanism is run on the truthful input, resulting in clicks $c_1, \ldots, c_n$ for each bidder. Then, for any $\epsilon > 0$ there is a pure-strategy ϵ-Nash equilibrium of the GFP mechanism where each bidder receives $c_i \pm \epsilon$ clicks.*

4 Conclusions

In this paper we have given a truthful mechanism for assigning bidders to click-generating slots that respects budget and per-click price constraints. The mechanism also respects a scheduling constraint on the slots, using a classical result from scheduling theory to characterize (and compute) the possible allocations. We have also proved that the revenue-maximizing mechanism has an ϵ-Nash equilibrium whose outcome is arbitrarily close to our mechanism. This final result in some way suggests that our mechanism is the right one for this model. It would interesting to make this more formal; we conjecture that a general truthful mechanism cannot do better in terms of revenue.

Extensions. There are several natural generalizations of the *Online Ad Slot Scheduling* problem where it would be interesting to extend our results or apply the knowledge gained in this paper. We mention a few here. *(i) Click-through rates.* In sponsored search (e.g. [7]) it is common for each bidder to have a personal click-through-rate γ_i; in our model this would mean that a bidder i assigned to slot j for a time period of length α would receive $\alpha\gamma_i D_j$ clicks. All our results can be generalized to this setting by simply scaling the bids using $b'_i = b_i\gamma_i$. However, our mechanism in this case does not necessarily prefer more *efficient* solutions; i.e., ones that generate more overall clicks. It would be interesting to analyze a possible tradeoff between efficiency and revenue in this setting. *(ii) Multiple Keywords.* To model multiple keywords in our model, we could say that each query q had its own set of click totals $D_{q,1} \ldots D_{q,n}$, and each bidder is interested in a subset of queries. The greedy first-price mechanism is easily generalized to this case: maximally allocate clicks to bidders in order of their bid b_i (at price b_i) while respecting the budgets, the query preferences, and the click commitments to previous bidders. It would not be surprising if there was an equilibrium of this extension of the greedy mechanism that could be computed directly with a generalization of the PS mechanism. *(iii) Online queries, uncertain supply.* In sponsored search, allocations must be made online in response to user queries, and some of the previous literature has focused on this aspect of the problem (e.g., [19,17]). Perhaps the ideas in this paper could be used to help make online allocation decisions using (unreliable) estimates of the supply, a setting considered in [17], with game-theoretic considerations.

Acknowledgments. We thank Cliff Stein and Yishay Mansour for helpful discussions and the anonymous reviewers for their suggestions.

References

1. Abrams, Z., Mendelevitch, O., Tomlin, J.: Optimal delivery of sponsored search advertisements subject to budget constraints. In: ACM Conference on Electronic Commerce (2007)
2. Aggarwal, G., Goel, A., Motwani, R.: Truthful auctions for pricing search keywords. In: ACM Conference on Electronic Commerce (EC) (2006)
3. Arrow, K., Debreu, G.: Existence of an equilibrium for a competitive economy. Econometrica 22, 265–290 (1954)
4. Borgs, C., Chayes, J.T., Immorlica, N., Mahdian, M., Saberi, A.: Multi-unit auctions with budget-constrained bidders. In: ACM Conference on Electronic Commerce (EC) (2005)
5. Deng, X., Papadimitriou, C., Safra, S.: On the complexity of equilibria. In: Proceedings of ACM Symposium on Theory of Computing (2002)
6. Devanur, N., Papadimitriou, C., Saberi, A., Vazirani, V.: Market equilibrium via a primal-dual-type algorithm (2002)
7. Edelman, B., Ostrovsky, M., Schwarz, M.: Internet advertising and the generalized second price auction: Selling billions of dollars worth of keywords. In: Second workshop on sponsored search auctions (2006)
8. Eisenberg, E., Gale, D.: Consensus of subjective probabilities: The pari-mutuel method. Annals Of Mathematical Statistics 30(165) (1959)
9. Feldman, J., Nikolova, E., Muthukrishnan, S., Pál, M.: A truthful mechanism for offline ad slot scheduling (2007), http://arxiv.org/abs/0801.2931
10. Gonzalez, T., Sahni, S.: Preemptive scheduling of uniform processing systems. Journal of the ACM 25, 92–101 (1978)
11. Graham, R.L., Lawler, E.L., Lenstra, J.K., Rinnooy Kan, A.H.G.: Optimization and approximation in deterministic sequencing and scheduling: a survey. Ann. Discrete Math. 4, 287–326 (1979)
12. Horvath, E.C., Lam, S., Sethi, R.: A level algorithm for preemptive scheduling. J. ACM 24(1), 32–43 (1977)
13. Johari, R., Tsitsiklis, J.N.: Efficiency loss in a network resource allocation game. Mathematics of Operations Research 29(3), 407–435 (2004)
14. Kelly, F.: Charging and rate control for elastic traffic. European Transactions on Telecommunications 8, 33–37 (1997)
15. Lahaie, S.: An analysis of alternative slot auction designs for sponsored search. In: ACM Conference on Electronic Commerce (EC) (2006)
16. Lahie, S., Pennock, D., Saberi, A., Vohra, R.: Sponsored Search Auctions. In: Algorithmic Game Theory, pp. 699–716. Cambridge University Press, Cambridge (2007)
17. Mahdian, M., Nazerzadeh, H., Saberi, A.: Allocating online advertisement space with unreliable estimates. In: ACM Conference on Electronic Commerce (2007)
18. Mahdian, M., Saberi, A.: Multi-unit auctions with unknown supply. In: ACM conference on Electronic commerce (EC) (2006)
19. Mehta, A., Saberi, A., Vazirani, U., Vazirani, V.: AdWords and generalized online matching. In: FOCS (2005)
20. Nguyen, T., Tardos, E.: Approximately maximizing efficiency and revenue in polyhedral environments. In: ACM Conference on Electronic Commerce (EC) (2007)
21. Varian, H.: Position auctions. International Journal of Industrial Organization 25(6), 1163–1178 (2007)

Alternatives to Truthfulness Are Hard to Recognize*

Vincenzo Auletta[1], Paolo Penna[1], Giuseppe Persiano[1], and Carmine Ventre[2,**]

[1] Dipartimento di Informatica ed Applicazioni, Università di Salerno, Italy
{auletta,penna,giuper}@dia.unisa.it
[2] Computer Science Department, University of Liverpool, UK
Carmine.Ventre@liverpool.ac.uk

Abstract. The central question in mechanism design is how to implement a given social choice function. One of the most studied concepts is that of *truthful* implementations in which truth-telling is always the best response of the players. The Revelation Principle says that one can focus on truthful implementations without loss of generality (if there is no truthful implementation then there is no implementation at all). Green and Laffont [1] showed that, in the scenario in which players' responses can be *partially verified*, the revelation principle holds only in some particular cases.

When the Revelation Principle does not hold, non-truthful implementations become interesting since they might be the only way to implement a social choice function of interest. In this work we show that, although non-truthful implementations may exist, they are hard to find. Namely, it is NP-hard to decide if a given social choice function can be implemented in a non-truthful manner, or even if it can be implemented at all. This is in contrast to the fact that truthful implementability can be recognized efficiently, even when partial verification of the agents is allowed. Our results also show that there is no "simple" characterization of those social choice functions for which it is worth looking for non-truthful implementations.

1 Introduction

Social choice theory deals with the fact that individuals (agents) have different preferences over the set of possible alternatives or outcomes. A social choice function maps these preferences into a particular outcome, which is not necessarily the one preferred by the agents. The main difficulty in implementing a social choice function stems from the fact that agents can *misreport* their preferences. Intuitively speaking, a social choice function can be implemented if there is a method for selecting the desired outcome which cannot be manipulated by

* Research funded by Università di Salerno and by the EU through IP AEOLUS.
** The author is also supported by DFG grant Kr 2332/1-2 within Emmy Noether Program.

B. Monien and U.-P. Schroeder (Eds.): SAGT 2008, LNCS 4997, pp. 194–205, 2008.

rational agents. By 'desired outcome' we mean the one specified by the social choice function applied to the *true* agents' preferences.

More precisely, each agent has a *type* which specifies the utility he derives if some outcome is selected. When agents are also endowed with payments, we consider agents with quasi linear utility: the type specifies the gross utility and the agent's utility is the sum of gross utility and payment received. In either case, a rational agent reports a type so to maximize his own utility and the reported type must belong to a *domain* consisting of all possible types. In the case of *partially verifiable* information, the true type of an agent further restricts the set of types that he can possibly report [1].

One of the most studied solution concepts is that of *truthful* implementations in which agents always maximize their utilities by truthfully reporting their types. The *Revelation Principle* says that one can focus on truthful implementations without loss of generality: A social choice function is implementable if and only if it has a truthful implementation. Green and Laffont [1] showed that, in the case of partially verifiable information, the Revelation Principle holds only in some particular cases. When the Revelation Principle does not hold, non-truthful implementations become interesting since they might be the only way to implement a social choice function of interest. Although a non-truthful implementation may induce some agent to misreport his type, given that he reports the type maximizing his utility, it is still possible to compute the desired outcome "indirectly". Singh and Wittman [3] observed that the Revelation Principle fails in several interesting cases and show sufficient conditions for the existence of non-truthful implementations.

1.1 Our Contribution

In this work, we study the case in which the declaration of an agent can be partially verified. We adopt the model of Green and Laffont [1] in which the ability to partially verify the declaration of an agent is encoded by a *correspondence* function M: $M(t)$ is the set of the possible declarations of an agent of type t. Green and Laffont [1] characterized the correspondences for which the Revelation Principle holds; that is, correspondences M for which a social choice function is either truthfully implementable or not implementable at all.

We show that although non-truthful implementations may exist, they are hard to find. Namely, it is NP-hard to decide if a given social choice function can be implemented for a given correspondence in a non-truthful manner. This is in contrast to the fact that it is possible to efficiently decide whether a social choice function can be truthfully implemented for a given correspondence. Our results show that there is no "simple" characterization of those social choice functions that violate the Revelation Principle. These are the social choice functions for which it is worth looking for non-truthful implementations since this might be the only way to implement them.

We prove these negative results for a very restricted scenario in which we have only one agent and at most two possible outcomes, and the given function does not have truthful implementations. We give hardness proofs both for the case in which payments are not allowed and the case in which payments are allowed and the agent has quasi linear utility.

In general payments are intended as a tool for enlarging the class of social choice functions that can be implemented. We find that there is a rich class of correspondences for which it is NP-hard to decide if a social choice function can be implemented *without* payments, while for the same correspondences it is trivial to test truthful implementable with payments via the approach in [3]. Finally, we complement our negative results by showing a class of correspondences for which there is an efficient algorithm for deciding whether a social choice function can be implemented.

We note that the characterization of Green and Laffont [1] has no direct implication in our results. Indeed, the property characterizing the Revelation Principle can be tested efficiently. Moreover, when the Revelation Principle does not hold, we only know that there exists *some* social choice function which is only implemented in a non-truthful manner. Hence, we do not know if the social choice function of interest can be implemented or not. Note that this question can be answered efficiently when the Revelation Principle holds since testing the existence of truthful implementations is computationally easy.

Road map. We introduce the model with partial verification by Green and Laffont [1] in Section 2. The case with no payments is studied in Section 3. Section 4 presents our results for the case in which payments are allowed and the agent has quasi linear utility. We draw some conclusions in Section 5.

2 The Model

The model considered in this work is the one studied by Green and Laffont [1] who considered the so called principal-agent scenario. Here there are two players: the agent, who has a type t belonging to a domain D, and the principal who wants to compute a social choice function $f : D \to \mathcal{O}$, where $\mathcal{O}$ is the set of possible of outcomes. The quantity $t(X)$ denotes the *utility* that an agent of type t assigns to outcome $X \in \mathcal{O}$.

The agent observes his type $t \in D$ and then transmits some message $t' \in D$ to the principal. The principal applies the outcome function $g : D \to \mathcal{O}$ to t' and obtains outcome $X = g(t')$. We stress that the principal fixes the outcome function g in advance and then the agent *rationally* reports t' so to maximize his utility $t(g(t'))$. Even though the principal does not exactly know the type of the agent, it is reasonable to assume that some *partial* information on the type of the agent is available. Thus the agent is restricted to report a type t' in a set $M(t) \subseteq D$, which is specified by a *correspondence* function $M : D \to 2^D$. We will only consider correspondences $M(\cdot)$ for which truth-telling is always an option; that is, for all $t \in D$, $t \in M(t)$. Notice that the case in which the principal has no information (no verification is possible) corresponds to setting $M(t) = D$ for all t.

Definition 1 ([1]). *A mechanism* (M, g) *consists of a correspondence* $M : D \to 2^D$ *and an* outcome function $g : D \to \mathcal{O}$. *The outcome function* g *induces a* *best response rule* $\phi_g : D \to D$ *defined by* $\phi_g(t) \in \arg\max_{t' \in M(t)}\{t(g(t'))\}$. *If* $t \in \arg\max_{t' \in M(t)}\{t(g(t'))\}$ *then we set* $\phi_g(t) = t$.

The correspondence M can be represented by a directed graph $\mathcal{G}_M$ (which we call the *correspondence graph*) defined as follows. Nodes of $\mathcal{G}_M$ are types in the domain D and an edge (t, t'), for $t \neq t'$, exists if and only if $t' \in M(t)$. We stress that the correspondence graph of M does not contain self-loops, even though we only consider correspondences M such that $t \in M(t)$ for all $t \in D$. We will often identify the correspondence M with its correspondence graph $\mathcal{G}_M$ and say, for example, that a correspondence is acyclic meaning that its correspondence graph is acyclic. Sometimes it is useful to consider a weighted version of graph $\mathcal{G}_M$. Specifically, for a function $g : D \to \mathcal{O}$, we define $\mathcal{G}_{M,g}$ to be the weighted version of graph $\mathcal{G}_M$ where edge (t, t') has weight $t(g(t)) - t(g(t'))$.

We study the class of M-implementable social choice functions $f : D \to \mathcal{O}$.

Definition 2 ([1]). *An outcome function* $g : D \to \mathcal{O}$ M-implements *social choice function* $f : D \to \mathcal{O}$ *if for all* $t \in D$ $g(\phi_g(t)) = f(t)$ *where* $\phi_g(\cdot)$ *is the best response rule induced by* g. *A social choice function* $f : D \to \mathcal{O}$ *is* M-implementable *if and only if there exists an outcome function* $g : D \to \mathcal{O}$ *that* M*-implements* f.

The social choice functions that can be truthfully M-implemented are of particular interest.

Definition 3 ([1]). *An outcome function* $g : D \to \mathcal{O}$ truthfully M-implements *social choice function* $f : D \to \mathcal{O}$ *if* g M*-implements* f *and* $\phi_g(t) = t$ *for all* $t \in D$. *A social choice function* $f : D \to \mathcal{O}$ *is* truthfully M-implementable *if and only if there exists an outcome function* $g : D \to \mathcal{O}$ *that truthfully* M*-implements* f.

The classical notions of *implementation* and of *truthful implementation* are obtained by setting $M(t) = D$ for all $t \in D$. Actually in this case the two notions of implementable social choice function and of truthfully implementable social choice function coincide due to the well-known revelation principle.

Theorem 1 (The Revelation Principle). *If no verification is possible (that is,* $M(t) = D$ *for all* $t \in D$*), a social choice function is implementable if and only if it is truthfully implementable.*

The Revelation Principle does not necessarily hold for the notion of M-implementation and of truthful M-implementation. Green and Laffont [1] indeed give a necessary and sufficient condition on M for the revelation principle to hold. More precisely, a correspondence M satisfies the *Nested Range Condition* if the following holds: for any $t_1, t_2, t_3 \in D$ if $t_2 \in M(t_1)$ and $t_3 \in M(t_2)$ then $t_3 \in M(t_1)$.

Theorem 2 (Green-Laffont [1]). *If* M *satisfies the NRC condition then a social choice function* f *is* M*-implementable if and only if* f *is* M*-truthfully implementable. If* M *does not satisfy the NRC condition then there exists an* M*-implementable social choice function* f *that is not truthfully* M*-implementable.*

Besides its conceptual beauty, the Revelation Principle can also be used in some cases to decide whether a given social choice function f is M-implementable for a given correspondence M. Indeed, if the Revelation Principle holds for correspondence M, the problem of deciding M-implementability is equivalent to the problem of deciding truthful M-implementability which, in turn, can be efficiently decided.

Theorem 3. *There exists an algorithm running in time polynomial in the size of the domain that, given a social choice function f and a correspondence M, decides whether f is truthfully M-implementable.*

Proof. To test truthful M-implementability of f we consider graph $\mathcal{G}_{M,f}$ where edge (t, t') has weight $t(f(t)) - t(f(t'))$. Then it is obvious that f is M-truthful implementable if and only if no edge of $\mathcal{G}_{M,f}$ has negative weight. □

3 Hardness of the Implementability Problem

In this section we prove that the following problem is NP-hard.

Problem 1. The IMPLEMENTABILITY problem is defined as follows.

INPUT: domain D, outcome set $\mathcal{O}$, social choice function $f : D \to \mathcal{O}$ and correspondence M.

TASK: decide whether there exists an outcome function g that M-implements f.

The following lemma, whose proof is immediate, gives sufficient conditions for an outcome function g to M-implement social choice function f.

Lemma 1. *For outcomes $\mathcal{O} = \{T, F\}$, if the following conditions are satisfied for all $a \in D$ then outcome function g M-implements social choice function f.*

1. *If $f(a) = T$ and $a(T) < a(F)$ then, for all $v \in M(a)$, we have $g(v) = T$.*
2. *If $f(a) = F$ and $a(T) < a(F)$ then, there exists $v \in M(a)$ such that $g(v) = F$.*
3. *If $f(a) = T$ and $a(T) > a(F)$ then, there exists $v \in M(a)$ such that $g(v) = T$.*
4. *If $f(a) = F$ and $a(T) > a(F)$ then, for all $v \in M(a)$, we have $g(v) = F$.*

The reduction. We reduce from 3SAT. Let Φ a Boolean formula in 3-CNF over the variables $x_1, \cdots, x_n$ and let $C_1, \cdots, C_m$ be the clauses of Φ. We construct D, $\mathcal{O}$, M and $f : D \to \mathcal{O}$ such that f is M-implementable if and only if Φ is satisfiable. We set $\mathcal{O} = \{T, F\}$. We next construct a correspondence graph $\mathcal{G}_M$ representing M. We will use variable gadgets (one per variable) and clause gadgets (one per clause).

The variable gadget for the variable x_i is depicted in Figure 1(a). Each variable x_i of the formula Φ adds six new types to the domain D of the agent, namely, t_i, u_i, v_i, w_i, z_i^1 and z_i^2 satisfying the following relations:

$$t_i(F) > t_i(T), \tag{1}$$

$$u_i(F) > u_i(T), \tag{2}$$

$$v_i(T) > v_i(F), \tag{3}$$

$$w_i(T) > w_i(F). \tag{4}$$

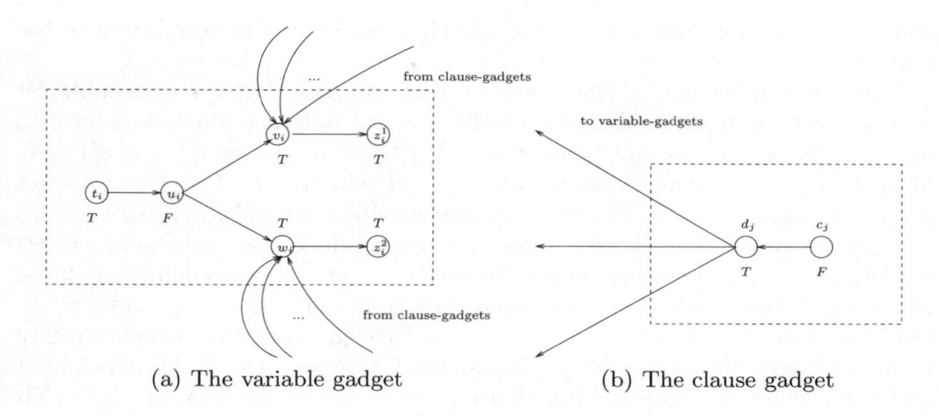

(a) The variable gadget (b) The clause gadget

Fig. 1. Gadgets used in the reduction

The labeling of the vertices defines the social choice function f; that is, $f(t_i) = T, f(v_i) = T, f(w_i) = T, f(z_i^1) = T, f(z_i^2) = T$, and $f(u_i) = F$. Directed edges of the gadget describe the correspondence M (rather the correspondence graph). Thus, for example, $M(t_i) = \{t_i, u_i\}$ and $M(u_i) = \{u_i, v_i, w_i\}$. Nodes v_i and w_i have incoming edges from the clause gadgets. The role of these edges will be clear in the following.

We observe that (1) implies that the social choice function f is not truthfully M-implementable. Indeed t_i prefers outcome $F = f(u_i)$ to $T = f(t_i)$ and $u_i \in M(t_i)$. Moreover, by Lemma 1, for any outcome function g implementing f we must have $g(t_i) = g(u_i) = T$. On the other hand, since $f(u_i) = F$ it must be the case that any g that M-implements f assigns outcome F to *at least* one node in $M(u_i) \setminus \{u_i\}$. Intuitively, the fact that every outcome function g that M-implements f must assign F to at least one between v_i and w_i corresponds to assigning "false" to respectively literal x_i and $\bar{x}_i$.

The clause gadget for clause C_j of Φ is depicted in Figure 1(b). Each clause C_j adds types c_j and d_j to the domain D of the agent such that

$$c_j(T) > c_j(F), \tag{5}$$

$$d_j(T) > d_j(F). \tag{6}$$

As before the labeling defines the social choice function f and we have $f(d_j) = T$ and $f(c_j) = F$. Moreover, directed edges encode correspondence M. Besides the directed edge (c_j, d_j), the correspondence graph contains three edges directed from d_j towards the three variable gadgets corresponding to the variables appearing in the clause C_j. Specifically, if C_j contains the literal x_i then d_j has an outgoing edge to node v_i. If C_j contains the literal $\bar{x}_i$ then d_j has an outgoing edge to node w_i. Similarly to the variable gadget, we observe that (5) implies that for any g M-implementing f it must be $g(d_j) = F$. Therefore, for g to M-implement f it must be the case that, for at least one of the neighbors a of d_j from a variable gadget, we have $g(a) = T$. We will see that this happens if and only if the formula Φ is satisfiable. This concludes the description of the reduction.

We next prove that the reduction is correct. Suppose that Φ is satisfiable, let τ be a satisfying truth assignment and let g be the outcome function defined as follows. For the i-th variable gadget we set $g(t_i) = g(u_i) = g(z_i^1) = g(z_i^2) = T$. Moreover, if x_i is true in τ, then we set $g(v_i) = T$ and $g(w_i) = F$; otherwise se set $g(v_i) = F$ and $g(w_i) = T$. For the j-th clause gadget, we set $g(d_j) = g(c_j) = F$.

Thus, to prove that the outcome function produced by our reduction M-implements f, it is sufficient to show for each type a the corresponding condition of Lemma 1 holds. We prove that conditions hold only for $a = u_i$ and $a = d_j$, the other cases being immediate. For u_i we have to verify that Condition 2 of Lemma 1 holds. Since τ is a truth assignment, for each i vertex u_i has a neighbor vertex for which the outcome function g gives F. For d_j we have to verify that Condition 3 of Lemma 1 holds. Since τ is a satisfying truth assignment, for each j there exists at least one literal of C_j that is true in τ; therefore, vertex d_j has a neighbor vertex for which the outcome function g gives T.

Conversely, consider an outcome function g which M-implements the social choice function f. This means that, for each clause C_j, d_j is connected to at least one node, call it a_j, from a variable gadget such that $g(a_j) = T$. Then the truth assignment that sets to true the literals corresponding to nodes $a_1, \cdots, a_m$ (and gives arbitrary truth value to the other variables) satisfies the formula.

The following theorem follows from the above discussion and from the observation that the reduction can be carried out in polynomial time and the graph we constructed is acyclic with maximum outdegree 3.

Theorem 4. *The* IMPLEMENTABILITY *Problem is NP-hard even for outcome sets of size* 2 *and acyclic correspondences of maximum outdegree* 3.

3.1 Corrrespondences with Outdegree 1

In this section, we study correspondences of outdegree 1.

We start by reducing the problem of finding g that M-implements f, for the case in which $\mathcal{G}_M$ is a line, to the problem of finding a satisfying assignment for a formula in 2CNF (that is every clause has at most 2 literals). We assume $D = \{t_1, \cdots, t_n\}$, $\mathcal{O} = \{o_1, \cdots, o_m\}$ and that, for $i = 2, \cdots, n$, $M(t_i) = \{t_i, t_{i-1}\}$ and $M(t_1) = \{t_1\}$. We construct a formula Φ in 2CNF in the following way. The formula Φ has the variables x_{ij} for $1 \leq i \leq n$ and $1 \leq j \leq m$. The intended meaning of variable x_{ij} being set to true is that $g(t_i) = o_j$. We will construct Φ so that every truth assignment that satisfies Φ describes g that M-implements f. We do so by considering the following clauses:

1. Φ contains clauses $(x_{if(t_i)} \vee x_{i-1f(t_i)})$, for $i = 2, \cdots, n$, and clause $x_{1f(t_1)}$. These clauses encode the fact that for g to M-implement f it must be the case that there exists at least one neighbor a of t_i in $\mathcal{G}_M$ such that $g(a) = f(t_i)$.
2. Φ contains clauses $(x_{ij} \rightarrow \overline{x}_{ik})$, for $i = 1, \cdots, n$ and for $1 \leq k \neq j \leq m$. These clauses encode the fact that g assigns at most one outcome to t_i.

3. Φ contains clauses $(x_{if(t_i)} \to \overline{x}_{i-1k})$ for all $i = 2, \cdots, n$ and for all k such that $t_i(o_k) > t_i(f(t_i))$.
   These clauses encode the fact that if g M-implements f and $g(t_i) = f(t_i)$ then agent of type t_i does not prefer $g(t_{i-1})$ to $g(t_i)$. Therefore, in this case t_i's best response is t_i itself.
4. Φ contains clauses $(x_{i-1f(t_i)} \to \overline{x}_{ik})$ for all $i = 2, \cdots, n$ and for all k such that $t_i(o_k) \geq t_i(f(t_i))$.
   These clauses encode the fact that if g M-implements f and $g(t_{i-1}) = f(t_i)$ then agent of type t_i does not prefer $g(t_i)$ to $g(t_{i-1})$. Therefore, in this case t_i's best response is t_{i-1}.

It is easy to see that Φ is satisfiable if and only if f is M-implementable. The above reasoning can be immediately extended to the case in which each node of $\mathcal{G}_M$ has outdegree at most 1 (that is $\mathcal{G}_M$ is a collection of cycles and paths). We thus have the following theorem.

Theorem 5. *The* IMPLEMENTABILITY *Problem can be solved in time polynomial in the sizes of the domain and of the outcome sets for correspondences of maximum outdegree* 1.

4 Implementability with Quasi Linear Utility

In this section we consider mechanisms with payments; that is, the mechanism picks an outcome and a payment to be transferred to the agent, based on the reported type of the agent. Therefore a mechanism is now a pair (g, p) where g is the outcome function and $p : D \to \mathbb{R}$ is the payment function. We assume that the agent has quasi linear utility.

Definition 4. *A mechanism (M, g, p) for an agent with quasi-linear utility is a triplet where $M : D \to 2^D$ is a correspondence, $g : D \to D$ is an outcome function, and $p : D \to \mathbb{R}$ is a payment function.*

The mechanism defines a best-response function $\phi_{(g,p)} : D \to D$ where $\phi_{(g,p)}(t) \in \arg\max_{t' \in M(t)}\{t(g(t')) + p(t')\}$. If $t \in \arg\max_{t' \in M(t)}\{t(g(t')) + p(t')\}$ then we set $\phi_g(t) = t$.

Definition 5. *The pair (g, p) M-implements social choice function $f : D \to \mathcal{O}$ for an agent with quasi-linear utility if for all $t \in D$, $g(\phi_{(g,p)}(t)) = f(t)$.*

The pair (g, p) truthfully M-implements social choice function f for an agent with quasi-linear utility if (g, p) M-implements f and, for all $t \in D$, $\phi_{(g,p)}(t) = t$.

In the rest of this section we will just say that (g, p) M-implements (or truthfully M-implements) f and mean that M-implementation is for agent with quasi-linear utility.

Testing truthful M-implementability of a social choice function f can be done in time polynomial in the size of the domain by using the following theorem that gives necessary and sufficient conditions. The proof is straightforward from the proof of [2] (see also [4]).

Theorem 6. *Social choice function f is truthfully M-implementable if and only if $\mathcal{G}_{M,f}$ has no negative weight cycle.*

Therefore, as in the previous case when payments were not allowed, if M has the NRC property then the Revelation Principle holds and the class of M-implementable social choice functions coincides with the class of truthfully M-implementable social choice functions. We next ask what happens for correspondences M for which the NRC property does not hold. Our answer is negative as we show that the following problem is NP-hard.

Problem 2. The QUASI-LINEAR IMPLEMENTABILITY problem is defined as follows.

INPUT: domain D, outcome set $\mathcal{O}$, social choice function $f : D \to \mathcal{O}$ and correspondence M.

TASK: decide whether there exists (g, p) that M-implements f.

We start with the following technical lemma.

Lemma 2. *Let M be a correspondence and let f be a social choice function for which correspondence graph has a negative-weight cycle $t \to t' \to t$ of length 2. If (g, p) M-implements f then*

$$\{\phi_{(g,p)}(t), \phi_{(g,p)}(t')\} \not\subseteq \{t, t'\}.$$

Proof. Let us assume for sake of contradiction that (g, p) M-implements f and that

$$\{\phi_{(g,p)}(t), \phi_{(g,p)}(t')\} \subseteq \{t, t'\}. \tag{7}$$

Since cycle $C := t \to t' \to t$ has weight

$$t(f(t)) - t(f(t')) + t'(f(t')) - t'(f(t)) < 0 \tag{8}$$

then $f(t) \neq f(t')$. Therefore, since (g, p) M-implements f, it holds $\phi_{(g,p)}(t) \neq \phi_{(g,p)}(t')$ and thus (7) implies that $\{\phi_{(g,p)}(t), \phi_{(g,p)}(t')\} = \{t, t'\}$.

Suppose that $\phi_{(g,p)}(t) = t'$ and thus $\phi_{(g,p)}(t') = t$. Then for (g, p) to M-implement f it must be the case that $g(t) = f(t')$, $g(t') = f(t)$. But then the payment function p must satisfy both the following:

$$p(t') + t(f(t)) \geq p(t) + t(f(t')),$$
$$p(t) + t'(f(t')) \geq p(t') + t'(f(t)),$$

which contradicts (8). The same argument can be used for the case $\phi_{(g,p)}(t) = t$ and $\phi_{(g,p)}(t') = t'$. □

The reduction. We are now ready to show our reduction from 3SAT to the QUASI-LINEAR IMPLEMENTABILITY problem. The reduction is similar in spirit to the one of the previous section. We start from a Boolean formula Φ in conjunctive normal form whose clauses contain exactly 3 literals and we construct a domain

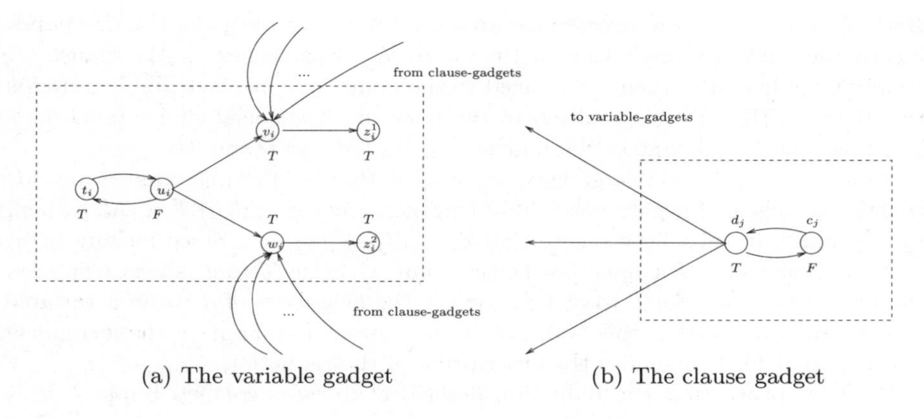

(a) The variable gadget (b) The clause gadget

Fig. 2. Gadgets used in the reduction

D, a set of outcomes $\mathcal{O}$, a social choice function f, and a correspondence M such that there exists (g, p) that M-implements f if and only if Φ is satisfiable.

We set $\mathcal{O} = \{T, F\}$ and fix constants $0 < \beta < \delta$. Let $x_1, \ldots, x_n$ be the variables and $C_1, \ldots, C_m$ be the clauses of Φ. The reduction uses two different gadgets: variable gadgets and clause gadgets.

We have one variable gadget for each variable; the gadget for x_i is depicted in Figure 2(a) where the depicted edges are edges of $\mathcal{G}_M$. Each variable x_i of the formula Φ adds six new types to the domain D: $t_i, u_i, v_i, w_i, z_i^1$, and z_i^2 satisfying the following:

$$t_i(T) - t_i(F) < u_i(F) - u_i(T), \tag{9}$$

$$u_i(T) - u_i(F) = \beta. \tag{10}$$

Nodes v_i and w_i have incoming edges from the clause gadgets. The role of these edges will be clear in the following. The labeling of the nodes describes the social choice function f to be implemented. More precisely, we have that $f(t_i) = f(v_i) = f(w_i) = f(z_i^1) = f(z_i^2) = T$ and $f(u_i) = F$.

We observe that, by (9), cycle $C := t_i \to u_i \to t_i$ has negative weight. Moreover, since $\phi_{(g,p)}(t_i) \in M(t_i) = \{t_i, u_i\}$, by Lemma 2, it must be the case that $\phi_{(g,p)}(u_i) \notin \{t_i, u_i\}$. Therefore, if (g, p) M-implements f then $g(\phi_{(g,p)}(u_i)) = f(u_i) = F$, and thus g assigns outcome F to *at least* one of the neighbors of u_i. Intuitively, the fact that the outcome function g assigns F to at least one between v_i and w_i corresponds to assigning "false" to literal x_i and $\bar{x}_i$.

We have one clause gadget for each clause; the gadget for clause C_j is depicted in Figure 2(b). Each clause C_j of Φ adds two new types to the domain D: c_j and d_j satisfying

$$c_j(F) - c_j(T) < d_j(T) - d_j(F), \tag{11}$$

$$d_j(T) - d_j(F) = \delta. \tag{12}$$

Node d_j has three edges directed towards the three variable gadgets corresponding to the variables appearing in the clause C_j. Specifically, if the clause C_j contains the literal x_i then d_j is linked to the node v_i. Conversely, if C_j contains the literal $\bar{x}_i$ then d_j is connected to the node w_i. The social choice function f is defined by the labeling of the nodes; that is, $f(d_j) = T$ and $f(c_j) = F$.

Similarly to the variable gadget, we observe that (11) implies that c_j and d_j constitute a cycle of negative weight of length 2. Since $\phi_{(g,p)}(c_j) \in \{c_j, d_j\}$, then, by Lemma 2, it must be the case that $\phi_{(g,p)}(d_j) \notin \{c_j, d_j\}$. Since for any (g,p) that M-implements f it must be the case that g assigns T to d_j's best response, then g assigns outcome T to at least one of the neighbors of d_j from a variable gadget. We will see that this happens for all clauses if and only if the formula Φ is satisfiable. This concludes the description of the reduction.

We next prove that the reduction described above is correct. Suppose Φ is satisfiable, let τ be a satisfying assignment for Φ, let γ be a constant such that $\beta < \gamma < \delta$ and consider the following pair (g,p). For $i = 1, \cdots, n$, we set $g(a) = T$ and $p(a) = 0$ for all nodes a of the variable gadget for x_i except for v_i and w_i. Then, if $\tau(x_i) = 1$, we set $g(v_i) = T$, $p(v_i) = 0$, $g(w_i) = F$ and $p(w_i) = \gamma$. If instead $\tau(x_i) = 0$, we set $g(v_i) = F$, $p(v_i) = \gamma$, $g(w_i) = T$ and $p(w_i) = 0$. For $j = 1, \cdots, m$, we set $g(c_j) = g(d_j) = F$ and $p(c_j) = p(d_j) = 0$.

We now show that (g,p) M-implements f. We show this only for types u_i from variable gadgets and types d_j from clause gadgets, as for the other types the reasoning is immediate. Notice that by definition, g assigns F to exactly one of v_i and w_i and T to the other. Thus, denote by a the vertex $a \in \{v_i, w_i\}$ such that $g(a) = F$ and by b the vertex $b \in \{v_i, w_i\}$ such that $g(b) = T$. We show that a is u_i's best response under (g,p). Observe that $u_i(g(a)) + p(a) = u_i(F) + \gamma > u_i(F) = u_i(g(t_i)) + p(t_i)$. Therefore t_i is not u_i's best response. On the other hand, we have $u_i(g(b)) + p(b) = u_i(T)$. But then, since $\gamma > \beta = u_i(T) - u_i(F)$, we have that a is u_i's best response under (g,p).

For d_j, we observe that, since τ satisfies clause C_j, there must exists at least one literal of C_j that is true under τ. By the definition of g, there exists at least one neighbor, call it a_j, of d_j from a variable gadget such that $g(a_j) = T$. We next show that a_j is d_j's best response. Notice that $p(a_j) = 0$. For all vertices b adjacent to d_j for which $g(b) = F$, we have $p(b) \leq \gamma$. But then, since $\gamma < \delta = d_j(T) - d_j(F)$ we have that a_j is d_j's best response under (g,p).

Conversely, consider an outcome function (g,p) that implements f and construct truth assignment τ as follows. Observe that, for any clause C_j, d_j and c_j constitute a cycle of negative weight and length 2. Moreover, c_j's best response is either c_j or d_j and thus, by Lemma 2, it must be the case that d_j's best response is a vertex, call it a_j, from a variable gadget such that $g(a_j) = T$. Then if $a_j = v_i$ for some i then we set $\tau(x_i) = 1$; if instead $a_j = w_i$ for some i we set $\tau(x_i) = 0$. Assignment τ (arbitrarily extended to unspecified variables) is easily seen to satisfy Φ.

The above discussion and the observation that the reduction can be carried out in polynomial time proves the following theorem.

Theorem 7. *The* QUASI-LINEAR IMPLEMENTABILITY *problem is NP-hard even for outcome sets of size* 2.

5 Conclusions

We have seen that is it NP-hard to decide if a given social choice function can be implemented even under the premise that the function does not admit a truthful implementation. Indeed, for these function it is NP-hard to decide if there is a non-truthful implementation, which in turn is the only way to implement them. An important factor here is the structure of the domain and the partial information, which we encode in the correspondence graph. In particular, we have the following results:

Correspondence Graph	No Payments	Payments and Quasi-linear Agent
Path	Polynomial [Th. 5]	Always implementable [3, Th. 4]
Directed acyclic	NP-hard [Th. 4]	Always implementable [3, Th. 4]
Arbitrary	NP-hard [Th. 4]	NP-hard [Th. 7]

Note that for directed acyclic graphs, the QUASI LINEAR IMPLEMENTABILITY Problem (where we ask implementability with payments) is trivially polynomial since all social choice functions are implementable wheras it is NP-hard to decide if an implementation without payments exists. So, it is also difficult to decide if payments are necessary or not for implementing a given function. Once again, this task becomes easy when restricting to truthful implementations [4].

Another interesting fact is that the problem without payments is not difficult because there are many possible outcomes, but because an agent may have several ways of misreporting his type. Indeed, the problem is easy if the agent has at most one way of lying (Theorem 5), but becomes NP-hard already for three (Theorem 4). The case of two remains open.

Finally, the fact that we consider the principal-agent model (the same as in [1]) only makes our negative results stronger since they obviously extend to the case of several agents (simply add extra agents whose corresponding function is $M(t) = \{t\}$).

On the other hand it remains open whether the positive result for graphs of outdegree at most 1 can be extended to many agents. Here the difficulty is the inter-dependance between the best response rules of the agents.

References

1. Jerry, R.: Green and Jean-Jacques Laffont. Partially Verifiable Information and Mechanism Design. The Review of Economic Studies 53, 447–456 (1986)
2. Rochet, J.-C.: A Condition for Rationalizability in a Quasi-Linear Context. Journal of Mathematical Economics 16, 191–200 (1987)
3. Singh, N., Wittman, D.: Implementation with partial verification. Review of Economic Design 6(1), 63–84 (2001)
4. Vohra, R.V.: Paths, cycles and mechanism design. Technical report, Kellogg School of Management (2007)

Distributed Algorithmic Mechanism Design and Algebraic Communication Complexity

Markus Bläser[1] and Elias Vicari[2,*]

[1] Computer Science, Saarland University
Postfach 151150, 66041 Saarbrücken, Germany
mblaeser@cs.uni-sb.de
[2] Institute of Theoretical Computer Science, ETH Zurich
8092 Zurich, Switzerland
vicariel@inf.ethz.ch

Abstract. In this paper, we introduce and develop the field of algebraic communication complexity, the theory dealing with the least number of messages to be exchanged between two players in order to compute the value of a polynomial or rational function depending on an input distributed between the two players. We define a general algebraic model, where the involved functions can be computed with the natural operations additions, multiplications and divisions and possibly with comparisons. We provide various lower bound techniques, mainly for fields of characteristic 0.

We then apply this general theory to problems from distributed mechanism design, in particular to the multicast cost sharing problem, and study the number of messages that need to be exchanged to compute the outcome of the mechanism. This addresses a question raised by Feigenbaum, Papadimitriou, and Shenker [9].

1 Introduction

Distributed algorithmic mechanism design is the art of analyzing, designing, and implementing mechanisms in a distributed setting. Feigenbaum, Papadimitriou, and Shenker [9] were the first to consider distributed aspects of mechanism design by investigating mechanisms for the multicast cost sharing problem in a distributed setting. Many mechanisms, like *marginal costs* and *Shaply value*, can be computed by arithmetic operations ("+", "$*$", and "/"), equality tests ("="), and comparisons ("<"). These are exactly the operations that are considered in *algebraic complexity theory*, a well-developed area of complexity theory. In the multicast cost sharing problem, a provider, the root of a tree, wants to send a transmission to n players, residing at the nodes of the tree. Each player has a utility u_i for receiving the transmission. The provider has to pay some cost c_e when he uses a particular edge e of the tree for the transmission. The utilities u_i are private information whereas the edge costs c_e are public. The nodes are capable of doing multicast, so edges are shared by the users in the tree. A mechanism is sought that is strategy-proof, fulfills three natural side constraints, and either maximizes welfare

* The author is partially supported by the Swiss National Science Foundation under grant number $200021 - 105236/1$.

B. Monien and U.-P. Schroeder (Eds.): SAGT 2008, LNCS 4997, pp. 206–217, 2008.

or is budget-balanced. We assume that there is no centralized authority that knows all the values. Instead the computation is done at the nodes of the trees and values have to be sent over the links. The mechanisms maximizing welfare have efficient distributed implementations with only a constant number of messages per link.

Feigenbaum, Papadimitriou, and Shenker [9] investigate the communication complexity of computing budget balanced mechanisms with linear operations, that is, only multiplication with scalars but not with variables are allowed. This implies that all the messages sent are linear combinations of the values held by the players. In this setting, they show that the computation of any budget balanced mechanism requires a linear number of messages over a linear number of links, yielding an overall quadratic lower bound. Any function can be computed by this many messages by sending all inputs to one player, so this lower bound is optimal. They leave the extension of their results to non-linear operations, i.e., multiplications and divisions, as an open question.

Later, Feigenbaum et al. [8] used Boolean communication complexity to prove lower bounds for the bit complexity of distributed algorithms for budget balanced mechanisms. In the Boolean model, all numbers that occur are rational numbers (given as the quotient of two natural numbers in binary). The messages can be arbitrary bit strings. In this model, Feigenbaum et al. prove that any distributed Boolean algorithm has to send at least a linear number of *bits* over a linear number of links.

While this seems to solve the problem at a first glance, there is a flaw hidden: If we measure the bit complexity of the messages, we have to relate it to the *bit size* of the input. In particular, in order to prove the mentioned lower bound, Feigenbaum et al. consider the following simple scenario: There is one link from the provider to one node v at which $n/2$ players reside and there is one link from v to another node u at which another $n/2$ players reside. By instantiating the utilities of the players appropriately, it turns out that any distributed algorithm that computes a budget-balanced mechanism on this tree decides whether the sets of utilities at the nodes u and v are disjoint. If we now consider this to be a game between two players, one at u, the other at v, this becomes the well-known set disjointness problem, however with a small twist, the universe is not $\{1, \dots, n\}$. It is known that deciding the set disjointness problem in the Boolean setting where both players hold the characteristic vector of their subset of $\{1, \dots, n\}$ requires n bits of communication. The scenario that we get from the multicast cost sharing problem above is a generalization, so we get the lower bound of n bits as well. In this example, the utilities of the players are of the form C/i for some constant C and $1 \leq i \leq n$. Thus the overall input size $\ell = \Theta(n \log n)$, but the lower bound for the communication is only $\Omega(n^2) = \Omega(\ell^2 / \log^2 \ell)$. (Note that we have to send n bits. If we replace the one link between u and v by a path of length n, we get n messages over n links, yielding the quadratic lower bound.) While this is still bad enough, it is not quadratic in the input size.

We will show that in the general algebraic model, we can get a quadratic lower bound. We provide tools to investigate the communication complexity of distributed computations in a general algebraic setting. This means, that the player may perform arbitrary arithmetic operations and not only linear ones, together with comparisons. The messages are now arbitrary rational functions in the inputs of the players. We will show that in this setting, the computation of any budget balanced mechanism needs a linear number of messages over a linear number of links. This is now a tight lower bound.

The reduction by Feigenbaum et al. to a two-player communcation problem indicates that it is sufficient to understand two-player communication complexity. This is what we do here: *Algebraic communication complexity* deals with the problem of computing a multivariate function, when the input is distributed between different entities referred to as *players*. In particular, given a field k and a rational function $f \in k(X_1, \ldots, X_n, Y_1, \ldots, Y_m)$, we want to know how many messages have to be exchanged between the two players, until one of them has enough information at his/her disposal to be able to compute the function value. Here, one player is holding the X-indeterminates and the other one the Y-indeterminates.[1]

The boolean counterpart, where one deals with boolean functions and bit-strings as inputs, is well studied, thanks to its successful application in VLSI-design theory. [12,13] are excellent textbooks on Boolean communication complexity. Compared to the boolean setting, we work over infinite fields and treat the arising numbers as entities. So our results do not follow from discrete models with larger (but finite) alphabets.

2 Related Research and New Results

The Boolean model was proposed by Yao [18] and was typically motivated by VLSI design problems. But it quickly found other applications, see e.g. [12,13]. It is mostly of combinatoric nature, whereas our model is based on algebraic structures like fields in which the possible operations are the natural arithmetic operations.

Abelson [1,2] motivated and introduced continuous communication complexity theory over $\mathbb{R}$ assuming some differentiability properties of the involved functions. Luo and Tsitsiklis [16] improved Abelson's results in certain cases making use of algebraic tools, but they only consider computional problems and no decision problems. The reason for this is that they only consider smooth messages and smooth problems.

Other than these work, there are also studies leading to more specific directions, like optimization within an error of a sum of two distributed convex functions, where every player has access to a single function [15].

Our aim is to develop a model based on an algebraic structure, in order to take advantage of the powerful tools of algebra and algebraic complexity theory. Even though this restricts the power of the messages the players can exchange and the family of the multivariate objective functions, we strongly believe that this model of computation fits better in the context of possible applications, since the power of the involved messages is realistically bounded. Moreover, in our setting we can introduce new features in a natural way. For instance by allowing equality tests and comparisons, we actually can deal with non-continuous functions. We can speak about *decision problems*, too, where the goal is not to compute the function, but decide whether a given input lies in its zero-set. Or we can introduce *nondeterminism* in a flavor recalling the usual nondeterminism of computational complexity.

[1] One can define the multi-party case accordingly. Due to space limitations, we solely deal with the two player case in this paper.

Feigenbaum, Papadimitriou, and Shenker [9] investigated the multicost cost sharing problem in a restricted algebraic setting. They only allow messages that are linear forms in the inputs (i.e. bids) of the players. They leave the proof of lower bounds in a general model as an open problem. We here prove a quadratic lower bound for the communication complexity of budget-balanced mechanisms for the multicast cost sharing problem in such a general model.

In particular, we show the following results:

- We completely solve the one-way case for the computation problem and give a characterization in terms of the transcendence degree.
- We almost completely solve the one-way case for the decision problem, where "almost" means up to one message.
- We provide several lower bounds techniques for the two-way case:
  - a method that is inspired by the substitution method known from algebraic complexity theory
  - a dimension bound (which is solely suited for homogeneous problems)
- We extend our methods to decision problems with equality tests and with comparisons.
- We apply our methods to distributed mechanism design problems, like the multicast cost sharing problem and auctions with single minded bidders.

In the full version, we also introduce nondeterminism and provide an interesting link to the decision and the computation problem in the one-way case.

Recently and independently, Grigoriev [11] introduced a model similar to ours, but he only deals with the one-way communication complexity. His model is actually an extension of the one-way model presented here: both parties send messages to some referee who then makes the decisions. We will use one of his results, which can be readily generalized to the two-way model, to deal with comparisons. Grigoriev allows randomization, too, but he mainly uses rank-based lower bounds.

3 Model of Computation and Notation

Throughout this paper, k is a field of characteristic zero which we often assume to be algebraically closed. Our mechanism design problems will be reduced to a two-player algebraic communication problem; so we define and analyze this setting first. We give the players names: Alice (A) and Bob (B). Alice usually holds an input denoted by $X = (x_1, \ldots, x_n)$ and Bob holds $Y = (y_1, \ldots, y_m)$ and their aim is to compute a rational function $f : k^n \times k^m \to k$. Each player may send messages that are rational functions in his/her inputs and the messages (s)he has received from the other player so far. In the end, one of them has to be able to compute $f(X, Y)$ (*computation problem*), to decide if the value is zero or not (*decision problem*), or to prove that indeed the input lies in the zero-set of f (*nondeterminism*).

We denote, as usual in the literature, by $k[X, Y]$ the ring of polynomials with variables $(x_1, \ldots, x_n, y_1, \ldots, y_m)$ and coefficients in k and by $k(X, Y)$ the field of rational functions over the same set of variables. Further, we denote by $M_{A \to B}$ and $M_{A \leftarrow B}$ the index-set of messages sent by Alice to Bob and vice versa, respectively. The network where messages are sent is completely reliable: there is no data loss and transmissions are error-free. In particular we assume that we can send field-elements as they are, meaning that every message counts as one, no matters how large the number is (for example as real number) and how we encode it. What is really important in this framework is the total number of messages sent during a protocol, neglecting the amount of computation performed by the players. In fact, our model relies on Luo and Tsitsiklis' model [16], but the assumptions on the involved functions are different: we deal solely with polynomial/rational functions. Later, we introduce equality tests and comparisons. In this extended model, we can also compute non-continuous functions and investigate decision problems, too.

A protocol P for computing a rational function $f \in k(X,Y)$ is a list of instructions telling the players the form of the messages they have to send and in which order, so to let a player to be able to compute $f(X,Y)$.

Definition 1 (Protocol). *A* two-way protocol *P for computing f consists of:*

1. *Disjoint inputs X, Y distributed between player A and player B, respectively.*
2. *A collection of messages $m_1, \dots, m_r$ belonging to some field extensions of k, sent in this order, with the following property: for each $1 \le i \le r$, we have:*
   - *if $i \in M_{A \to B}$, then $m_i \in k(X, m_1, \dots, m_{i-1})$*
   - *if $i \in M_{A \leftarrow B}$, then $m_i \in k(Y, m_1, \dots, m_{i-1})$*
3. *We have either $f \in k(X, m_1, \dots, m_r)$ or $f \in k(Y, m_1, \dots, m_r)$.*

P is called one-way *if in addition $M_{A \leftarrow B} = \emptyset$ and $f \in k(Y, m_1, \dots, m_r)$.*

Definition 2. *The* two-way communication complexity *of f is defined as*

$$\mathrm{C(f)} := \min_{\mathrm{P}} \mathrm{r(P)}$$

where the minimum is taken over the set of all protocols P for f and $r(P)$ is the number of messages sent in P. Similarly we define the one-way communication complexity $\mathrm{C}^{\rightarrow}(f)$.

A message m is said to be *feasible* in step i, if the second property in the definition of a protocol holds. If a message m is feasible, this exactly means that m can be computed by additions, multiplications, and divisions from the inputs of the particular player and all the messages he received so far.

We will also speak about divisionfree protocols. In this case, m is feasible if $m \in k[X, m_1, \dots, m_{i-1}]$ or $m \in k[Y, m_1, \dots, m_{i-1}]$, respectively. With such protocols, we can of course only compute polynomials.

Finally, we can also extend the notion of a protocol and communication complexity to a set of functions $f_1, \dots, f_\ell$. We just require $f_1, \dots, f_\ell \in k(X, m_1, \dots, m_r)$ or $f_1, \dots, f_\ell \in k(Y, m_1, \dots, m_r)$ in the third item of the definition of a protocol.

Example 1. Let

$$f_1(X,Y) = (y_1 + x_1 y_2 + x_1^2 y_3) x_1 + (y_1 + x_1 y_2 + x_1^2 y_3)^2 x_2 + (y_1 + x_1 y_2 + x_1^2 y_3)^3 x_3.$$

Then $\mathrm{C(f_1)} \le 2$. Indeed, it is easy to see, that if Alice sends to Bob the value of x_1, then he is able to compute the polynomial $y_1 + x_1 y_2 + x_1^2 y_3$, which, in turn, enables Alice to compute the whole function. One can easily show with the results of this paper, that in fact $\mathrm{C(f_1)} = 2$ but $\mathrm{C}^{\rightarrow}(f_1) = 3$.

4 One-Way Communication

In this section we deal exclusively with the one-way communication model. One-way communication corresponds to distributed one-pass algorithms. For the example of multicast pricing, this means that the result has to be computed in one sweep over the multicast tree. We provide a technique to compute exactly the value $\mathrm{C}^{\rightarrow}(f)$ for *every* rational function $f \in k(X,Y)$.

For such an $f \in k(X,Y)$ we denote by $\mathrm{Coeff}_Y\, f$ the field extension over k generated by adding the coefficients of f seen as function of Y. In particular, $\mathrm{Coeff}_Y\, f \subseteq k(X)$. Obviously, Bob is able to compute the value of f from the messages $m_1,\ldots,m_r$ received from Alice if and only if $\mathrm{Coeff}_Y\, f \subseteq k(m_1,\ldots,m_r)$.

Theorem 1 (Transcendence degree bound). *For every field k and rational function* $f \in k(X,Y)$*, we have* $\mathrm{C}^{\rightarrow}(f) \geq \mathrm{tr\,deg}_k\, \mathrm{Coeff}_Y\, f$ *where* tr deg *denotes the transcendence degree of the field extension* $\mathrm{Coeff}_Y\, f/k$.

Proof. We proceed by induction on $q := \mathrm{tr\,deg}_k\, \mathrm{Coeff}_Y\, f$. For $q = 0$, we have $f \in k(Y)$, since any X-variable cannot be algebraic over k.

For $q > 0$, let m_i be the first message such that $\mathrm{tr\,deg}_{k(m_1,\ldots,m_{i-1})}\, k(m_1,\ldots,m_i) = 1$. Then we have $\mathrm{tr\,deg}_{k(m_1,\ldots,m_i)}\, \mathrm{Coeff}_Y(f) = \mathrm{tr\,deg}_k\, \mathrm{Coeff}_Y(f) - 1$ and the induction hypothesis applies. Hence Alice sends at least $q - 1 + i \geq q$ messages. □

It is easy to prove with the Primitive Element Theorem from algebra (see [3]), that $\mathrm{tr\,deg}_k\, \mathrm{Coeff}_Y\, f + 1$ is an upper bound for $\mathrm{C}^{\rightarrow}(f)$: Alice simply sends a complete transcendence basis of $\mathrm{Coeff}_Y\, f$ over k to Bob using $\mathrm{tr\,deg}_k\, \mathrm{Coeff}_Y\, f$ messages. Furthermore, since fields of characteristic zero are separable, the Primitive Element Theorem for algebraic extensions assures that at most one more message makes the extended field equal to $\mathrm{Coeff}_Y\, f$. The following lemma helps us to strengthen this result.

Lemma 1. *Let* $q = \mathrm{tr\,deg}_k\, k(f_1,\ldots,f_r)$*, then there exist* $g_1,\ldots,g_q \in k(f_1,\ldots,f_r)$ *with the property that* $k(f_1,\ldots,f_r) \subseteq k(g_1,\ldots,g_q)$.

This lemma assures that Alice can send a transcendence basis $\{m_1,\ldots,m_q\}$ with the property that $\mathrm{Coeff}_Y\, f \subseteq k(m_1,\ldots,m_q)$, with $q = \mathrm{tr\,deg}_k\, \mathrm{Coeff}_Y\, f$, so that we do not need the Primitive Element Theorem anymore. This establishes $\mathrm{tr\,deg}_k\, \mathrm{Coeff}_Y\, f$ as the correct number of messages for every optimal protocol in the one-way communication model. It can be easily computed as the rank of the matrix, whose columns are given by the gradient of the coefficients of f, see [16].

5 Two-Way Communication

Next we study the two-way model. Here, we do not have a tight characterization as in the one-way case, but we provide several lower bound techniques that show tight bounds for some specific functions, in particular for the ones arising from our mechanism design problem.

5.1 Substitution Method

The idea of the substitution method is to make the first message trivial by adjoining it to the ground field. Since we deal with several ground fields in this section, we write occasionally $\mathrm{C(f; k)}$ instead of $\mathrm{C(f)}$ to stress the underlying ground field.

Lemma 2. *Let P be some protocol for computing a rational function f over some field k that uses r messages. Then there is an extension field $k' \supset k$ with $\operatorname{tr}\deg_k k' \leq 1$ and either $k' \subseteq k(X)$ or $k' \subseteq k(Y)$ and there is a protocol P' for computing f with $r-1$ messages over k'.*

Proof. Let m_1 be the first message of the protocol. Set $k' = k(m_1)$. Then we do not need to send the first message, since it is now known to both players in advance. The claim on the transcendence degree is obvious and also that either $k' \subseteq k(X)$ or $k' \subseteq k(Y)$, since m_1 is the first message. □

Example 2. Assume that Alice and Bob both hold n and m input values. We want to know whether they have at least one value in common. This is called the *set disjointness problem*. It is modelled by the following function

$$\mathrm{Disj}_{n,m}(X, Y) = \prod_{i=1}^{n} \prod_{j=1}^{m} (x_i - y_j).$$

It is now possible to show the following lower bound by induction.

Theorem 2. *For every rational function $g = p/q$ such that q is coprime with* $\mathrm{Disj}_{n,m}$, $\mathrm{C(g \cdot Disj_{n,m})} \geq \mathrm{min\{m, n\}}$.

Proof. The proof is by induction on $\min\{m, n\}$. Since q is coprime with $\mathrm{Disj}_{n,m}$, we need at least one message. If $\min\{m, n\}$ is 1, then the claim is trivial. Otherwise we want to apply Lemma 2. We treat the case $k' \subseteq k(X)$, the other case is symmetric. By renaming variables, we can assume that $x_1, \ldots, x_{n-1}, m_1$ are algebraically independent, where m_1 is the message substituted. $x_1, \ldots, x_{n-1}$ and $y_1, \ldots, y_m$ are algebraically independent over $k' = k(m_1)$ We can write $\mathrm{Disj}_{n,m} = g' \cdot \mathrm{Disj}_{n-1,m}$ where g' consists of all terms that contains x_n.

By the inductions hypothesis $\mathrm{C(gg' \cdot Disj_{n-1,m}; k')} \geq \mathrm{min\{n-1, m\}}$ and by Lemma 2, $\mathrm{C(g \cdot Disj_{n,m}; k)} \geq \mathrm{C(gg' \cdot Disj_{n-1,m}; k')} + 1$. This proves the claim. □

5.2 Dimension Bound

In this section, we show lower bounds using methods from algebraic geometry and in particular, we relate results about the dimension of a variety to our problem.

We here consider only homogeneous polynomials $f_1, \ldots, f_\ell$ and we want to decide whether the inputs of Alice and Bob lie in the zero set $V(f_1, \ldots, f_\ell)$ of all of them. In this setting, it makes sense to restrict oneself to *projective protocols*, i.e., protocols in which all messages are homogeneous polynomials. Now we can work over the projective space $\mathbb{P}^{2n} := \left(k^{2n+1} \setminus \{0\}\right) / \sim$, where $a \sim b$ iff there exists a $0 \neq \lambda \in k$ such that

$a = \lambda b$. The inputs of the players are embedded canonically into the projective space by adding a 1 as the last coordinate. The input $(x_1, \ldots, x_n, y_1, \ldots, y_n)$ is mapped to the $(x_1, \ldots, x_n, y_1, \ldots, y_n, 1)$. Whenever we speak of the point $(0, 0)$, we mean the corresponding point $(0, \ldots, 0, 0, \ldots, 0, 1)$ in $\mathbb{P}^{2n}$. Since we always work in the affine subspace of $\mathbb{P}^{2n}$ of points with the last coordinate $\neq 0$, we will usually omit the last 1.

Theorem 3. *Let k be an algebraically closed field and $f : k^n \times k^n \to k^\ell$ be a homogeneous polynomial mapping. Assume that $\dim V(f(0, Y)) = q < n$. Then deciding whether $(X, Y) \in V(f) = \{(X, Y) \in k^n \times k^n \mid f(X, Y) = 0\}$ requires at least $n - q$ homogeneous messages even in the two-way model.*

Proof. Let P be a homogeneous protocol where w.l.o.g. Alice is able to eventually decide whether $f(X, Y) = 0$ for $(X, Y) \in k^{2n}$. We embed (X, Y) into $\mathbb{P}^{2n}$ by adding a component 1 as described above. For contradiction, we assume that $n - q - 1$ messages are enough for Alice to decide whether $f(X, Y) = 0$; we denote them by $m_1, \ldots, m_{n-q-1}$. Since the messages are homogeneous, it follows in particular that $m_i(0, 0) = 0$, $1 \leq i \leq n - q - 1$. Also $M := V(m_1, \ldots, m_{n-q-1})$ is a projective variety in $\mathbb{P}^{2n}$ with $\dim M \geq n + q + 1$.

Consider the variety $E := V(x_1, \ldots, x_n)$ of dimension n. By the properties of the projective space it follows that $\dim M \cap E \geq q + 1$. On the other hand, $\dim V(f(0, Y)) = q$, so we can find a point $(0, b)$ in the intersection $M \cap E$ with $f(0, b) \neq 0$. If we run P on the inputs $(0, 0)$ and $(0, b)$, we notice that Alice has the same input and receives vanishing messages in both instances. Hence she is unable to distinguish between $f(0, 0) = 0$ and $f(0, b) \neq 0$; this is a contradiction. □

Example 3. This theorem applies in a straightforward manner to the *equality problem*. We want to decide whether the two inputs are identical. The problem is modelled by the n functions $f_1 = x_1 - y_1, \ldots, f_n = x_n - y_n$. Let $f(X, Y) = (f_1(X, Y), \ldots, f_n(X, Y))$. Of course we have $\dim V(f(0, Y)) = 0$, since only $(0, 0)$ belongs to it. Thus any homogeneous protocol for deciding whether the two inputs are the same requires n messages.

Remark 1. This argument is interesting in another important context as well. If the theorem should extend to the equality problem over every field k of characteristic zero not necessarily closed, then this would rule out the existence of an injective polynomial $p \in k[X, Y]$, which is still an open problem for general fields. On the other hand note that the equality problem can be solved with one message on certain ground sets. For instance, the polynomial $(a, b) \mapsto \frac{1}{2}\big((a + b)^2 + 3a + b\big)$ is a bijection over $\mathbb{N}^2 \to \mathbb{N}$.

6 Decision Problems

Now we allow the players to perform equality tests and comparisons, the latter only over $\mathbb{R}$. In this setting, players do not only compute and send the same messages for every input, but they are allowed to go through a decision tree, where at every node they check whether some two functions of their input are equal. Every internal node in the tree has two successors, one corresponds to the outcome of the fact that the compared functions are equal, one to the outcome that they are not equal. Depending on the result of the test, the players follow the respective branch in the decision tree. When they reach a leaf, they send a message accordingly. Furthermore, we do not want to compute f but it is sufficient to decide whether the input of the players lies in the zero set $V(f)$ of f or not.

If we allow comparisions, then we can also decide semi-algebraic sets. Every test now has three outcomes, $<$, $=$, or $>$, and every node in the decision tree has three successors. The formal definitions can be found in the full version of this paper.

The communication complexities $\mathrm{C}^{\rightarrow}_{\mathrm{dec}}(f)$ and $\mathrm{C}_{\mathrm{dec}}(f)$ are defined in the same way as $\mathrm{C}^{\rightarrow}(f)$ and C(f), but over the larger class of protocols with equality tests. Therefore, it is clear that $\mathrm{C}^{\rightarrow}_{\mathrm{dec}}(f) \leq \mathrm{C}^{\rightarrow}(f)$ and $\mathrm{C}_{\mathrm{dec}}(f) \leq \mathrm{C(f)}$ for all rational functions f. In the same way, we define $\mathrm{C}^{\rightarrow}_{\mathrm{dec},<}(S)$ and $\mathrm{C}_{\mathrm{dec},<}(S)$ for semi-algebraic sets S. $\mathrm{C}^{\rightarrow}_{\mathrm{dec},<}(f)$ and $\mathrm{C}_{\mathrm{dec},<}(f)$ shorthand $\mathrm{C}^{\rightarrow}_{\mathrm{dec},<}(V(f))$ and $\mathrm{C}_{\mathrm{dec},<}(V(f))$.

Intuitively, one would expect that these additions would indeed decrease the number of messages needed between the two players. We will see that often, this decrease is very modest.

6.1 Equality Tests

Lemma 3. *For every irreducible polynomial $f \in k[X,Y]$ over an algebraically closed field, there is a rational function $h = p/q$ with q and f being coprime, such that $\mathrm{C}^{\rightarrow}_{\mathrm{dec}}(f) \geq \mathrm{C}^{\rightarrow}(hf)$ and $\mathrm{C}_{\mathrm{dec}}(f) \geq \mathrm{C(hf)}$.*

Proof. Consider a protocol P, one-way or two-way, deciding the membership of the inputs in $\Omega := V(f)$, the variety defined by f. Since the possible inputs are infinite (k is closed), almost every input (in the Zariski sense) follows the same path π_0. Let π be the typical path of an element from Ω and let ν be the node where π and π_0 separate for the first time. Following the path π_0 up to ν, we find rational functions $g_1^{(1)}, g_1^{(2)}, \ldots, g_r^{(1)}, g_r^{(2)}$ such that $g_i^{(1)} \stackrel{?}{=} g_i^{(2)}$ is tested by some player. Obviously $g_i := g_i^{(1)} - g_i^{(2)}$ is not identically zero and because we follow the path taken by most inputs, $g_1, \ldots, g_{r-1}$ do not vanish on the given input when we follow π or π_0. Since Ω is a closed set in the Zariski topology, it follows that the elements of Ω reaching ν lie also in $V(g_r)$. Altogether, for an input (X,Y), from $f(X,Y) = 0$, we necessarily have that at least one of $g_1(X,Y), \ldots, g_r(X,Y)$ vanishes, in other words $g := g_1 \cdot \ldots \cdot g_r$ vanishes on (X,Y). Therefore, applying the Nullstellensatz on the numerator of g and noting that $\mathrm{rad}(f) = (f)$ (since f is irreducible), we have $g = h \cdot f$, for a rational function $h = p/q$, q coprime with f. Thus from the protocol P that decides $V(f)$, we get a protocol of the same type that computes hf. □

For the one-way case, we get an almost tight characterisation meaning that performing equality tests does not bring any significant help in the communication task.

Corollary 1. *Given an irreducible polynomial $f \in k[X,Y]$ over an algebraically closed field k, $\mathrm{C}^{\rightarrow}_{\mathrm{dec}}(f) \geq \mathrm{tr\,deg}_k\, \mathrm{Coeff}_Y\, f - 1$.*

Proof. We have $\mathrm{tr\,deg\,Coeff}_Y\, g \geq \mathrm{tr\,deg\,Coeff}_Y\, f - 1$, see [6]. □

We cannot apply Lemma 3 to $\mathrm{Disj}_{n,m}$, since $\mathrm{Disj}_{n,m}$ is not irreducible.

Lemma 4. *Let $\ell_1, \ldots, \ell_t$ be linear forms such that any two of them are linearly independent. Let $L = \ell_1 \cdots \ell_t$. Then there is a rational function $h = p/q$ with L and q being coprime, such that* $\mathrm{C_{dec}}(L) \geq \mathrm{C(hL)}$.

Proof. Consider a protocol P for deciding L and let π_0 be the path taken by almost all inputs. Let π_τ be the path taken by almost all inputs in $V(\ell_\tau)$. Let v_τ be the node where these two paths separate for the first time. Let $g_\tau = p_\tau/q_\tau$ be the rational function tested at v_τ. By Gauss' Lemma, $\ell_\tau | p_\tau$. Thus $g_1 \cdots g_t$ can be written as hL. □

Corollary 2. $\mathrm{C_{dec}}(\mathrm{Disj}_{n,m}) \geq \min\{n, m\}$.

6.2 Comparisons

Consider the lexicographic monomial ordering on $y_n, \ldots, y_1, x_n, \ldots, x_1$, i.e., a monomal m is smaller than an other one m' if the exponent vector of m is smaller than that of m' in the lexicographic ordering (with variables in the order as above). For a polynomial p, $\mathrm{lt}(p)$ denotes the least term with respect to the chosen monomial ordering. Grigoriev [11] essentially proves the following result.

Lemma 5. *Let $\ell_1, \ldots, \ell_n$ be linear forms such that $x_1, \ldots, x_n, \ell_1, \ldots \ell_n$ are linearly independent. Let V be the union of some hyperplanes, among them $V(\ell_1), \ldots, V(\ell_n)$. Then there is a polynomial f such that* $\mathrm{C_{dec,<}}(V) \geq \mathrm{C(f)}$ *and $\ell_1, \ldots, \ell_n$ divide* $\mathrm{lt}(f)$. *The same holds for the seminalgebraic set S defined by $\ell_1 > 0, \ldots, \ell_t > 0$.*

But now the substitution method from Section 5.2 allows us to get rid of one message for each linear form ℓ_i that divides $\mathrm{lt}(f)$ and get a bound for $\mathrm{C}^{\rightarrow}_{\mathrm{dec},<}(\ell_1, \ldots, \ell_n)$ in this way. Note that every indeterminate is substituted by a polynomial, so the $\mathrm{lt}(f)$ does not change and we can perform induction.

Corollary 3. $\mathrm{C_{dec,<}}(\mathrm{Disj}_{n,m}) = \min\{n, m\}$.

Proof. $\mathrm{Disj}_{n,m}$ is a product of linear forms, among them $x_1 - y_1, \ldots, x_n - y_n$. From the lemma above, we get that these linear forms divide $\mathrm{lt}(f)$. Now we can apply the substitution method to f in the same manner as we did before to $\mathrm{Disj}_{n,m}$. □

In the same way, we can show that the set defined by $x_1 > y_1, \ldots, x_n > y_n$ has communication complexity equal to n. Grigoriev shows the last two bounds using rank based methods.

7 Applications

Using the reduction by Feigenbaum et al. [8] and applying Corollary 3, we get the following lower bound for the multicast cost sharing problem.

Theorem 4. *There is a tree with n players such that every algebraic algorithm with comparisons and equality tests that computes a strategyproof and budget-balanced mechanism for the multicast cost sharing problem sends at least n messages over a linear number of links.*

As a second example, we consider a distributed version of a combinatorial auction with single minded bidders, see e.g. [14,4]. We have n players and a collection of objects. There is a partial order on these objects. Each player bids on exactly one of the objects. A selection algorithm gets the bids of the players and selects those players that get an object. Such an algorithm is called monotone if a selected player that now makes a higher or equal bid on a lesser or equal object still is selected (assuming that all other players bids are the same). From such an algorithm, one can construct a truthful mechanism via the critical value scheme, see [14,4] for details. We now assume that the players are distributed in a network. The players can send their bids along the links and there has to be one (arbitrary) node at which the selection process takes places. Consider the following scenario: we have a network of two subgraphs connected by one link. Each subgraph contains n players. We have n objects and one player of each subgraph bids on one of them. In particular, such a mechanism can decide whether the bids in one subgraphs are all larger than the corresponding bids in the other subgraph. But this is exactly the semialgebraic set defined by $x_1 > y_1, \dots, x_n > y_n$, which has communication complexity n.

Theorem 5. *Any distributed truthful critical value scheme for a combinatorial auction with single minded bidders needs to send at least n messages over linearly many links in the worst case.*

8 Conclusions and Open Problems

From the point of distributed mechanism design, it would be nice to find more examples to which our techniques can be applied.

From the point of algebraic complexity theory, the most interesting open problem is to relate the complexity of f to the complexity of $g \cdot f$ in the two-way model. This would directly show that any lower bound for the computation problem is also a lower bound for the decision problem (Lemma 3). For instance, if the rank bound of Abelson [2] yields a lower bound of q for the complexity of f, then it yields a lower bound for $q-2$ for $g \cdot f$, under some strict assumptions on the functions. However, there are cases where the rank bound is not tight at all. It is not clear to us how to prove the general case. We conclude with the following conjectures.

Conjecture 1. For all irreducible polynomials f over algebraically closed fields,

1. $\mathrm{C(f)} \geq \mathrm{C(gf)} - 1$ for every coprime polynomial $g \neq 0$ and
2. $\mathrm{C(f)} \geq \mathrm{C(f^j)} - 1$ for every $j \geq 1$.

If both conjectures were true, then the decision complexity and the nondetermistic complexity (see the full version for definitions and results) would be closely related to the communicitation complexity of computing f. The case $f = x_1 y_1 + x_1^2 x_2 y_2$, $g = x_2$ shows that $\mathrm{C(f)} \geq \mathrm{C(gf)}$ is in general not true.

Acknowledgments

We would like to thank L. Shankar Ram and Andreas Meyer for carefully reading an early draft of this paper and in particular the latter for the help provided in proving Lemma 1.

References

1. Abelson, H.: Towards a theory of local and global in computation. Theoret. Comput. Sci. 6(1), 41–67 (1978)
2. Abelson, H.: Lower bounds on information transfer in distributed computations. J. Assoc. Comput. Mach. 27(2), 384–392 (1980)
3. Bosch, S.: Algebra, 3rd edn. Springer, Heidelberg (1999)
4. Briest, P., Krysta, P., Vöcking, B.: Approximation techniques for utilitarian mechanism design. In: Proc. ACM Symp. on Theory of Computing (2005)
5. Bürgisser, P., Clausen, M., Amin Shokrollahi, M.: Algebraic Complexity Theory. Springer, Heidelberg (1997)
6. Bürgisser, P., Lickteig, T.: Test complexity of generic polynomials. J. Complexity 8, 203–215 (1992)
7. Chen, P.: The communication complexity of computing differentiable functions in a multicomputer network. Theoret. Comput. Sci. 125(2), 373–383 (1994)
8. Feigenbaum, J., Krishnamurthy, A., Sami, R., Shenker, S.: Hardness results for Multicast Cost Sharing. In: Agrawal, M., Seth, A.K. (eds.) FSTTCS 2002. LNCS, vol. 2556, pp. 133–144. Springer, Heidelberg (2002)
9. Feigenbaum, J., Papadimitriou, C.H., Shenker, S.: Sharing the cost of a multicast transmission. J. Comput. Sys. Sci. 63, 21–41 (2001)
10. Feigenbaum, J., Shenker, S.: Distributed algorithmic mechanism design: Recent results and future directions. In: Proc. 6th Int. Workshop on Discr. Alg. and Methods for Mobile Comput. and Communic., pp. 1–13 (2002)
11. Grigoriev, D.: Probabilistic communication complexity over the reals (preprint, 2007)
12. Hromkovič, J.: Communication Complexity and Parallel Computation. Springer, Heidelberg (1998)
13. Kushilevitz, E., Nisan, N.: Communication Complexity. Cambridge University Press, Cambridge (1997)
14. Lehmann, D., O'Callaghan, L., Shoham, Y.: Truth revelation in approximately efficient combinatorial auctions. In: Proc. ACM Conference on Electronic Commerce (2003)
15. Luo, Z.-Q., Tsitsiklis, J.N.: Communication complexity of convex optimization. J. Complexity 3, 231–243 (1987)
16. Luo, Z.-Q., Tsitsiklis, J.N.: On the communication complexity of distributed algebraic computation. J. Assoc. Comput. Mach. 40(5), 1019–1047 (1993)
17. Shafarevich, I.R.: Basic algebraic geometry 1 – Varieties in projective space, 2nd edn. Springer, Heidelberg (1994)
18. Yao, A.C.: Some complexity questions related to distributed computing. In: Proc. of 11th ACM Symp. on Theory of Comput., pp. 209–213 (1979)

The Price of Anarchy of a Network Creation Game with Exponential Payoff

Nadine Baumann[1,⋆] and Sebastian Stiller[2,⋆⋆]

[1] Universität Dortmund, Fachbereich Mathematik, 44221 Dortmund, Germany
nadine.baumann@math.uni-dortmund.de
[2] Technische Universität Berlin, Fakultät II, Straße des 17. Juni 136, 10623 Berlin, Germany
stiller@math.tu-berlin.de

Abstract. We analyze a graph process (or network creation game) where the vertices as players can establish mutual relations between each other at a fixed price. Each vertex receives income from every other vertex, exponentially decreasing with their distance. To establish an edge, both players have to make a consent acting selfishly. This process has originially been proposed in economics to analyse social networks of cooperation. Though the exponential payoff is a desirable principle to model the benefit of distributed systems, it has so far been an obstacle for analysis.

We show that the process has a positive probability to cycle. We reduce the creation rule with payoff functions to graph theoretic criteria. Moreover, these criteria can be evaluated locally. This allows us to thoroughly reveal the structure of all stable states. In addition, the question for the price of anarchy can be reduced to counting the maximum number of edges of a stable graph. This together with a probabilistic argument allows to determine the price of anarchy exactly.

1 Introduction

A fundamental Graph Process. Graph processes and network creation games help to understand the structure of real-world networks. Though these tools often fall short of a detailed modeling, their analysis elates by linking a simple and intuitive, creative principle to typical features of huge real-world networks. In this way, e.g., the preferential attachment model (cf. [6] for details) explains the scale-free structure formed by the pages of the WWW and their links.

Several such models have been proposed mostly in an economic context. We consider one which seems fundamental among these. The network created is a simple, undirected graph $G(V,E)$. It is created step by step. In each step a pair of vertices $\{u,v\}$ is chosen with no respect to whether their common edge already exists or not. For the edge to exist at the end of the step both vertices u and v have to benefit from it. In case at least one of them disapproves, the edge will not be present at the end of the step. A vertex benefits from an edge e, if the current graph with e gives that vertex a higher

⋆ This work was supported by DFG Focus Program 1126, "Algorithmic Aspects of Large and Complex Networks", grant no. SK 58/4-1 and SK 58/5-3.

⋆⋆ This work was partially supported by the ARRIVAL project, within the 6th Framework Programme of the European Commission under contract no. FP6-021235-2.

B. Monien and U.-P. Schroeder (Eds.): SAGT 2008, LNCS 4997, pp. 218–229, 2008.

payoff than the current graph without e. The costs are *local*, namely every vertex pays a factor c times its degree, but the vertex enjoys income *globally*, namely from every other vertex exponentially declining with the distance to the other vertex. Whether to be at distance 2 or 3 is a much greater difference than whether to be at distance 100 or 101. These intuitions are not limited to economics. Think of the network as a means to send information from any vertex to any other vertex along paths. Unfortunately, at every edge used a $(1-\delta)$-portion of the information sent into the edge is lost. (Equivalently, each edge can have a probability of temporary failure of $(1-\delta)$.) Hence, every vertex v will send its deliveries to w via a shortest path and only a portion of $\delta^{\mathrm{dist}(v,w)}$ of the total amount sent from v to w and vice versa will reach its destination.

This creation rule can be understood as a game, where the vertices as players create the edges myopicly, selfishly, and non-cooperatively, though each edge requires both its end-vertices to agree. Obviously, it depends on the order of steps which networks are created. One is interested in certain equilibria or stable states, i.e., situations which every player would leave unchanged, if it was her turn now. The notion of stability suitable to this model is called *pairwise stability* (cf. [10]), i.e., a graph is stable, if it will stay unchanged, no matter which *pair* of vertices is chosen for the next step. Alternatively, one can define a probability distribution according to which the next pair of players is chosen. Then the game becomes a random graph process, i.e., a sequence of random variables $(G_i)_{i\in\mathbb{N}}$, each one representing a network. Again the same stable states and the possibility for the process to cycle are of primary interest. For the graph process, we accept every distribution that assigns a positive probability to every pair of vertices.

Besides the stable graphs, one is interested in graphs maximizing the sum of the payoff, i.e., the total throughput of information minus the total edge costs. These graphs are called *efficient graphs* or *system optima*. The smallest ratio between the total payoff of a stable graph and that of a system optimum is called the *price of anarchy* of the graph process and is of high interest in network creation games since its first mentioning by Koutsoupias and Papadimitriou [12]. The classical notion of Nash-equilibrium (cf. [14]) is not adequate for bilateral games. Note that pairwise stability is a stricter notion of equilibrium as a player is not allowed to overhaul his whole strategy without the other players reacting to his steps (cf. also [7]).

Related Results. The huge number of network creation games proposed in the literature shows the great interest for these explanatory tools. See Jackson [10] for a survey article. During the previous decade, also the interest in the analysis especially of the price of anarchy and Nash equilibria of those network creation games increased. The first to treat the price of anarchy were Koutsoupias and Papadimitriou in their seminal paper [12]. A relevant but simple network creation game is the unilateral game with linear payoff function as proposed by Fabrikant et al. [9]. Their model is the same as ours except for two features: For them an edge is used mutually, but only one of its end-vertices pays for it. Second, the income from other vertices decreases *linearly* with their distance. Recently, Albers et al. [1] give the best known upper bound on the price of anarchy for this model. Moreover, they disprove a structural conjecture for stable states made by Fabrikant et al. [9] when they show that graphs with cycles can be stable.

Corbo and Parkes [7] analyzed a bilateral consent-driven variant of the model by Fabrikant et al. and determined lower and upper bounds of the price of anarchy. Among other improvements Demaine et al. [8] lifted the lower bound to match the upper bound. Further, they compared these values with the unilateral model. As already stated by Corbo and Parkes, Nash equilibria are not appropriate for bilateral games. They therefore introduced the alleviated notion of *pairwise Nash* which is equivalent to pairwise stability.

The model we consider uses a more elaborate payoff function and the bilateral approach for sharing the costs of the edge. It was proposed by Jackson and Wolinsky [11] and interpreted as a graph process by Watts [15]. We state their results in Section 2. Though in comparison to those of [9] and [1] as well as [7] their results may appear limited to peculiar cases and immediate from the definition, the model of [11] and [15] is more convincing for two reasons: They give a consistent interpretation of mutual relations. And the income decreases exponentially with the distance.

Another model using bilateral cost sharing is given by Melendez-Jiminez [13]. Models using cost sharing principles are for example Bala and Goyal [5]. Anshelevich et al. [4] and [3] establish a near optimal solution for selfish players and determine the price of anarchy in a model with fair costs.

Our Contribution. It would be desirable to achieve the same level of mathematical insight for the process of Jackson, Wolinsky [11] and Watts [15] as provided for the process of Fabrikant et al. [9] by Albers et al. [1], namely some structural knowledge of the stable states and bounds on the price of anarchy. We achieve even more. Our process depends on two parameters, c and δ. First, we show that this process behaves equally whenever c is in $(\delta-\delta^2,\delta-\delta^3)$. For these cases we show that the process has positive probability to cycle.

Further, we provide for thorough structural insight to stable states. On the one hand, this is of interest in its own. On the other hand, it allows for our main theorem: We give an explicit formula in c, δ, and the number of vertices for the exact price of anarchy for all $c\in(\delta-\delta^2,\delta-\delta^3)$. We argue by reducing the creation rule and payoff functions to local, graph theoretic criteria. In particular, the price of anarchy can be reduced to the number of edges in a maximum stable graph.

For $c>\delta-\delta^3$ we indicate how and to which extent our methods can be carried over. Further, we point out how an analogon to our main result is linked to extremal graph theory.

2 Preliminaries

For a graph G as usual $V(G)$ and $E(G)$ stand for its vertex and its edge set. For a pair of vertices $e=\{u,v\}$, we use $G+e$ and $G-e$, no matter whether $e\in E(G)$ or not, to denote the graph G with or without the edge e. The neighborhood of a vertex v will be symbolized by $N(v)$, its degree by $d(v)$ $(=|N(v)|)$ and the distance between two vertices u and v by $\operatorname{dist}(u,v)$, i.e., the minimum number of edges of a path connecting u and v. If there does not exist a path, we say $\operatorname{dist}(u,v)=\infty$.

Formally we define a *graph process* (or a game) to be a triple of the cost coefficient, the income basis and the number of vertices, $(c,\delta,n)\in\mathbb{R}\times(0,1)\times\mathbb{N}$.

For every vertex v the *income* is a function of $E(G)$ given by $\sum_{u\in V\setminus\{v\}} \delta^{\mathrm{dist}(u,v)}$. The *costs* of a vertex v are $c\cdot d(v)$, and its *payoff* is its income minus its costs. The total (or social) payoff is the sum of all vertices' payoff: $\sum_{v\in V}\left(\sum_{u\in V\setminus\{v\}} \delta^{\mathrm{dist}(u,v)} - c\cdot d(v)\right)$. A graph maximizing the total payoff is called *system optimum*.

A *situation* is a graph G together with a pair $\{u,v\}$ of its vertices. Every situation defines a polynomial in δ for each of the pair's vertices u and v, expressing the change in income (not yet in payoff) for that vertex between $G-e$ and $G+e$, with $e=\{u,v\}$. In these terms the creation rule reads as follows: When the (possible) edge $e=\{u,v\}$ is evaluated given the graph G, the *decision* will be positive, i.e., $G+e$ will be the resulting graph, if both polynomials are bigger than c, and negative if at least one is smaller, i.e., $G-e$ results. We do not consider cases where a polynomial can equal c, i.e., a vertex is indifferent about an edge. It would be possible to extend the model and results to these cases, but it would also be tedious.

Observe that if an edge is inserted by the process, this can only increase the total payoff. The deletion of an edge by the process can be locally advantageous but decrease the total payoff.

A graph G is called *stable*, if G together with any $e=\{u,v\}\in E(G)$ is a situation with positive decision, and G together with any $e=\{u,v\}\notin E(G)$ is a situation with negative decision. The *price of anarchy* is defined as the maximum ratio $\frac{\text{Total Payoff of a System Optimum}}{\text{Total Payoff of } G}$ over all stable graphs G.

Expressed in these terms, Jackson and Wolinsky [11] and Watts [15] observe that for $c<\delta-\delta^2$ the complete graph is the only stable graph and the unique system optimum, because, no matter what the graph looks like, every further edge is beneficial. Trees have the least total cost among all connected graphs. In a star all not directly connected pairs of vertices are at distance 2. Therefore, if the cost factor c is high enough to draw any attention to the costs the star is optimal, namely for $c\in(\delta-\delta^2,\delta+\frac{n-2}{2}\delta^2)$. Beyond that limit for the costs, even the star's payoff becomes negative and the empty graph is the system optimum. Notably, the star is a stable graph for $c\in(\delta-\delta^2,\delta)$. Beyond that the empty graph is a stable state (though not the only one).

Our first lemma links stability and the choice of c and δ to a structural property.

Lemma 1. *If a graph G is a stable state of a graph process with $c<\delta-\delta^{k+1}, k\in\mathbb{N}$, then G has diameter less than or equal to k.*

Proof. Assume to the contrary that there are two vertices u and v at distance greater than k. The (non-existing) edge $\{u,v\}$ would improve the income for u from at least v, i.e., the increase in income is greater than or equal to $\delta-\delta^{k+1}$. As the analogon holds for v, the edge would be inserted, which is to say, the graph is unstable.

3 The Graph Process for $c\in(\delta-\delta^2,\delta-\delta^3)$

In this main section we restrict to the case $c\in(\delta-\delta^2,\delta-\delta^3)$. The restriction has the nice property that all graph processes in those cases are identical. We say that a set of graph processes $\{(c,\delta,n)\}$ is *identical*, if and only if for every situation the decision is the same for all processes in the set.

Theorem 1. *For fixed n the set of graph processes* $\{(c, \delta, n) \mid c \in (\delta - \delta^2, \delta - \delta^3)\}$ *is identical.*

Proof. Let G be the graph and $e = \{u, v\}$ be the edge of an arbitrary situation. If $\text{dist}_{G-e}(u, v) \geq 3$, then the income in $G + e$ is for u and for v at least $\delta - \delta^3$ higher than in $G - e$, because this is the minimal improvement in income from u for v and vice versa. As no distance gets longer by inserting an edge this lower bounds the total increase in income. As $c < \delta - \delta^3$ the change in payoff is positive.

Assume now $\text{dist}_{G-e}(u, v) = 2$ and the decision to be in favor of e. The gain of the edge for v from u (and vice versa) is $\delta - \delta^2$,i.e., less than its cost. Thus, for both u and v further vertices must be closer in $G + e$ than in $G - e$. For some $x \neq v$ *any* shortest path from u to x in $G + e$ must use the edge e, i.e., is of the form: $(u, v, \ldots, x)$. This implies, that at least for one *neighbor* y of v all shortest path from u to y in $G + e$ go via v (and are shorter than those in $G - e$). We can conclude that the change in income for u from y is at least $\delta^2 - \delta^3$, as $\text{dist}_{G-e}(u, y) \geq 3$. Therefore, the total change in income for u is at least that from v plus that from y, so at least $\delta - \delta^3$. An analogon holds for v. Thus, a situation that is positive for one graph process with $c < \delta - \delta^3$ is positive for of those.

As all graph processes on n vertices are identical under the restriction $c \in (\delta - \delta^2, \delta - \delta^3)$, we also speak of *the* graph process. The argument of the theorem allows for all graph processes with $c \in (\delta - \delta^2, \delta - \delta^3)$ to reduce a decision to a graph theoretical set of rules. An edge $e = \{u, v\}$ will be kept or inserted, if and only if at least one of the following conditions holds in the graph without e.

1. We have $\text{dist}_{G-e}(u, v) > 2$.
2. The end-vertex u has a neighbor x with $\text{dist}_{G-e}(v, x) = 3$, and the end-vertex v has a neighbor y with $\text{dist}_{G-e}(u, y) = 3$.

Cycling. The price of anarchy is defined with reference to the stable graphs. But there is an infinite series of pairs of vertices, that never leads to a stable graph. Apply the graph rules above to the sequence depicted in Figure 1 to check that it cycles. In the first two situations the dashed edge is inserted because the two marked vertices are too far from each other. In the third situation the dashed edge is removed, because the two marked vertices have an alternative short connection.

Theorem 2. *The graph process for parameters c and* δ *with* $c \in (\delta - \delta^2, \delta - \delta^3)$ *can cycle.*

The Price of Anarchy In order to determine the price of anarchy, we need to establish criteria for stable graphs for the considered graph process.

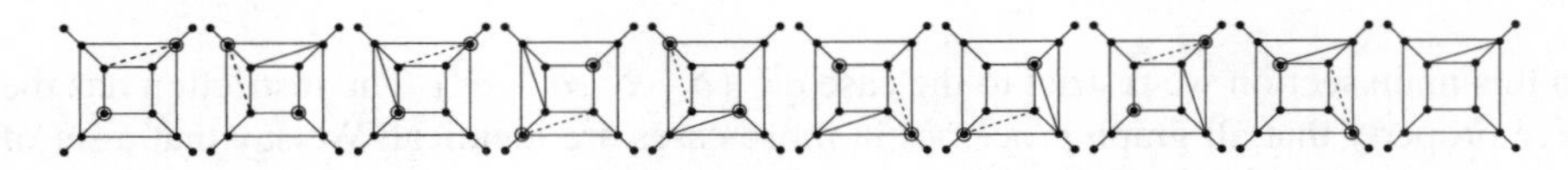

Fig. 1. Cycling sequence of graphs, to be read from the left to the right

For the graph process Lemma 1 amounts to say that every stable graph must have diameter exactly 2, as the complete graph obviously is not stable. Consequently, *the star is the only stable tree*. Further, we can give a sufficient condition for a graph to be stable.

Theorem 3. *A graph G is stable, if G has diameter 2 and contains no triangles.*

Proof. For stability, on the one hand, we have to show that no edge in G will be removed. The graph G contains no triangles. Hence the shortest path between the end-vertices u and v of a currently present edge e in the graph without that edge, $G-e$, has length greater than or equal to 3. Thus, by the same argument as in Lemma 1, the edge is beneficial for both its endpoints and therefore kept.

On the other hand, no further edge will be inserted as the diameter suitable to the parameters is already reached: For any edge $e = \{u,v\}$ not present in G, calculating the payoff for one of its end-vertices u in $G+e$, is the same as in G except that the other end-vertex v will change from distance two to distance one. Consequently, the change in income by inserting e is exactly $\delta - \delta^2$ and therefore it is not beneficial to insert e.

By the above observations, we reduced the decisions of any situation to graph theoretic considerations. In fact, we can do the same for the price of anarchy, or equivalently, the total cost of a stable graph. For a stable graph we know all distances to be less than or equal to 2. Consequently, we can rewrite the total payoff of such a graph $\sum_{u\in V}\left(\sum_{v\in V\setminus\{u\}}\delta^{\mathrm{dist}(u,v)} - d(u)c\right) = \sum_{\{u,v\}\in E}(\delta - c) + \sum_{\substack{\{u,v\}\in V\times V,\\ \mathrm{dist}(u,v)=2}}\delta^2 = m(\delta - c) + \left(\binom{n}{2} - m\right)\delta^2$, where m denotes the number of edges. As $\delta - c - \delta^2 < 0$ by the choice of the parameters, it directly follows that the payoff of a stable graph is the bigger the less edges it has.

Lemma 2. *Let G and G' be stable graphs. The total payoff of G is greater than that of G' if and only if $|E(G)| < |E(G')|$.*

This together with the description of a stable graph in Theorem 3 provides for a lower bound on the price of anarchy. Recall that the star is the unique system optimum. It is well known that the graph $K_{\lfloor\frac{n}{2}\rfloor,\lceil\frac{n}{2}\rceil}$ maximizes the number of edges in a triangle free graph. For the lower bound of the price of anarchy as the ratio of this graph to the star, we need the stability of the graph $K_{\lfloor\frac{n}{2}\rfloor,\lceil\frac{n}{2}\rceil}$. But as the graph $K_{\lfloor\frac{n}{2}\rfloor,\lceil\frac{n}{2}\rceil}$ also has diameter equal to 2, we get from Theorem 3 that the maximum bipartite graph $K_{\lfloor\frac{n}{2}\rfloor,\lceil\frac{n}{2}\rceil}$ is stable.

Corollary 1. *The price of anarchy of the graph process with n vertices is bounded from below by*

$$\frac{(n-1)((\frac{n}{2}-1)\delta^2 + \delta - c)}{\mu((\frac{n^2-n}{2\mu} - 1)\delta^2 + \delta - c)} \tag{1}$$

where $\mu := \lceil\frac{n}{2}\rceil\lfloor\frac{n}{2}\rfloor$ is the number of edges in $K_{\lfloor\frac{n}{2}\rfloor,\lceil\frac{n}{2}\rceil}$.

To show that Corollary 1 exactly states the price of anarchy for the graph process, we need to show that $K_{\lfloor\frac{n}{2}\rfloor,\lceil\frac{n}{2}\rceil}$ maximizes the number of edges among *all* stable graphs with n vertices. That would be easy if the converse of Theorem 3 was true, i.e., all stable graphs had no triangles. This is not the case as will be shown in the next part. There, we analyze the occurrence of triangles in detail. By the end of that part we will show (Theorem 6) that all stable graphs that contain at least one triangle have less edges than $K_{\lfloor \frac{n}{2} \rfloor, \lceil \frac{n}{2} \rceil}$. Therefore, we can conclude our main result:

Theorem 4. *For all graph processes with n vertices and* $c \in (\delta - \delta^2, \delta - \delta^3)$*, the maximum price of anarchy equals* (1) *and is produced by the maximum bipartite graph* $K_{\lfloor \frac{n}{2} \rfloor, \lceil \frac{n}{2} \rceil}$ *against the star* $K_{1,n-1}$.

For $c \searrow (\delta - \delta^2)$ the expression (1), i.e., the price of anarchy, tends to 1 for every n and $\delta \in (0,1)$, whereas it converges to 2 for $c \nearrow (\delta - \delta^3)$, $n \nearrow \infty$, and $\delta \searrow 0$.

Triangles. We have seen situations where a triangle is closed in the course of the graph process. There the graph process cycles and therefore does not reach a stable state. Nonetheless, there are stable graphs containing triangles. The three graphs depicted left in Figure 2 are stable and contain a triangle. The black vertices in the second graph form the leftmost graph. The white vertices in the second can be added one by one, such that for every number of vertices strictly greater than 6 stable graphs with at least one triangle are possible.

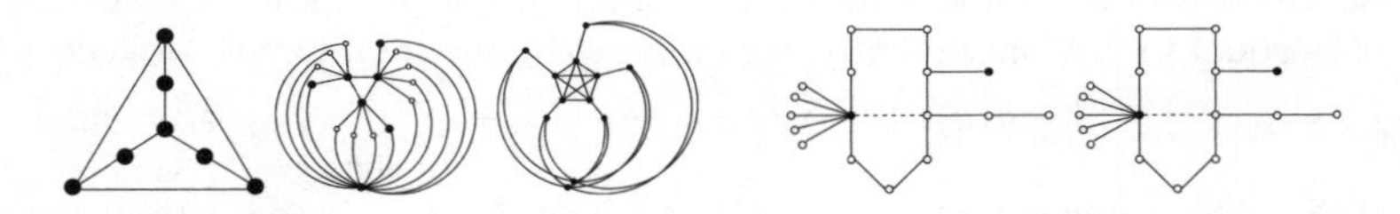

Fig. 2. Stable graphs that contain a triangle and situations with intersecting polynomials

Next, we show a structural result for stable graphs with triangles that is of interest in itself though not necessary for the proof of Theorem 4.

Theorem 5. *If G is a stable state of the graph process and the vertex set* $\{a,b,c\}$ *forms a triangle in G, then there exists at least one* $v \in V(G)$ *with distances* $\mathrm{dist}(a,v) = \mathrm{dist}(b,v) = \mathrm{dist}(c,v) = 2$.

In order to show Theorem 5, we first need some lemmata.

Lemma 3. *Let G be stable and contain a triangle* $\{a,b,c\}$*. Then for every* $i,j \in \{a,b,c\}$, $i \neq j$ *there is a vertex in* $N(i) \setminus \{a,b,c\}$ *that has no edge to neither any vertex in* $N(j)$ *nor j itself.*

Proof. Assume the claim of the lemma is false. Then j can reach all neighbors of i via one of its own neighbors or directly. Moreover, it can reach i itself in two steps via the two other arcs of the triangle. That is as good as anything i can offer to j. Hence j would drop the arc $\{i,j\}$ contradicting the stability of G.

Note that the premises of the following Lemmata 4 and 5 are assumptions to be falsified to prove Theorem 5. The proofs rest on the fact that without a vertex at distance 2 to all triangle vertices, those behave totally jealous towards their other neighbors.

Lemma 4. *Let G be stable, contain a triangle $\{a,b,c\}$, and have no vertex v with* $\operatorname{dist}(i,v) = 2$*, for all $i \in \{a,b,c\}$. Then the neighborhoods of the triangle vertices form a disjoint partition of $V(G) \setminus \{a,b,c\}$.*

Proof. First, it holds that $\bigcup_{i\in\{a,b,c\}} N(i) = V(G)$ because there is no vertex at distance 2 to the triangle and every stable graph has diameter less than 3. Assume vertex v to be in $N(i) \cap N(j)$, $j,i \in \{a,b,c\}, j \neq i, v \notin \{a,b,c\}$. Remove the arc $\{i,v\}$. One figures out quickly that i can still reach any vertex including v within two steps because even in G there is no vertex at distance 2 to all of the vertices a,b,c. Therefore, the assumption contradicts the stability of G.

Lemma 5. *Let G be stable, contain a triangle $\{a,b,c\}$, and have no vertex v with* $\operatorname{dist}(i,v) = 2$*, for all $i \in \{a,b,c\}$. Then the neighborhoods of the triangle vertices are independent sets.*

Proof. Assume $v,w \in N(i) \setminus \{a,b,c\}$ for some $i \in \{a,b,c\}$ with $\{v,w\} \in E(G)$. When removing the edge $\{i,v\}$, considerations like in the previous proof show that the triangle vertex i can still reach every vertex within 2 steps. Hence G was not stable.

Proof (Proof of Theorem 5). Assume the claim of the theorem not to be true for G, containing a triangle and being stable. Let x be a vertex in $N(a)$ as guaranteed to exist by Lemma 3 that has neither a connection to b nor to one of its neighbors, and y be a vertex in $N(b)$ that has neither a connection to c nor to one of its neighbors. As G is stable, its diameter is 2. Hence, x and y must have a common neighbor, as they cannot be adjacent by definition of x. By definition of y, such a neighbor is neither c nor one of c's neighbors. By Lemma 4 it is neither a nor b and by Lemma 5 it is neither a neighbor of b nor of a. Thus, we have a contradiction.

Using Theorem 5 and the insights of Lemmata 3-5 we get that every stable graph that contains a triangle has at least 7 vertices. Further, the number of triangles in a stable graph does not need to be small, as the middle graph in Figure 2 shows, where the clique can consist of $\lceil \frac{n-1}{2} \rceil$ vertices. One figures out quickly that such a graph features the biggest clique in a stable graph of n vertices.

In order to determine the price of anarchy exactly, by Lemma 2 one has to look for the stable graph with the maximum number of edges. Intuitively, stable graphs with triangles should have more edges than triangle free stable graphs. We show that all stable graphs with triangles have less edges than some without triangles, namely the $K_{\lfloor \frac{n}{2} \rfloor, \lceil \frac{n}{2} \rceil}$.

Theorem 6. *For a number n of vertices, the maximum bipartite graph $K_{\lfloor \frac{n}{2} \rfloor, \lceil \frac{n}{2} \rceil}$ has the maximum number of edges among all stable graphs on n vertices.*

Proof. We need to show that $K_{\lfloor \frac{n}{2} \rfloor, \lceil \frac{n}{2} \rceil}$ has more edges than any stable graph on n vertices containing at least one triangle.

For every graph G we define a random variable based on the uniform distribution over the vertices of G as follows: Pick a vertex uniformly at random and sum up the degrees of its neighbors. Denote the expectation of that random variable by $\Phi(G) = \frac{1}{|V(G)|}\sum_{v\in V(G)}\sum_{u\in N(v)} d(u)$. We denote by $\mu := \lfloor \frac{n}{2} \rfloor \cdot \lceil \frac{n}{2} \rceil$ the number of edges in $K_{\lfloor \frac{n}{2} \rfloor, \lceil \frac{n}{2} \rceil}$. In order to show the statement of the theorem we prove two claims.

Claim 1. Let G be a stable graph with n vertices. Then $\Phi(G) \leq \Phi(K_{\lfloor \frac{n}{2} \rfloor, \lceil \frac{n}{2} \rceil})$.
Claim 2. Let G be a graph with n vertices and μ edges. Then $\Phi(K_{\lfloor \frac{n}{2} \rfloor, \lceil \frac{n}{2} \rceil}) \leq \Phi(G)$.

These claims yield the statement of the theorem. Assume a stable graph G on n vertices with more edges than the maximum bipartite graph $K_{\lfloor \frac{n}{2} \rfloor, \lceil \frac{n}{2} \rceil}$. Arbitrarily remove edges from G until the resulting graph G' has exactly as many edges as $K_{\lfloor \frac{n}{2} \rfloor, \lceil \frac{n}{2} \rceil}$. Observe that removing an edge reduces the value of Φ by definition. Hence, $\Phi(G') < \Phi(G)$. By the second claim $\Phi(K_{\lfloor \frac{n}{2} \rfloor, \lceil \frac{n}{2} \rceil}) \leq \Phi(G')$. This implies $\Phi(K_{\lfloor \frac{n}{2} \rfloor, \lceil \frac{n}{2} \rceil}) < \Phi(G)$, which contradicts the first claim. Hence there is no stable graph with more than μ edges, which proves the theorem.

It remains to prove Claims 1 and 2. In addition, we will prove an even stronger version of the first claim, namely that every stable graph containing at least one triangle has a strictly smaller value of Φ than the complete bipartite graph and consequently less edges.

Proof of Claim 1: For the maximum bipartite graph we have $\Phi(K_{\lfloor \frac{n}{2} \rfloor, \lceil \frac{n}{2} \rceil}) = \mu$. Now, consider an arbitrary, stable graph G. For a randomly chosen vertex $v \in V(G)$ let $N(v)$ be all neighboring vertices and $b := |N(v)|$. Partition the edges incident to the vertices in $N(v)$ into three sets: first the edges incident to v, second those within $N(v)$ and third the edges to other vertices. The first and the last set together contain at most μ edges, because of the bipartiteness of the subgraph formed by these edges. Every edge in the second set belongs to a triangle containing v. Denote the number of vertices in $N(v)$ that belong to at least one triangle with v by ℓ. Then there are at most $\frac{\ell^2-\ell}{2}$ edges in the second set. As Φ counts the degrees of v's neighbors, each edge in the second set counts twice.

Assume a vertex u to be part of a triangle with v. Why will v be interested in the edge $\{u, v\}$? There must be at least one vertex that v can reach within two steps only via u, or $\text{dist}(u, v) \geq 3$ if the edge $\{u, v\}$ was removed. The latter is wrong, as u and v are in a triangle. Thus there exists a neighbor $w \neq v$ of u that is not connected to any vertex in $(N(v) \cup \{v\}) \setminus \{u\}$. In other words, u has an exclusive attraction to v in the sense that no other vertex in $N(v) \cup \{v\}$ is connected to w. Therefore we have to subtract $(b-1)$ from the number of possible edges in the third set for each of the vertices in $N(v)$ that participate in a triangle with v. Altogether, we get that the sum over the degrees of neighbors of v is at most $\sum_{u\in N(v)} d(u) \leq \mu + (\ell^2 - \ell) - \ell(b-1)$, if ℓ neighbors of v participate in triangles with v. As $\ell \leq b$ and the preceeding inequality holds for all vertices $v \in V(G)$, we get

$$\Phi(G) \leq \mu. \tag{2}$$

This proves Claim 1 as stated above. Moreover, a graph containing a triangle does not achieve equality in Inequality (2). To show this, observe that in case of equality for each vertex the number ℓ must be in the set $\{0, b\}$, and if $\ell = b$ for a vertex v, then $N(v) \cup \{v\}$ is a clique. In other words, the neighborhood of a vertex v either contains no triangle with v or forms a clique together with the vertex v. For at least one vertex v with $d(v) > 1$ the latter must be true (otherwise G does not contain a triangle). A neighbor u of such a vertex v is not interested in its edge to v because u has a direct edge to all of v's neighbors and can reach v itself via another neighbor of v ($d(v) > 1$) within two steps. To report accurately, a graph G can fulfill at most two of the following three properties:

1. G is stable.
2. G achieves equality in Inequality (2).
3. G has a triangle.

Proof of Claim 2: Rewriting the counting function $\Phi(G)$ with $\sum_{v \in V(G)} \sum_{u \in N(v)} d(u) = \sum_{\{u,v\} \in E(G)} d(u) + d(v) = \sum_{v \in V(G)} d^2(v)$ gives $\Phi(G) = \frac{1}{|V(G)|} \sum_{v \in V(G)} d^2(v)$. It is easy to see that among all multiset of n natural numbers $s_i, 1 \leq i \leq n$ with $\sum_i s_i = 2m$ the degree sequence d_i of $K_{\lfloor \frac{n}{2} \rfloor, \lceil \frac{n}{2} \rceil}$ minimizes $\sum_i s_i^2$, which yields the claim.

4 Further Choices of c and δ

The model of Jackson, Wolinsky and Watts is rather a family of different models in their own right. The methods we used to analyze the case where $c \in (\delta - \delta^2, \delta - \delta^3)$ to a certain extent can be carried over to other cases.

Every situation gives rise to two polynomials in δ defined over the $(0,1)$ interval and mapping to the positive reals determined by the change in income of the two vertices in question. We can also interpret the set $(0,1) \times \mathbb{R}^+$ in which the graphs (here: graphs of functions) of those polynomials live as the set of all graph processes for a fixed number of vertices n, because they are defined by a δ coordinate in $(0,1)$ and a c coordinate in $\mathbb{R}^+$. In this picture a decision is positive for exactly those graph processes that are below the polynomials of both end-vertices. This visualizes how the polynomials separate the set of all processes in those for which they are positive and those for which they are negative. Restricting to $c \in (\delta - \delta^2, \delta - \delta^3)$ we exploit the nature of $\delta - \delta^2$ and $\delta - \delta^3$ as *threshold functions*. We call a function $f : (0,1) \to \mathbb{R}^+$ a threshold function if for every polynomial p that stems from a situation we have $\exists x_0 \in (0,1) : p(x_0) < f(x_0) \Rightarrow p(x) < f(x) \forall x \in (0,1)$ and $\exists x_0 \in (0,1) : p(x_0) > f(x_0) \Rightarrow p(x) > f(x) \forall x \in (0,1)$. One may carry on looking for threshold functions and redo our analyze for these cases. The next theorem gives the next threshold function.

Theorem 7. *For fixed n all graph process with $c \in (\delta - \delta^3, \delta - \delta^4)$ are identical.*

Proof. We show, that a situation is either positive for all graph processes in the interval or for none. Classify all situations by the distance $a := \text{dist}(u,v)$ of the considered pair $\{u,v\}$. For $a \geq 4$ the increase in income is automatically bigger than $\delta - \delta^4 > c$, thus all decisions positive. For $a = 3$ the end-vertices u and v must attract each other with vertices not located on a shortest path between u and v to have a gain greater than $c > \delta - \delta^3$. Therefore they at least have an increase of $\delta - \delta^3 + \delta^2 + \delta^3$. This in turn always suffices. For $a = 2$ at least two further vertices outside the shortest paths are required for a positive decision. All positive situations for $c \in (\delta - \delta^3, \delta - \delta^4)$ contain

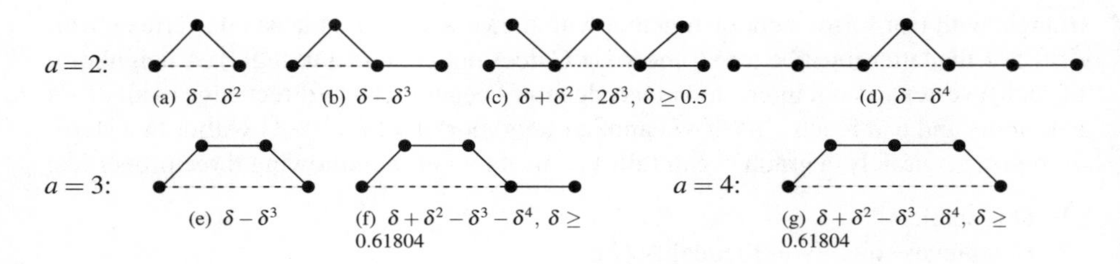

Fig. 3. Subnetworks that need to be considered in the proof of Theorem 7. The caption determines the benefit of the dashed edge. The lower bounds on δ in Figures 3(f) and 3(g) guarantee that $c < \delta$ which makes the initial insertion of edges into an empty graph possible.

Figure 3(d) or Figure 3(c) as subgraphs. The increase in income in Figure 3(d) is the smallest one and bigger than c.

One may suppose that all polynomials stemming from situations are threshold functions. Yet, there are situations yielding polynomials that intersect each other in the open interval $(0,1)$. Consider the two situations depicted to the right in Figure 2. In both cases the black end-vertex has so many neighbors that the opposite end-vertex of the dashed edge will endorse the insertion of the dashed edge. The polynomial for the black vertex in the left situation minus that of the black vertex to the right gives: $\delta^3 - \delta^5 - (\delta^4 - \delta^7)$. For some values of $\delta \in (0,1)$ this is negative, and for some it is positive. In other words, some processes will insert the edge in the left situation but not in the right situation, whereas other processes will in both situations do the contrary. This implies that for predicting the process' behavior it is no longer sufficient to specify one parameter by bounds of the other.

Nevertheless, general results, e.g., concerning the price of anarchy, might be achievable along the lines of this work. Choose a graph process, where the star is optimal. Every stable graph will have diameter less than k if $c < \delta - \delta^{k-1}$. What can be said about graphs that have the maximum number of edges among all those containing no cycle smaller than $k-1$, i.e., in graph theoretic terms, which have *girth* $k-1$? They are not necessarily stable, but none of their present edges will currently be removed. The graph of our main result $K_{\lfloor \frac{n}{2} \rfloor, \lceil \frac{n}{2} \rceil}$ is an extremal graph in the sense that it maximizes the number of edges for girth 4. We conjecture that in general the graphs with maximum number of edges for girth $k-1$ are a good approximation for those maximizing the price of anarchy. Most astonishingly the known upper bounds [2] for this long standing problem of extremal graph theory would imply that the price of anarchy becomes *constant*. For simplicity account for the price of anarchy as the greatest fraction between the number of edges in a stable and in an optimal graph. If our conjecture holds that the maximum graph of girth $k-1$ approximates the price of anarchy for graph processes with $\delta - \delta^{k-1} < c < \delta - \delta^k$, this means that this price of anarchy would drop from worst ($\mathcal{O}(n)$) to best ($\mathcal{O}(1)$) as k growth. In other words, the users form an optimal network if the costs are very low ($c < \delta - \delta^2$) and an almost optimal network if the costs are very high. The worst outcome is then caused by costs $c \in (\delta - \delta^2, \delta - \delta^3)$, which is the case we have analyzed.

References

1. Albers, S., Eilts, S., Even-Dar, E., Mansour, Y., Roditty, L.: On Nash equilibria for a network creation game. In: Proceedings of the 17th ACM-SIAM Symposium on Discrete Algorithms (SODA 2006), pp. 89–98 (2006)
2. Alon, N., Hoory, S., Linial, N.: The Moore bound for irregular graphs. Graphs and Combinatorics 18(1), 53–57 (2002)
3. Anshelevich, E., Dasgupta, A., Kleinberg, J., Tardos, E., Wexler, T., Roughgarden, T.: The price of stability for network design with fair cost alloccation. In: Proceedings of the 45th Annual IEEE Symposium on Foundations of Computer Science (FOCS 2004), pp. 295–304 (2004)
4. Anshelevich, E., Dasgupta, A., Tardos, E., Wexler, T.: Near-optimal network design with selfish agents. In: Proceedings of 35th Annual ACM Symposium on Theory of Computing (STOC 2003), pp. 511–520 (2003)
5. Bala, V., Goyal, S.: A non-cooperative model of network formation. Journal of Econometrics 68(5), 1181–1229 (2000)
6. Barabási, A.-L., Albert, R.: Emergence of scaling in random networks. Science 286, 509–512 (1999)
7. Corbo, J., Parkes, D.C.: The price of selfish behavior in bilateral network formation. In: Proceedings of the 24th ACM Symposium on Principles of Distributed Computing (PODC 2005) (2005)
8. Demaine, E.D., Hajiaghayi, M.T., Mahini, H., Zadimoghaddam, M.: The price of anarchy in network creation games. In: Proceedings of the 26th Annual ACM SIGACT-SIGOPS Symposium on Principles of Distributed Computing (PODC 2007) (2007)
9. Fabrikant, A., Luthra, A., Maneva, E., Papadimitriou, C.H., Shenker, S.: On a network creation game. In: Proceedings of the 22nd annual Symposium On Principles of Distributed Computing (PODC 2003), pp. 347–351 (2003)
10. Jackson, M.O.: A survey of models of network formation: Stability and efficiency. In: Demange, G., Wooders, M. (eds.) Group Formation in Economics: Networks, Clubs and Coalitions, ch. 1, pp. 11–57. Cambridge University Press, Cambridge (2004)
11. Jackson, M.O., Wolinsky, A.: A strategic model of social and economic networks. Journal of Economic Theory 71, 44–74 (1996); reprinted in Networks and Groups: Models of Strategic Formation, edited by Dutta and Jackson, Springer, Heidelberg (2003)
12. Koutsoupias, E., Papadimitriou, C.H.: Worst-case equilibria. In: Meinel, C., Tison, S. (eds.) STACS 1999. LNCS, vol. 1563, pp. 404–413. Springer, Heidelberg (1999)
13. Melendez-Jiminez, M.A.: Network formation and coordination: Bargaining the division of link costs. In: Presented at 57th European Meeting of the Econometric Society (August 2002), http://www.webdeptos.uma.es/THEconomica.net1.pdf
14. Nash, J.F.: Non-cooperative games. Annals of Mathematics 54(286–295), 2951
15. Watts, A.: A dynamic model of network formation. Games and Economic Behavior 34, 331–341 (2001)

A Hierarchical Model for Cooperative Games

Ulrich Faigle[1] and Britta Peis[2]

[1] Zentrum für Angewandte Informatik Köln (ZAIK), Weyertal 80, 50931 Köln, Germany
faigle@zpr.uni-koeln.de
[2] Technische Universität Berlin, Strasse des 17. Juni 136, 10623 Berlin, Germany
peis@math.tu-berlin.de

Abstract. Classically, a cooperative game is given by a normalized real-valued function v on the collection of all subsets of the set N of players. Shapley has observed that the core of the game is non-empty if v is a non-negative convex (a.k.a. supermodular) set function. In particular, the Shapley value of a convex game is a member of the core. We generalize the classical model of games such that not all subsets of N need to form feasible coalitions. We introduce a model for ranking individual players which yields natural notions of Weber sets and Shapley values in a very general context. We establish Shapley's theorem on the nonemptyness of the core of monotone convex games in this framework. The proof follows from a greedy algorithm that, in particular, generalizes Edmonds' polymatroid greedy algorithm.

1 Introduction

A *classical cooperative game* on a (finite) set N of players is given by a characteristic function $v : 2^N \to \mathbb{R}$ describing the gain $v(S)$ the set S of players can achieve by cooperating. A *solution concept* is a mathematical model for a "fair" allocation of the value $v(N)$ to the players in N. Among the most attractive solution concepts in cooperative game theory are the core and the Shapley value (see below).

However, cooperation is often only feasible for certain subsets $F \subseteq N$ of players. Collecting these into the family $\mathcal{F}$ of *feasible* coalitions, we call $\Gamma = (\mathcal{F}, v)$ a *cooperative game with restricted cooperation* (cf. [10]), where $v : \mathcal{F} \to \mathbb{R}$ is a given function. The basic question of fair allocations to the individual players poses itself also in this more general context.

Related results. Many generalizations of classical cooperative games have been investigated. For example, multi-choice cooperative games were studied in [22] with respect to the Shapley value. These games are special cases of cooperative games under precedence constraints on the players in the sense of [11], where not all coalitions are necessarily feasible (see also [20,4,18,19]). Moreover, the precedence structure suggests a natural approach to the Shapley value based on the "hierarchical strength" on a player. The latter model was carried over to coalition structures that are even more general and constitute, from a combinatorial point of view, so-called convex geometries and antimatroids (*cf.* [1,2,3]).

B. Monien and U.-P. Schroeder (Eds.): SAGT 2008, LNCS 4997, pp. 230–241, 2008.

The question whether the core of a cooperative game is non-empty can be attacked as a problem in combinatorial optimization. In fact, Shapley's characterization of cores of convex games corresponds to the characterization of so-called polyhedral matroids by the greedy algorithm (*cf.* [7,9]). Extending this approach to cooperative games with precedence structures on the players, a generalization of Shapley's theorem was derived in [11,13,14,17,18,19]. A very general polyhedral model for matroid-type optimization problems is given by the lattice polyhedra of [21]. While the general algorithmic properties of lattice polyhedra are unclear, some progress has been obtained in [16] with respect to a model with certain decreasing supermodular constraint functions, which also includes Shapley's convex game model as a special case.

Our results. We introduce *hierarchical games*, a very far-reaching model for cooperative games with restricted cooperation. The model is based on formalized selection schemes that allow rankings of players. The previous models of multi-choice games, games with precedence constraints on the players, antimatroids *etc.* are all subsumed under our general framework. Employing a ranking procedure for the players, we show that the hierarchical strength of a player is meaningfully defined in our model and yields a notion of a Shapley value, of which the classical Shapley value is a special case. We furthermore study the core of hierarchical games. In particular, the appropriate theory of "hierarchically convex" games is established and it is shown that the Shapley value of a general convex hierarchical game lies in the core (*cf.* Section 5). The proof follows from a new (poly-)matroid-type greedy algorithm for the constructions of weighted ranking vectors.

Our results extend the scope of "Shapley's theorem for convex cooperative games" considerably in that there are cooperative games that are convex within the hierarchical model (and hence have a non-empty core) although none of their standard extensions to classical cooperative games is convex–so that the classical Shapley Theorem does not apply (see Ex. 1).

Before going into generalizations, we recall the definition of core and Shapley value in classical cooperative game theory.

Classical core and Shapley value. Given a classical cooperative game $\Gamma = (2^N, v)$, an allocation vector $x \in \mathbb{R}^N_+$ is said to lie in the *core* if x does not allocate more than $v(N)$ in total and each coalition S receives at least its value $v(S)$, i.e.if $x(N) \leq v(N)$ and $x(S) \geq v(S)$ holds for all $S \subseteq N$, where we use the standard notation $x(S) = \sum_{p \in S} x_p$.

A *ranking* of the players is a linear arrangement (permutation) $\pi = p_1 \dots p_n$ of the elements of N. The *marginal contribution* of p_i w.r.t. π is

$$x^\pi(p_i) = v(\{p_1, \dots, p_i\}) - v(\{p_1, \dots, p_{i-1}\}).$$

The individual marginal contributions yield the marginal allocation vector x^π. Letting Π denote the collection of all linear arrangements of N, the *Shapley value* is the average marginal allocation vector $\Phi(v) = \frac{1}{n!} \sum_{\pi \in \Pi} x^\pi$.

While the core may be empty, the Shapley value exists by definition (*cf.* Shapley [25]). Shapley [26] observed that a (classical) monotone game is *convex*, i.e, satisfies

$$v(S) + v(T) \leq v(S \cup T) + v(S \cap T) \quad \text{for all } S, T \subseteq N,$$

if and only if all marginal allocation vectors lie in the core. Since the core is a convex subset of $\mathbb{R}^N$, one finds that the core of any non-negative convex game is non-empty and contains the Shapley value of the game.

Cooperative games with restricted cooperation. In the subsequent discussion of cooperative games $\Gamma = (\mathcal{F}, v)$ with restricted cooperation, we will always view such games as *profit games* and thus interpret $v(F)$ as the *gain* the feasible coalition $F \in \mathcal{F}$ can achieve.

While a classical cooperative profit game is essentially the same as a cooperative *cost game*, this is no longer true for a cooperative game with restricted cooperation $\Gamma = (\mathcal{F}, v)$. Cost games are briefly discussed in Section 6.

For any subset $X \subseteq N$, we use the notation

$$\mathcal{F}(X) = \{F \in \mathcal{F} \mid F \subseteq X\}$$

for the *restriction* of $\mathcal{F}$ to X. For technical convenience, we assume $\emptyset \in \mathcal{F}$ and thus, in particular, that $\mathcal{F}$ is non-empty. Moreover, we assume v to be 0-*normalized* in the sense $v(\emptyset) = 0$. Since N is not necessarily a feasible coalition, we allow the players to form pairwise disjoint coalitions to generate a total benefit

$$v^* = \max\{\sum_i v(F_i) \mid F_i \in \mathcal{F}, F_i \cap F_j = \emptyset \text{ if } i \neq j\},$$

which we call the *value* of the game Γ. A natural extension of the notion of the core now considers all non-negative allocation vectors x that do not allocate more than v^* in total and satisfy all feasible coalitions. So we define

$$\text{core}(\Gamma) = \{x \in \mathbb{R}^N_+ \mid x(N) \leq v^*, x(F) \geq v(F) \text{ for all } F \in \mathcal{F}\}$$

to be the *core* of the game Γ. (Note that the inequality $x(N) \leq v^*$ can be replaced by the equality $x(N) = v^*$ in the definition of the core.)

An appropriate extension of the Shapley value to cooperative games with restricted cooperation is less obvious. Standard approaches extend $\Gamma = (\mathcal{F}, v)$ first to a classical cooperative game and then compute the corresponding Shapley value (see the null-extension and the core-extension of Γ below). However, this approach might lead to a non-convex classical game such that Shapley's Theorem cannot be applied (see Ex. 1). We propose another approach that is based on the combinatorial structure of $\mathcal{F}$ via *rankings* (see Section 5) and properly generalizes the classical model.

Game Extensions. The *null-extension* of $\Gamma = (\mathcal{F}, v)$ is the cooperative game $\Gamma^0 = (2^N, v^0)$ with $v^0(S) = 1$ if $S \in \mathcal{F}$ and $v^0(S) = 0$ otherwise.

The *core-extension* $\tilde{\Gamma} = (2^N, \tilde{v})$ assesses the maximal benefit the players in an arbitrary subset S can secure by forming pairwise disjoint feasible coalitions, i.e.,

$$\tilde{v}(S) = \max\{\sum_i v(F_i) \mid F_i \in \mathcal{F}(S),\ F_i \cap F_j = \emptyset \text{ for } i \neq j\}.$$

Thus, $v^* = \tilde{v}(N)$ is the value of the game Γ. It is straightforward to verify that $\tilde{v}$ is non-negative, monotone increasing and superadditive on the subsets of N.

Observe that the cores of the game extensions coincide:

$$\begin{aligned}\text{core}(\Gamma) &= \{x \in \mathbb{R}^N_+ \mid x(N) \leq v^*,\ x(S) \geq v^0(S) \text{ for all } S \subseteq N\}\\ &= \text{core}(\tilde{\Gamma}).\end{aligned}$$

2 Selections and Rankings

Generalizing the classical model for choice functions (see, e.g., [24,23,17]), we call $(\mathcal{D}, \mu)$ a *selection scheme* for $\mathcal{F}$ if $\mathcal{D}$ is a family of subsets of N and $\mu : \mathcal{D} \to \mathcal{F}$ a function such that

(0) $N \in \mathcal{D}$,
(i) $\mu(D) \subseteq D$,
(ii) $\mu(D) \neq \emptyset$ if $\mathcal{F}(D) \neq \{\emptyset\}$,
(iii) $D \setminus \{p\} \in \mathcal{D}$ for all $p \in \mu(D)$.

Given the selection scheme $(\mathcal{D}, \mu)$, an ordered selection $\pi = p_1 \dots p_k$ of players is a *ranking* if the players $p_k, p_{k-1}, \dots$ are selected in turn and, in case of ties, preference is given to players that have been eligible for selection in a previous turn. In other words, π is constructed according to the following procedure:

(R0) $X \leftarrow N$; $\pi \leftarrow \Box$; $w(p) \leftarrow 1$ for all $p \in N$;
(R1) Choose $p \in \mu(X)$ of minimal weight $w(p)$;
(R2) Update $w(p') \leftarrow [w(p') - w(p)]$ for all $p' \in \mu(X)$;
Update $X \leftarrow X \setminus \{p\}$; $\pi \leftarrow p\pi$;
(R3) If $\mu(X) = \emptyset$, stop and output π. Return to (R1) otherwise;

We denote by M_i the set $\mu(X)$ from which player p_i is selected during the ranking procedure and thus obtain with $\pi = p_1 \dots p_k$ the *Monge family*

$$\mathcal{M}^\pi = \{M_0, M_1, \dots, M_k\} \quad (\text{with } M_0 = \emptyset).$$

In the case $\mathcal{D} = \mathcal{F} = 2^N$, the choice $\mu(D) = D$ yields exactly the permutations of the elements in N as rankings. We give more examples.

Ordered Sets of Players. Let $P = (N, \leq)$ be a partial order. An *antichain* of P is a set $A \subseteq N$ of pairwise incomparable elements. An *ideal* of P is a set $I \subseteq N$ such that $p \in I$ implies $q \in I$ for all $q \leq p$.

Let $\mathcal{A}$ be the collection of antichains of P. Then $(2^N, \mu_{\mathcal{A}})$ yields a selection scheme for $\mathcal{A}$ when $\mu_{\mathcal{A}}(X)$ denotes the set of maximal elements of X (with respect to P). A selection scheme for the collection $\mathcal{I}$ of ideals of P is $(2^N, \mu_{\mathcal{I}})$, where $\mu_{\mathcal{I}}(X)$ denotes the maximal ideal contained in X. Note that the two selection schemes lead to generally *different* rankings π of players.

Antimatroids and Convex Geometries. An *antimatroid* is a union-closed family $\mathcal{I}$ of subsets of N such that $\emptyset, N \in \mathcal{I}$ and for each $I, J \in \mathcal{I}$ one has the so-called Steinitz augmentation property

(S) $|I| < |J| \quad \Longrightarrow \quad I \cup p \in \mathcal{I}$ for some $p \in J \setminus I$.

A selection scheme for the antimatroid $\mathcal{I}$ is $(2^N, \mu_{\mathcal{I}})$ with

$$\mu_{\mathcal{I}}(X) = \text{the maximal member of } \mathcal{I} \text{ contained in } X.$$

A *convex geometry* is a family $\mathcal{C}$ of subsets of N such that the complementary family $\mathcal{C}^c = \{N \setminus C \mid C \in \mathcal{C}\}$ is an antimatroid. $\mathcal{C}$ is thus intersection-closed and

$$\overline{X} = \bigcap\{C \mid X \subseteq C\} \in \mathcal{C}$$

is well-defined for every $X \subseteq N$. Using the augmentation property (S) of the antimatroid $\mathcal{C}^c$, it is not hard to prove:

Lemma 1. *Let $\mathcal{C}$ be a convex geometry on N and $X \subseteq N$ arbitrary. Then there exists a unique minimal subset $\mu_{\mathcal{E}}(X) \subseteq X$ such that $\overline{X} = \overline{\mu_{\mathcal{E}}(X)}$.* ◇

$\mu_{\mathcal{E}}(X)$ is the set of *extreme points* of $\overline{X}$. Letting $\mathcal{E} = \{\mu_{\mathcal{E}}(C) \mid C \in \mathcal{C}\}$, we thus obtain the selection scheme $(2^N, \mu_{\mathcal{E}})$ for the family $\mathcal{E}$ of extremal sets relative to an underlying convex geometry.

Strong Hierarchies. The preceding examples can be cast into the following model. We assume that $\mathcal{F}$ can be equipped with a partial order relation $(\mathcal{F}, \preceq)$ such that

(SH1) every restriction $(\mathcal{F}(X), \preceq)$ has a unique maximal element $\mu(X)$,
(SH2) $(2^N, \mu)$ is a selection scheme for $\mathcal{F}$.

We refer to $(\mathcal{F}, \preceq)$ as a *strong hierarchy* if (SH1) and (SH2) holds.

Proposition 1. *Let $\pi = p_1 \ldots p_k$ be a ranking of the strong hierarchy $(\mathcal{F}, \preceq)$. Then for each feasible coalition $F \in \mathcal{F} \setminus \emptyset$, there is some ranked player $p_i \in \pi$ with $p_i \in F$.*

Proof. Suppose $F \in \mathcal{F}$ is a counterexample to the Proposition. Letting $\mathcal{M}^\pi = \{M_0 \preceq \ldots \preceq M_k\}$ be the Monge set of π, choose i minimal such that $F \preceq M_i$. So $F \neq M_i$ and $F \not\preceq M_{i-1}$ holds by assumption. Consequently, $\mu(F \cup M_{i-1})$ is larger than M_i and satisfies

$$\mu(F \cup M_{i-1}) \cap \{p_k, p_k - 1, \ldots, p_i\} = \emptyset,$$

which contradicts the choice of $M_{i-1} = \mu(N \setminus \{p_k, p_{k-1}, \ldots, p_i\})$ in the ranking procedure. ◇

3 Weber Sets and the Shapley Property

Let $(\mathcal{F}, v)$ be a cooperative game with restricted cooperation and $(\mathcal{D}, \mu)$ a selection scheme for $\mathcal{F}$. Then we call $\Gamma = (\mathcal{F}, v, \mathcal{D}, \mu)$ a *hierarchical game.* We denote by Π the collection of all rankings of Γ.

3.1 Ranking Vectors

Let $\pi = p_1 \dots p_k \in \Pi$ be a ranking of the hierarchical game Γ with Monge set $\mathcal{M}^\pi = \{M_0, M_1, \dots, M_k\}$. The key to our analysis is the following observation.

Theorem 1. *There is a unique vector $h^\pi \in \mathbb{R}^N$ with the properties*

(a) $h^\pi(M_i) = v(M_i)$ for all i.
(b) $h^\pi(p) = 0$ for all $p \in N$ that do not occur in π.
(c) $h^\pi(N) \leq v^*$.

Proof. The unique existence of h^π is recursively established. We set $h^\pi(p_1) = v(M_1)$ and then iterate

$$h^\pi(p_i) = v(M_i) - \sum \{p_j \mid j < i, p_j \in M_i\} \quad (i = 2, \dots, k).$$

We briefly sketch the proof of (c). Let $\mathcal{M}' \subseteq \mathcal{M}^\pi$ be the collection of all those feasible coalitions $M_i \in \mathcal{M}^\pi$ such that $p_i \in M_i$ has weight $w(p_i) = 1$ when it is selected in the ranking procedure. It is not difficult to see that the members of $\mathcal{M}'$ must be pairwise disjoint, whence we conclude from (a):

$$h^\pi(N) = \sum_{i=1}^{k} h^\pi(p_j) = \sum_{M_i \in \mathcal{M}'} h^\pi(M_i) = \sum_{M_i \in \mathcal{M}'} v(M_i) \leq v^*.$$

◇

The ranking vectors h^π are the analogues of the marginal contribution vectors in classical games. The inequality (c) says that h^π is a "reasonable" allocation in the sense that it distributes not more to the players than the value $\tilde{v}(N)$ the players can generate by forming pairwise disjoint feasible coalitions.

The Weber Set. Generalizing the Weber set of classical cooperative games (see [27]), we define the *Weber set* of the hierarchical game Γ as the convex hull of its ranking vectors:

$$W(\Gamma) = \text{conv}\{h^\pi \mid \pi \in \Pi\}.$$

So $W(\Gamma)$ is a convex polytope in $\mathbb{R}^N$. Note that $W(\Gamma)$ depends on the particular selection scheme $(\mathcal{D}, \mu)$ according to which rankings of players are determined.

The Shapley Value. It is now natural to allocate to each player p_i his average ranking value in the hierarchical game $\Gamma = (\mathcal{F}, v, \mathcal{D}, \mu)$. So we define the *Shapley value* of Γ as the allocation vector

$$\Phi(\Gamma) = \frac{1}{|\Pi|} \sum_{\pi \in \Pi} h^\pi.$$

The definition implies $\Phi(\Gamma) \in W(\Gamma)$. Moreover, the feasibility of the allocation $\Phi(\Gamma)$ is guaranteed:

$$\sum_{p \in N} \Phi(\Gamma)_p = \sum_{p \in N} \frac{1}{|\Pi|} \sum_{\pi \in \Pi} h^\pi(p) \leq v^*.$$

3.2 The Shapley Property

The cooperative game $\Gamma = (\mathcal{F}, v)$ with restricted cooperation is said to have the *Shapley property* if there exists a selection scheme $(\mathcal{D}, \mu)$ for Γ with collection Π of rankings such that $W(\Gamma) \subseteq \text{core}(\Gamma)$ or, equivalently,

$$h^\pi \in \text{core}(\Gamma) \quad \text{for all } \pi \in \Pi.$$

Recall that $h^\pi(N) = v^*$ necessarily holds for any ranking vector h^π that satisfies $h^\pi \in \text{core}(\Gamma)$. We will exhibit sufficient conditions for the Shapley property later and now state a general form of a Shapley-type theorem (cf. [26]):

Theorem 2 ("Shapley's Theorem"). *The cooperative game $\Gamma = (\mathcal{F}, v)$ has the Shapley property if and only if there exists a selection scheme $(\mathcal{D}, \mu)$ for $\mathcal{F}$ such that*

$$\text{core}(\Gamma) = W(\Gamma)$$

holds for the associated Weber set.

We prove Theorem 2 *via* a greedy algorithm (see Section 4 and Theorem 3) for linearly weighted ranking vectors that actually optimizes a linear function over $\text{core}(\Gamma)$ if $W(\Gamma) = \text{core}(\Gamma)$ holds. So also $\text{core}(\Gamma) \subseteq W(v)$ must be true, which yields the claim of the Theorem.

We have remarked before that in the classical case the Shapley property is equivalent with the game to be non-negative convex. The next example shows that a hierarchical game Γ may possess the Shapley property although neither of its extensions Γ^0 and $\tilde{\Gamma}$ are convex.

Example 1. Consider $N = \{a, b, c\}$ together with the strong hierarchy $(\mathcal{F}, \preceq) = \{\emptyset \prec \{a, b\} \prec \{b, c\}\}$ and let $v(a, b) = v(b, c) = 1$. It is easy to check that each ranking vector relative to $(\mathcal{F}, \preceq)$ lies in the core of $\Gamma = (\mathcal{F}, v)$. So Γ has the Shapley property. However, neither the null-extension v^0 nor the core-extension $\tilde{v}$ are convex on the collection of all subsets of N since

$$\begin{aligned} 2 = v(a,b) + v(b,c) &= v^0(a,b) + v^0(b,c) &= \tilde{v}(a,b) + \tilde{v}(b,c) \\ 1 = \tilde{v}(b) + \tilde{v}(a,b,c) &> v^0(b) + v^0(a,b,c) &= 0. \end{aligned}$$

4 The Greedy Algorithm

Let $c \in \mathbb{R}^N$ be an arbitrary vector of weights for the elements in N. Given the game $\Gamma = (\mathcal{F}, v, \mathcal{D}, \mu)$ with collection Π of rankings, we are interested in finding the optimal weighted ranking vector, i.e., the discrete optimization problem

$$\max_{\pi \in \Pi} \sum_{p \in N} c_p h^{\pi}(p). \tag{1}$$

Passing to the weighting $\gamma = -c$, problem (1) becomes equivalent with

$$\min_{\pi \in \Pi} \sum_{p \in N} \gamma_p h^{\pi}(p). \tag{2}$$

We will investigate the problem in its minimization version (2) and consider the following *Greedy Algorithm* relative to a weight vector w':

(G0) $X \leftarrow N$; $\pi \leftarrow \Box$; $w(p) \leftarrow w'(p)$ for all $p \in N$;
(G1) Choose $p \in \mu(X)$ of minimal weight $w(p)$;
(G2) Update $w(p') \leftarrow [w(p') - w(p)]$ for all $p' \in \mu(X)$;
Update $X \leftarrow X \setminus \{p\}$; $\pi \leftarrow p\pi$;
(G3) If $\mu(X) = \emptyset$, stop and output π. Return to (G1) otherwise;

If we choose the weights $w'(p) = 1$, the greedy algorithm reduces to the ranking algorithm (and thus produces a ranking π for Γ). For general weights $w'(p)$, however, the greedy algorithm may end with some π that is not a ranking. We solve this problem, by adding a "large" weight $\lambda > 0$ to each of the original weights:

Lemma 2. *Choose some parameter $\lambda > 3\sum_{p \in N} |\gamma_p|$ and define $w'(p) = \gamma_p + \lambda$ for all $p \in N$. Then the greedy algorithm with input w' will produce a ranking π for the hierarchical game Γ.*

Proof. The greedy algorithm will choose in the first step some $p_k \in M_k = \mu(N)$

and then reduce all the weights in M_k by the amount $\gamma_p - \lambda$. Afterwards, all weights in M_k will be "small" while all the other weights are still "large". When $p_{k-1} \in M_{k-1} = \mu(N \setminus p_k)$ is to be selected in the next step, preference will thus be given to elements from M_k *etc.*, which is exactly the procedure of the ranking algorithm. $\diamond$

We call the greedy algorithm for problem (2) with the modified weights $w'(p) = \gamma_p + \lambda$ the *modified greedy algorithm.* One of our main results which can be proved using linear programming duality is

Theorem 3. *Let $(\mathcal{D}, \mu)$ be a selection scheme for the game $\Gamma = (\mathcal{F}, v)$ such that every ranking vector h^{π} lies in the core of Γ. Then the modified greedy algorithm solves problem (2) optimally. Moreover, the greedy vector h^{π} is an optimal solution for the problem*

$$\max_{x \in \text{core}(\Gamma)} c^T x = \sum_{p \in N} c_p x_p. \tag{3}$$

Consequently, the core and the Weber set of Γ coincide.

5 Convex Hierarchical Games

We now discuss sufficient conditions for the cooperative game $\Gamma = (\mathcal{F}, v)$ to have the Shapley property. We restrict ourselves to the case where $(\mathcal{F}, \preceq)$ is a strong hierarchy (with associated selection scheme $(2^N, \mu)$). We furthermore assume that $(\mathcal{F}, \preceq)$ satisfies the *dominant player axiom* for all feasible coalitions $F_1, F_2, F_3 \in \mathcal{F}$ and players $p \in F_3 \setminus F_2$:

(DP) $F_1 \preceq F_2 \preceq F_3 \quad \Longrightarrow \quad p \notin F_1$.

We say that a player $p \in F_3$ is F_2*-dominant* if $F_2 \preceq F_3$ and $p \notin F_2$. So (DP) says that any F_2-dominant player p is also F_1-dominant if $F_1 \preceq F_2$. Note that the examples in Section enjoy property (DP).

For any $F \in \mathcal{F}$, we define the *excess* of the ranking $\pi = p_1 \ldots p_k$ on F as the difference between the corresponding π-allocation of F and the value of F:

$$e_\pi(F) = \sum_{p \in F} h^\pi(p) - v(F).$$

Where $\mathcal{M}^\pi = \{M_0 \prec M_1 \prec \ldots \prec M_k\}$ is the Monge chain of π, the function $e_\pi : \mathcal{F} \longrightarrow \mathbb{R}$ has the property

$$e_\pi(M_i) = \sum_{p \in M_i} h^\pi(p) - v(M_i) = 0 \quad (i = 1, \ldots, k).$$

Given the ranking π of $(\mathcal{F}, \preceq)$, we call an arbitrary function $f : \mathcal{F} \to \mathbb{R}$ π*-convex* if the following holds:

(PC) For all $F \in \mathcal{F} \setminus \mathcal{M}^\pi$ such that $F \prec M_i$ and $F \npreceq M_{i-1}$, there is some $F' \in \mathcal{F}$ with the property $F' \preceq F$, $F' \preceq M_{i-1}$ and

$$f(F') + f(M_i) \geq f(F) + f(M_{i-1}).$$

Say that $\Gamma = (\mathcal{F}, \preceq, v)$ is π*-convex* if the excess function e_π is π-convex.

Lemma 3. *Let π be a ranking of the hierarchical game $\Gamma = (\mathcal{F}, \preceq, v)$ and assume that Γ is π-convex. Then we have for all $F \in \mathcal{F}$,*

$$e_\pi(F) \geq 0 \,, \quad \text{i.e.} \,, \quad h^\pi(F) \geq v(F) \,.$$

Proof. Suppose the Lemma is false for the ranking π with associated chain $\mathcal{M}^\pi$ and there is some $F \in \mathcal{F}$ with $e_\pi(F) < 0$. Choose F to be as small as possible relative to the partial order $(\mathcal{F}, \preceq)$. In view of $e_\pi(M_j) = 0$ for all $M_j \in \mathcal{M}^\pi$, F cannot be a member of $\mathcal{M}^\pi$. Because $\mathcal{M}^\pi$ includes the maximal element of $(\mathcal{F}, \preceq)$, there exists a (unique) coalition $M_i \in \mathcal{M}^\pi$ such that $F \preceq M_i$ and $M_{i-1} \npreceq F$. Then π-convexity guarantees some $F' \in \mathcal{F}$ with $F' \preceq F$ and $F' \preceq M_{i-1}$ such that

$$e_\pi(F) \geq e_\pi(F') + e_\pi(M_{i-1}) - e_\pi(M_i) = e_\pi(F').$$

Note that $F' \preceq M_{i-1}$ implies $F' \neq F$. So the choice of F as a minimal counterexample yields $0 \leq e_\pi(F') \leq e_\pi(F)$. This contradiction establishes the claim of the Lemma. ◇

Since $h^\pi(N) \leq v^*$ holds for every ranking π of $(\mathcal{F}, \preceq)$, Lemma 3 immediately provides sufficient conditions for the non-emptyness of core(v):

Proposition 2. *Assume that $\Gamma = (\mathcal{F}, \preceq, v)$ is π-convex. Then*

$$h^\pi \in \text{core}(v) \quad \Longleftrightarrow \quad h^\pi \geq 0. \qquad \diamond$$

It follows from the construction and property (DP) that $h^\pi \geq 0$ is satisfied whenever v is monotone:

Lemma 4. *Let $\Gamma = (\mathcal{F}, \preceq, v)$ be a hierarchical game and assume that v is monotone increasing with respect to $(\mathcal{F}, \preceq)$. Then every ranking π yields:*

$$h^\pi(p) \geq 0 \quad \textit{for all } p \in N. \qquad \diamond$$

We illustrate the notions of this Section with the following three applications:

Example 2 ("Shapley's Theorem"). The excess functions of a classical convex game $\Gamma = (2^N, \subseteq, v)$ satisfy the inequality

$$e_\pi(S \cup T) + e_\pi(S \cap T) \geq e_\pi(S) + e_\pi(T).$$

If $F \subseteq N$ is namely such that $F \subseteq M_i$ and $F \not\subseteq M_{i-1}$ holds, we have $M_i = F \cup M_i$. So the choice $F' = F \cap M_{i-1}$ exhibits Γ as π-convex with respect to any permutation π of N. h^π is non-negative (and hence lies in core(v)) if and only if v is monotone on the chain $\mathcal{M}^\pi$. It follows that monotone convex games have a non-empty core (*cf.* [26]).

Example 3 (Distributive Games). If $P = (N, \leq)$ is a partial order on the players, the selection scheme $(2^N, \mu_{\mathcal{I}^+})$, where $\mu_{\mathcal{I}^+}(X)$ contains the elements of maximal height (with respect to P) in $\mu_{\mathcal{I}}(X)$ (i.e., the maximal ideal contained in X), leads to the core and Weber set of distributive games proposed in [18] and [19] (see also [13]).

Example 4 (Chains). Let $\mathcal{F}$ be and $v : \mathcal{F} \to \mathbb{R}_+$ any normalized function. Order $\mathcal{F}$ linearly so that v is monotone and note the resulting chain $(\mathcal{F}, \preceq)$ is strong hierarchy. Hence, if the dominant player property (DP) is satisfied, the core of the game $(\mathcal{F}, v)$ is non-empty.

Example 5 (Lattice Hierarchies). Assume $\wedge, \vee : \mathcal{F} \times \mathcal{F} \to \mathcal{F}$ are binary operations with $F_1 \wedge F_2 \preceq F_1, F_2 \preceq F_1 \vee F_2$ and

$$(F_1 \wedge F_2) \cup (F_1 \vee F_2) \subseteq F_1 \cup F_2.$$

Then $(\mathcal{F}, \preceq)$ is a strong hierarchy. (For example, any inclusion-wise ordered union-closed family $(\mathcal{F}, \subseteq)$ is a strong hierarchy and satisfies (DP)). Provided property (DP) holds for $(\mathcal{F}, \preceq)$ and the profit function v is monotone increasing and supermodular with respect to $\wedge, \vee$, one can show that v is π-convex for every ranking π. So $\Gamma = (\mathcal{F}, v)$ has the Shapley property.

6 Final Remarks

From a modeling point of view, it is interesting to recall that in the classical context every cooperative profit game is equivalent to an associated *cost* game in the sense that both games have the same core. The present hierarchical games generalize profit games. It seems less straightforward to exhibit natural cost games with the same core in the wider model of hierarchies.

It appears reasonable to associate with a cost game $\Gamma = (\mathcal{F}, c)$ the cost parameter

$$c^* = \min\{\sum_{i=1}^{\ell} c(S_i) \mid N \subseteq \bigcup_{i=1}^{\ell} S_i, S_i \in \mathcal{F}\}$$

and to consider the polyhedron

$$\text{core}(c) = \{x \in \mathbb{R}_+^N \mid \sum_{p \in N} x(p) \geq c^*, \sum_{p \in S} x(s) \leq c(S) \text{ for all } S \in \mathcal{F}\}.$$

While a Shapley value $\Phi(c)$ of the cost game Γ may be formally defined in complete analogy with the Shapley value of Section 3, it seems to be less clear to what an extent an analogue of Theorem 2 is true. For some results on greedy algorithms in this direction, we refer the interested reader to [15].

References

1. Algaba, E., Bilbao, J.M., van den Brink, R., Jiménez-Losada, A.: Cooperative games on antimatroids. Discr. Mathematics 282, 1–15 (2004)
2. Bilbao, J.M., Jiménez, N., Lebrón, E., López, J.J.: The marginal operators for games on convex geometries. Intern. Game Theory Review 8, 141–151 (2006)
3. Bilbao, J.M., Lebrón, E., Jiménez, N.: The core of games on convex geometries. Europ. J. Operational Research 119, 365–372 (1999)
4. Derks, J., Gilles, R.P.: Hierarchical organization structures and constraints in coalition formation. Intern. J. Game Theory 24, 147–163 (1995)
5. Dietrich, B.L., Hoffman, A.J.: On greedy algorithms, partially ordered sets, and submodular functions. IBM J. Res. & Dev. 47, 25–30 (2003)
6. Danilov, V., Koshevoy, G.: Choice functions and extending operators (preprint, 2007)
7. Edmonds, J.: Submodular functions, matroids and certain polyhedra. In: Proc. Int. Conf. on Combinatorics (Calgary), pp. 69–87 (1970)
8. Edelman, P.H., Jamison, R.E.: The theory of convex geometries. Geometriae Dedicata 19, 247–270 (1985)
9. Fujishige, S.: Submodular Functions and Optimization. 2nd edn.; Ann. Discrete Mathematics 58 (2005)
10. Faigle, U.: Cores of games with restricted cooperation. Methods and Models of Operations Research 33, 405–422 (1989)
11. Faigle, U., Kern, W.: The Shapley value for cooperative games under precedence constraints. Intern. J. Game Theory 21, 249–266 (1992)
12. Faigle, U., Kern, W.: Submodular linear programs on forests. Math. Programming 72, 195–206 (1996)

13. Faigle, U., Kern, W.: On the core of ordered submodular cost games. Math. Programming 87, 483–489 (2000)
14. Faigle, U., Kern, W.: An order-theoretic framework for the greedy algorithm with applications to the core and Weber set of cooperative games. Order 17, 353–375 (2000)
15. Faigle, U., Peis, B.: Two-phase greedy algorithms for some classes of combinatorial linear programs. In: SODA 2008 (accepted, 2008)
16. Frank, A.: Increasing the rooted-connectivity of a digraph by one. Math. Programming 84, 565–576 (1999)
17. Fujishige, S.: Dual greedy polyhedra, choice functions, and abstract convex geometries. Discrete Optimization 1, 41–49 (2004)
18. Grabisch, M., Xie, L.J.: The core of games on distributive lattices (working paper)
19. Grabisch, M., Xie, L.J.: A new investigation about the core and Weber set of multichoice gamse. Mathematical Methods of Operations Research (to appear)
20. Gilles, R.P., Owen, G., van den Brink, R.: Games with permission structures: the conjunctive approach. Intern. J. Game Theory 20, 277–293 (1992)
21. Hoffman, A.J., Schwartz, D.E.: On lattice polyhedra. In: Hajnal, A., Sós, V.T. (eds.) Proc. 5th Hungarian Conference in Combinatorics, pp. 593–598. North-Holland, Amsterdam (1978)
22. Hsiao, C.-R., Raghavan, T.E.S.: Shapley value for multi-choice cooperative games. Games and Economic Behavior 5, 240–256 (1993)
23. Koshevoy, G.: Choice functions and abstract convex geometries. Math. Soc. Sci. 38, 35–44 (1999)
24. Moulin, H.: Choice functions over a finite set: a summary. Soc. Choice Welfare 2, 147–160 (1985)
25. Shapley, L.S.: A value for n-person games. In: Kuhn, H.W., Tucker, A.W. (eds.) Contributions to the Theory of Games, Ann. Math. Studies, vol. 28, pp. 307–317. Princeton University Press, Princeton (1953)
26. Shapley, L.S.: Cores of convex games. Intern. J. Game Theory 1, 12–26 (1971)
27. Weber, R.J.: Probabilistic values for games. In: Roth, A.E. (ed.) The Shapley Value, pp. 101–120. Cambridge University Press, Cambridge (1988)

Strategic Characterization of the Index of an Equilibrium

Arndt von Schemde and Bernhard von Stengel

Department of Mathematics, London School of Economics, London WC2A 2AE, United Kingdom
schemde@gmail.com, stengel@nash.lse.ac.uk

Abstract. We prove that an equilibrium of a nondegenerate bimatrix game has index $+1$ if and only if it can be made the unique equilibrium of an extended game with additional strategies of one player. The main tool is the "dual construction". A simplicial polytope, dual to the common best-response polytope of one player, has its facets subdivided into best-response regions, so that equilibria are completely labeled points on the surface of that polytope. That surface has dimension $m-1$ for an $m \times n$ game, which is much lower than the dimension $m+n$ of the polytopes that are classically used.

1 Introduction

The index of a Nash equilibrium is an integer that is related to notions of "stability" of the equilibrium. In this paper, we only consider nondegenerate bimatrix games; "generic" (that is, almost all) bimatrix games are nondegenerate. A bimatrix game is nondegenerate if any mixed strategy with support of size k has at most k pure best responses [15]; the support of a mixed strategy is the set of pure strategies that are played positive probability. Nondegeneracy implies that the two strategies of a mixed equilibrium have supports of equal size. The index has the following elementary definition due to Shapley [13].

Definition 1. *Let (x,y) be a Nash equilibrium of a nondegenerate bimatrix game (A,B) with positive payoff matrices A and B, and let L and J be the respective supports of x and y, with corresponding submatrices A_{LJ} and B_{LJ} of the payoff matrices A and B. Then the* index *of (x,y) is defined as*

$$(-1)^{|L|+1}\text{sign}\ (\det(A_{LJ})\ \det(B_{LJ})). \tag{1}$$

The index has the following properties, which require that its sign alternates with the parity of the support size as in (1).

Proposition 2. *In a nondegenerate bimatrix game,*
(a) *the index of an equilibrium is $+1$ or -1;*
(b) *any pure-strategy equilibrium has index $+1$;*
(c) *the index only depends on the payoffs in the support of the equilibrium strategies;*
(d) *the index does not depend on the order of a player's pure strategies;*

B. Monien and U.-P. Schroeder (Eds.): SAGT 2008, LNCS 4997, pp. 242–254, 2008.

(e) *the endpoints of any Lemke–Howson path have opposite index;*
(f) *the sum of the indices over all equilibria is* $+1$*;*
(g) *in a* 2×2 *game with two pure equilibria, the mixed equilibrium has index* -1.

Condition (a) holds because payoff-submatrices A_{LJ} or B_{LJ} that do not have full rank $|L|$ can only occur for degenerate games. The simple property (g) applies to, say, a coordination game and easily follows from (1) or (f). It is one indication that, as suggested by Hofbauer [7], equilibria of index $+1$ are in many respects "sustainable" according to Myerson [10], who discusses ways to refine or select equilibria in "culturally familiar games". Hofbauer [7] also shows that only equilibria of index $+1$ can be stable under any "Nash dynamics", that is, a vector field on the set of mixed strategy profiles whose rest points are the Nash equilibria [6][4]. Such dynamics may represent evolutionary or learning processes.

The most interesting computational property is (e), proved by Shapley [13]. The Lemke–Howson (LH) algorithm [9] (for an exposition see [15]) defines a path which can either start at a trivial "artificial equilibrium" with empty support, or else at any Nash equilibrium, and which ends at another equilibrium. The equilibria of the game, plus the artificial equilibrium, are therefore the endpoints of the LH paths. By (1), the artificial equilibrium has index -1. Consequently, the game has one more equilibrium of index $+1$ than of index -1, and (f) holds.

Equilibria as endpoints of LH paths provide a "parity argument" that puts the problem of finding one Nash equilibrium of a bimatrix game into the complexity class PPAD [11]. This stands for "polynomial parity argument with direction", where the direction of the path is provided by the index (which can also be determined locally at any point on the path).

The index of an equilibrium can also be defined for general games (which may be degenerate and have more than two players) as the degree of a topological map that has the Nash equilibria as fixed points, like the mentioned "Nash dynamics" [6][4].

The index is a relatively complicated topological notion, essentially a geometric orientation of the equilibrium. In this paper, we prove the following theorem, first conjectured in [7], which characterizes the index in much simpler strategic terms.

Theorem 3. *A Nash equilibrium of a nondegenerate* $m \times n$ *bimatrix game* G *has index* $+1$ *if and only if it is the unique equilibrium of a game* G' *obtained from* G *by adding suitable strategies. It suffices to add* $3m$ *strategies for the column player.*

The equilibrium of G in Theorem 3 is re-interpreted as an equilibrium of G', so none of the added strategies is used in the equilibrium; their purpose is to eliminate all other equilibria. Unplayed strategies do not matter for the index of an equilibrium by Prop. 2(c). By (f), a unique equilibrium has index $+1$, so only equilibria with positive index can be made unique as stated in Theorem 3; the nontrivial part is therefore to show that this is always possible.

We prove Theorem 3 using a novel geometric-combinatorial method that we call the *dual construction*. It allows to visualize all equilibria of an $m \times n$ game in a diagram of dimension $m-1$. For example, all equilibria of a $3 \times n$ game are visualized with a diagram (essentially, of suitably connected $n+3$ points) in the plane. This should provide new insights into the geometry of Nash equilibria.

A better understanding of the geometry of Nash equilibria may also be relevant algorithmically, and we think the index is relevant apart from providing the "D" in "PPAD". Recent results on the complexity of finding one Nash equilibrium of a bimatrix game have illustrated the difficulty of the problem: it is PPAD-complete [2], and LH paths may be exponentially long [12]. Even a sub-exponential algorithm for finding one Nash equilibrium is not in sight. In designing any such algorithm, for example incremental or by divide-and-conquer, it is important that the information carried to the next phase of the algorithm does not describe the entire set of equilibria, because questions about that set, for example uniqueness of the Nash equilibrium, tend to be NP-hard [5][3]. On the other hand, Nash equilibria with additional properties (for example, a minimum payoff) may not exist, or give rise to NP-complete decision problems. However, it is always possible to restrict the search to an equilibrium with index $+1$; whether this is of computational use remains speculative.

The dual construction has first been described in the first author's PhD dissertation, published in [14]. Some steps of the construction are greatly simplified here, and the constructive proof outlined in Section 5 is new.

2 Dualizing the First Player's Best Response Polytope

We use the following notation. All vectors are column vectors. If $d \in \mathbb{R}^k$ and $s \in \mathbb{R}$, then ds is the vector d scaled with s, as the product of a $k \times 1$ with a 1×1 matrix. If $s = 1/t$, we write d/t for ds. The vectors $\mathbf{0}$ and $\mathbf{1}$ have all components equal to 0 and 1, respectively. Inequalities like $d \geq \mathbf{0}$ between vectors hold for all components. A matrix C with all entries scaled by s is written as sC. We write $C = [c_1 \cdots c_k]$ if C is a matrix with columns $c_1, \ldots, c_k$. The transpose of C is $C^\top$.

The index of an equilibrium is defined in (1) via the sign of determinants. We recall some properties of determinants. Exchanging any two rows or any two columns of a square matrix $C = [c_1 \cdots c_k]$ changes the sign of $\det(C)$, which implies Prop. 2(d). The determinant is multilinear, so that, for any $d \in \mathbb{R}^k$, $s \in \mathbb{R}$ and $1 \leq i \leq k$,

$$\begin{aligned} \det[c_1 \cdots c_{i-1}\ c_i s\ c_{i+1} \cdots c_k] &= s \det(C), \\ \det[c_1 \cdots c_{i-1}\ (c_i + d)\ c_{i+1} \cdots c_k] &= \det(C) + \det[c_1 \cdots c_{i-1}\ d\ c_{i+1} \cdots c_k]. \end{aligned} \tag{2}$$

Let $s_1, \ldots, s_k$ be scalars, which we add to the columns of C. Repeated application of (2) gives

$$\det(C + [\mathbf{1}s_1 \cdots \mathbf{1}s_k]) = \det(C) + \sum_{i=1}^{k} s_i \det[c_1 \cdots c_{i-1}\ \mathbf{1}\ c_{i+1} \cdots c_k]. \tag{3}$$

The right-hand side of (3) is linear in $(s_1, \ldots, s_k)$. In particular, if $s_1 = \cdots s_k = s$, then the expression $\det(C + s[\mathbf{1} \cdots \mathbf{1}])$ is linear in s and changes its sign at most once.

We first explain why the matrices A and B are assumed to be positive in Def. 1. Consider an equilibrium (x, y), and discard for simplicity all pure strategies that are not in the support of x or y, so that A and B are the matrices called A_{LJ} and B_{LJ} in (1), which have full rank. Then the equilibrium payoffs to the two players are u and v, respectively, with $Ay = \mathbf{1}u$ and $B^\top x = \mathbf{1}v$. We want that always $u > 0$ and $v > 0$; this clearly holds if A and B have positive entries, although this not a necessary condition. Adding any constant t to all payoffs of A does not change the equilibria of the game, but does change the equilibrium payoff from u to $u+t$. Consequently, we could achieve $Ay = \mathbf{0}$ (with $t = -u$), or $Ay < \mathbf{0}$. However, $Ay = \mathbf{0}$ implies $\det(A) = 0$, and consequently a change of the sign of u implies a change of the sign of $\det(A)$. Because the sign of $\det(A)$ changes only once, that sign is unique whenever A is positive. Similarly, the sign of $\det(B)$ is unique, so (1) defines the index uniquely.

For the rest of the paper, we consider a nondegenerate $m \times n$ bimatrix game (A,B) so that the best-response payoff to any mixed strategy of a player is always positive, for example by assuming that A and B are positive. The following polytopes can be used to characterize the equilibria of (A,B) [15]:

$$\begin{aligned} P &= \{x \in \mathbb{R}^m \mid \ x \geq \mathbf{0},\ B^\top x \leq \mathbf{1}\}, \\ Q &= \{y \in \mathbb{R}^n \mid Ay \leq \mathbf{1}, \quad y \geq \mathbf{0}\}. \end{aligned} \tag{4}$$

Any $(x,y) \in P \times Q$ with $x \neq \mathbf{0}$, $y \neq \mathbf{0}$ represents a mixed strategy pair with best-response payoffs scaled to one; normalizing x and y as probability vectors re-scales these payoffs. The inequalities in P and Q are *labeled* with the numbers $1,\ldots,m+n$ to indicate the pure strategies of player 1 (labels $1,\ldots,m$) and player 2 (labels $m+1,\ldots,m+n$). Given $x \in P$ and $y \in Q$, each *binding* inequality (which holds as an equation) defines a facet of P or Q (by nondegeneracy, it cannot be a lower-dimensional face [15]). The corresponding label defines an unplayed own pure strategy or best response of the other player. An equilibrium of (A,B) corresponds a pair (x,y) of $P \times Q$ where each pure strategy $1,\ldots,m+n$ appears as a label of x or y. The artificial equilibrium is given by $(x,y) = (\mathbf{0},\mathbf{0})$.

The first step of our construction is to dualize the polytope P by considering its *polar* polytope [16]. Suppose R is a polytope defined by inequalities that has $\mathbf{0}$ it its interior, so that it can be written as $R = \{z \in \mathbb{R}^m \mid c_i^\top z \leq 1,\ 1 \leq i \leq k\}$. Then the polar polytope is defined as $R^\triangle = \operatorname{conv}\{c_i \mid 1 \leq i \leq k\}$, that is, as the convex hull of the normal vectors c_i of the inequalities that define R. The face lattice of $R^\triangle$ is that of R upside down, so $R^\triangle$ and R contain the same combinatorial information about the face incidences. More precisely, assuming that R has full dimension m, any face of R of dimension h given by $\{z \in R \mid c_i^\top z = 1 \text{ for } i \in K\}$ (with maximal set K) corresponds to the face $\operatorname{conv}\{c_i \mid i \in K\}$ of dimension $m-1-h$. So facets (irredundant inequalities) of R correspond to vertices of $R^\triangle$, and vertices of R correspond to facets of $R^\triangle$. If R is simple, that is, has no point that lies on more that m facets, then $R^\triangle$ is simplicial, that is, all its facets are simplices.

Because the game is nondegenerate, the polytope P is simple, and any binding inequality of P defines either a facet or the empty face, the latter corresponding to a dominated strategy of player 2 that can be omitted from the game. In particular, player 2 has no weakly dominated strategy, which would define a lower-dimensional face of P.

Because P does not have $\mathbf{0}$ in its interior, we consider instead the polytope

$$\begin{aligned} P_\varepsilon &= \{x \in \mathbb{R}^m \mid \qquad\quad x \geq -\mathbf{1}\varepsilon,\ B^\top x \leq \mathbf{1}\}, \\ &= \{x \in \mathbb{R}^m \mid -1/\varepsilon \cdot Ix \leq \mathbf{1}, \quad B^\top x \leq \mathbf{1}\}, \end{aligned} \tag{5}$$

where $\varepsilon > 0$ and I is the $m \times m$ identity matrix. For sufficiently small ε, the polytopes P and P_ε are combinatorially equivalent, because the simple polytope P allows small perturbations of its facets. Moreover, nondegeneracy crucially forbids weakly dominated strategies, which would be "better" than the dominating strategy under the "negative probabilities" x_i allowed in P_ε, and hence define facets of P_ε but not of P. Then

$$P_\varepsilon^\triangle = \mathrm{conv}(\{-e_i/\varepsilon \mid 1 \le i \le m\} \cup \{b_j \mid 1 \le j \le n\}), \tag{6}$$

where e_i is the ith unit vector in $\mathbb{R}^m$ and $B = [b_1 \cdots b_n]$. That is, $P_\varepsilon^\triangle$ is the convex hull of sufficiently largely scaled negative unit vectors and of the columns b_j of the payoff matrix B of player 2. We will later add points, which are just additional payoff columns; this is the reason why we perturb, rather than translate, P.

Any facet F of $P_\varepsilon^\triangle$ is a simplex, given as the convex hull of m vertices $-e_i/\varepsilon$ for $i \in K$ and b_j for $j \in J$, where $|K| + |J| = m$. We write

$$F = F(K,J) = \mathrm{conv}(\{-e_i/\varepsilon \mid i \in K\} \cup \{b_j \mid j \in J\}). \tag{7}$$

Then the vertices of the facet $F(K,J)$ define *labels* i and $m+j$ which represent unplayed strategies $i \in K$ of player 1 and best responses $j \in J$ of player 2. These labels are the labels of the facets of P_ε, and hence of P, that correspond to the vertices of $F(K,J)$.

The facet $F(K,J)$ itself corresponds to a vertex x_ε of P_ε. Namely, because $P_\varepsilon^{\triangle\triangle} = P_\varepsilon$ [16], we have $F(K,J) = P_\varepsilon \cap \{z \in \mathbb{R}^m \mid x_\varepsilon^\top z = 1\}$, where $x_\varepsilon^\top z \le 1$ holds for all $z \in P_\varepsilon$. The vertex x_ε of P_ε corresponds to a vertex x of P, which is determined from K and J by the linear equations $x_i = 0$ for $i \in K$ and $\sum_{i \notin K} b_{ij} x_i = 1$ for $j \in J$. The corresponding equations for x_ε are $(x_\varepsilon)_i = -\varepsilon$ for $i \in K$ and $\sum_{i=1}^m b_{ij}(x_\varepsilon)_i = 1$ for $j \in J$, so $x_\varepsilon \to x$ as $\varepsilon \to 0$.

In summary, the normal vectors of facets $F(K,J)$ of $P_\varepsilon^\triangle$ correspond to mixed strategies x of player 1. The vertices of such facets represent player 1's unplayed strategies $i \in K$ and best responses $j \in J$ of player 2. A similar representation of mixed strategies and best responses is considered by Bárány, Vempala and Vetta [1], namely the polyhedron defined as the intersection of the halfspaces with nonnegative normal vectors that contain the points $b_1, \ldots, b_n$. Our polytope $P_\varepsilon^\triangle$ approximates that polyhedron when it is intersected with the halfspace with supporting hyperplane through the m points $-e_i/\varepsilon$ for $1 \le i \le m$.

The facet $F_0 = F(\{1,\ldots,m\},\emptyset)$ of $P_\varepsilon^\triangle$ whose vertices are the m points $-e_i/\varepsilon$ for $1 \le i \le m$ corresponds to the vertex $\mathbf{0}$ of P. The surface of $P_\varepsilon^\triangle$ can be projected to F_0, giving a so-called Schlegel diagram [16]. A suitable projection point is $-\mathbf{1}/\varepsilon$. The Schlegel diagram is a subdivision of the simplex F_0 into simplices that correspond to the other facets of $P_\varepsilon^\triangle$. All these simplices have dimension $m-1$, so for $m = 3$ one obtains a subdivided triangle. An example is the left picture in Fig. 1 for the matrix B of the 3×4 game

$$A = \begin{bmatrix} 0 & 10 & 0 & 10 \\ 10 & 0 & 0 & 0 \\ 8 & 0 & 10 & 8 \end{bmatrix}, \qquad B = \begin{bmatrix} 0 & 10 & 0 & -10 \\ 0 & 0 & 10 & 8 \\ 10 & 0 & 0 & 8 \end{bmatrix}. \tag{8}$$

In that picture, the labels $i = 1,2,3$ correspond to the scaled negative unit vectors $-e_i/\varepsilon$, the labels $m+j = 4,5,6,7$ to the columns b_j of B. The nonpositive entries of A and B are allowed because a player's best-response payoff is always positive.

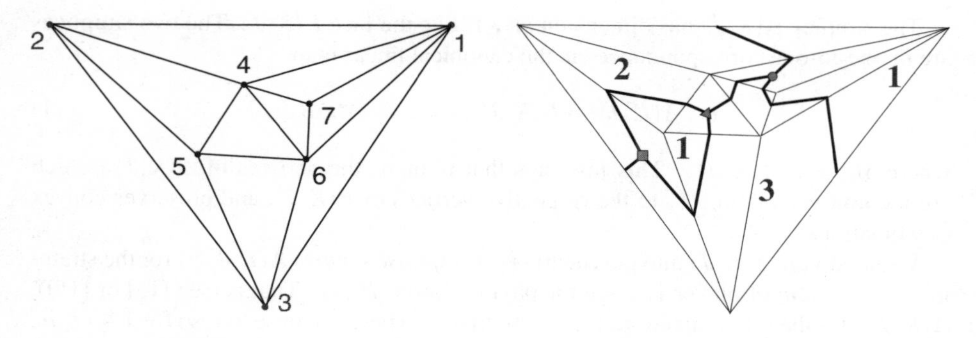

Fig. 1. Left: Schlegel diagram of $P_\varepsilon^\triangle$ for the example (8). Right: Subdivision into best-response regions of player 1, which completes the dual construction.

3 Subdividing Simplices into Best-Response Regions

The second step of our construction is the incorporation of player 1's best responses into the surface of $P_\varepsilon^\triangle$. Let $F(K,J)$ be a facet of $P_\varepsilon^\triangle$ as in (7). Consider the $m \times m$ matrix

$$[I_K\, A_J] = [e_{i_1} \cdots e_{i_k}\, a_{j_1} \cdots a_{j_{m-k}}] \qquad \text{if } K = \{i_1, \ldots, i_k\},\ J = \{j_1, \ldots, j_{m-k}\}, \quad (9)$$

that is, the columns of $[I_K\, A_J]$ are the columns e_i of the identity matrix I for $i \in K$ and the columns a_j of player 1's payoff matrix A for $j \in J$, where $A = [a_1 \cdots a_n]$. We write the unit simplex $\text{conv}\{e_1, \ldots, e_m\}$ in $\mathbb{R}^m$ as

$$\Delta(K,J) = \{z \in \mathbb{R}^K \times \mathbb{R}^J \mid z \geq \mathbf{0},\ \mathbf{1}^\top z = 1\}. \quad (10)$$

Proposition 4. *Let $(x,y) \in P \times Q - \{(\mathbf{0},\mathbf{0})\}$. Then (x,y) is a Nash equilibrium of (A,B) if and only if the vertex x of P corresponds to a facet $F(K,J)$ of $P_\varepsilon^\triangle$ so that $[I_K\, A_J]z = \mathbf{1}u$ for some $z = (z_K, z_J) \in \Delta(K,J)$ and some $u > 0$, and $y_J = z_J/u$, where y_J is y restricted to its support J.*

Proof. Because the game is nondegenerate, only vertices x of P can represent equilibrium strategies. Let $F(K,J)$ be the facet of $P_\varepsilon^\triangle$ that corresponds to x, where $K = \{i \mid x_i = 0\}$ and J is the set of best responses to x. Then y is a best response to x if and only if the support of y is J; suppose this holds, so that $Ay = A_J y_J$. In turn, x is a best response to y if and only if $(Ay)_i = 1$ whenever $i \notin K$, because $Ay \leq \mathbf{1}$. This is equivalent to $I_K s_K + A_J y_J = \mathbf{1}$ for a suitable slack vector $s_K \in \mathbb{R}^K$, $s_K \geq \mathbf{0}$. With $u = 1/(\sum_{i \in K} s_i + \sum_{j \in J} y_j)$ and $z = (s_K u, y_J u)$ this is equivalent to $z \in \Delta(K,J)$ and $[I_K\, A_J]z = \mathbf{1}u$ as claimed. □

Given a facet $F(K,J)$ of $P_\varepsilon^\triangle$ that corresponds to a vertex x of P, Prop. 4 states that x is part of a Nash equilibrium (x,y) if and only if there is a mixed strategy $z = (z_K, z_J) \in \Delta(K,J)$ so that all pure strategies of player 1 are best responses against z when the payoff matrix to player 1 is $[I_K\, A_J]$. Suitably scaled, z_K is a vector of slack variables, and z_J represents the nonzero part y_J of player 2's mixed strategy y. Nondegeneracy implies that z is completely mixed and hence in the interior of $\Delta(K,J)$.

The simplex $\Delta(K,J)$ has dimension $m-1$, like the face $F(K,J)$. The two simplices are in one-to-one correspondence via the canonical linear map

$$\alpha : \Delta(K,J) \to F(K,J), \qquad z \mapsto [M_K\ B_J]z, \tag{11}$$

where $M_K = (-1/\varepsilon \cdot I)_K$. This just says that α maps the vertices of $\Delta(K,J)$ (which are the unit vectors in $\mathbb{R}^m$) to the respective vertices of $F(K,J)$, and preserves convex combinations.

We subdivide $\Delta(K,J)$ into polyhedral *best response regions* $\Delta(K,J)(i)$ for the strategies $i = 1,\ldots,m$ of player 1, using the payoff matrix $[I_K\ A_J]$. That is (see [13] or [15]), $\Delta(K,J)(i)$ is the set of mixed strategies z so that i is a best response to z, so for $1 \le i \le m$,

$$\Delta(K,J)(i) = \{z \in \Delta(K,J) \mid ([I_K\ A_J]z)_i \ge ([I_K\ A_J]z)_k \text{ for all } k = 1,\ldots,m\}. \tag{12}$$

We say z in $\Delta(K,J)$ has *label* i if $z \in \Delta(K,J)(i)$, and correspondingly $\alpha(z)$ in $F(K,J)$ has label i if z has label i.

This *dual construction* [14] labels every point on the surface of $P_\varepsilon^\triangle$. The labeling is unique because the payoffs to player 1, and the map α, only depend on the vertices of the respective facets, so the labels agree for points that belong to more than one facet. For the game in (8), this labeling is shown in the right picture of Fig. 1.

As a consequence of Prop. 4, we obtain that the equilibria of the game correspond to the points on the surface of $P_\varepsilon^\triangle$ that have all labels $1,\ldots,m$. We call such points *completely labeled*. The three equilibria of the game (8) are marked by a small square, triangle and circle in Fig. 1. Not all facets of $P_\varepsilon^\triangle$ contain such a completely labeled point, if the corresponding vertex x of P is not part of a Nash equilibrium.

"Completely labeled" now means "all strategies of player 1 appear as labels". What happened to the strategies of player 2? They correspond to the vertices of $P_\varepsilon^\triangle$. They are automatically best responses when considering the facets of $P_\varepsilon^\triangle$, and they are the only strategies that player 2 is allowed to use, apart from the slacks $i \in K$, when subdividing a facet $F(K,J)$ into the labeled regions $\alpha(\Delta(K,J)(i))$ for the labels $i = 1,\ldots,m$.

4 Visualizing the Index and the LH Algorithm

The described dual construction, the labeled subdivision of the surface of $P_\varepsilon^\triangle$, visualizes all equilibria of an $m \times n$ game in a geometric object of dimension $m-1$. Figure 2 also shows the index of an equilibrium as the orientation in which the labels $1,\ldots,m$ appear around the point representing the equilibrium, here counterclockwise for index $+1$, and clockwise for index -1. The artificial equilibrium is the completely labeled point $M(\mathbf{1}/m)$ (see (11) with $J = \emptyset$) of the facet F_0 of $P_\varepsilon^\triangle$, which has negative index. This facet should be seen as the underside of the "flattened" view of $P_\varepsilon^\triangle$ given by the Schlegel diagram, so the dashed border of F_0 in Fig. 2 is to be identified with the border of the large triangle.

Our goal is to formalize the orientation of a completely labeled point in the dual construction, and to show that it agrees with the index in Def. 1. A nonsingular $m \times m$ matrix C has *positive orientation* if $\det(C) > 0$, otherwise *negative orientation*.

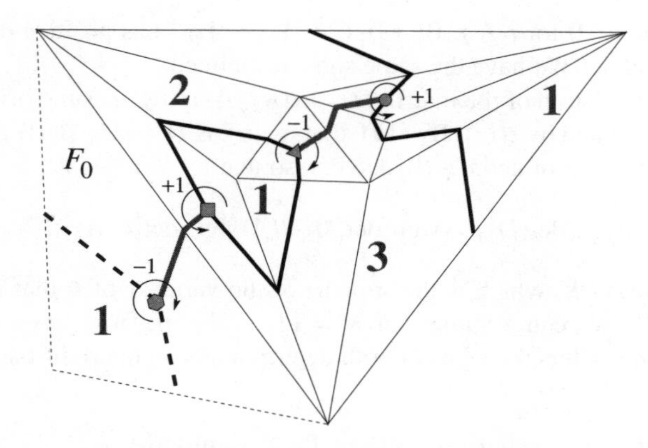

Fig. 2. Indices of equilibria as positive or negative orientations of the labels, and LH paths for missing label 1. The facet on the left with dashed border indicates the flapped-out "underside" of the Schlegel diagram, the facet F_0 of $P_\varepsilon^\triangle$.

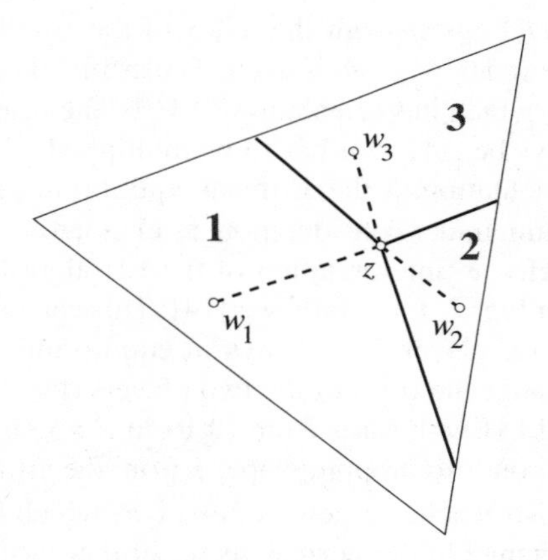

Fig. 3. Points w_1, w_2, w_3 in $\Delta(K,J)$, here for $K = \emptyset$, so that (13) holds

Let $\alpha(z)$ be a completely labeled point of a facet $F(K,J)$ of $P_\varepsilon^\triangle$. We first consider points $w_1, \ldots, w_m$ so that w_i belongs only to the best-response region $\Delta(K,J)(i)$ for $1 \le i \le m$. More specifically, we want that for suitable $s_i \ge 0$, $t_i > 0$,

$$[I_K\ A_J]w_i = \mathbf{1}s_i + e_i t_i, \tag{13}$$

that is, player 1's payoff against w_i is $s_i + t_i$ for his pure strategy i, and a smaller constant s_i for all other pure strategies $k \neq i$. Such points w_i exist, by extending the line through the completely labeled point z defined by the $m-1$ labels $k \neq i$ into the region $\Delta(K,J)(i)$, as shown in Fig. 3. For $i \in K$, we can simply choose $w_i = e_i$ to obtain (13), a case that is not shown in Fig. 3.

Let $W = [w_1 \cdots w_m]$. We want to show that W has the same orientation as $[I_K\ A_J]$. Because of (13), $[I_K\ A_J]W = C + [\mathbf{1}s_1 \cdots \mathbf{1}s_m]$ for the diagonal matrix C with entries

$c_{ii} = t_i > 0$ and $c_{ij} = 0$ for $i \neq j$. By (3), $C + [\mathbf{1}s_1 \cdots \mathbf{1}s_m]$ has positive determinant, so that $\det[I_K\ A_J]$ and $\det(W)$ have the same sign, as claimed.

We take the orientation of the matrix $D = [\alpha(w_1) \cdots \alpha(w_m)]$ as the orientation of the equilibrium represented by $\alpha(z)$. By (11), that matrix is $D = [M_K\ B_J]\,W$. Its orientation $\mathrm{sign}(\det(D))$ is the sign of $\det[M_K\ B_J]\det(W)$, so that

$$\mathrm{sign}(\det(D)) = \mathrm{sign}(\det[M_K\ B_J])\,\mathrm{sign}(\det[I_K\ A_J]). \tag{14}$$

Let $L = \{1,\ldots,m\} - K$, which is the support of the vertex x of P that corresponds to the facet $F(K,J)$. We can assume that $K = \{1,\ldots,k\}$, because any transposition of player 1's strategies alters the signs of both determinants on the right-hand side of (14). Then

$$\begin{aligned}\mathrm{sign}(\det(D)) &= \mathrm{sign}((-1/\varepsilon)^k \det(B_{LJ}))\ \mathrm{sign}(\det(A_{LJ}))\\ &= (-1)^{m-|L|}\mathrm{sign}(\det(B_{LJ}))\ \mathrm{sign}(\det(A_{LJ}))\\ &= (-1)^{m-1}(-1)^{|L|+1}\mathrm{sign}(\det(B_{LJ}))\ \mathrm{sign}(\det(A_{LJ})).\end{aligned}$$

Consequently, $\mathrm{sign}(\det(D))$ agrees with the index of the equilibrium when m is odd, and is the negative of the index when m is even. The artificial equilibrium corresponds to the center point of F_0, which has orientation $(-1)^m$. The orientation of the artificial equilibrium should always be -1, so it has to be multiplied with -1 when m is even. Hence, relative to the orientation of the artificial equilibrium, $\mathrm{sign}(\det(D))$ is exactly the index of the equilibrium under consideration, as claimed.

We mention very briefly an interpretation of the LH algorithm with the dual construction, as illustrated in Fig. 2; for details see [14]. This can only be done for missing labels of player 1, because player 2 is always in equilibrium. For missing labels of player 2 one has to exchange the roles of the two players (the dual construction works either way). The original LH path starts from $(\mathbf{0},\mathbf{0})$ in $P \times Q$ by dropping a label, say label 1, in P. This corresponds to dropping label 1 from the artificial equilibrium given by the center of F_0. It also reaches a new vertex of P, which in $P_\varepsilon^\triangle$ means a change of facet. This means a change of the normal vector of that facet, which is an invisible step in the dual construction because the path is at that point on the joint face of the two facets. Traversal of an edge of Q is represented by traversing the line segment in a face $P_\varepsilon^\triangle$ that has all $m-1$ labels except for the missing label. That line segment either ends at an equilibrium, or else reaches a new facet of $P_\varepsilon^\triangle$. The path then (invisibly) changes to that facet, followed by another line segment, and so on. Algorithmically, the LH pivoting steps are just like for the path on $P \times Q$, so nothing changes.

Figure 2 also illustrates why the endpoints of LH paths have opposite index: Along the path, the $m-1$ labels that are present preserve their orientation around the path, whereas the missing label is in a different direction at the beginning and at the end of the path. In Fig. 2, an LH path from a -1 to a $+1$ equilibrium with missing label 1 has label 2 on the left and label 3 on the right. This intuition of Shapley's result Prop. 2(e) [13] can be given without the dual construction (see [12, Fig. 1]), but here it is provided with a figure in low dimension.

5 Proof Sketch of Theorem 3

In this section, we give an outline of the proof of Theorem 3 with the help of the dual construction. We confine ourselves to an equilibrium (x,y) of index $+1$ that uses all m strategies of player 1, which captures the core of the argument. Hence, the facet $F(K,J)$ of $P_\varepsilon^\triangle$ that corresponds to the fully mixed strategy x of player 1 has $K=\emptyset$. The m best responses of player 2 are $j \in J$, which define the payoff vectors b_j as points in $\mathbb{R}^m$, and an $m \times m$ submatrix B_J of B.

For player 1 and player 2, we will construct three $m \times m$ matrices A', A'', A''' and B', B'', B''', respectively, so that the extended game G' in Theorem 3 is defined by the two $m \times (n+3m)$ payoff matrices $[A\, A'\, A''\, A''']$ and $[B\, B'\, B''\, B''']$.

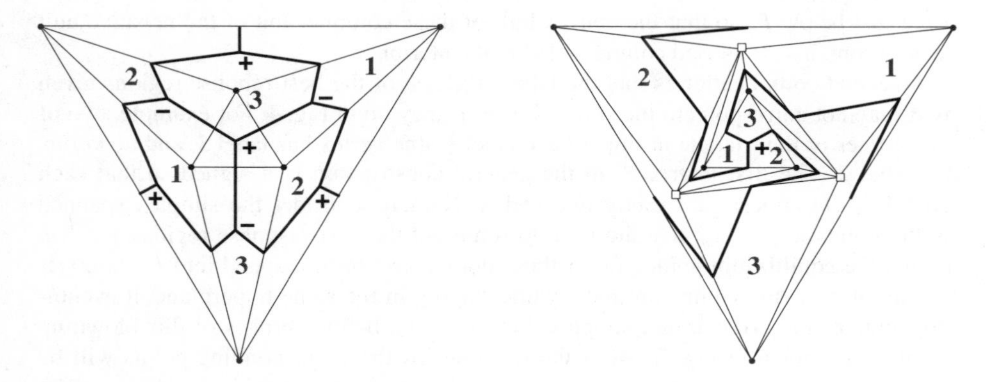

Fig. 4. Left: Dual construction for a 3×3 coordination game, which has four equilibria with index $+1$ and three with index -1. Right: Adding three strategies for the column player (small white squares) so that only the completely mixed equilibrium remains, see (15).

An example is the 3×3 coordination game where both players' payoffs are given by the identity matrix. The game has seven equilibria: three pure-strategy equilibria and the completely mixed equilibrium, all of index $+1$, and three equilibria where each player mixes two strategies, with index -1. The left picture in Fig. 4 shows the dual construction for this game. The index of each equilibrium point is indicated by its sign, given by the orientation of the labels $1,2,3$ around the point. The completely mixed equilibrium is on the central triangle with facet F whose vertices are the three columns of B. We want to make this equilibrium unique by adding strategies. In this case, we need only the matrices A'' and B'', for example as follows:

$$[AA''] = \begin{bmatrix} 1 & 0 & 0 & 0 & 0 & 1 \\ 0 & 1 & 0 & 1 & 0 & 0 \\ 0 & 0 & 1 & 0 & 1 & 0 \end{bmatrix}, \quad [BB''] = \begin{bmatrix} 11 & 10 & 10 & 12 & 8.9 & 10 \\ 10 & 11 & 10 & 10 & 12 & 8.9 \\ 10 & 10 & 11 & 8.9 & 10 & 12 \end{bmatrix}. \qquad (15)$$

The dual construction for the game in (15) is shown on the right in Fig. 4. As desired, only the original facet F has an equilibrium point, which is now unique. It is also clear that its index must be $+1$, because otherwise it would not be possible to "twist" the best response regions $1,2,3$ outwards to meet the labels at the outer vertices.

In this example, the columns of B'' span a simplex (with vertices indicated by small white squares in Fig. 4), whose projection to F_0 in the Schlegel diagram contains the original facet F as a subset. In fact, the columns of B'' are first constructed as points in the hyperplane defined by F so that they define a larger simplex than F. Subsequently, these new points are moved slightly to the origin, so that F re-appears in the convex hull: Note that in (15), the normal vector for the hyperplane through the columns of B'' is $\mathbf{1}$, but its scalar product with these columns is 30.9 and not 31 like for the columns of B (the matrix B is the identity matrix with 10 added to every entry).

In the general construction, several complications have to be taken care of. First, the original game may have additional columns that are not played in the considered equilibrium. The example makes it clear that this is a minor complication: Namely, the simplex spanned by the columns of B'' can be magnified, while staying in the hyperplane just below F, so that the convex hull of these columns and of the negative unit vectors contains all unused columns of B in its interior.

A second complication is that the labels $1,\ldots,m$ of the best-reponse regions given by A may not correspond to the vertices of F as they do in Fig. 4. For example, two of the vertices of the triangle in Fig. 3 have label 1, one vertex has label 3, and no vertex has label 2. The first matrix B' in the general construction is designed so that each label $1,\ldots,m$ appears at exactly one vertex. Namely, consider the simplex spanned by the points $w_1,\ldots,w_m$ that are used to represent the best-response regions $1,\ldots,m$ around the equilibrium point z (after these points have been mapped into F via α). If this simplex is "blown up" around z while staying in the same hyperplane, it eventually contains the original unit simplex. Let $v_1,\ldots,v_m$ be the vertices of this blown-up simplex, as shown in Fig. 5. After the mapping α, the corresponding points will be

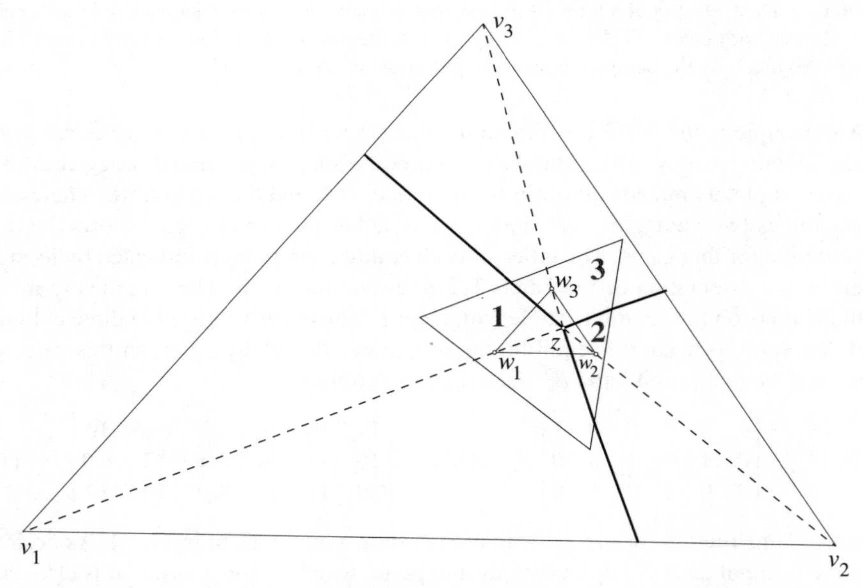

Fig. 5. Points $v_1,\ldots,v_m$ as expanded points $w_1,\ldots,w_m$ along the dashed lines around z in the same hyperplane, so that $\text{conv}\{v_1,\ldots,v_m\}$ contains the original simplex $\Delta(K,J)$

in the hyperplane defined by F and define a simplex that contains F as a subset. We merely move these points $\alpha(v_1),\ldots,\alpha(v_m)$ slightly towards to the origin, which defines the matrix B'. The corresponding payoffs A' to player 1 are given by the diagonal matrix $[e_1t_1\cdots e_mt_m]$ with the payoffs $t_1,\ldots,t_m$ given as in (13). We could add an arbitrary constant $\mathbf{1}s_i$ to the ith column of A' (for each i) without changing the subdivision into best-response regions of the simplex defined by B'. Hence, if B' was still in the same hyperplane as F, that subdivision would coincide with the subdivision of F, which it still does after moving B' slightly inwards. From then on, we consider the simplex by spanned B' instead of F, which then looks essentially like in the special case of Fig. 15 because the corresponding matrix A' is a diagonal matrix.

We also use two increasingly larger simplices defined by B'' and B''', with identity matrices A'' and A'''. In the resulting construction, each pair of matrices (M,B'''), (B''',B''), (B'',B') and (B',B_J) (where the columns of M and B_J are the vertices of F_0 and F, respectively) defines a pair of simplices whose convex hull is a "distorted prism". These distorted prisms are stacked on top of each other, with the points of intermediate layers spread outwards in parallel hyperplanes to maintain a convex set. In the projected Schlegel diagram, each simplex is contained in the next.

The missing matrices B'' and B''' are constructed using the following theorem of [8]: *Every matrix with positive determinant is the product of three P-matrices.* A P-matrix is a matrix P such that every principal minor P_{SS} of P (where S is an arbitray set of rows and the same set of columns) has positive determinant. The P-matrices are useful for "stacking distorted prisms" because of the following property:

Proposition 5. *Let $P=[p_1\cdots p_m]$ be a P-matrix where each column p_i is scaled such that $\mathbf{1}^\top p_i = 2$. Let X be the convex hull of the vectors p_i and the unit vectors e_i for $1\le i\le m$. Assume X is a simplicial polytope, if necessary by slightly perturbing P. Let a facet of X have label i if it has p_i or e_i has a vertex. Then the only facets of X that have all labels $1,\ldots,m$ are those spanned by $p_1,\ldots,p_m$ and by $e_1,\ldots,e_m$.*

Proposition 5 may be a novel observation about P-matrices. It can be proved using a parity argument: Additional completely labeled facets would have to come in pairs of opposite orientation, and a negatively oriented facet contradicts the P-matrix property.

Consequently, a distorted prism X defined by the columns of the identity matrix I and a P-matrix P (scaled as in Prop. 5) has no completely labeled facet other than its two "end" facets defined by I and P. If Q is another such P-matrix, the same observation holds for I and Q, and consequently for P and PQ because it does not change under affine transformations. Finally, for another P-matrix R, we see that PQ and PQR define prisms that have no completely labeled "side" facets, either. According to said theorem of [8], PQR can represent an arbitrary matrix with positive determinant.

The stack of prisms that we want to generate should start at the convex hull M of the negative unit vectors used in (11). We move these vectors in the direction $\mathbf{1}$ until it crosses the origin, so that the resulting matrix, which we call N, has opposite orientation to M. As shown in Section 4, N has therefore the same orientation as the matrix D in (14) and hence as B'. Therefore, $N^{-1}B'$ has positive determinant and can be represented as a product PQR of three P-matrices, so that $NPQR = B'$. Then the additional matrices are given by $B'' = NPQ$ and $B''' = NP$.

We have to omit details for reasons of space. We conclude with the following intuition why we use two matrices B'' and B''' rather than just a single one. In Fig. 4, the columns of B'', which is the only matrix used, are indicated by the white squares, but these show only the projection in the Schlegel diagram. Their (invisible) distances from the origin are very important because they determine the facets spanned by the columns of B'' and of B. Essentially, B'' represents a single intermediate step when "twisting" F by 180 degrees towards the outer triangle, and A'' is a matrix of suitably ordered unit vectors. It is not clear if this can be done in higher dimensions. With two intermediate sets of points B'' and B''', their exact relative position is not very relevant when using P-matrices, due to Prop. 5, and one can use identity matrices A'' and A'''.

References

1. Bárány, I., Vempala, S., Vetta, A.: Nash equilibria in random games. In: Proc. 46th FOCS, pp. 123–131 (2005)
2. Chen, X., Deng, X.: Settling the complexity of 2-player Nash-equilibrium. In: Proc. 47th FOCS, pp. 261–272 (2006)
3. Conitzer, V., Sandholm, T.: Complexity results about Nash equilibria. In: Proc. 18th International Joint Conference on Artificial Intelligence (IJCAI), pp. 765–771 (2003)
4. Demichelis, S., Germano, F.: On the indices of zeros of Nash fields. Journal of Economic Theory 94, 192–217 (2000)
5. Gilboa, I., Zemel, E.: Nash and correlated equilibria: some complexity considerations. Games and Economic Behavior 1, 80–93 (1989)
6. Govindan, S., Wilson, R.: Uniqueness of the index for Nash equilibria of two-player games. Economic Theory 10, 541–549 (1997)
7. Hofbauer, J.: Some thoughts on sustainable/learnable equilibria. In: 15th Italian Meeting on Game Theory and Applications, Urbino, Italy, July 9–12 (2003) (accessed on October 29, 2007), `http://www.econ.uniurb.it/imgta/PlenaryLecture/Hofbauer.pdf`
8. Johnson, C.R., Olesky, D.D., van den Driessche, P.: Matrix classes that generate all matrices with positive determinant. SIAM J. Matrix Anal. Appl. 25, 285–294 (2003)
9. Lemke, C.E., Howson Jr., J.T.: Equilibrium points of bimatrix games. Journal of the Society for Industrial and Applied Mathematics 12, 413–423 (1964)
10. Myerson, R.B.: Sustainable equilibria in culturally familiar games. In: Albers, W., et al. (eds.) Understanding Strategic Interaction: Essays in Honor of Reinhard Selten, pp. 111–121. Springer, Heidelberg (1997)
11. Papadimitriou, C.H.: On the complexity of the parity argument and other inefficient proofs of existence. Journal of Computer and System Sciences 48, 498–532 (1994)
12. Savani, R., von Stengel, B.: Exponentially many steps for finding a Nash equilibrium in a bimatrix game. In: Proc. 45th FOCS, pp. 258–267 (2004)
13. Shapley, L.S.: A note on the Lemke–Howson algorithm. Mathematical Programming Study 1: Pivoting and Extensions, pp. 175–189 (1974)
14. von Schemde, A.: Index and Stability in Bimatrix Games. Lecture Notes in Economics and Mathematical Systems, vol. 1853. Springer, Berlin (2005)
15. von Stengel, B.: Computing equilibria for two-person games. In: Aumann, R.J., Hart, S. (eds.) Handbook of Game Theory, vol. 3, pp. 1723–1759. North-Holland, Amsterdam (2002)
16. Ziegler, G.M.: Lectures on Polytopes. Springer, New York (1995)

The Local and Global Price of Anarchy of Graphical Games

Oren Ben-Zwi[1] and Amir Ronen[2,*]

[1] Haifa University, Haifa 31905 Israel
nbenzv03@cs.haifa.ac.il

[2] Faculty of Industrial Engineering and Management, Technion, Israel
amirr@ie.technion.ac.il

Abstract. This paper initiates a study of connections between local and global properties of graphical games. Specifically, we introduce a concept of local price of anarchy that quantifies how well subsets of agents respond to their environments. We then show several methods of bounding the global price of anarchy of a game in terms of the local price of anarchy. All our bounds are essentially tight.

Keywords: Graphical games, price of anarchy, local global properties.

1 Introduction

The model of graphical games [10], is a recent representation method of games in which the dependencies among the agents are represented by a graph. In a graphical game, each agent is identified by a vertex, and its utility is determined solely by its own action and the actions of its graph neighbors. Note that every game can be represented by a graphical game with a complete graph. Yet, often, a much more succinct representation is possible. While the original motivation of defining graphical games was computational, we believe that an important property of the model is that it enables an investigation of many natural structural properties of games.

In this work we investigate connections between local and global properties of graphical games. Specifically, we study the Price of Anarchy (PoA) which is the ratio between the welfare of a worst Nash equilibrium and the optimal possible welfare [11].

We introduce a novel notion of a *local* price of anarchy which quantifies how well subsets of agents respond to their environments. We then study the relations between this local measure and the global price of anarchy of the game. We provide several methods of bounding the global price of anarchy in terms of the local price of anarchy, and demonstrate the tightness of these methods.

One possible interpretation of our results is as follows: if a decentralized system is comprised of smaller, well behaved units, with small overlap between them,

* The author was supported in part by grant number 969/06 from the Israel Science Foundation.

B. Monien and U.-P. Schroeder (Eds.): SAGT 2008, LNCS 4997, pp. 255–266, 2008.

then the whole system behaves well. This holds independently of the size of the small units, and even when the small units only behave well on average. This phenomenon may have implications, for example, on organizational theory. From a computational perspective, the price of anarchy of large games is likely to be extremely hard to compute. However, computing the local price of anarchy of small units is relatively easy since they correspond to much smaller games. Once these are computed, our methods can be invoked to bound the price of anarchy of the overall game.

1.1 Related Work

The model of graphical games was introduced in [10]. The original motivation for the model was computational as it permitted a succinct representation of many games of interest. Moreover, for certain graph families, there are properties that can be computed efficiently. For example, although computing a Nash equilibrium is usually a hard task [5,4], it can be computed efficiently for graphical games on graphs with maximum degree 2 [6]. Rather surprisingly, the proofs of the hardness of computing Nash equilibria of normal form games are conducted via reductions to graphical games [5].

[8] investigates the structure of equilibria of graphical games under some symmetry assumptions on the utility of the agents. It shows that in these games, there always exists a pure strategy equilibrium. For such games of incomplete information, [8] shows that there is a monotone relationship between the degree of a vertex and its payoff, and investigates further the connections between the level of information the game possesses and the monotonicity of the players' degree in equilibria. Several works coauthored by Michael Kearns explore economic and game theoretic properties which are related to structure (e.g. [9]). The questions addressed in these works are very different from the ones we address here.

The price of anarchy [11] is a natural measure of games. After the discovery of fundamental results regarding the price of anarchy of congestion games [15,2], the price of anarchy and the price of stability[1] [1] have become almost standard methods for evaluating games.

[12] investigates deductions that can be made on global properties of graphs after examining only local neighborhoods. It shows that for any graph G, where $V[G] = n$, and a function $f : V \to \Re^+ \cup \{0\}$, if the local average of f over every ball of positive radius less or equal to r in G is greater or equal to α, the global average of f is at least $\frac{\alpha}{n^{O(1/\log r)}}$. [12] also demonstrates the tightness of this bound.

In this work we make an extensive use of graph covers, but we do not introduce a method for finding them. Algorithms that find good covers can be found, for example, in [3] and [13]. Due to the game theoretic nature of our setup, these algorithms cannot be applied to it directly.

In general, the field of property testing in computer science examines the connections between local and global properties of combinatorial objects (see,

[1] The price of stability is the ratio between the *best* Nash equilibrium and the optimum of the game.

for example, [7] for a survey). For many properties it is known that if an object satisfies a property in a local sense, then it is "not too far" from satisfying it globally. As we shall see, the additivity of the welfare function, enables even stronger connections between the local and the global perspectives in our setup.

2 Preliminaries

We denote the utility of a player $i \in [n] = \{1, 2, \dots, n\}$ by u_i. In most of this work we focus on games where the utility of each player is non-negative, i.e. $u_i \geq 0$. Every player wishes to maximize its own utility.

Definition 1. *[10]* **(graphical game)** *An n players* graphical game *is a pair (G, M), where G is an undirected graph on n vertices and M is an n players finite game. Every player i is associated with the node v_i and its utility is determined only by its action and by the actions of its neighbors.*

Since we only discuss graphical representations of games in this paper, the terms game and graphical game are treated as synonyms. We note again that every game can be represented as a graphical game.

Definition 2. (welfare) *Let s be a vector of agent strategies for a game G. The* welfare $|s|$ *is the sum of the agents' utilities resulting from s, i.e. $|s| = \sum_i u_i(s)$.*

The welfare of a game is a common measure of the aggregation of the agents' utilities. It is by no means the only aggregation method. In this paper we focus on maximizing the welfare as the sole criterion of how good a game is. Our results can immediately be generalized to any measure of the form $\sum_i \psi_i(u_i(\cdot))$ where $\psi_i : \Re^+ \to \Re^+$ are non decreasing functions.

Definition 3. *[14]* **(Nash equilibrium)** *A strategy where no player can unilaterally divert from and increase its utility is called a* Nash equilibrium.

A Nash equilibrium does not have to be unique. In this work we are only interested in *global worst Nash equilibria*, i.e. in equilibria that obtain the minimal welfare of the whole game. A global worst Nash equilibrium is not necessarily unique but its value is fixed for a game and we can just pick one such arbitrary strategy vector to work with. We denote a global worst Nash equilibrium by $GWNE$, and its value by $|GWNE|$. From compactness and continuity considerations, a global worst Nash equilibrium always exists.

We denote by $U_{opt}(G)$ (or just U_{opt} when the context is clear) a strategy vector that achieves the optimal welfare of the game, and by $|U_{opt}|$ the optimal welfare.

Definition 4. *[11]* **(price of anarchy)** *For a game G, the ratio between the welfare of a worst Nash equilibrium and the optimal welfare is called the* price of anarchy *(PoA). That is: $PoA = \frac{|GWNE|}{|U_{opt}|}$.*

Note that the price of anarchy is always between 0 and 1. A price of anarchy of 1 means that all Nash equilibria are optimal[2]. It is natural to define the price of anarchy of sub-games as well. We thus denote the PoA of the whole game by GPoA (global price of anarchy).

The following combinatorial definitions are standard.

Definition 5. (cover, disjoint cover) *A* cover $\mathcal{S} = \{S_1, S_2, \ldots, S_l\}$ *of a graph* G *is a collection of subsets of* $V[G]$ *such that for every vertex* $v_j \in V[G]$*, there exists a set* $S_i \in \mathcal{S}$*, where* $v_j \in S_i$*. We say that* $\mathcal{S}$ *is a* disjoint cover *if for every vertex* $v_j \in V[G]$*, there exists a distinct set* $S_i \in \mathcal{S}$*, such that* $v_j \in S_i$*.*

Definition 6. (width) *Let* $\mathcal{S}$ *be a cover of* $V[G]$*. The* width *of* $v \in V[G]$ *is the number of sets that contain it. The* width *of a cover* $\mathcal{S}$ *is* β*, if* β *is the maximum width of a vertex in* $V[G]$*. That is:* $\beta = \max_{v \in V[G]} |\{S_i \mid v \in S_i\}|$

For a set S, we let $S^{(-)}$ denote the set S minus its internal boundary (i.e. $S^{(-)}$ contains only nodes that do not have neighbors outside S). We let $S^{(+)}$ denote the set S plus its external boundary (its neighbors). For a collection of sets $\mathcal{S} = \{S_1, S_2, \ldots, S_l\}$, we let $\mathcal{S}^{(-)}$ denote the collection $\{S_1^{(-)}, S_2^{(-)}, \ldots, S_l^{(-)}\}$, and $\mathcal{S}^{(+)} = \{S_1^{(+)}, S_2^{(+)}, \ldots, S_l^{(+)}\}$.

Observation 1. *Let* $\mathcal{S}$ *be a disjoint cover of a graph* G *of max degree* d*, then* $\mathcal{S}^{(+)}$ *is a cover of width* $d+1$ *at the most.*

Next we introduce a basic definition of the local price of anarchy. Note that if $S \subseteq V[G]$ is a set of players in a graphical game G, the utility of S depends only on the actions of the players in S and the actions of S's neighbors. Therefore every set of strategies of the neighbors of S induces a *sub-game* on S.

Definition 7. (local price of anarchy) *Let* G *be a graphical game. The* local price of anarchy *of a set of players* S_i *is* α*, if for every set of actions of its neighbors, the PoA of the induced sub-game is at least* α*. Let* $\mathcal{S} = \{S_1, S_2, \ldots, S_l\}$ *be a cover of* $V[G]$*. We say that the* local PoA *of* G *with respect to* $\mathcal{S}$ *(*$LPoA_{\mathcal{S}}(G)$*) is* α*, if the local price of anarchy of every subset* $S_i \in \mathcal{S}$ *is at least* α*.*

Intuitively, a high local price of anarchy means that every set S_i in the cover responds well to its neighbors' actions. The local price of anarchy of the set of all players, equals the global PoA of the game. Note that we could focus only on neighbors' actions which are part of a global Nash equilibrium and still obtain all the results in this paper.

We denote by $\widehat{u}(S) = \max_s \sum_i u_i(s)$ the maximum welfare that a set S can achieve (over all the possible vectors of strategies of $S^{(+)}$). We let $u(S)$ denote the sum of utilities of S when the game is in a global worst Nash equilibrium (the equilibrium is always clear from the context so we suppress it from the notation). Similarly, we let $u(i)$ denote the utility of player i in this equilibrium. Note that if the game is in a global Nash equilibrium then *all* the subsets are also in local Nash equilibria (i.e. all the induced sub-games are in equilibrium).

[2] For utility games, the PoA is sometimes defined by $\frac{|U_{opt}|}{|GWNE|}$, and then the game is better if the $PoA \geq 1$ is smaller.

3 A Basic Bound on the Price of Anarchy

In this section we introduce a basic lower bound of the global price of anarchy in terms of the local price of anarchy.

Definition 8. (**(α, β)-cover**) *Let G be a graph. A cover $\mathcal{S} = \{S_1, S_2, ..., S_l\}$ of $V[G]$ is called an* (α, β)-cover *if the following hold:*

1. *$LPoA_{\mathcal{S}}(G) \geq \alpha$*
2. *$\mathcal{S}$ is of width at most β*
3. *The collection of interiors $\mathcal{S}^{(-)}$ is also a cover*

Intuitively, an (α, β)-cover is good when $0 \leq \alpha \leq 1$ is high, and $\beta \geq 1$ is low. It is possible to view an (α, β)-cover in the following manner: every set S_i is well behaved, that is, reacts well to its external conditions, and the interaction (overlap) between sets is limited. The requirement that $\mathcal{S}^{(-)}$ is also a cover, is crucial. Without it, the next theorem could not be established.

Theorem 2. *Let $\mathcal{S}$ be an (α, β)-cover for G a graphical game. Then,*

$$GPoA(G) \geq \alpha/\beta$$

Proof. We will see that the fact that the local price of anarchy is α helps us in achieving an α factor between every Nash equilibrium of S_i and the optimal welfare of $S_i^{(-)}$. The requirement for $\mathcal{S}^{(-)}$ to be a cover will thus be used to bound the global optimum, and the width of $\mathcal{S}$ will generate the $1/\beta$ factor.

Consider a global worst Nash equilibrium.

Claim. Let $S_i \in \mathcal{S}$ be a set in the cover. Then, $u(S_i) \geq \alpha \widehat{u}(S_i^{(-)})$.

Proof. The welfare of $S_i^{(-)}$ only depends on the actions of $S_i^{(-)}$ and the neighbors of $S_i^{(-)}$, that is, the welfare only depends on S_i. Thus, there exists a vector of actions of S_i that gives $S_i^{(-)}$ a welfare of $\widehat{u}(S_i^{(-)})$. Since the utilities are always non-negative, this vector guarantees S_i at least $\widehat{u}(S_i^{(-)})$. In other words, the utility $u^{Nash}(S_i^{best})$, when the neighbors play worst Nash and the set S_i plays the optimum for this induced game, is $u^{Nash}(S_i^{best}) \geq \widehat{u}(S_i^{(-)})$. Since the local price of anarchy of S_i is at least α, $\frac{u(S_i)}{\widehat{u}(S_i^{(-)})} \geq \frac{u(S_i)}{u^{Nash}(S_i^{best})} \geq \alpha$.

Summing the former over all the sets in the cover $\mathcal{S}$ yields:

$$\sum_{S_i \in \mathcal{S}} u(S_i) \geq \alpha \sum_{S_i \in \mathcal{S}} \widehat{u}(S_i^{(-)})$$

Claim. $\sum_{S_i \in \mathcal{S}} \widehat{u}(S_i^{(-)}) \geq |U_{opt}|$

Proof. Since $\mathcal{S}^{(-)}$ is a cover, the subsets $H_i^{(-)} = S_i^{(-)} - \bigcup_{j<i} S_j^{(-)}$ compose a disjoined cover $H^{(-)}$ of the graph. Since the utilities are non negative, $U_{opt}(H_i^{(-)}) \leq U_{opt}(S_i^{(-)})$ for all i. Since $H^{(-)}$ is a cover we get that

$$|U_{opt}| = \sum_i U_{opt}(H_i^{(-)}) \leq \sum_i U_{opt}(S_i^{(-)}) \leq \sum_{S_i \in \mathcal{S}} \widehat{u}(S_i^{(-)})$$

where the last inequality is due to the optimality of $\widehat{u}(S_i^{(-)})$.

Claim. $\sum_{i \in V} u(i) \geq \frac{1}{\beta} \sum_{S_i \in \mathcal{S}} u(S_i)$

Proof. $\mathcal{S}$ is of width β at the most. Therefore every element on the left hand side appears at most β times in the sum on the right hand side.

Putting all together we get:

$$|GWNE| = \sum_{i \in V} u(i) \geq \frac{1}{\beta} \sum_{S_i \in \mathcal{S}} u(S_i) \geq \frac{\alpha}{\beta} \sum_{S_i \in \mathcal{S}} \widehat{u}(S_i^{(-)}) \geq \frac{\alpha}{\beta} |U_{opt}|$$

Remarks. While the local and global price of anarchies refer to Nash equilibria, it is possible to obtain a similar bound for many solution concepts (e.g. correlated or strong equilibria). We note that we do not know how to show an analog of this theorem for the price of stability. The width parameter is purely combinatorial and can be interpreted as a measure of interaction between the sub-games (subsets). The α parameter is a measure of how well the small subsets behave. Later, we will average these parameters and also study the effects of other local parameters on the global price of anarchy. An interesting algorithmic issue is how to decompose a large game into small units such that the resulting bound on the global price of anarchy is as tight as possible, i.e. how to find a good cover.

3.1 Covers by Balls

One of the most natural ways of obtaining a cover is by taking all balls of a certain radius.

Definition 9. **(r-LPoA)** *Let G be a game. We say that the r-LPoA of G is at least α, if every ball B of radius r has a local price of anarchy of at least α.*

Corollary 1. *Let G be a game with maximum degree d. If the 1-LPoA of G is at least α, then the $GPoA(G) \geq \frac{\alpha}{d+1}$.*

Proof. Consider the cover $\mathcal{S}$ of all balls of radius 1. Since this cover is of width $\beta \leq d+1$ and $\mathcal{S}^{(-)}$ is also a cover, it is an $(\alpha, d+1)$-cover. Now we can apply Theorem 2, and the corollary follows.

Open problem. Interestingly, an $r - LPoA \geq \alpha$ only guarantees a bound of $O(\frac{\alpha}{r^{d+1}})$ on the $GPoA$. On the other hand it is natural to conjecture that the right bound is $\Theta(\frac{\alpha}{d})$. We leave this as an interesting open problem. When the game is relatively balanced, we can show that indeed $\Theta(\frac{\alpha}{d})$ is the right'.

4 Examples

Before refining Theorem 2, let us consider a few simple examples. The star-of-cliques game demonstrates how to use the theorem, the example of covering a torus by grids demonstrates the need for refining the basic bound, and the biased consensus game shows that the theorem is tight, i.e. that in the general case, it is not possible to improve the α/β bound.

4.1 Star-of-Cliques

The following example demonstrates the use of the theorem.

Definition 10. **((k,l) star-of-cliques graph)** *A (k,l)* star-of-cliques *is a graph $G = (V, E)$, where*

$$V = \{w, v_1, v_2, \ldots, v_k, x_1^1, x_1^2, \ldots, x_1^{l-1}, x_2^1, x_2^2, \ldots, x_2^{l-1}, x_3^1, \ldots, x_k^{l-1}\}$$

$$E = \{\{\{w, v_i\} | \forall 1 \le i \le k\} \cup \{\{v_i, x_i^j\} | \forall i, j\} \cup \{\{x_i^{j_1}, x_i^{j_2}\} | \forall i, j_1, j_2\}\}$$

That is, the vertices $w, v_1, \ldots, v_k$ form a star with w in the center, and for each i, the l vertices $v_i, x_i^1, x_i^2, \ldots, x_i^{l-1}$ form a clique.

The game has three types of players, $w, v,$ *and* x. Each agent has two available strategies, 0 and 1. For each vertex y we denote by $N_a(y)$ the number of y's neighbors of type $a \in \{w, v, x\}$ that play the same strategy as y, and by $\overline{N_a(y)}$ the number of y's neighbors of type $a \in \{w, v, x\}$ that play the opposite of y. We denote by $Maj(y)$ the strategy of the majority of y's neighbors.

Example 1. **((k,l) star-of-cliques game)** A (k,l) *star-of-cliques game*, is an n-player game, where the graph of the game is a (k,l)-star-of-cliques, and the utilities of the players are given by:

$$u_y = \begin{cases} N_x(y) + (l-2)\overline{N_v(y)} & \text{if } y \text{ is of type } x \\ \overline{N_x(y)} + (l-1)N_w(y) & \text{if } y \text{ is of type } v \\ 1 & \text{if } y \text{ is of type } w \text{ and its strategy is } Maj(y) \\ 0 & \text{if } y \text{ is of type } w \text{ and its strategy is not } Maj(y) \end{cases}$$

We discuss this game farther in the full version of this work, where we prove the following:

Proposition 1. *Let G be a (k,l)-star-of-cliques game. Then, $GPoA(G) \ge \frac{1}{2(k+1)}$.*

4.2 Covering a Torus by Grids

Example 2. Let G be an $m \times m$ torus, and let k such that k divides m. We let $\mathcal{S}^{(-)}$ be a disjoint cover of $k \times k$ grids.

Consider the case of $k > 2$. Here, $\beta = 3$, $\forall i$, $|S_i| = k^2 + 4k$, and $|S_i^{(-)}| = k^2$. When k is large, almost all the vertices have a width of 1. An immediate conclusion of Theorem 2 is as follows.

Corollary 2. *For the torus example, if $LPoA_{\mathcal{S}}(G) = \alpha$, then $\alpha/3 \leq GPoA$.*

In other words, an LPoA of α implies a GPoA of about $\alpha/3$. The example however demonstrates the need for refining the basic theorem: While the width of the cover is 3, almost all the vertices have a width of 1 and are therefore counted only once in $\sum_i u(S_i)$. Thus, typically, one should expect a GPoA of around α and not $\alpha/3$. This can be addressed by the various refinements of the basic theorem shown in this paper.

4.3 A Tight Example

Our final example shows that in general, the α/β bound of the basic theorem is essentially tight. For simplicity we focus on pure Nash equilibria.

Example 3. **(biased consensus game)** Let $\gamma > 1$ be a parameter. In a *biased consensus game* the agent strategies are taken from $\{0, 1\}$. The utility of each agent i is defined by:

$$u_i = \begin{cases} 1 & \text{if } i \text{ and all its neighbors choose 1} \\ 1/\gamma & \text{if } i \text{ and some of its neighbors choose 1} \\ 1/\gamma & \text{if } i \text{ and all its neighbors choose 0} \\ 0 & \text{otherwise} \end{cases}$$

In other words, if a player has a neighbor playing 1 it should play 1 as well, if all its neighbors are playing 0, it should play 0 too. We assume that the graph of the game is connected.

The following proposition shows the tightness of Theorem 2. We omit the proof from this extended abstract.

Proposition 2. *For every $\epsilon > 0$, there exists a graphical game G, and an (α, β)-cover $\mathcal{S}$, where:*

$$\frac{\alpha}{\beta} \leq GPoA \leq (1 + \epsilon)\frac{\alpha}{\beta}$$

Proof. Consider the biased consensus game played on a d-regular graph and the cover by all balls of radius 1.

By an observation omitted from this abstract, the local PoA of such a ball S, is obtained when all its neighbors are playing 0. In this case, the worst local Nash equilibrium occurs when all the players in the ball are playing 0. This yields a utility of $1/\gamma$ to every member of S. In the optimal strategy for S all its members play 1. This strategy vector results in a utility of $1/\gamma$ for each of the d boundary nodes of S, and a utility of 1 for the inner node. Therefore, the local price of anarchy of S equals: $\alpha_S = \frac{\frac{1}{\gamma}(d+1)}{1+d/\gamma} = \frac{d+1}{d+\gamma}$. Since G is d-regular, $\alpha = \frac{d+1}{d+\gamma}$ as all the balls have $d + 1$ nodes. Since $\beta = d + 1$, we have that $\alpha/\beta = \frac{1}{(d+\gamma)}$. Thus, if we set $\gamma > d/\epsilon$, we get that

$$(1 + \epsilon)\frac{\alpha}{\beta} = \frac{1 + \epsilon}{d + \gamma} > \frac{1 + \epsilon}{\gamma\epsilon + \gamma} = 1/\gamma = GPoA(G)$$

5 Refinements of the Basic Bound

5.1 Averaging the Parameters

In the biased consensus game (Example 3), all the induced sub-games of the cover have the same local price of anarchy. Most games do not possess this property, and the basic theorem is thus often wasteful (as $LPoA_{\mathcal{S}}(G)$ is the minimum local PoA of the sets in the cover). Similarly, β is the maximum width. For this purpose we generalize the definitions of the local price of anarchy and the width to be an average instead of the minimum and maximum, respectively. We introduce improved bounds on the global price of anarchy using the new definitions.

Definition 11. (average local price of anarchy) *Let G be a graphical game and let $\mathcal{S} = \{S_1, S_2, \ldots, S_l\}$ be a cover of $V[G]$ such that $\mathcal{S}^{(-)}$ is also a cover. Let α_i be the local PoA of S_i. The* average local price of anarchy *of G w.r.t. $\mathcal{S}$, $\overline{LPoA_{\mathcal{S}}(G)}$, is the average of α_i by the maximum utilities of $S_i^{(-)}$, that is $\overline{LPoA_{\mathcal{S}}(G)} = \frac{\sum_{i=1}^{l} \alpha_i \hat{u}(S_i^{(-)})}{\sum_{i=1}^{l} \hat{u}(S_i^{(-)})}$.*

Theorem 3. *Let G be a graphical game and let $\mathcal{S} = \{S_1, S_2, \ldots, S_l\}$ be a cover of $V[G]$ such that $\mathcal{S}^{(-)}$ is also a cover and $\mathcal{S}$ is of width β. Let $\alpha = \overline{LPoA_{\mathcal{S}}(G)}$, then $GPoA \geq \alpha/\beta$.*

Proof (sketch): The proof resembles the one of Theorem 2, and we thus only sketch it.
Let $S_i \in \mathcal{S}$. If we follow the steps of the proof of Claim 3 in the proof of Theorem 2, with the new definition of α_i, we will get: $u(S_i) \geq \alpha_i \hat{u}(S_i^{(-)})$. Now:

$$\sum_{i=1}^{l} u(S_i) \geq \sum_{i=1}^{l} \alpha_i \hat{u}(S_i^{(-)}) = \alpha \sum_{i=1}^{l} \hat{u}(S_i^{(-)})$$

where the 2nd equality is due to the definition of $\overline{LPoA_{\mathcal{S}}(G)}$, and the first is just a summation of the former.

Like in Claim 3, since $\mathcal{S}$ is of width β we have that $\beta \sum_{i=1}^{n} u(i) \geq \sum_{i=1}^{l} u(S_i)$. Since $\mathcal{S}^{(-)}$ is a cover we have that $\sum_{i=1}^{l} \hat{u}(S_i^{(-)}) \geq |U_{opt}|$ (Similarly to Claim 3 in the proof of Theorem 2).

Putting all together we conclude that:

$$\sum_{i=1}^{n} u(i) \geq 1/\beta \sum_{i=1}^{l} u(S_i) \geq 1/\beta \sum_{i=1}^{l} \alpha_i \hat{u}(S_i^{(-)}) \geq \alpha/\beta \sum_{i=1}^{l} \hat{u}(S_i^{(-)}) \geq \frac{\alpha}{\beta} |U_{opt}|$$

The above refinement is also interesting for the algorithmic task of finding a good cover. This is because one can look for sub-games with high average PoA instead of a cover with a high minimum PoA.

Next, we consider a weighted version of the width parameter.

Theorem 4. *Let G be a graphical game and let $\mathcal{S} = \{S_1, S_2, \ldots, S_l\}$ be a cover for $V[G]$ such that $\mathcal{S}^{(-)}$ is a cover, and the width of node $i \in V[G]$ in $\mathcal{S}$ is β_i. Define $\overline{\beta}$ as the average of β_i weighted by the agent utilities in a global worst Nash equilibrium, that is $\overline{\beta} = \frac{\sum_{i=1}^n \beta_i u(i)}{\sum_{i=1}^n u(i)}$. Let $\alpha = \overline{LPoA_{\mathcal{S}}(G)}$. Then*

$$GPoA(G) \geq \alpha/\overline{\beta}$$

Proof (sketch): By definitions $\overline{\beta} \sum_{i=1}^n u(i) = \sum_{i=1}^n \beta_i u(i) = \sum_{i=1}^l u(S_i)$. We can then proceed according to the proof of Theorem 3.

Going back to the star-of-cliques (Example 1), one can see now that in this case $\overline{\beta} = 1 + \epsilon$ for a small $\epsilon = \epsilon(k, l)$ whereas $\beta = k + 1$ is the non weighted width. This is because, in the proposed cover, the center w is of width $k + 1$, the k vertices of type v are of width 2, and all the $k(l - 1)$ vertices of type x are of width 1, and the weights are roughly the same. Thus, Theorem 4 yields a bound of $GPoA(G) \geq \frac{1}{2(1+\epsilon)}$, instead of the much weaker bound of $\frac{1}{2(k+1)}$ of the basic theorem. As we noted before, it can be shown that the actual global price of anarchy is slightly greater than 1/2, so the above bound is tight.

Note that in the last theorem we took the average according to the utilities of the agents in the global equilibrium. Therefore, averaging the β parameter may sometimes be less constructive.

5.2 Nash Expansion

We now introduce a different local parameter that sometimes helps analyzing games which are not well addresses by the previous theorems. This parameter resembles graph expansion parameters but refers directly to the equilibrium welfare so it cannot be deduced solely from the graph.

Definition 12. (Nash expansion of a set) *Let G be a graphical game and $S \subseteq V[G]$. We say that the* Nash expansion *of S is ξ if, for all sets of strategies for the neighbors of S, for all Nash equilibria of S*

$$\xi \leq \frac{\sum_{j \in S^{(-)}} u_j}{\sum_{j \in S} u_j}$$

In other words, in every Nash equilibrium, the ratio between the welfare of $S^{(-)}$ and the welfare of its (external) boundary is bounded by $\frac{\xi}{1-\xi}$.

Definition 13. (Nash expansion) *Let G be a graphical game and $\mathcal{S}$ a cover. We say that the* Nash expansion *of $\mathcal{S}$ is $\xi = \xi_G(\mathcal{S})$ if for all $S_i \in \mathcal{S}$ the Nash expansion of the set S_i is greater or equal ξ.*

Observation 5. *Let G be a graphical game and $\mathcal{S}$ a cover. If the Nash expansion of $\mathcal{S}$ is at least $\xi = \xi_G(\mathcal{S})$ then:*

$$\frac{\sum_{S_i} u(S_i^{(-)})}{\sum_{S_i} u(S_i)} \geq \xi$$

Note that ξ is a local parameter that can be obtained by examining each subset S_i in isolation. It is possible to show that if a cover $\mathcal{S}$, where $\mathcal{S}^{(-)}$ is disjoint, has a Nash expansion of ξ, then its weighted width $\overline{\beta}$ is bounded by $1/\xi$ as well. This yields the following theorem:

Theorem 6. *Let G be a graphical game. Let $\mathcal{S} = \{S_1, S_2, \ldots, S_l\}$ be a cover with $\alpha = \overline{LPoA_{\mathcal{S}}(G)}$ and a Nash expansion ξ, such that $\mathcal{S}^{(-)}$ is a disjoint cover. Then $GPoA(G) \geq \alpha\xi$*

□

It is also possible to average ξ and obtain a similar theorem. In the longer version of this work we discuss the properties of the expansion parameter further. Specifically, we show that if we can bound the maximum ratio between pairs of players' utilities in a global worst Nash equilibrium, then we can replace the Nash expansion parameter by a simple **combinatorial** parameter.

6 Discussion and Future Research

In real life, almost every game is embedded in a larger game and players are likely to be able to consider only their close vicinity. Thus, we view the investigation of the relations between local and global properties of games as a basic issue in the understanding of large games. This paper demonstrates that at least from the perspective of the price of anarchy, a good local behavior of a game implies a good global behavior. The converse is not necessarily true, and there are many non-trivial questions which are related to bounding the price of anarchy of graphical games. Of course, it is natural to investigate questions, similar to the ones which are studied here, in the context of other properties of games.

In general, we believe that models like graphical games provide an excellent opportunity to introduce many structural properties into games. We believe that such properties arise naturally in many contexts and can give rise to a lot of fruitful research on the border of game theory, combinatorics, and computer science.

Acknowledgment. We thank Michal Feldman, Yishai Mansour, and Ehud Lehrer for useful discussions. We thank Oded Lachish and Inbal Ronen for commenting on a previous draft of this paper.

References

1. Anshelevich, E., Dasgupta, A., Kleinberg, J., Tardos, E., Wexler, T., Roughgarden, T.: The Price of Stability for Network Design with Fair Cost Allocation. In: The 45th Annual IEEE Symposium on Foundations of Computer Science (FoCS), pp. 59–73 (2004)
2. Awerbuch, B., Azar, Y., Epstein, A.: The Price of Routing Unsplittable Flow. In: The 37th Annual ACM Symposium on Theory of Computing (SToC) (2005)
3. Awerbuch, B., Peleg, D.: Sparse Partitions. In: The 31st Annual IEEE Symposium on Foundations of Computer Science (FoCS), pp. 503–513 (1990)

4. Chen, X., Deng, X.: Settling the Complexity of 2-Player Nash-Equilibrium. In: The 47th Annual IEEE Symposium on Foundations of Computer Science (FoCS) (2006)
5. Daskalakis, C., Goldberg, P.W., Papadimitriou, C.: The Complexity of Computing a Nash Equilibrium. In: The 38th Annual ACM Symposium on Theory of Computing (SToC) (2006)
6. Elkind, E., Goldberg, L.A., Goldberg, P.W.: Nash Equilibria in Graphical Games on Trees Revisited. In: The 7th ACM conf. on Electronic Commerce (EC 2006) (2006)
7. Fischer, E.: The Art of Uninformed Decisions: A Primer to Property Testing. The Bulletin of the European Association for Theoretical Computer Science 75, 97–126 (2001)
8. Galeotti, A., Goyaly, S., Jackson, M.O., Vega-Redondox, F., Yariv, L.: Network Games (unpublished, 2006), `http://www.stanford.edu/~jacksonm/networkgames.pdf`
9. Kakade, S., Kearns, M., Ortiz, L., Pemantle, R., Suri, S.: Economic Properties of Social Networks. In: The 18th Annual Conference on Neural Information Processing Systems (NIPS) (2004)
10. Kearns, M., Littman, M., Singh, S.: Graphical Models for Game Theory. In: The 17th Conference in Uncertainty in Artificial Intelligence (UAI) (2001)
11. Koutsoupias, E., Papadimitriou, C.: Worst Case Equilibria. In: Meinel, C., Tison, S. (eds.) STACS 1999. LNCS, vol. 1563, Springer, Heidelberg (1999)
12. Linial, N., Peleg, D., Rabinovich, Y., Saks, M.: Sphere Packing and Local Majorities in Graphs. In: The 2nd Israel Symp. on Theory of Computing Systems (1993)
13. Linial, N., Saks, M.: Low Diameter Graph Decompositions. Combinatorica Issue Volume 13, 441–454 (1993)
14. Nash, J.F.: Non-Cooperative Games. Annals of Mathematics 54, 286–295 (1951)
15. Roughgarden, T., Tardos, E.: How Bad is Selfish Routing? In: The 41st Annual IEEE Symposium on Foundations of Computer Science (FoCS) (2000)

Approximate Nash Equilibria for Multi-player Games*

Sébastien Hémon[1,2], Michel de Rougemont[1,3], and Miklos Santha[1]

[1] CNRS-LRI, Univ. Paris-Sud, F-91405 Orsay
[2] LRDE-EPITA F-94276 Le Kremlin-Bicetre
[3] Univ. Paris II F-75005 Paris

Abstract. We consider games of complete information with $r \geq 2$ players, and study approximate Nash equilibria in the additive and multiplicative sense, where the number of pure strategies of the players is n. We establish a lower bound of $\sqrt[r-1]{\frac{\ln n - 2\ln\ln n - \ln r}{\ln r}}$ on the size of the support of strategy profiles which achieve an ε-approximate equilibrium, for $\varepsilon < \frac{r-1}{r}$ in the additive case, and $\varepsilon < r - 1$ in the multiplicative case. We exhibit polynomial time algorithms for additive approximation which respectively compute an $\frac{r-1}{r}$-approximate equilibrium with support sizes at most 2, and which extend the algorithms for 2 players with better than $\frac{1}{2}$-approximations to compute ε-equilibria with $\varepsilon < \frac{r-1}{r}$. Finally, we investigate the sampling based technique for computing approximate equilibria of Lipton et al.[12] with a new analysis, that instead of Hoeffding's bound uses the more general McDiarmid's inequality. In the additive case we show that for $0 < \varepsilon < 1$, an ε-approximate Nash equilibrium with support size $\frac{2r\ln(nr+r)}{\varepsilon^2}$ can be obtained, improving by a factor of r the support size of [12]. We derive an analogous result in the multiplicative case where the support size depends also quadratically on g^{-1}, for any lower bound g on the payoffs of the players at some given Nash equilibrium.

1 Introduction

Classical games of complete information with r players model situations where r decision makers interact and pursue well-defined objectives. A Nash equilibrium describes strategies for each player such that no player has any incentive to change her strategy. The algorithmic study of Nash equilibria started with the work of Lemke and Howson [11] in the 1960's, for the case of two players. This classical algorithm is exponential in the number of strategies (see [15]). Computing a Nash equlibrium is indeed not an easy task. It was proven recently that this computation is complete for the class PPAD, first for $r \geq 4$ in [7], then for $r \geq 3$ in [6] and [2], and finally for $r \geq 2$ in [4]. Therefore it is unlikely to be feasible in polynomial time.

* Research supported by the ANR Blanc AlgoQP grant of the French Research Ministry.

B. Monien and U.-P. Schroeder (Eds.): SAGT 2008, LNCS 4997, pp. 267–278, 2008.

Approximate Nash equilibria have been studied both in the additive and the multiplicative models of approximation. An ε-approximate Nash equilibrium describes strategies for each player such that by changing her strategy unilaterally, no player can improve her gain by more than ε. Lipton et al. [12] studied additive approximate Nash equilibria for r-player games by considering small-support strategies, and obtained an approximation scheme which computes an ε-approximate equilibrium in the additive sense, in time $n^{O(\frac{\ln n}{\varepsilon^2})}$, where n is the maximum number of pure strategies. It is known that there is no Fully Polynomial Time Approximation Scheme (FPTAS) for this problem [3], but it is open to decide if there is a PTAS. Daskalakis at al. [8] gave a simple algorithm for computing an additive $\frac{1}{2}$-approximate equilibirum in 2-player games, using strategies with support at most 2. Feder et al. [10] showed that the factor $\frac{1}{2}$ was optimal when the size of the support could not exceed $\log n - 2\log\log n$. Breaking the $\frac{1}{2}$ barrier required approximation strategies with larger support size. In [9] Papadimitriou et al. have exhibited an additive $\frac{3-\sqrt{5}}{2}$-approximate polynomial time algorithm, using linear programming. Further improvements for the approximation of the equilibrium in 2-player game were obtained by Bosse et al. [1] and Tsaknakis et al. [16], but the case of polynomial time approximation in games with more than 2 players was not investigated. The case of the multiplicative approximation has been studied by Chien and Sinclair [5] for dynamic strategies.

Here we study approximate Nash equilibria for r-player games, where the number of pure strategies of the players is n. First we extend the lower bounds on the factors of approximations for strategies with small support size. In Theorem 1 we prove that no ε-approximate equilibrium can be achieved with strategy profiles of support size less than $\sqrt[r-1]{\frac{\ln n - 2\ln\ln n - \ln r}{\ln r}}$ if $\varepsilon < \frac{r-1}{r}$ in the additive case, and $\varepsilon < r-1$ in the multiplicative case.

Then we exhibit polynomial time algorithms for additive approximation. Our results are based on the algorithm of Theorem 2 which extends approximations for r-player games to approximations for $(r+1)$-player games. As a consequence, we design in Corollary 3 a polynomial time algorithm which computes an $\frac{r-1}{r}$-approximate equilibrium with support size at most 2, and in Corollary 4 extend the algorithms breaking the $\frac{1}{2}$-approximation threshold in 2-player games into algorithms breaking the $\frac{r-1}{r}$ approximation threshold in r-player games.

Finally, we investigate the sampling based technique for computing approximate additive equilibria of Lipton et al.[12]. We propose a new analysis of this technique that instead of the Hoeffding's bound uses the more general McDiarmid's inequality [13] which enables us to bound the deviation of a function of independent random variables from its expectation. In Theorem 4 we show that for $0 < \varepsilon < 1$, an ε-approximate Nash equilibrium with support size $\frac{2r\ln(nr+r)}{\varepsilon^2}$ can be obtained, improving by a factor r the support size of [12]. We also establish a result analogous to the additive case in Theorem 5, where we show that for $0 < \varepsilon < 1$, a multiplicative ε-approximate Nash equilibrium with support size $\frac{9r\ln(nr+r)}{2g^2\varepsilon^2}$ can be achieved where g is a lower bound on the payoffs of the players at some given Nash equilibrium. In Remark 2 we argue that some dependence on g is necessary if we want the support of the approximate equilibrium to be included in the support of the given Nash equilibrium.

2 Preliminaries

For a natural number n, we denote by $[n]$ the set $\{1, \ldots, n\}$. For an integer $r \geq 2$, an *r-player game in normal form* is specified by a set of *pure strategies* S_p, and a *utility* or *payoff* function $u_p : S \rightarrow \mathbb{R}$, for each player $p \in [r]$, where $S = S_1 \times \cdots \times S_r$ is the set of *pure strategy profiles.* For $s \in S$, the value $u_p(s)$ is the payoff of player p for pure strategy profile s. Let $S_{-p} = S_1 \times \cdots \times S_{p-1} \times S_{p+1} \times \cdots \times S_r$, the set of all pure strategy profiles of players other than p. For $s \in S$, we set the *partial* pure strategy profile s_{-p} to be $(s_1, \ldots s_{p-1}, s_{p+1}, \ldots, s_r)$, and for s' in S_{-p}, and $t_p \in S_p$, we denote by (s'_{-p}, t_p) the *combined* pure strategy profile $(s'_1, \ldots, s'_{p-1}, t_p, s'_{p+1}, \ldots, s'_r) \in S$. Let $B = \{e_1, \ldots, e_n\}$ be the canonical basis of the vector space $\mathbb{R}^n$. We will suppose that each player has n pure strategies and that $S_p = B$, for all $p \in [r]$, and therefore $S = B^r$.

A *mixed strategy* for player p is a probability distribution over S_p, that is a vector $x_p = (x_p^1, \ldots x_p^n)$ such that $x_p^i \geq 0$, for all $i \in [n]$, and $\sum_{i \in [n]} x_p^i = 1$. We define $\text{supp}(x_p)$, the *support* of the mixed strategy x_p, as the set of indices i for which $x_p^i > 0$. Following [12], a mixed strategy x_p is called *k-uniform*, for some $k \in [n]$, if for every $i \in [n]$, there is an integer $0 \leq l \leq k$ such that $x_p^i = \frac{l}{k}$. Obviously, the size of the support of a k-uniform strategy is at most k. We denote by Δ_p the set of mixed strategies for p, and we call $\Delta = \Delta_1 \times \cdots \times \Delta_r$ the set of *mixed strategy profiles.* For a mixed strategy profile $x = (x_1, \ldots, x_r)$ we set $\text{supp}(x) = \text{supp}(x_1) \times \cdots \times \text{supp}(x_r)$, and $\text{size}(x) = \max\{|\text{supp}(x_p)| : p \in [r]\}$. For a mixed strategy profile $x = (x_1, \ldots, x_r)$ and pure strategy profile $s \in S$, the product $x_s = x_1^{s_1} x_2^{s_2} \cdots x_r^{s_r}$ denotes the probability of s in x. We will consider the multilinear extension of the payoff functions from S to Δ defined by $u_p(x) = \sum_{s \in S} x_s u_p(s)$. The set Δ_{-p}, the partial mixed strategy profile x_{-p} for $x \in \Delta$, and the combined mixed strategy profile (x', x_p) for $x' \in \Delta_{-p}$ and $x_p \in \Delta_p$ are defined analogously to the pure case. The pure strategy s_p is a *best response* for player p against the partial mixed strategy profile x_{-p} if it maximizes $u_p(x_{-p}, \cdot)$. We will denote by $\text{br}(x_{-p})$ the set of best responses against x_{-p}.

A *Nash equilibrium* is a mixed strategy profile x^* such that for all $p \in [r]$, and for all $x_p \in \Delta_p$,

$$u_p(x^*_{-p}, x_p) \leq u_p(x^*).$$

An equivalent condition is $u_p(x^*_{-p}, s_p) \leq u_p(x^*)$ for every $s_p \in \text{br}(x^*_{-p})$. Nash has shown [14] that for games with a finite number of players there exists always an equilibrium. It is immediate that the set of Nash equilibria is invariant by translation and positive scaling of the utility functions. Therefore we will suppose that they take values in the interval $[0, 1]$.

Several relaxations of the notion of equilibrium have been considered in the form of additive and multiplicative approximations. Let $\varepsilon > 0$. An *additive*

ε-approximate equlibrium is a mixed strategy profile x^* such that for all $p \in [r]$, and for all $x_p \in \Delta_p$,

$$u_p(x^*_{-p}, x_p) \leq u_p(x^*) + \varepsilon.$$

A *multiplicative ε-approximate equlibrium* is a mixed strategy profile x^* such that for all $p \in [r]$, and for all $x_p \in \Delta_p$,

$$u_p(x^*_{-p}, x_p) \leq (1 + \varepsilon) u_p(x^*).$$

Since by our convention $0 \leq u_p(x^*) \leq 1$, a multiplicative ε-approximate equilibrium is always an additive ε-approximate equilibrium, but the converse is not necessarily true.

The input of an r-player game is given by the description of rn^r rational numbers. Here we will consider the computational model where arithmetic operations and comparisons have unit cost.

3 Inapproximability Results for Small Support Size

In [10] Feder, Nazerzadeh and Saberi have shown that there are 2-player games where for $\varepsilon < 1$, no multiplicative ε-approximation can be achieved with support size less than $\ln n - 2 \ln \ln n$. We generalize this result for r-player games in both models of approximation.

Theorem 1. *For $r \in o(n)$ there exists an r-player game such that no mixed strategy profile x with size$(x) < \sqrt[r-1]{\frac{\ln n - 2 \ln \ln n - \ln r}{\ln r}}$ can be an additive ε-approximate equilibrium for $\varepsilon < \frac{r-1}{r}$, or a multiplicative ε-approximate equilibrium for $\varepsilon < r - 1$.*

Proof. We use the probabilistic method and will show that a random game from an appropriately chosen probabilistic space satisfies the claimed properties with positive probability. The space is defined as follows: for every pure strategy profile $s = (s_1, \ldots, s_r) \in S$, choose a uniformly random $p \in [r]$ and set $u_p(s) = 1$ and $u_q(s) = 0$ for all $q \neq p$. This defines a random r-player 0/1 game with constant sum 1.

Fix $k < \sqrt[r-1]{\frac{\ln n - 2 \ln \ln n - \ln r}{\ln r}}$, and set $S^{\leq k} = \{K_1 \times \cdots \times K_r \subseteq S : |K_p| \leq k \text{ for } p \in [r]\}$. Clearly size$(x) \leq k$ exactly when supp$(x) \in S^{\leq k}$. We define $S^{\leq k}_{-p}$ analogously. The event E_p is defined as follows: For all $K \in S^{\leq k}_{-p}$, there exists a pure strategy $t_p \in B$ such that for all $s' \in K$, we have $u_p(s', t_p) = 1$. Let $E = \bigwedge_{p \in [r]} E_p$. When E is realized, then for every x with size$(x) \leq k$, each player can increase her payoff to 1 by changing her strategy. Since the total payoff of the players is 1, at least one player has payoff at most $1/r$, and therefore x is not an additive ε-approximate equilibrium for $\varepsilon < \frac{r-1}{r}$, nor a multiplicative ε-approximate equilibrium for $\varepsilon < r - 1$.

We will prove that $\Pr[\overline{E}_p] < 1/r$ for all $p \in [r]$, and therefore $\Pr[E] > 0$. For fixed $K \in S_{-p}^{\leq k}$ and $t_p \in B$, the probability that there exists $s' \in K$ with $u_p(s', t_p) = 0$ is

$$1 - \Pr[\forall s' \in K \;\; u_p(s', t_p) = 1] \leq 1 - \frac{1}{r^{k^{r-1}}}.$$

Since the payoff functions are set independently, using the union bound we get

$$\Pr[\overline{E}_p] \leq \binom{n}{k}^{r-1} \left(1 - \frac{1}{r^{k^{r-1}}}\right)^n.$$

To prove the bound on $\Pr[\overline{E}_p]$ as claimed we bound the logarithm of the right hand side of the above inequality. This is at most

$$k(r-1)\ln n \; - \; \frac{n}{2r^{k^{r-1}}},$$

which can easily seen to be no more than $-\ln r$ for the chosen value of k by rearranging, and taking logarithms again.

Corollary 1. *For $r \in O(1)$ there exists an r-player game such that for some constant $c > 0$, no mixed strategy profile x with size$(x) < c \sqrt[r-1]{\ln n}$ can be an additive ε-approximate equilibrium for $\varepsilon < \frac{r-1}{r}$, or a multiplicative ε-approximate equilibrium for $\varepsilon < r - 1$.*

How essential are the restrictions on r and ε in Theorem 1? As we will show in the next section, for r fixed, the bound on ε is optimal in the case of additive approximation. The optimality of the bound for the multiplicative error remains open, and we don't know either if the restriction $r \in o(n)$ is necessary. Observe, however, that the case $r \geq n$ is anyhow of limited interest, since the uniform distribution on the pure strategies, for each players, is clearly an additive $\frac{n-1}{n}$-approximation, and a multiplicative $(n-1)$-approximation.

4 Polynomial Time Additive Approximations

We know from the previous section that no strategy profile of constant support size can achieve a better than $\frac{r-1}{r}$-approximate additive Nash equilibrium. We will prove here on the other hand that there exists an additive $\frac{r-1}{r}$-approximate Nash equilibrium of constant support size, and that it can be computed in polynomial time. It is also shown that there are polynomial time computable additive η-approximate equilibria for some $\eta < \frac{r-1}{r}$. These results are based on an algorithm which extends any additive approximation for r-player games to an approximation for $(r+1)$-player games.

Theorem 2. *Given an algorithm $\mathcal{A}$ that computes in time $q(r,n)$ an additive ε-approximate equilibrium for r-player games, there exists an algorithm $\mathcal{A}'$ that computes in time $q(r,n) + O(n^{r+1})$ an additive $\frac{1}{2-\varepsilon}$-approximate equilibrium for $(r+1)$-player games. Moreover, in algorithm $\mathcal{A}'$, the support of the last player is of size at most 2, and the sizes of the supports of the first r players are respectively the same as in algorithm $\mathcal{A}$.*

Proof. Let s_{r+1} an arbitrary pure strategy of player $r + 1$. This induces an r-player game for the other players, assuming that player $p + 1$ is restricted to s_{r+1}. Algorithm $\mathcal{A}$ finds for the induced game an additive ε-approximate equilibrium, say $x = (x_1, \ldots, x_r)$. Compute now in time $O(n^{r+1})$ a pure strategy t_{r+1} for the last player which is in $\mathrm{br}(x_1, \ldots, x_r)$. Let us define the mixed strategy $x_{r+1} = \frac{1}{2-\varepsilon} s_{r+1} + \frac{1-\varepsilon}{2-\varepsilon} t_{r+1}$. We claim that $x^* = (x, x_{r+1})$ is an $\frac{1}{2-\varepsilon}$-approximate equilibrium.

Consider any of the first r players. She can earn an additional payoff at most ε when player $r + 1$ plays s_{r+1}, and an additional payoff at most 1 when the chosen strategy is t_{r+1}. Therefore the overall gain by changing strategy is at most $\frac{\varepsilon}{2-\varepsilon} + \frac{1-\varepsilon}{2-\varepsilon} = \frac{1}{2-\varepsilon}$.

The last player has no way to increase her payoff when she plays her best response strategy t_{r+1}. Therefore her overall gain by changing strategy is at most $\frac{1}{2-\varepsilon}$.

Corollary 2. *Given an algorithm $\mathcal{A}$ that computes in time $q(n)$ an additive ε-approximate equilibrium for 2-player games, there exists an algorithm that computes for any $r \geq 3$, in time $q(n) + O(n^r)$ an additive $\frac{(r-2)-(r-3)\varepsilon}{(r-1)-(r-2)\varepsilon}$-approximate equilibrium for r-player games. Moreover, the supports of all but the first two players are of size at most 2, and the support sizes of the first two players are respectively the same as in algorithm $\mathcal{A}$.*

Proof. We apply Theorem 2 inductively. Let ε_l be the approximation obtained for l-player games. Then $\varepsilon_2 = \varepsilon$ and $\varepsilon_{l+1} = \frac{1}{2-\varepsilon_l}$. Solving the recursion gives the result.

Corollary 2 never returns a better than $\frac{r-2}{r-1}$-approximate Nash equilibrium. And, the procedure yields for r players an additive ε-approximation with $\varepsilon \geq \frac{r-2}{r-1}$ only if the original two-player algorithm $\mathcal{A}$ computes an additive η-approximation with $\eta \leq \frac{(r-2)-(r-1)\varepsilon}{(r-2)\varepsilon-(r-3)}$.

Corollary 3. *There exists an algorithm which computes an additive $\frac{r-1}{r}$-approximate equilibrium for r-player games in time $O(n^r)$. Moreover the support of all players is of size at most 2.*

Proof. We apply Corollary 2 to the algorithm of [8] which computes in time $O(n^2)$ an additive $\frac{1}{2}$-approximation for 2-player games with support size at most 2.

Let us stress here that though the complexity of the algorithm of Corollary 3 is exponential in r, it is sublinear in the input size.

Corollary 4. *There exist algorithms which in polynomial time compute an additive ε-approximate equilibrium for r-player games for some constant $\varepsilon < \frac{r-1}{r}$.*

Proof. Apply Corollary 2 to any of the polynomial time algorithms for 2-player games, such as [9], [1] or [16], which obtain an additive η-approximate equilibrium for some $\eta < \frac{1}{2}$.

□

5 Subexponential Time Additive and Multiplicative Approximation

In one of the most interesting works on approximate equilibria, Lipton, Markakis and Mehta [12] have shown that for r-player games, for every $0 < \varepsilon < 1$, there exists a k-uniform additive ε-approxima- tion whenever $k > \frac{3r^2 \ln(r^2 n)}{\varepsilon^2}$. The result is proven by averaging, for all players, independent samples of pure strategies according to any Nash equilibrium.

Here we improve their bound by a factor r by showing that for $0 < \varepsilon < 1$, an additive ε-approximation exists already when $k > \frac{2r \ln(rn+r)}{\varepsilon^2}$. We also establish an analogous result for multiplicative ε-approximation when $k > \frac{9r \ln(rn+r)}{2g^2 \varepsilon^2}$, where g is a lower bound on the payoffs of the players at the equilibrium.

The proof is based on the probabilistic method and is analogous to the one given in [12]. The main difference is that instead of the Hoeffding's bound, we use the more general Mc Diarmid's inequality [13] which bounds the deviation of a function of several independent random variables from its expectation. It specializes to the Hoeffding's bound when the function is the sum of the variables. It is stated as follows:

Theorem 3 (McDiarmid). *Let $Y_1, \ldots, Y_m$ be independent random variables on a finite set A, and let $f : A^m \longrightarrow \mathbb{R}$ be a function with the property that there exist real numbers $c_1, \ldots, c_m$ such that for all $(a_1, \ldots, a_m, b) \in A^{m+1}$ and $1 \leq l \leq m$:*

$$|f(a_1, \ldots, a_l, \ldots, a_m) - f(a_1, \ldots, b, \ldots, a_m)| \leq c_l \ .$$

Then, for every $\varepsilon > 0$,

$$\Pr[f(Y_1, \ldots Y_m) - \mathbb{E}[f(Y_1, \ldots Y_m)] > \varepsilon] \leq e^{-\frac{2\varepsilon^2}{\sum_l c_l^2}} \ .$$

Theorem 4. *For all $0 < \varepsilon < 1$, and for all $k > \frac{2r \ln(rn+r)}{\varepsilon^2}$, there exists a k-uniform additive ε-approximate equilibrium.*

Proof. For every $p \in [r]$, let $X_p^1, \ldots X_p^k$ be k copies of the random variable that takes the pure strategy $e_i \in B$ with probability x_p^i. We define $\mathcal{X}_p = \frac{1}{k} \sum_{j=1}^k X_p^j$, and let $\mathcal{X} = (\mathcal{X}_1, \ldots, \mathcal{X}_r)$. Observe that $\mathbb{E}[u_p(\mathcal{X})] = u_p(x)$. For $p \in [r]$ and $i \in [n]$, we consider the events

$$E_p : |u_p(\mathcal{X}) - u_p(x)| < \frac{\varepsilon}{2} \ ,$$
$$F_p^i : |u_p(\mathcal{X}_{-p}, e_i) - u_p(x_{-p}, e_i)| < \frac{\varepsilon}{2} \ ,$$

and we define E as the conjunction of all them.

For every $p \in [r]$ and $i \in [n]$, the event F_p^i, the fact that x is a Nash equilibrium, and the event E_p imply that

$$|u_p(\mathcal{X}_{-p}, e_i) - u_p(\mathcal{X})| < \varepsilon.$$

Therefore, when E is realized, $\mathcal{X}$ is an additive ε-approximate Nash equilibrium.

We prove that event E occurs with strictly positive probability. We start by bounding the probability of $\overline{E}_p$. We use McDiarmid's inequality with $m = rk$, when A is the canonical basis B, and the function f is defined as

$$f(a_1^1, \ldots, a_1^k, \ldots, a_r^1, \ldots, a_r^k) = u_p\left(\frac{1}{k}\sum_{j=1}^{k} a_1^j, \ldots, \frac{1}{k}\sum_{j=1}^{k} a_r^j\right).$$

Observe that $f(X_1^1, \ldots, X_1^k, \ldots, X_r^1, \ldots, X_r^k) = u_p(\mathcal{X}_1, \ldots, \mathcal{X}_r)$ and therefore

$$\mathbb{E}[f(X_1^1, \ldots, X_1^k, \ldots, X_r^1, \ldots, X_r^k)] = u_p(x)\ .$$

We claim that the values c_p^j can be chosen as $1/k$. Let a_p^j for $j \in [k]$ and $p \in [r]$ be some pure strategies. Fix $j \in [k]$, $p \in [r]$ and let b_p^j be a pure strategy. For $q \neq p$, we define the mixed strategies $\alpha_q = \frac{1}{k}\sum_{j=1}^{k} a_q^j$. Then, using the multilinearity of u_p, we have

$$\begin{aligned}
&f(a_1^1, \ldots, a_p^j, \ldots, a_r^k) - f(a_1^1, \ldots, b_p^j, \ldots, a_r^k) \\
&= \frac{1}{k}\left(u_p(\alpha_1, \ldots, a_p^j, \ldots, \alpha_r) - u_p(\alpha_1, \ldots, b_p^j, \ldots, \alpha_r)\right)\ .
\end{aligned}$$

This implies the claim, because u_p takes values in $[0, 1]$. Since $\sum_{j,p}(c_p^j) = kr\frac{1}{k} = \frac{r}{k}$, by McDiarmid's inequality we have

$$\Pr[\overline{E}_p] \leq e^{-\frac{\varepsilon^2 k}{2r}}\ .$$

For bounding from above the probability of $\overline{F}_p^i$, just observe that McDiarmid's inequality can be applied analogously for a function defined with $(r-1)k$ variables. This gives

$$\Pr[\overline{F}_p^i] \leq e^{-\frac{\varepsilon^2 k}{2(r-1)}}\ ,$$

and it follows from the union bound that

$$\Pr[\overline{E}] \leq r(n+1)e^{-\frac{\varepsilon^2 k}{2r}}.$$

The right side of this inequality is smaller than 1 when $k > \frac{2r\ln(rn+r)}{\varepsilon^2}$.

□

Theorem 5. *Let x be a Nash equilibrium for an r-player game and let $g > 0$ be a lower bound on the payoff of each player at the equilibrium. Then, for all $0 < \varepsilon < 1$, and for all $k > \frac{9r\ln(rn+r)}{2g^2\varepsilon^2}$, there exists a k-uniform multiplicative ε-approximate equilibrium.*

Proof. The proof is a slight modification of the previous one. The random variable $\mathcal{X}$ is defined identically. We set $\eta = 1 - \frac{1}{\sqrt{1+\varepsilon}}$ and $\zeta = \sqrt{1+\varepsilon} - 1$. The events E_p and F_p^i are defined as

$$E_p : |u_p(\mathcal{X}) - u_p(x)| < \eta\ u_p(x) ,$$
$$F_p^i : |u_p(\mathcal{X}_{-p}, e_i) - u_p(x_{-p}, e_i)| < \zeta\ u_p(x) ,$$

and E as the conjunction of all them.

Recursively applying E_p, we get for every integer $m > 0$,

$$u_p(x) < \eta^m u_p(x) + u_p(\mathcal{X}) \sum_{l<m} \eta^l .$$

Therefore, using that $\frac{1}{1-\eta} = 1 + \zeta$, we have

$$u_p(x) \leq (1+\zeta) u_p(\mathcal{X}) .$$

The event F_p^i and the fact that x is a Nash equilibrium imply that

$$u_p(\mathcal{X}_{-p}, e_i) < (1+\zeta) u_p(x) .$$

Since $(1+\zeta)^2 = 1 + \varepsilon$, it follows from the last two inequalities that $\mathcal{X}$ is a multiplicative ε-approximate Nash equilibrium when E is realized.

Using that g is a lower bound for $u_p(x)$, by McDiarmid's inequality we get

$$\Pr[\overline{E}_p] \leq e^{-\frac{2g^2\eta^2 k}{r}} ,$$

and

$$\Pr[\overline{F}_p^i] \leq e^{-\frac{2g^2\zeta^2 k}{r-1}} .$$

As $\eta = \frac{\zeta}{1+\zeta}$ we have that $\eta < \zeta$. Also, it is not hard to see that $0 < \varepsilon < 1$ implies $\frac{\varepsilon}{3} < \eta$. Therefore

$$\Pr[\overline{E}] \leq r(n+1) e^{-\frac{2g^2\varepsilon^2 k}{9r}} ,$$

and $\Pr[E] > 0$ when $k \geq \frac{9r \ln(rn+r)}{2g^2\varepsilon^2}$.

□

Remark 1. The condition $\varepsilon < 1$ in Theorem 5 is not a real restriction, since when $\varepsilon \geq 1$ then $\eta > \frac{1}{4}$, and therefore there exists a k-uniform multiplicative ε-approximate equilibrium for $k > \frac{8r \ln(rn+r)}{g^2}$.

Remark 2. If we require in Theorem 5 that the support of the approximate equilibrium is a subset of the support of the Nash equilibrium, the dependence on g is indeed necessary. Consider the following two players game given in the standard bimatrix representation where the number of the pure strategies of the players is $2n$:

$$M_1 = \begin{pmatrix} O_n & \frac{1}{2}I_n \\ \frac{1}{n}I_n & A_{n\times n} \end{pmatrix} \quad M_2 = \begin{pmatrix} I_n & O_n \\ O_n & I_n \end{pmatrix} .$$

Here, O_n denotes the $n \times n$ matrix with everywhere 0's, I_n is the $n \times n$ identity matrix, and $A_{n\times n}$ the is the $n \times n$ matrix with everywhere $1/n$ except on its diagonal where all entries are 0. The game has a Nash equilibrium $x = (x_1, x_2)$ where $x_1 = x_2 = \frac{1}{n}\sum_{i=1}^{n} e_{n+i}$. The payoffs of the first and second player are respectively $u_1(x, y) = \frac{1}{n} - \frac{1}{n}$ and $u_2(x_1, x_2) = \frac{1}{n}$ and therefore, the minimum of the payoffs is $g = \Theta(1/n)$. Let $0 < \varepsilon < 1$, and let $y = (y_1, y_2)$ be a multiplicative ε-approximate Nash equilibrium. Let k denote the size of supp(y_2), we claim that $k \geq \frac{n}{2(1+\varepsilon)}$. For this, observe first that $u_1(y_1, y_2) \leq 1/n$. Since supp$(y_2) \subseteq \{n+1, \ldots, 2n\}$, there exists $i \in [n]$ such that $y_2^{n+i} \geq 1/k$, and therefore $u_1(e_i, y_2) \geq \frac{1}{2k}$. Since y is a multiplicative ε-approximate equilibrium, we have that $\frac{1}{2k} \leq \frac{1+\varepsilon}{n}$ and the statement follows. Observe on the other hand that there exists, for any $\varepsilon > \frac{2}{n-2}$, multiplicative ε-approximate equilibria with support size only 2 if we let the support be outside this of the Nash equilibrium.

In [12] it was already observed that when the number of players is constant, the sampling method yields an additive ε-approximation, for all constant $\varepsilon > 0$, in time $n^{O(\ln n)}$. When $g = \Omega(1)$, Theorem 5 implies a similar result for the multiplicative approximation. This condition is satisfied for example if all the utility functions are bounded from below by a constant.

Corollary 5. *If in an r-player game, where r is constant, there exists a Nash-equilibrium at which all the players payoffs are bounded from below by a constant then for all constant $\varepsilon > 0$, a multiplicative ε-approximation can be found in time $n^{O(\ln n)}$.*

It can be interesting to compare the complexities of the two additive approximation algorithms based on the Lipton, Markakis and Mehta sampling technique. Let $\mathcal{A}(r)$ be the additive ε-approximation r-player algorithm based on Theorem 4 which searches exhaustively trough all the $\frac{2r^2\ln(nr+r)}{\varepsilon^2}$-uniform strategies. Let $\mathcal{B}(r)$ be the r-player algorithm we obtain by applying the iterative construction technique of Corollary 2 to $\mathcal{A}(2)$. Since by this technique we will never obtain a better than $\frac{r-2}{r-1}$-approximation, let us fix some $\varepsilon > \frac{r-2}{r-1}$. The overall running time of $\mathcal{A}(r)$ is $O(n^{\frac{2r\ln(rn+r)}{\varepsilon^2}})$ since the search is applied to the r players independently. The complexity of algorithm $\mathcal{B}(r)$ is $O(n^{8\frac{r\ln(rn+r)}{\zeta^2}} + n^r)$ where $\zeta = \frac{(r-2)-(r-1)\varepsilon}{(r-2)\varepsilon-(r-3)}$. A simple computation shows that for all $r \geq 3$, algorithm $\mathcal{B}(r)$ has a smaller complexity.

Obviously, $\mathcal{A}(r)$ and $\mathcal{B}(r)$ are not the fastest algorithms when Corollary 4 yields a polynomial time procedure for computing an additive approximate Nash equilibrium. A simple analysis shows that for each two-player polynomial time η-approximation, when $\eta > 0$, Corollary 4 gives a polynomial time algorithm computing an additive $(\frac{r-2}{r-1}+\eta')$-approximate equilibrium in an r-player game, for some $\eta' > 0$. This means that, at least for the time being, when ε is in some right neighborhood of $\frac{r-2}{r-1}$, the algorithm $\mathcal{B}(r)$ is the most efficient known procedure for computing an additive ε-approximate Nash equilibrium in an r-player game.

6 Conclusion and Open Problems

In this paper, we have started the study of approximate Nash equilibria for r-player games when $r \geq 2$. The main open problem, just as in the two-player case, is the existence of a PTAS. We enumerate a few other, possibly much simpler, problems left also open:

1. Does the lower bound of Theorem 1 on the size of the strategy profiles hold also in the case when $r = cn$, for a constant $c \leq 1$?
2. Can we reduce the gap on the support size between the lower bound of Theorem 1 and the upper bound of Theorems 4 and 5 ? For example, when $r = \Theta(1)$, the lower bound is $\Omega(\sqrt[r-1]{\ln n})$ and the upper bound is $O(\ln n)$. When $r = 2$, these bounds are tight.
3. Is there a polynomial time algorithm which computes a multiplicative $(r-1)$-approximate Nash equilibrium?

References

1. Bosse, H., Byrka, J., Markakis, E.: New algorithms for approximate Nash equilibria in bimatrix games. In: International Workshop on Internet and Network Economics (2008)
2. Chen, X., Deng, X.: 3-NASH is PPAD-complete. Electronic Colloquium in Computational Complexity (134) (2005)
3. Chen, X., Deng, X.: Computing Nash equilibria: approximation and smoothed complexity. In: IEEE Symposium On Foundations of Computer Science, pp. 603–612 (2006)
4. Chen, X., Deng, X.: Settling the complexity of two-player Nash equilibrium. In: IEEE Symposium on Foundations of Computer Science, pp. 261–272 (2006)
5. Chien, S., Sinclair, A.: Convergence to approximate Nash equilibria in congestion games. In: ACM-SIAM Symposium On Discrete Algorithms, pp. 169–178 (2007)
6. Daskalakis, C., Goldberg, P., Papadimitriou, C.: Three players games are hard. Electronic Colloquium on Computational Complexity (139) (2005)
7. Daskalakis, C., Goldberg, P., Papadimitriou, C.: The complexity of computing a Nash equilibrium. In: ACM Symposium on Theory of Computing, pp. 71–78 (2006)
8. Daskalakis, C., Mehta, A., Papadimitriou, C.: A note on approximate Nash equilibria. In: International Workshop on Internet and Network Economics, pp. 297–306 (2006)
9. Daskalakis, C., Mehta, A., Papadimitriou, C.: Progress in approximate Nash equilibria. In: ACM Conference on Electronic Commerce, pp. 355–358 (2007)
10. Feder, T., Nazerzadeh, H., Saberi, A.: Approximating Nash equilibria using small-support strategies. In: ACM Conference on Electronic Commerce, pp. 352–354 (2007)

11. Lemke, C.E., Howson Jr., J.T.: Equilibrium points of bimatrix games. Journal of the Society of Industrial and Applied Mathematics, 413–423 (1964)
12. Lipton, R., Markakis, E., Mehta, A.: Playing large games using simple strategies. In: ACM Conference on Electronic Commerce, pp. 36–41 (2003)
13. McDiarmid, C.: On the method of bounded differences. In: Surveys in combinatorics, London Math. Soc. Lectures Notes, vol. 141, pp. 148–188. Cambridge Univ. Press, Cambridge (1989)
14. Nash, J.F.: Non-cooperative games. Annals of mathematics 54, 286–295 (1951)
15. Savani, R., von Stengel, B.: Exponentially many steps for finding a Nash equilibrium in a bimatrix game. In: IEEE Symposium on Foundations Of Computer Science, pp. 258–267 (2004)
16. Tsaknakis, H., Spirakis, P.: An optimization approach for approximate Nash equilibria. In: International Workshop on Internet and Network Economics, pp. 42–56 (2008)

Subjective *vs.* Objective Reality — The Risk of Running Late

Amos Fiat and Hila Pochter

School of Computer Science, Tel Aviv University
{fiat,hilapoch}@tau.ac.il

Abstract. We study selfish agents that have a "distorted view" of reality. We introduce a framework of *subjective vs. objective reality*. This is very useful to model risk averse behavior. Natural quality of service issues can be cast as special cases thereof.

In particular, we study two applicable variants of the price of anarchy paradigm, the *subjective price of anarchy* where one compares the "optimal" subjective outcome to the outcome that arises from selfish subjective reality agents, and the *objective price of anarchy* where one compares the optimal objective outcome to that derived by selfish subjective agents.

1 Introduction

In this paper we consider the problem arising from selfish agents seeking to minimize server delay[1], where agents may have different risk sensitivities and also possibly different assumptions on the underlying risks.

We introduce a framework of "distorted reality" where we assume some underlying *objective reality* and where agents have a distorted view of objective truth, *i.e.*, agents have their own *subjective reality*. We introduce criteria to measure the degree of distortion (M, $\tilde{M}$, hereinafter), and study the impact of this distortion on the outcome.

This framework of distorted reality allows us to model risk averse behavior and to study the impact of risk aversion on the price of anarchy, in both subjective and objective terms. The distortion setting allows us to consider many variants of the problem, encompassing both agent-specific risk sensitivity and also different agent-specific assumptions on inherent underlying risks.

More specifically we consider these problems in the context of server delay and take the social welfare to be the makespan (objective or subjective).

Perhaps not surprisingly, it suffices for little distortion of reality (say 1%) (analogously recast as a small degree of risk aversion) to result in a dramatic change in the price of anarchy, super polynomial in the number of servers, at least for some server models.

[1] Note that we use the terminology of congestion games (delay), rather than the nomenclature of machine load balancing (load).

B. Monien and U.-P. Schroeder (Eds.): SAGT 2008, LNCS 4997, pp. 279–290, 2008.

1.1 Risk Neutral Server Delay Problems

The problem of minimizing server delay is that of selfish weighted tasks that wish to assign their task to a server with minimal delay. There are several well studied versions of this problem, including the equal, related, and unrelated models.

In general, for every task i and server s we have a positive real weight $w_{i,s}$. An assignment A is a mapping of n tasks to m servers. Given an assignment A, the delay of a server s is $\sum_i w_{i,s}$ over all tasks i assigned to server s. The delay of task i in assignment A is the delay of the server A_i to which it is assigned.

The problem variants (equal, related, unrelated) reflect different assumptions on the weights $w_{i,s}$.

1. The most general case, *unrelated servers*, assumes no restrictions on the $w_{i,s}$ values. The unrelated server model is applicable in cases where different servers are suitable for different tasks, to varying degrees.
2. The equal server variant assumes that $w_{i,s} = w_{i,s'}$ for all servers s, s', and all tasks i, *i.e.*, all servers are (inherently) equally suitable for all tasks. Thus, for equal servers, one can omit the subscript s, and use w_i in place of $w_{i,s}$.
3. For related servers, one associates a positive real value a_s with every server s, and requires that $w_{i,s}/a_s = w_{i,s'}/a_{s'}$, for all servers s, s'. This can be interpreted as meaning that different servers have different speeds/throughput which are proportional to $1/a_s$. Thus, by normalizing the minimal a_s to be 1, we can associate a weight, w_i with task i, and a speed, $1/a_s$ with server s, so that $w_{i,s} = a_s w_i$. [2]

The makespan of an assignment is the maximal delay over all servers s. The price of anarchy (POA) is the ratio between the highest makespan of a Nash equilibrium and the optimal makespan (the minimal makespan of any assignment). We denote the price of anarchy in the risk-free model by $\mathrm{POA}_{\mathrm{neut}}$.

Let t be the maximal ratio between task weights (which aren't infinity) on any two servers. The price of anarchy for the various server delay problems is 2, $\Theta(\log m/\log\log m)$ [4], and $\Theta(t + \frac{\log m}{\log(1+\log(\log m/t))})$ [3], for equal, related and unrelated servers, respectively.

1.2 Related Work: Agent-Specific Utility Functions

Agent-specific utility functions were first considered by Milchtaich [8] in the context of congestion games where he showed that unweighed agents with non decreasing latency functions and singleton congestion games always have a pure Nash equilibrium. Additional studies on agent specific utilities and conditions for the existence of pure Nash equilibria are given in [7,6,1].

Most relevant to our work herein is that of [6], in which Gairing, Monien and Tiemann consider the price of anarchy for linear player-specific utility functions. They consider what we call herein the *subjective price of anarchy*, and they take the social welfare function to be the sum of agent utilities.

---

[2] One can extend the notion of related machines to allow delay of the form $a_s W_s + b_s$ for some $b_s \geq 0$, b_s can be interpreted as a setup delay. It is not hard to extend previous results (bounds on the price of anarchy for related machines) in this model.

1.3 Subjective Agents

Whatever the context (equal, related, unrelated), let A be some assignment of tasks to servers. Let A_i be the server to which agent i is assigned under assignment A and let $W_s(A)$ be the objective delay on server s under assignment A. *I.e.*, $W_s(A) = \sum_{i.A_i=s} w_{i,s}$. Given an assignment A, we define the i-subjective delay for server s as $W_s(A)r_{i,s}$. *I.e.*, $r_{i,s}$, indexed by agent i and server s, determines how different agents view reality. As noted above, agent subjective delays, albeit without a corresponding objective delay, have been addressed before under the title of *player specific payoff functions* in a sequence of papers including [8] and [6].

However, we seek to distinguish between two different outcome *measures* :

1. Subjective outcomes (viewing the outcome from the perspective of the risk averse agents), or
2. Objective outcomes, viewing the outcome from the perspective of a risk neutral (external) observer.

Thus, for a specific Nash equilibrium Q, where Q_i is the server to which task i is assigned, we consider the following social cost functions:

1. Subjective makespan, denoted $\mathrm{SC}_{\mathrm{sub}}(Q) = \max_i W_{Q_i}(Q) r_{i,Q_i}$.
2. Objective makespan, denoted $\mathrm{SC}_{\mathrm{obj}}(Q) = \max_i W_{Q_i}(Q)$.

Similarly, one can define subjective and objective "optimal" solutions. The optimal objective makespan, denoted $\mathrm{OPT}_{\mathrm{obj}}$, minimizes the objective makespan, $\mathrm{SC}_{\mathrm{obj}}$. The optimal subjective makespan, denoted, $\mathrm{OPT}_{\mathrm{sub}}$, minimizes the maximal *subjective makespan*, $\mathrm{SC}_{\mathrm{sub}}$.

Next, we consider two versions of the price of anarchy , dealing with objective and subjective outcomes:

$$\mathrm{POA}_{\mathrm{obj}} = \sup_{\text{Nash equilibria } Q} \frac{\mathrm{SC}_{\mathrm{obj}}(Q)}{\mathrm{OPT}_{\mathrm{obj}}};$$

$$\mathrm{POA}_{\mathrm{sub}} = \sup_{\text{Nash equilibria } Q} \frac{\mathrm{SC}_{\mathrm{sub}}(Q)}{\mathrm{OPT}_{\mathrm{sub}}};$$

1.4 Modeling Server Risk: $r_{i,s} = 1 + k_i x_{i,s}$

Consider a set of servers S and a set of n agents. Agent i, seeking a quality of service guarantee, seeks a server of minimal expected delay T, subject to the restriction that task i completes within time T with probability at least $1 - \epsilon_i$.

Fix an assignment A. To study uncertainty and risk with regard to server performance we want to recast the notion of server delay, $W_s(A)$, as the expected (risk-neutral) delay. We do this as follows:

1. Associated with server s and an agent i is a random variable $X_s^i > 0$, with expectation $\mu(X_s^i) = 1$ and standard deviation $\sigma(X_s^i)$, let $x_{i,s} = \sigma(X_s^i)$. We will refer to $x_{i,s}$ as the risk belief of agent i for server s.

2. The delay of server s is the random variable $X_s^i W_s(A)$. Note that the expectation of $X_s^i W_s(A)$ is $W_s(A)$. What we've done here is identify the uncertainty with regard to server performance with the properties of the distribution for X_s^i.
3. Associate with every task i, a probability, ϵ_i, and let $k_i = 1/\sqrt{\epsilon_i}$. We will refer to k_i as the risk aversion of agent i.

Now, let $r_{i,s} = 1 + k_i x_{i,s} = 1 + \frac{1}{\sqrt{\epsilon_i}}\sigma(X_s^i)$. Let A be some assignment where task i is assigned to server s and let A' be the same as A except that task i is now assigned to server s'. Assume that agent i chose to use server s'.

Applying the Chernoff bound $Pr(|Y - \mu(Y)| \geq t\sigma(Y)) \leq \frac{1}{t^2}$, to $Y = X_{s'}^i W_{s'}(A')$, and setting $t = 1/\sqrt{\epsilon_i}$, we get

$$Pr\left[|X_{s'}^i W_{s'}(A') - \mu(X_{s'}^i W_{s'}(A'))| \geq 1/\sqrt{\epsilon_i}\sigma(X_{s'} W_{s'}(A'))\right] < \epsilon_i,$$

so

$$Pr\left[X_{s'}^i W_{s'}(A) \geq (1 + k_i x_{i,s'}) W_{s'}(A)\right] < \epsilon_i,$$

which means that the probability to experience delay larger then $(1+k_i x_{i,s'})W_{s'}(A)$ is less then ϵ_i.

This implies that the probability that task i will regret the choice of server s' (over server s) is no more than ϵ_i, and this is the best result she could have achieved, having no other apriori knowledge on the distribution of X_s^i and $X_{s'}^i$.

Thus, we have modelled this problem of quality of service sensitive agents, in terms of a subjective delay function.

We remark that we could have used other measures besides the standard deviation, and obtain tighter bounds on the probability of not meeting the deadline. In particular, using the standard deviation penalizes both upside and downside volatility equally (as does the Sharpe ratio [9] used to measure risk in equities), and one could alternately emulate the Sortino [10] ratio and penalize only delay (or equity returns) falling above (below) a user-specified target. For further risk related reading see [2,5].

1.5 Beliefs and Risk

Agents may have inconsistent vision of the inherent risk of different servers as well as different risk sensitivity. This gives us 4 different models:

1. II: Individual risk beliefs, Individual risk aversion. This is the most general model. Every agent may have different beliefs regarding server risk, and the risk aversion of the agents may vary. In this model $r_{i,s} = 1 + k_i x_{i,s}$
2. CI: Common risk beliefs, Individual risk aversion. All agents have the same beliefs regarding server risk, but the risk aversion of the agents vary. In this model $r_{i,s} = 1 + k_i x_s$, since for all $1 \leq i, j \leq n$, $x_{i,s} = x_{j,s}$.
3. IC: Individual risk beliefs, Common risk aversion. All agents have the same risk aversion, but different beliefs regarding server risk. In this model $r_{i,s} = 1 + k x_{i,s}$, since for all $1 \leq i, j \leq n$, $k_i = k_j$.
4. CC: Common risk beliefs, Common risk aversion. All agents have the same risk aversion, and the same beliefs regarding server risk. In this model $r_{i,s} = 1 + k x_s$.

Obviously, the CI, IC, and CC models are special cases of the II model. In fact, we can map any instance of II (k_i, $x_{i,s}$), to an instance of IC ($k'_i = k'_j = k'$, $x'_{i,s}$), choose $k' = 1$, and set $x'_{i,s} = k_i x_{i,s}$. This implies that insofar as regarding the price of anarchy, the II and IC models are equivalent.

We define two parameters of the system depending on the risk beliefs of the agents

$$M = \max_{i,s} r_{i,s} / \min_{i,s} r_{i,s}, \qquad \tilde{M} = \max_i (\max_s r_{i,s} / \min_s r_{i,s})$$

Lemma 1. $\mathrm{POA_{sub}}/M \leq \mathrm{POA_{obj}} \leq M \cdot \mathrm{POA_{sub}}$.

These results holds for all models $(\{II, IC, CI, CC\} \times \{$equal servers,related servers,unrelated servers$\})$.

Notice that if $\tilde{M} = 1$, the problem is uninteresting, therefore our results are for $\tilde{M} \neq 1$.

The results for the three models are shown in Figures 1 and 2. Proofs are in Section 2.

For the risk sensitive agents that hold common risk beliefs, and assuming related servers, we give significantly better bounds. In this case, the bound on the objective price of anarchy drops dramatically from being exponential in $\tilde{M}$

Model	$\mathrm{POA_{obj}}$	$\mathrm{POA_{sub}}$
Equal	$2\tilde{M}$	$2\tilde{M}M$
Related	$\tilde{M}^{O\left(\sqrt{\log m/\log \tilde{M}}\right)}$	$M\tilde{M}^{O\left(\sqrt{\log m/\log \tilde{M}}\right)}$
Unrelated	$\tilde{M}^{O\left(\sqrt{t^2+\log m/\log \tilde{M}}\right)}$ [3]	$M\tilde{M}^{O\left(\sqrt{t^2+\log m/\log \tilde{M}}\right)}$

Fig. 1. Upper bounds on the price of anarchy for pure Nash equilibria in the II model. (The same bounds hold for the CI and IC models).

Model	$\mathrm{POA_{obj}}$	$\mathrm{POA_{sub}}$
Equal	$2\tilde{M}$	$\Theta(\frac{\log m}{\log\log m})$
Related	$\Theta\left(\frac{\log(m)}{\log\log(m)}\tilde{M}\right)$	$\Theta\left(\frac{\log(m)}{\log\log(m)}\right)$
Unrelated	$O(t\tilde{M} + \frac{\log m}{\log(1+\log(\log m/t\tilde{M}))})\tilde{M}$	$\Theta(t\tilde{M} + \frac{\log m}{\log(1+\log(\log m/t\tilde{M}))})$

Fig. 2. Upper Bounds on the price of anarchy for pure Nash equilibria in the CC model. All bounds are tight up to a constant, except the $\mathrm{POA_{obj}}$ for the unrelated model.

[3] We also give a lower bound of $\tilde{M}^{\Omega\left(\sqrt{\log m/\log \tilde{M}}\right)}$, see Section 2.1.

to being linear. We also give better bounds on the subjective price of anarchy. See Section 3.

2 Bounds on the Price of Anarchy

2.1 Individual Risk Beliefs, Individual Risk Aversion

In this subsection we will present upper bounds in the equal, related and unrelated servers models, under the II assumption. Obviously, the results apply to the IC, CI and CC cases as well. These proofs closely follow the earlier risk free results [3,4]. We sketch some of these proofs below.

Lemma 2. *For equal servers, we have that* $\mathrm{POA}_{\mathrm{obj}} \leq 2\tilde{M}$.

Proof (Sketch). This is a minor modification of the proof for risk-free equal servers. Agents with with a distorted view of reality ($r_{i,s}$ factors) will seek to place their task on a the server that *they view* as having minimal latency. This distortion of reality introduces a factor of $\tilde{M}$ into the price of anarchy.

Theorem 1. *The objective price of anarchy for related servers,* $\mathrm{POA}_{\mathrm{obj}} \leq \tilde{M}^{O\left(\sqrt{\log m/\log \tilde{M}}\right)}$.

Proof. This proof, too, is a distorted version of the proof of [4], which we sketch here.

Let the servers be sorted according to their speed a_i, so that $a_1 \leq a_2 \leq \ldots \leq a_m$. We denote the maximal objective delay in the Nash equilibrium Q by $W_{\max} = \max_s W_s(Q)$. We define $s_{\max} = \mathrm{argmax_s W_s(Q)}$. For $p \geq 1$, define Z_p to be the smallest index $p \in \{0, 1, \ldots, m\}$ s.t. $W_{s_{Z_p+1}}(Q) < p \cdot \tilde{M}^{p-1} \cdot OPT$ or $Z_p = m$, if no such index exists. We define $p_{\max}$ to obey $W_{s_1}(Q) = p_{\max} \cdot \tilde{M}^{p_{\max}-1} \cdot OPT$.

We now consider two cases:

Case 1: If $p_{\max} \leq 3$, then $\mathrm{POA}_{\mathrm{obj}} \leq 3\tilde{M}^3 + \tilde{M}$.

Claim. $W_{\max} \leq (W_{s_1}(Q) + \mathrm{OPT}) \cdot \tilde{M}$.

Proof. Look at an agent i assigned to server $s_{\max}$ in Q. Since it is a pure Nash equilibrium, the agent doesn't want to switch servers which means that $W_{\max} r_{i,s_{\max}} \leq (W_{s_1}(Q) + a_1 w_i) r_{i,s_1}$. But as s_1 is the fastest server available, $a_1 w_i \leq \mathrm{OPT}$. So we get that for each agent i assigned to server $s_{\max}$ in Q we have $W_{\max} \cdot r_{i,s_{\max}} \leq (W_{s_1}(Q) + \mathrm{OPT}) \cdot r_{i,s_1}$. Bounding from above and below we get that $W_{\max} \cdot \min_s r_{i,s} \leq (W_{s_1}(Q) + \mathrm{OPT}) \cdot \max_s r_{i,s}$. Which means that $W_{\max} \leq (W_{s_1}(Q) + \mathrm{OPT}) \frac{\max_s r_{i,s}}{\min_s r_{i,s}}$, *i.e.*, $W_{\max} \leq (W_{s_1}(Q) + \mathrm{OPT}) \cdot \tilde{M}$.

According to the definitions, if $p_{\max} \leq 3$ then $W_{s_1}(Q) \leq 3\mathrm{OPT} \cdot \tilde{M}^2$. It now follows that $W_{\max} \leq (W_{s_1}(Q) + \mathrm{OPT}) \cdot \tilde{M} \leq (3\mathrm{OPT} \cdot \tilde{M}^2 + \mathrm{OPT}) \cdot \tilde{M}$, which leads to $\mathrm{POA}_{\mathrm{obj}} \leq 3\tilde{M}^3 + \tilde{M}$.

Case 2: If $p_{\max} > 3$, then $\mathrm{POA}_{\mathrm{obj}} \le \tilde{M}^{O\left(\sqrt{\log m/\log \tilde{M}}\right)}$.

We argue that $\tilde{M}^{(p_{\max}-1)+(p_{\max}-2)+\ldots+1} \cdot p_{\max}! \le Z_1$. Let S be the set of servers $s_1 \ldots s_{Z_{p+1}}$.

Claim. If in a pure Nash equilibrium Q, agent i is assigned to a server in S, an optimal strategy does not assign i to a server s_r such that $r > Z_p$.

Proof omitted. Distorted version of [4].

Claim. If an optimal strategy assigns tasks from group S to servers $s_1, \ldots, s_{Z_p}$ then $Z_p \ge (p+1)\tilde{M}^p \cdot Z_{p+1}$.

Proof omitted. Distorted version of [4].

Since $Z_{p_{\max}} \ge 1$ we have that $p_{\max}! \cdot M^{(p_{\max}-1)+(p_{\max}-2)+\ldots+1} \le Z_1 \le m$. Solving the equation we get that $p_{\max} \le O\left(\sqrt{\frac{\log m}{\log \tilde{M}}}\right)$. Substituting we get that

$$\mathrm{POA}_{\mathrm{obj}} \le \tilde{M}^{O\left(\sqrt{\frac{\log m}{\log \tilde{M}}}\right)}.$$

Theorem 2. *For unrelated servers* $\mathrm{POA}_{\mathrm{obj}} \le O\left(\tilde{M}^{O\left(\sqrt{t^2+\log m/\log \tilde{M}}\right)}\right)$, *where t is the maximal ratio between task weights (not including infinity).*

Proof (Sketch). Let Q be a pure Nash equilibrium. Without loss of generality, we assume that the servers are sorted in decreasing order according to their objective delay $W_s(Q)$. For $p \ge 1$, define Z_p to be the smallest index $p \in \{0, 1, \ldots, m\}$ s.t. $W_{s_{Z_p+1}}(Q) < p \cdot M^p \cdot OPT$ or $Z_p = m$, if no such index exists. Let S be the set of servers $s_1 \ldots s_{Z_{p+1}}$. The following two claims give us the required result.

1. If in a pure Nash equilibrium Q, agent i is assigned to a server in S, an optimal strategy does not assign i to a server s_r such that $r > Z_p$.
2. For every $p \ge 1$, $\frac{Z_p}{Z_{p+1}} \ge \frac{(p+1)\tilde{M}^{p+1}}{t}$.

Lemma 3. *For unrelated servers,* $\mathrm{POA}_{\mathrm{obj}} = \Omega\left(\tilde{M}^{\sqrt{\frac{\log m}{\log \tilde{M}}}}\right)$.

Proof. We prove that for every k and every $r_{\max} = \max_{i,s} r_{i,s}$ and $r_{\min} = \min_{i,s} r_{i,s}$, we can construct a game in which $\mathrm{POA}_{\mathrm{obj}} = \Omega\left(\tilde{M}^{\sqrt{\frac{\log m}{\log \tilde{M}}}}\right)$. Consider $p+1$ groups of servers indexed $0 \ldots p$ (p will be defined later). Group j contains n_j servers, where

* $n_j = n_{j-1} \cdot \tilde{M}^{p-j}$
* $n_0 = 1$

Now, consider p sets of tasks indexed $0..p-1$. Set j contains $n_j \cdot \tilde{M}^{p-1-j}$ tasks. Let the tasks from set i have weight $\tilde{M}$ on servers from group i, weight 1 on servers from group $i+1$ and infinity on the rest of the servers. The tasks from set i have risk adjustment factor $r_{\min}$ on servers from group i and factor $r_{\max}$ on the rest of the servers.

Let's look at a pure Nash equilibrium Q, in which all the tasks from set i are equally divided on the servers of group i. The subjective delay experienced by the tasks from the i-th set is $(\tilde{M} \cdot \tilde{M}^{p-1-i}) \cdot r_{\min} = (\tilde{M}^{p-i}) \cdot r_{\min}$. Since an agent from set i can only choose to switch to a server from group $i+1$ (otherwise she will experience a delay of infinity), and the subjective delay after the move would be $(\tilde{M}^{p-i-1} + 1) \cdot r_{\max} > \tilde{M}^{p-i-1} \cdot r_{\max} = \tilde{M}^{p-i} \cdot r_{\min}$, the agent has no incentive to switch servers. The objective makespan is achieved on the servers from group 0, so we have $SC_{\text{obj}}(Q) = \tilde{M}^p$.

Clearly the objective optimal solution is achieved when every agent from the l-th set is assigned to a distinct server in the $(l+1)$-th group, leading to an objective makespan of 1. Therefore $\text{POA}_{\text{obj}} \geq \tilde{M}^p$. $m = \tilde{M}^{\theta(p^2)}$ implies that $p = \theta\left(\sqrt{\frac{\log m}{\log \tilde{M}}}\right)$ and $\text{POA}_{\text{obj}} = \Omega\left(\tilde{M}^{\sqrt{\frac{\log m}{\log \tilde{M}}}}\right)$.

2.2 Common Risk Beliefs, Common Risk Aversion

In this subsection we give bounds on the price of anarchy when all agents are equally risk averse and have common risk beliefs ($\forall s\ \forall i,j\ r_{i,s} = r_{j,s} = r_s$).

Lemma 4. *Under the CC assumption in the equal server model, the subjective price of anarchy is equal to the price of anarchy in the game without risk with adjusted weights* $w'_{i,s} = w_{i,s} r_s$.

From Lemma 4 and Lemma 1 we conclude that

Lemma 5. *Under the CC assumption in the related server model,* $\text{POA}_{\text{obj}} = \tilde{M} \cdot \text{POA}_{\text{neut}} = O\left(\frac{\log m}{\log\log m} \tilde{M}\right)$.
For unrelated servers, $\text{POA}_{\text{obj}} = O\left(\left(t\tilde{M} + \frac{\log m}{\log(1+\log(\log m / t\tilde{M}))}\right) \tilde{M}\right)$.

3 Bounds on the Price of Anarchy for Related Servers, Common Risk Beliefs, Individual Risk Aversion, and Equal Weights

In this section we bound the price of anarchy under the CI assumption where all the tasks have equal weights , $w_i = 1$. For the rest of this section let $x_s = x_{i,s}$ (common for all agents). Let $M' = \max_s \frac{\max_i r_{i,s}}{\min_i r_{i,s}}$.

Theorem 3. *Under the CI assumption with related servers, where* $\forall i, w_i = 1$, *it holds that: 1.* $\text{POA}_{\text{obj}} \leq \tilde{M}$ *and 2.* $\text{POA}_{\text{sub}} \leq M'$.

Let G be a game with individual risk aversion and $G_{\max}$ the corresponding game with common risk aversion $k_{\max} = \max_i k_i$.

Lemma 6. POA_{obj} *is bounded by* POA_{obj} *in* $G_{\max}$.

Proof. First we prove that the objective makespan of the worst Nash equilibrium in G is bounded by the objective makespan of the worst Nash equilibrium in $G_{\max}$. For this we will need the following 2 claims:

Claim (1). Assume that in an assignment A, agent 1 is playing her best response on server s_1. If the risk aversion of agent 1 is changed to $k > k_1$ then her new best response (server s_2) satisfies $x_{s_2} \leq x_{s_1}$ and $W_{s_2}(A') \geq W_{s_1}(A)$, where A' is the same as A after the task i was allocated to s_2 . Likewise, her new subjective delay is not lower than the subjective delay before the change of k_1.

Proof (of Claim). If the agent does not want to switch servers after we change her risk aversion then the claim holds with equality. Therefore, we assume that the agent has switched positions from server s_1 to server s_2. Since the agent didn't want to switch places before the change of k_1 but does want to switch now, we get

$$\begin{aligned} W_{s_1}(A)(1+k_1x_{s_1}) &\leq (W_{s_2}(A)+a_{s_2}w_1)(1+k_1x_{s_2}) = W_{s_2}(A')(1+k_1x_{s_2}) \\ &< W_{s_2}(A')(1+kx_{s_2}) < W_{s_1}(A)(1+kx_{s_1}). \end{aligned}$$

So we get that $\frac{1+kx_{s_2}}{1+kx_{s_1}} < \frac{W_{s_1}(A)}{W_{s_2}(A')} \leq \frac{1+k_1x_{s_2}}{1+k_1x_{s_1}}$. We see the there is an inverse correlation between the risk beliefs $r_{i,s}$ and the objective delay. Now let's define $f(k) = \frac{1+kx_{s_2}}{1+kx_{s_1}}$. Under the assumptions we see that $f(k)$ is monotonically decreasing and $f'(k) = \frac{-x_{s_1}+x_{s_2}}{(1+kx_{s_1})^2} < 0$. Therefore, $x_{s_1} > x_{s_2}$, and $\frac{W_{s_1}(A)}{W_{s_2}(A')} < 1$ which completes our proof. □

Claim (2). If we start from a given a Nash equilibrium in G and in each step we change one agent's risk aversion k_i to $k_{\max}$ and let the agent switch servers according to the best response. Then all the agents with risk aversion $k_{\max}$ have no new best response.

Corollary 1. *The process of changing the risk aversion k_i of the agents to $k_{\max}$, one agent at a time (in arbitrary order), and re-adjusting that agent's best response, culminates with a Nash equilibrium.*

Proof. By induction on the number of steps.

Basis - In step 0, since this is a Nash equilibrium, all the agents which have risk aversion $k_{\max}$ are in a best response position.

Now we assume that after n steps, each agent j that has $k_j = k_{\max}$ is playing her best response, and prove that this stays true after step $n+1$. Let's change agent p's risk aversion k_p to $k_{\max}$, and let her switch to the server which is her new best response.

Now look at all the other agents with risk aversion $k_{\max}$. According to the induction assumption, before the last agent switched servers they were playing their best response. If the agent does not switch, nothing has changed and the induction claim holds. Now assume that agent p switches from server a to server b. The agent is playing her best response, so we only have to prove that we didn't affect the other agents with risk aversion $k_{\max}$.

Let's look at the agents using server a. Their subjective delays are lower (because of the decrease in the objective delay), so if they were playing their best response before the switch, they are definitely playing their best response now. Now look at the agents with risk aversion $k_{\max}$ on server b. Their new subjective delays are not lower than before but the agent that switched to server b has the same w_i and k_i (and of course the same risk beliefs $x_{i,s} = x_s$) , which means that there is no better server for them either. *i.e.*., they are still playing their best response. Now look at an agent with $k_{\max}$ that uses server $c \neq a, b$. The objective delay for using server b just got higher, so the agent doesn't want to move there. The objective delay for using server a got lower, but agent p, which has the same w_i and k_i preferred to switch to server b, and b was not better than c in the last stage, (because the agent was playing her best response). This means that the agent's current location is still her best response. □

We now conclude the proof of Lemma 6. We proved that by starting with any Nash equilibrium and changing the agents' risk aversion k_j to $k_{\max}$ we arrive at a Nash equilibrium. Now we show an order of changing the agents risk aversion such that the Nash equilibrium we arrive at, in $G_{\max}$, has objective makespan at least as high as the Nash equilibrium we started with in G. We do this by successively choosing an agent j, on the server of highest objective delay, with $k_j < k_{\max}$.

Let's look at the agents using the server s with the highest objective delay. When we change one of the agent's (agent p) risk aversion, the agent can only switch to a server in which the objective delay will be at least as high as the objective delay on s (according to Claim 1). Now we treat 2 cases:

1. *Case 1.* Agent p does not switch servers. This means that we can change the risk aversion of all the agents using this server to $k_{\max}$ one by one, and their best response will be to remain on s. By Claim 2, we can continue in any order and get a Nash equilibrium with at least this server's objective delay.
2. *Case 2.* Agent p chooses to switch to another server s'. We look at the best possible server the agent wants to switch to. If it contains an agent q with $k_q < k_{\max}$, then if we changed k_q first, agent q would remain on this server. So we return agent p to her original server and only change agent q's risk aversion. We do that to all the agents using s' - they are still in best response after their risk aversion changed. Now we move agent p to s' and continue in any order and get a Nash equilibrium with at least server s''s objective.

As we can see from the two cases above, an agent from the original server of highest objective delay, never switches servers, unless the agent wants to switch to a new server that all the agents that use it, have the maximal risk aversion $k_{\max}$. In this case, after the agent switch servers, all the agents that are assigned to the new server have the maximal risk aversion, and from Claim 1, we can see that the server's objective delay is higher or equal to the objective delay of the original server of highest objective delay. In any case, we get an objective makespan higher or equal to the original Nash equilibrium.

This means that the objective makespan of the worst Nash equilibrium in G is not larger than the one of the worst Nash equilibrium in $G_{\max}$. Therefore, since they share an objective optimum, and the objective makespan of the optimum is the same in both games, we get that the objective price of anarchy in G is bounded by the objective price of anarchy in $G_{\max}$. (end of Lemma 6). □

Corollary 2. $\mathrm{POA}_{\mathrm{obj}} \leq \tilde{M}$

Proof. In the related server game with common risk aversion, common risk belief and equal weights, we have that $\mathrm{POA}_{\mathrm{obj}} \leq \tilde{M}$ as in this setting we trivially have that $\mathrm{POA}_{\mathrm{sub}} = 1$. The corollary follows immediately.

We would have liked to use the same technique as in Lemma 6 to bound the general weighted case. The technique will not work as in the weighted case as there exists a 2 server counter example to the theorem in the weighted setting.

Lemma 7. $\mathrm{POA}_{\mathrm{sub}}$ *in* G *is bounded by* $M' \cdot POA_{\mathrm{sub}}$ *in* $G_{\max}$*.*

Proof. First we prove that the subjective makespan of the worst pure Nash equilibrium in G, is bounded by the one of the worst pure Nash equilibria in $G_{\max}$. Then we prove that $OPT_{\max}$, the subjective makespan of the optimal solution in $G_{\max}$ is bounded by $OPT \cdot M'$. Combining the two give use the required result.

Claim. The subjective makespan of a pure Nash equilibrium in G, is bounded by the subjective makespan in $G_{\max}$

Proof. The proof is similar to the proof of Lemma 6, but we start with the agent that experience the highest subjective delay. The only thing to notice is that if agent p switches servers (after raising her risk aversion), she does so to a server on which her subjective delay is not lower than her original subjective delay.

Let Q^* be a subjective optimal assignment in G and $Q^*_{\max}$ to be a subjective optimal assignment in $G_{\max}$. For an assignment A, denote by $\mathrm{SC}_{\mathrm{sub}}(A)$ the subjective makespan of A in G, and by $\mathrm{SC}^k_{\mathrm{sub}}(A)$ the subjective makespan in $G_{\max}$. Let us look at the assignment Q^* under $G_{\max}$. Obviously $\mathrm{SC}^k_{\mathrm{sub}}(Q^*) \geq SC^k_{\mathrm{sub}}(Q^*_{\max})$. Now we prove that $\mathrm{SC}^k_{\mathrm{sub}}(Q^*) \leq M' \cdot SC_{\mathrm{sub}}(Q^*)$. Examining the subjective delay of agent i in Q^* under G we get $W_{Q^*_i}(Q^*)(1 + k_i x_{Q^*_i}) \leq \mathrm{SC}_{\mathrm{sub}}(Q^*)$. Now examine the agent's subjective delay when $k_i = k_{\max}$

$$
\begin{aligned}
W_{Q^*_i}(Q^*)(1 + k_i x_{Q^*_i} + (k_{\max} - k_i)x_{Q^*_i}) &\leq \mathrm{SC}_{\mathrm{sub}}(Q^*) + W_{Q^*_i}(Q^*)(k_{\max} - k_i)x_{Q^*_i} \\
&\leq \mathrm{SC}_{\mathrm{sub}}(Q^*) + \frac{SC(Q^*)}{1 + k_i x_{Q^*_i}}(k_{\max} - k_i)x_{Q^*_i} \\
&\leq \mathrm{SC}_{\mathrm{sub}}(Q^*)\left(1 + \frac{(k_{\max} - k_i)x_{Q^*_i}}{1 + k_i x_{Q^*_i}}\right) \leq \mathrm{SC}_{\mathrm{sub}}(Q^*)\left(\frac{1 + k_{\max} x_{Q^*_i}}{1 + k_i x_{Q^*_i}}\right) \\
&\leq \mathrm{SC}_{\mathrm{sub}}(Q^*)\left(\frac{1 + \max_i k_i \max_s x_s}{1 + \min_i k_i \max_s x_s}\right) = \mathrm{SC}_{\mathrm{sub}}(Q^*)M'.
\end{aligned}
$$

Where the last inequality holds because the function $\frac{1+\max_i k_i \cdot x}{1+\min_i k_i \cdot x}$ is monotonically increasing. Since the subjective makespan is the maximal subjective delay of an agent we have that $M' \cdot \mathrm{SC}_{\mathrm{sub}}(Q^*) \geq \mathrm{SC}^k_{\mathrm{sub}}(Q^*) \geq \mathrm{SC}^k_{\mathrm{sub}}(Q^*_{\max})$ which means that $M' \cdot OPT \geq OPT_{\max}$. □

Corollary 3. *As* $\mathrm{POA}_{\mathrm{sub}}$ *in* G *is bounded by 1 it follows that* $POA_{\mathrm{sub}} \leq M'$.

References

1. Ackermann, H., Röglin, H., Vöcking, B.: Pure nash equilibria in player-specific and weighted congestion games. In: Spirakis, P.G., Mavronicolas, M., Kontogiannis, S.C. (eds.) WINE 2006. LNCS, vol. 4286, pp. 50–61. Springer, Heidelberg (2006)
2. Altman, E., Saunders, A.: Credit risk measurement: Developments over the last 20 years. Journal of Banking & Finance 21, 1721–1742 (1998)
3. Awerbuch, B., Azar, Y., Richter, Y., Tsur, D.: Tradeoffs in worst-case equilibria. In: Solis-Oba, R., Jansen, K. (eds.) WAOA 2003. LNCS, vol. 2909, pp. 41–52. Springer, Heidelberg (2004)
4. Czumaj, A., Vöcking, B.: Tight bounds for worst-case equilibria. In: SODA, pp. 413–420 (2002)
5. Elton, E., Gruber, M.: Modern portfolio theory, 1950 to date. Journal of Banking & Finance 21, 1743–1759 (1997)
6. Gairing, M., Monien, B., Tiemann, K.: Routing (un-) splittable flow in games with player-specific linear latency functions. In: Bugliesi, M., Preneel, B., Sassone, V., Wegener, I. (eds.) ICALP 2006. LNCS, vol. 4051, pp. 501–512. Springer, Heidelberg (2006)
7. Mavronicolas, M., Milchtaich, I., Monien, B., Tiemann, K.: Congestion games with player-specific constants. In: Kučera, L., Kučera, A. (eds.) MFCS 2007. LNCS, vol. 4708, pp. 633–644. Springer, Heidelberg (2007)
8. Milchtaich, I.: Congestion games with player-specific payoff functions. Games and Economic Behavior 13, 111–124 (1996)
9. Sharpe, W.: Mutual fund performance. Journal of Business 39(1), 119–138 (1966)
10. Sortino, F., Price, L.: Performance measurement in a downside risk framework. The Journal of Investing, 59–65

On the Hardness and Existence of Quasi-Strict Equilibria*

Felix Brandt and Felix Fischer

Institut für Informatik, Universität München
80538 München, Germany
{brandtf,fischerf}@tcs.ifi.lmu.de

Abstract. This paper investigates the computational properties of quasi-strict equilibrium, an attractive equilibrium refinement proposed by Harsanyi, which was recently shown to always exist in bimatrix games. We prove that deciding the existence of a quasi-strict equilibrium in games with more than two players is NP-complete. We further show that, in contrast to Nash equilibrium, the support of quasi-strict equilibrium in zero-sum games is unique and propose a linear program to compute quasi-strict equilibria in these games. Finally, we prove that every symmetric multi-player game where each player has two actions at his disposal contains an efficiently computable quasi-strict equilibrium which may itself be asymmetric.

1 Introduction

Perhaps the most ubiquitous solution concept in non-cooperative game theory is Nash equilibrium—a strategy profile that does not permit beneficial unilateral deviation. One of the main drawbacks of this concept is its potential multiplicity: While Nash equilibria are guaranteed to exist in finite games, there may be many of them, which causes uncertainty among the players which one to choose. For this reason, a number of concepts that single out particularly reasonable Nash equilibria—so-called equilibrium refinements—have been proposed over the years.

An important result by Norde et al. [23] has cast doubt upon this strand of research. Norde et al. [23] have shown that Nash equilibrium can be completely characterized by *utility maximization in one-player games*, *consistency*,[1] and *existence*. As a consequence, all common equilibrium refinements either violate consistency or existence. In particular, all refinements that are guaranteed to

* This material is based upon work supported by the Deutsche Forschungsgemeinschaft under grant BR 2312/3-2.

[1] Consistency as introduced by Peleg and Tijs [25] is defined as follows. Let S be a solution of game G and let G' be a reduced game where a subset of players are assumed to invariably play the strategies prescribed by S. A solution concept is *consistent* if the solution S' in which all of the remaining players still play according to S is a solution of G'.

B. Monien and U.-P. Schroeder (Eds.): SAGT 2008, LNCS 4997, pp. 291–302, 2008.

exist such as *perfect*, *proper*, and *persistent* equilibria suffer from inconsistency while other refinements such as *quasi-strict*, *strong*, and *coalition-proof* equilibria may not exist. Since consistency is a very intuitive and appealing property, its failure may be considered more severe than possible non-existence. Harsanyi's quasi-strict equilibrium, which refines the Nash equilibrium concept by requiring that every action in the support yields strictly more payoff than actions not in the support, has been shown to always exist in bimatrix games [22] and generic n-player games (and thus in "almost every" game) [12]. Furthermore, Squires [28] has shown that quasi-strict equilibrium is very attractive from an axiomatic perspective as it satisfies the Cubitt and Sugden axioms [6], a strengthening of a similar set of axioms by Samuelson. This result can be interpreted so that the existence of quasi-strict equilibrium is sufficient to justify the assumption of common knowledge of rationality. In fact, Quesada [26] even poses the question whether the existence of quasi-strict equilibrium is sufficient for *any* reasonable justification theory. Finally, isolated quasi-strict equilibria satisfy almost all desirable properties defined in the refinements literature. They are essential, strongly stable, regular, proper, and strictly perfect [see, *e.g.*, 14, 29, 30].[2]

In recent years, the computational complexity of a number of equilibrium refinements in various classes of games such as extensive-form games, congestion games, or graphical games has come under increasing scrutiny [see, *e.g.*, 11, 19, 20, 27]. In this paper, we study the computational properties of quasi-strict equilibrium in zero-sum games, general normal-form games, and certain classes of symmetric or anonymous games. The remainder of the paper is structured as follows. In the next section, we introduce classes of strategic games and the solution concepts of Nash equilibrium and quasi-strict equilibrium. Section 3 focuses on two-player games. We show that quasi-strict equilibria of zero-sum games, unlike Nash equilibria, possess a unique support, and propose linear programs that characterize the quasi-strict equilibria in non-symmetric and symmetric zero-sum games. In Section 4 we turn to games with more than two players. We first distinguish multi-player games where a quasi-strict equilibrium is guaranteed to exist and can be found efficiently from those where existence is not guaranteed. An example of the former class are symmetric games where every player has two actions. We then move on to show that deciding the existence of a quasi-strict equilibrium in games with more than two players is NP-complete in general. This is in contrast to the two-player case, where the decision problem is trivial due to the existence result by Norde [22].

2 Preliminaries

An accepted way to model situations of strategic interaction is by means of a normal-form game [see, *e.g.*, 17].

[2] Using the framework of Peleg and Tijs [25] and Norde et al. [23], quasi-strict equilibria could easily be axiomatically characterized by consistency and *strict* utility maximization in one-player games.

Definition 1 (normal-form game). *A* game in normal-form *is a tuple* $\Gamma = (N, (A_i)_{i \in N}, (p_i)_{i \in N})$ *where* $N = \{1, \ldots, n\}$ *is a set of* players *and for each player* $i \in N$, A_i *is a nonempty set of* actions *available to player* i, *and* $p_i : (\times_{i \in N} A_i) \to \mathbb{R}$ *is a function mapping each action profile of the game (*i.e., *combination of actions) to a real-valued* payoff *for player* i.

A vector $s \in \times_{i \in N} A_i$ of actions is also called a profile of *pure strategies*. This concept can be generalized to *(mixed) strategy profiles* $s \in S = \times_{i \in N} S_i$ by letting players randomize over their actions. We have S_i denote the set of probability distributions over player i's actions, or *(mixed) strategies* available to player i. We further write s_i and s_{-i}, respectively, for the strategy of player i and the strategy profile for all players but i. For $a \in A_i$, we denote by $s(a)$ the probability with which player i plays a in strategy profile s. A game with two players is often called a bimatrix game. A bimatrix game is a *zero-sum game* if $p_1(a) + p_2(a) = 0$ for all $a \in A$.

Our results on symmetric and anonymous games will be based on the taxonomy introduced by Brandt et al. [4].[3] A common aspect of games in all classes of the taxonomy is that players cannot, or need not, distinguish between the other players. A lattice of four classes of symmetric games is then defined by considering two additional properties: *identical payoff functions* for all players and the ability to *distinguish oneself* from the other players.

Definition 2 (symmetries). *Let* $\Gamma = (N, (A_i)_{i \in N}, (p_i)_{i \in N})$ *be a normal-form game and* A *a set of actions such that* $A_i = A$ *for all* $i \in N$. *For any permutation* $\pi : N \to N$ *of the set of players, let* $\pi' : A^N \to A^N$ *be the permutation of the set of action profiles given by* $\pi'((a_1, \ldots, a_n)) = (a_{\pi(1)}, \ldots, a_{\pi(n)})$. Γ *is called*

- anonymous *if* $p_i(s) = p_i(\pi'(s))$ *for all* $s \in A^N$, $i \in N$ *and all* π *with* $\pi(i) = i$,
- symmetric *if* $p_i(s) = p_j(\pi'(s))$ *for all* $s \in A^N$, $i, j \in N$ *and all* π *with* $\pi(j) = i$,
- self-anonymous *if* $p_i(s) = p_i(\pi'(s))$ *for all* $s \in A^N$, $i \in N$, *and*
- self-symmetric *if* $p_i(s) = p_j(\pi'(s))$ *for all* $s \in A^N$, $i, j \in N$.

It is easily verified that the class of self-symmetric games equals the intersection of symmetric and self-anonymous games, and that both of these are strictly contained in the class of anonymous games. Anonymous multi-player games admit a compact representation when the number of actions is bounded.

One of the best-known solution concepts in game theory is Nash equilibrium [21]. In a Nash equilibrium, no player is able to increase his payoff by unilaterally changing his strategy.

Definition 3 (Nash equilibrium). *Let* $\Gamma = (N, (A_i)_{i \in N}, (p_i)_{i \in N})$ *be a normal-form game. A strategy profile* $s \in S$ *is called* Nash equilibrium *if for all* $i \in N$, $a \in A_i$,

$$p_i(s) \geq p_i(s_{-i}, a).$$

[3] However, the terminology has been adjusted to coincide with that of Daskalakis and Papadimitriou [7].

The solution concept of quasi-strict equilibrium proposed by Harsanyi [12] refines the Nash equilibrium concept by requiring that actions played with positive probability must yield *strictly* more payoff than actions played with probability zero.[4]

Definition 4 (quasi-strict equilibrium). *Let $\Gamma = (N, (A_i)_{i\in N}, (p_i)_{i\in N})$ be a normal-form game. A Nash equilibrium $s \in S$ is called* quasi-strict *if for all $i \in N$ and all $a, b \in A_i$ with $s_i(a) > 0$ and $s_i(b) = 0$, $p_i(s_{-i}, a) > p_i(s_{-i}, b)$.*

Quasi-strict equilibrium is a very natural concept in that it requires *all* best responses to be played with positive probability.

3 Two-Player Zero-Sum Games

It has been shown by a rather elaborate construction using Brouwer's fixed point theorem that quasi-strict equilibrium always exists in two-player games [22]. Since every quasi-strict equilibrium is also a Nash equilibrium, the problem of finding a quasi-strict equilibrium is intractable unless $P = PPAD$ [5]. The same is true for symmetric two-player games, because the symmetrization of Gale et al. [9] preserves quasi-strictness [15]. For the restricted class of zero-sum games, however, quasi-strict equilibria, like Nash equilibria, can be found efficiently by linear programming. In contrast to Nash equilibria, the support of quasi-strict equilibria in zero-sum games turns out to be unique.

Theorem 1. *Quasi-strict equilibria in two-player zero-sum games possess a unique support and can be computed using the linear program given in Figure 2.*

Proof. It is known from the work of Jansen [13] that every bimatrix game with a convex equilibrium set, and thus every two-player zero-sum game, possesses a quasi-strict equilibrium. We first establish that the support of a quasi-strict equilibrium must contain every action that is played with positive probability in some equilibrium of the game. Assume for contradiction that (s_1, s_2) is a quasi-strict equilibrium with value v and $a \in A_1$ is an action with $s_1(a) = 0$. It follows from the definition of quasi-strict equilibrium that $p_1(a, s_2) < v$. Now, if a is in the support of some Nash equilibrium, the exchangeability of equilibrium strategies in zero-sum games requires that $p_1(a, s_2) = v$, a contradiction. As a consequence, the support of any quasi-strict equilibrium is unique and consists precisely of those actions that are played with positive probability in some equilibrium.

In order to compute quasi-strict equilibria, consider the two standard linear programs for finding the minimax strategies for player 1 and 2, respectively, given in Figure 1 [see, *e.g.*, 17]. It is well-known from the minimax theorem [31], and also follows from LP duality, that the value v of the game is identical and

[4] Harsanyi originally referred to quasi-strict equilibrium as "quasi-strong". However, this term has been dropped to distinguish the concept from Aumann's strong equilibrium [1].

maximize v
subject to

$$\begin{array}{lll} \sum_{a \in A_1} s_1(a)\, p(a,b) & \geq v & \forall b \in A_2 \\ s_1(a) & \geq 0 & \forall a \in A_1 \\ \sum_{a \in A_1} s_1(a) & = 1 & \end{array}$$

minimize v
subject to

$$\begin{array}{lll} \sum_{b \in A_2} s_2(j)\, p(a,b) & \leq v & \forall a \in A_1 \\ s_2(b) & \geq 0 & \forall b \in A_2 \\ \sum_{b \in A_2} s_2(b) & = 1 & \end{array}$$

Fig. 1. Primal/dual linear programs for computing minimax strategies in zero-sum games

maximize ε
subject to

$$\begin{array}{lll} \sum_{b \in A_2} s_2(b)\, p(a,b) & \leq v & \forall a \in A_1 \\ s_2(b) & \geq 0 & \forall b \in A_2 \\ \sum_{b \in A_2} s_2(b) & = 1 & \\ s_1(a) + v - \sum_{b \in A_2} s_2(b)\, p(a,b) - \varepsilon & \geq 0 & \forall a \in A_1 \\ \sum_{a \in A_1} s_1(a)\, p(a,b) & \geq v & \forall b \in A_2 \\ s_1(a) & \geq 0 & \forall a \in A_1 \\ \sum_{a \in A_1} s_1(a) & = 1 & \\ s_2(b) + v - \sum_{a \in A_1} s_1(a)\, p(a,b) - \varepsilon & \geq 0 & \forall b \in A_2 \end{array}$$

Fig. 2. Linear program for computing quasi-strict equilibria in zero-sum games

unique in both cases. We can thus construct a linear feasibility program that computes equilibrium strategies for both players by simply merging the sets of constraints and omitting the minimization and maximization objectives. Now, quasi-strict equilibrium requires that action a yields strictly more payoff than action b if and only if a is in the support and b is not. For a zero-sum game with value v this can be achieved by requiring that for every action $a \in A_1$ of player 1, $s_1(a) + v > \sum_{b \in A_2} s_2(a)\ p(a,b)$. If $s_1(a) = 0$ (a is not in the support), action a yields strictly less payoff than the game's value. If, on other hand, $s_1(a) > 0$ (a is in the support), these constraints do not impose any restrictions if the strategy profile is indeed an equilibrium with value v, which is ensured by the remaining constraints. Since strict inequalities are not allowed in linear programs, we introduce another variable ε to be maximized. Due to the existence of at least one quasi-strict equilibrium, we will always find a solution with positive ε, turning the weak inequality into a strict one. The resulting linear program is given in Figure 2. □

We proceed by showing that every symmetric zero-sum game contains a symmetric quasi-strict equilibrium. This result should be contrasted to Theorem 3 in Section 4 which establishes that this is *not* the case for symmetric two-player games in general.

Theorem 2. *Every symmetric two-player zero-sum game contains a symmetric quasi-strict equilibrium that can be computed using the linear program given in Figure 3.*

$$
\begin{array}{lll}
\multicolumn{3}{l}{\textbf{maximize } \varepsilon} \\
\multicolumn{3}{l}{\textbf{subject to}} \\
\sum_{b \in A_2} s(b)\, p(a,b) & \leq 0 & \forall a \in A_1 \\
s(b) & \geq 0 & \forall b \in A_2 \\
\sum_{b \in A_2} s(b) & = 1 & \\
s(a) - \sum_{b \in A_2} s(b)\, p(a,b) - \varepsilon & \geq 0 & \forall a \in A_1
\end{array}
$$

Fig. 3. Linear program for computing quasi-strict equilibria in symmetric zero-sum games

Proof. According to Theorem 1, the support of any quasi-strict equilibrium contains all actions that are played with positive probability in some equilibrium. Clearly, in symmetric games, these actions coincide for both players and any minimax probability distribution over these actions constitutes a symmetric equilibrium. Since both players can enforce their minimax value using the same strategy in a symmetric zero-sum game, the value of the game has to be zero. Given that the equilibrium strategies (s, s) have to be symmetric and that the value of the game is known, the linear program in Figure 2 can be significantly simplified, resulting in the linear program given in Figure 3. □

The linear program in Figure 3 can be used to directly compute the *essential set* of a dominance graph. The essential set is defined as the set of actions played with positive probability in some Nash equilibrium of the zero-sum game given by the (symmetric) adjacency matrix of a directed graph [8]. It follows from Theorem 1 that this is exactly the unique support of all quasi-strict equilibria.

4 Multi-Player Games

In games with three or more players the existence of a quasi-strict equilibrium is no longer guaranteed. However, there are very few examples in the literature for games without quasi-strict equilibria.[5]. An important question is of course which natural classes of games always contain a quasi-strict equilibrium. It has already been shown that this is not the case for the class of *single-winner* games which require that all outcomes are permutations of the payoff vector $(1, 0, \ldots, 0)$ [3].

In the following, we will analyze whether *symmetric* and *anonymous* games always admit a quasi-strict equilibrium. It turns out that self-anonymous games, and thus also anonymous games, need *not* possess a quasi-strict equilibrium. For this, consider the following three-player single-loser game where players Alice, Bob, and Charlie independently and simultaneously are to decide whether to raise their hand or not (for instance, in order to decide who has to take out the garbage). Alice loses if exactly one player raises his hand, whereas Bob loses if exactly two players raise their hands, and Charlie loses if either all or no players raise their hand. The matrix form of this self-anonymous game is depicted in

[5] To the best of our knowledge, there are four examples, all of which involve three players with two actions each [29, 16, 6, 3].

Figure 4. The game exhibits some peculiar phenomena, some of which may be attributed to the absence of quasi-strict equilibrium. For example, the security level of all players is 0.5 and the expected payoff in the *only* Nash equilibrium (which has Alice raise her hand and Charlie randomize with equal probability) is $(0.5, 0.5, 1)$. However, the minimax strategies of Alice and Bob are different from their equilibrium strategies, *i.e.*, they can *guarantee* their equilibrium payoff by *not* playing their respective equilibrium strategies.[6] Furthermore, the unique equilibrium is not quasi-strict, *i.e.*, Alice and Bob could as well play any other action without jeopardizing their payoff.

$(1,1,0)$	$(0,1,1)$
$(0,1,1)$	$(1,0,1)$

$(0,1,1)$	$(1,0,1)$
$(1,0,1)$	$(1,1,0)$

Fig. 4. Self-anonymous game without a quasi-strict equilibrium. Players 1, 2, and 3 choose rows, columns, and matrices, respectively. In the only Nash equilibrium of the game player 1 plays his second action, player 2 plays his first action, and player 3 randomizes over both his actions.

For symmetric and self-symmetric games, on the other hand, the picture appears to be different. Self-symmmetric games are a subclass of common-payoff games, where the payoff of all players is identical in every outcome. Starting from an outcome with maximum payoff p for all players, a quasi-strict equilibrium can be found by iteratively adding actions to the support by which a player, and thus all players, can obtain the same payoff p. We will extend this result by showing that the existence of quasi-strict equilibria also holds for *symmetric* games where each player has only two actions at his disposal. It follows from a theorem by Nash [21] that every symmetric game has a *symmetric* Nash equilibrium, *i.e.*, a Nash equilibrium where all players play the same strategy. Perhaps surprisingly, it may be the case that all quasi-strict equilibria of a symmetric game are *asymmetric*.

Theorem 3. *Every symmetric game with two actions for each player has a quasi-strict equilibrium. Such an equilibrium can be found in polynomial time.*

Proof. Let $\Gamma = (N, \{0,1\}^N, (p_i)_{i\in N})$ be a symmetric game. By Definition 2, there exist $2(n-1)$ numbers $p_{m\ell} \in \mathbb{R}$ such that for all i, $p_i(s) = p_{m\ell}$ whenever $s_i = \ell$ and m is the number of players playing action 1 in s_{-i}. We can further assume w.l.o.g. that $p_{00} = p_{01}$ and $p_{n-1,0} \geq p_{n-1,1}$, and that $p_{m0} \neq p_{m1}$ for some m. To see this, observe that Γ must possess a symmetric equilibrium s [21], which we can assume to be the pure strategy profile where all players play action 0 with probability 1. If all players played both of their actions with positive probability, this would directly imply quasi-strictness of s. Now, if one of the former two equations was not satisfied, then one of the two symmetric pure strategy

[6] Similar phenomena were also observed by Aumann [2].

profiles would be a quasi-strict equilibrium. If the latter condition were not to hold, there would exist a quasi-strict equilibrium where all players randomize between their actions.

We will now distinguish two different cases according to the relationship between p_{m0} and p_{m1} for the different values of m. First assume that there exists m such that $p_{m0} > p_{m1}$ and for all $m' < m$, $p_{m'0} = p_{m'1}$. We claim that in this case any strategy profile s in which $m-1$ players randomize between both actions and the remaining $n-m+1$ players play action 0 is a quasi-strict equilibrium. To see this, consider first any player $i \in N$ who randomizes between both of his actions. It is easily verified that for every action profile a which is played with positive probability in s_{-i} and in which exactly m' players play action 1, it must hold that $p_{m'0} = p_{m'1}$. On the other hand, consider any player $i \in N$ who plays action 0 with probability 1. Then, for any action profile a which is played with positive probability in s_{-i} and in which exactly m' players play action 1, it must hold that $p_{m'0} \geq p_{m'1}$, and there exists one such action profile for which the inequality is strict.

Now assume that there exists m such that $p_{m0} < p_{m1}$, and choose m' such that for all m'', $m < m'' < m'$, $p_{m''0} = p_{m''1}$, and either $p_{m'0} > p_{m'1}$ or $m' = n$. We claim that in this case any strategy profile where $n-m'$ players play action 0, m players play action 1, and the remaining $m'-m$ players randomize between both of their actions is a quasi-strict equilibrium of Γ. For this, again consider any player $i \in N$ who plays both actions with positive probability. It is easily verified that for every action profile a which is played with positive probability in s_{-i} and in which exactly m' players play action 1, it must hold that $p_{m'0} = p_{m'1}$. On the other hand, for any player $i \in N$ who plays action 0 with probability 1 and any action profile a which is played with positive probability in s_{-i} and in which exactly m' players play action 1, it must hold that $p_{m'0} \geq p_{m'1}$, and there exists one such action profile for which the inequality is strict. Finally, for any player $i \in N$ who plays action 1 with probability 1 and any action profile a which is played with positive probability in s_{-i} and in which exactly m' players play action 1, it must hold that $p_{m'0} \leq p_{m'1}$, and there exists one such action profile for which the inequality is strict.

Since a symmetric equilibrium of a symmetric game with a constant number of actions can be found in polynomial time [24], and since the proof of the first part of the theorem is constructive, the second part follows immediately. □

We leave it as an open problem whether all symmetric games contain a quasi-strict equilibrium. If the symmetrization procedure due to Gale et al. [9] can be extended to multi-player games while still preserving quasi-strictness, a counter-example could be constructed from one of the known examples of games without quasi-strict equilibria. Of course, in light of Theorem 3, the number of actions per player in such a counter-example has to be greater than two (and may very well be substantially greater than that).

We conclude by showing that deciding whether a given normal-form game contains a quasi-strict equilibrium is NP-complete.

	b_1 $\cdots$ $b_{\lvert V\rvert}$	b_0
a_1		$(0,0,0)$
$\vdots$	$(m_{ij}, e_{ij}, m_{ij})_{i,j\in V}$	$\vdots$
$a_{\lvert V\rvert}$		$(0,0,0)$
a_0	$(0,0,0)$ $\cdots$ $(0,0,0)$	$(0,1,0)$

c_1

	b_1	$\cdots$	$b_{\lvert V\rvert}$	b_0
a_1	$(0,0,K)$	$\cdots$	$(0,0,K)$	$(0,0,0)$
$\vdots$	$\vdots$	$\ddots$	$\vdots$	$\vdots$
$a_{\lvert V\rvert}$	$(0,0,K)$	$\cdots$	$(0,0,K)$	$(0,0,0)$
a_0	$(1,0,0)$	$\cdots$	$(1,0,0)$	$(0,0,0)$

c_2

Fig. 5. Three-player game Γ used in the proof of Theorem 4. Players 1, 2, and 3 choose rows columns, and matrices, respectively.

Theorem 4. *Deciding whether a game in normal-form possesses a quasi-strict equilibrium is NP-complete, even if there are just three players and a constant number of payoffs.*

Proof. Membership in NP is obvious. We can simply guess a strategy profile and verify that it is an equilibrium and that the payoff is strictly lower for all actions that are not played.

For *hardness*, we provide a reduction from the NP-complete problem CLIQUE [see, *e.g.*, 10] reminiscent to a construction used by McLennan and Tourky [18] to give simplified NP-hardness proofs for various problems related to Nash equilibria in bimatrix games. Given an undirected graph $G = (V, E)$ and a positive integer $k \leq |E|$, CLIQUE asks whether G contains a clique of size at least k, *i.e.*, a subset $V' \subseteq V$ such that $|V'| \geq k$ and for all $v, w \in V'$, $(v, w) \in E$. Given a particular CLIQUE instance $((V, E), k)$ with $V = \{1, \ldots, m\}$, we construct a game Γ with three players, actions $A_1 = \{\, a_i \mid i \in V \,\} \cup \{a_0\}$, $A_2 = \{\, b_i \mid i \in V \,\} \cup \{b_0\}$ and $A_3 = \{c_1, c_2\}$, and payoffs p_i illustrated in Figure 5. If player 3 plays c_1 and players 1 and 2 play a_i and b_j, respectively, for $i, j \in V$, payoffs are given by a matrix $(m_{ij})_{i,j\in V}$ defined according to G, and by the identity matrix $(e_{ij})_{i,j\in V}$, where

$$m_{ij} = \begin{cases} 1 & \text{if } (i,j) \in E \\ 0 & \text{if } i = j \\ -1 & \text{otherwise} \end{cases} \quad \text{and} \quad e_{ij} = \begin{cases} 1 & \text{if } i = j \\ 0 & \text{otherwise.} \end{cases}$$

If player 3 instead plays c_2, he obtains a payoff of $K = (2k-3)/2k$. We claim that Γ possesses a quasi-strict equilibrium if and only if there exists a clique of size at least k in G.

Assume there exists a clique $V' \subseteq V$, $|V'| \geq k$, and consider the strategy profile s with $s(c_1) = 1$, and $s(a_i) = s(b_i) = 1/|V'|$ if $i \in V'$ and $s(a_i) = s(b_i) = 0$ otherwise. By construction of Γ, for all $i \in V \cup \{0\}$, $p_2(s_{-2}, b_i) < p_2(s)$ whenever $s(a_i) = 0$. Furthermore, by maximality of V', $p_1(s_{-1}, a_i) < p_1(s)$ for all $i \notin V'$. Finally, $p_3(s) = (k-1)/k > (2k-3)/2k = p_3(s_{-3}, c_2)$. Thus, s is a quasi-strict equilibrium of Γ.

Now assume for contradiction that there is no clique of size at least k in G, and that s is a quasi-strict equilibrium of Γ. In equilibrium, for all $b, b' \in A_2$, we must have $p_2(s_{-2}, b) = p_2(s_{-2}, b')$ whenever $s(b) > 0$ and $s(b') > 0$, and thus, for all $a, a' \in A_1$, $s(a) = s(a')$ whenever $s(a) > 0$ and $s(a') > 0$. As a consequence, for s to be quasi-strict, $s(b_i) > 0$ whenever $s(a_i)$ for all $i \in V \cup \{0\}$. First consider the case where $s(c_1) > 0$. If $s(a_0) = s(b_0) = 1$, s cannot be quasi-strict for player 1. If on the other hand $s(a_i) > 0$ or $s(b_i) > 0$ for some $i \in V$, then there would have to be a set $V' \subseteq V$, $|V'| \geq k$, such that for all $i \in V$ with $s(a_i) > 0$ and all $j \in V'$, $j \neq i$, $p_1(a_i, b_j, c_1) = 1$. By construction of Γ, V' would be a clique of size at least k in G, a contradiction. Now consider the case where $s(c_2) = 1$. If $s(a_0) = 1$ or $s(b_0) = 1$, s is not quasi-strict for player 3. If, on the other hand, $s(a_i) > 0$ and $s(b_j) > 0$ for some $i, j \in V$, then player 1 could deviate to a_0 to get a higher payoff. This completes the proof. □

It follows that the problem of *finding* a quasi-strict equilibrium in games with more than two players is NP-hard (under polynomial-time Turing reductions), whereas no such statement is known for Nash equilibrium.

5 Conclusion

We investigated the computational properties of an attractive equilibrium refinement known as quasi-strict equilibrium. It turned out that quasi-strict equilibria in zero-sum games have a unique support and can be computed efficiently via linear programming. In games with more than two players, finding a quasi-strict equilibrium is NP-hard.

As pointed out in Section 1, classes of games that always admit a quasi-strict equilibrium, such as bimatrix games, are of vital importance to justify rational play based on elementary assumptions. We specifically looked at symmetric and anonymous games and found that self-anonymous games (and thus also anonymous games) may not contain a quasi-strict equilibrium while symmetric games with two actions for each player always possess a quasi-strict equilibrium. Other classes of multi-player games for which this question might be of interest include unilaterally competitive games, potential games, single-winner games where all players have positive security levels, and graphical games with bounded neighborhood.

References

1. Aumann, R.J.: Acceptable points in general n-person games. In: Tucker, A.W., Luce, R.D. (eds.) Contributions to the Theory of Games IV, Annals of Mathematics Studies, vol. 40, pp. 287–324. Princeton University Press, Princeton (1959)
2. Aumann, R.J.: On the non-transferable utility value: A comment on the Roth-Shafer examples. Econometrica 53(3), 667–678 (1985)
3. Brandt, F., Fischer, F., Harrenstein, P., Shoham, Y.: A game-theoretic analysis of strictly competitive multiagent scenarios. In: Veloso, M. (ed.) Proceedings of the 20th International Joint Conference on Artificial Intelligence (IJCAI), pp. 1199–1206 (2007)

4. Brandt, F., Fischer, F., Holzer, M.: Symmetries and the complexity of pure Nash equilibrium. In: Thomas, W., Weil, P. (eds.) STACS 2007. LNCS, vol. 4393, pp. 212–223. Springer, Heidelberg (2007)
5. Chen, X., Deng, X.: Settling the complexity of 2-player Nash-equilibrium. In: Proceedings of the 47th Symposium on Foundations of Computer Science (FOCS), pp. 261–272. IEEE Press, Los Alamitos (2006)
6. Cubitt, R., Sugden, R.: Rationally justifiable play and the theory of non-cooperative games. Economic Journal 104(425), 798–803 (1994)
7. Daskalakis, C., Papadimitriou, C.H.: Computing equilibria in anonymous games. In: Proceedings of the 48th Symposium on Foundations of Computer Science (FOCS), IEEE Computer Society Press, Los Alamitos (2007)
8. Dutta, B., Laslier, J.-F.: Comparison functions and choice correspondences. Social Choice and Welfare 16(4), 513–532 (1999)
9. Gale, D., Kuhn, H.W., Tucker, A.W.: On symmetric games. In: Kuhn, H.W., Tucker, A.W. (eds.) Contributions to the Theory of Games, vol. 1, pp. 81–87. Princeton University Press, Princeton (1950)
10. Garey, M.R., Johnson, D.S.: Computers and Intractability: A Guide to the Theory of NP-Completeness. W. H. Freeman, New York (1979)
11. Gottlob, G., Greco, G., Scarcello, F.: Pure Nash equilibria: Hard and easy games. Journal of Artificial Intelligence Research 24, 195–220 (2005)
12. Harsanyi, J.C.: Oddness of the number of equilibrium points: A new proof. International Journal of Game Theory 2, 235–250 (1973)
13. Jansen, M.J.M.: Regularity and stability of equilibrium points of bimatrix games. Mathematics of Operations Research 6(4), 530–550 (1981)
14. Jansen, M.J.M.: Regular equilibrium points of bimatrix points. OR Spektrum 9(2), 82–92 (1987)
15. Jurg, A.P., Jansen, M.J.M., Potters, J.A.M., Tijs, S.H.: A symmetrization for finite two-person games. ZOR – Methods and Models of Operations Research 36(2), 111–123 (1992)
16. Kojima, M., Okada, A., Shindoh, S.: Strongly stable equilibrium points of N-person noncooperative games. Mathematics of Operations Research 10(4), 650–663 (1985)
17. Luce, R.D., Raiffa, H.: Games and Decisions: Introduction and Critical Survey. Wiley, Chichester (1957)
18. McLennan, A., Tourky, R.: Simple complexity from imitation games (unpublished manuscript, 2005)
19. Miltersen, P.B., Sørensen, T.B.: Computing sequential equilibria for two-player games. In: Proceedings of the 17th Annual ACM-SIAM Symposium on Discrete Algorithms (SODA), pp. 107–116. SIAM, Philadelphia (2006)
20. Miltersen, P.B., Sørensen, T.B.: Fast algorithms for finding proper strategies in game trees. In: Proceedings of the 19th Annual ACM-SIAM Symposium on Discrete Algorithms (SODA), pp. 874–883. SIAM, Philadelphia (2008)
21. Nash, J.F.: Non-cooperative games. Annals of Mathematics 54(2), 286–295 (1951)
22. Norde, H.: Bimatrix games have quasi-strict equilibria. Mathematical Programming 85, 35–49 (1999)
23. Norde, H., Potters, J., Reijnierse, H., Vermeulen, D.: Equilibrium selection and consistency. Games and Economic Behavior 12(2), 219–225 (1996)
24. Papadimitriou, C.H., Roughgarden, T.: Computing equilibria in multi-player games. In: Proceedings of the 16th Annual ACM-SIAM Symposium on Discrete Algorithms (SODA), pp. 82–91. SIAM, Philadelphia (2005)
25. Peleg, B., Tijs, S.: The consistency principle for games in strategic form. International Journal of Game Theory 25, 13–34 (1996)

26. Quesada, A.: Another impossibility result for normal form games. Theory and Decision 52, 73–80 (2002)
27. Rozenfeld, O., Tennenholtz, M.: Strong and correlated strong equilibria in monotone congestion games. In: Spirakis, P.G., Mavronicolas, M., Kontogiannis, S.C. (eds.) WINE 2006. LNCS, vol. 4286, pp. 74–86. Springer, Heidelberg (2006)
28. Squires, D.: Impossibility theorems for normal form games. Theory and Decision 44, 67–81 (1998)
29. van Damme, E.: Refinements of the Nash Equilibrium Concept. Springer, Heidelberg (1983)
30. van Damme, E.: Stability and Perfection of Nash Equilibria, 2nd edn. Springer, Heidelberg (1991)
31. von Neumann, J.: Zur Theorie der Gesellschaftspiele. Mathematische Annalen 100, 295–320 (1928)

The Price of Stochastic Anarchy

Christine Chung[1], Katrina Ligett[2,⋆], Kirk Pruhs[1,⋆⋆], and Aaron Roth[2]

[1] Department of Computer Science, University of Pittsburgh,
{chung,kirk}@cs.pitt.edu
[2] Department of Computer Science, Carnegie Mellon University,
{katrina,alroth}@cs.cmu.edu

Abstract. We consider the solution concept of stochastic stability, and propose the *price of stochastic anarchy* as an alternative to the *price of (Nash) anarchy* for quantifying the cost of selfishness and lack of coordination in games. As a solution concept, the Nash equilibrium has disadvantages that the set of stochastically stable states of a game avoid: unlike Nash equilibria, stochastically stable states are the result of natural dynamics of computationally bounded and decentralized agents, and are resilient to small perturbations from ideal play. The price of stochastic anarchy can be viewed as a smoothed analysis of the price of anarchy, distinguishing equilibria that are resilient to noise from those that are not. To illustrate the utility of stochastic stability, we study the load balancing game on unrelated machines. This game has an unboundedly large price of Nash anarchy even when restricted to two players and two machines. We show that in the two player case, the price of stochastic anarchy is 2, and that even in the general case, the price of stochastic anarchy is bounded. We conjecture that the price of stochastic anarchy is $O(m)$, matching the price of strong Nash anarchy without requiring player coordination. We expect that stochastic stability will be useful in understanding the relative stability of Nash equilibria in other games where the worst equilibria seem to be inherently brittle.

1 Introduction

Quantifying the *price of (Nash) anarchy* is one of the major lines of research in algorithmic game theory. Indeed, one fourth of the authoritative algorithmic game theory text edited by Nisan et al. [20] is wholly dedicated to this topic. But the Nash equilibrium solution concept has been widely criticized [15,4,9,10]. First, it is a solution characterization without a road map for how players might arrive at such a solution. Second, at Nash equilibria, players are unrealistically assumed to be perfectly rational, fully informed, and infallible. Third, computing Nash equilibria is PPAD-hard for even 2-player, n-action games [6], and it is therefore considered very unlikely that there exists a polynomial time algorithm to compute a Nash equilibrium even in a centralized manner. Thus, it is unrealistic to assume that selfish agents in general games will converge

⋆ Partially supported by an AT&T Labs Graduate Fellowship and an NSF Graduate Research Fellowship.

⋆⋆ Supported in part by NSF grants CNS-0325353, CCF-0448196, CCF-0514058 and IIS-0534531.

B. Monien and U.-P. Schroeder (Eds.): SAGT 2008, LNCS 4997, pp. 303–314, 2008.

precisely to the Nash equilibria of the game, or that they will necessarily *converge* to anything at all. In addition, the price of Nash anarchy metric comes with its own weaknesses; it blindly uses the worst case over all Nash equilibria, despite the fact that some equilibria are more resilient than others to perturbations in play.

Considering these drawbacks, computer scientists have paid relatively little attention to if or how Nash equilibria will in fact be reached, and even less to the question of which Nash equilibria are more likely to be played in the event players do converge to Nash equilibria. To address these issues, we employ the stochastic stability framework from evolutionary game theory to study simple dynamics of computationally efficient, imperfect agents. Rather than defining a-priori states such as Nash equilibria, which might not be reachable by natural dynamics, the stochastic stability framework allows us to define a natural dynamic, and from it derive the stable states. We define the *price of stochastic anarchy* to be the ratio of the worst stochastically stable solution to the optimal solution. The stochastically stable states of a game may, but do not necessarily, contain all Nash equilibria of the game, and so the price of stochastic anarchy may be strictly better than the price of Nash anarchy. In games for which the stochastically stable states are a subset of the Nash equilibria, studying the ratio of the worst stochastically stable state to the optimal state can be viewed as a smoothed analysis of the price of anarchy, distinguishing Nash equilibria that are brittle to small perturbations in perfect play from those that are resilient to noise.

The evolutionary game theory literature on *stochastic stability* studies n-player games that are played repeatedly. In each round, each player observes her action and its outcome, and then uses simple rules to select her action for the next round based only on her size-restricted memory of the past rounds. In any round, players have a small probability of deviating from their prescribed decision rules. The state of the game is the contents of the memories of all the players. The *stochastically stable states* in such a game are the states with non-zero probability in the limit of this random process, as the probability of error approaches zero. The play dynamics we employ in this paper are the imitation dynamics studied by Josephson and Matros [16]. Under these dynamics, each player imitates the strategy that was most successful for her in recent memory.

To illustrate the utility of stochastic stability, we study the price of stochastic anarchy of the unrelated load balancing game [2,1,11]. To our knowledge, we are the first to quantify the loss of efficiency in any system when the players are in stochastically stable equilibria. In the load balancing game on unrelated machines, even with only two players and two machines, there are Nash equilibria with arbitrarily high cost, and so the price of Nash anarchy is unbounded. We show that these equilibria are inherently brittle, and that for two players and two machines, the price of stochastic anarchy is 2. This result matches the strong price of anarchy [1] without requiring coordination (at strong Nash equilibria, players have the ability to coordinate by forming coalitions). We further show that in the general n-player, m-machine game, the price of stochastic anarchy is bounded. More precisely the price of stochastic anarchy is upper bounded by the nmth n-step Fibonacci number. We also show that the price of stochastic anarchy is at least $m + 1$.

Our work provides new insight into the equilibria of the load balancing game. Unlike some previous work on dynamics for games, our work does not seek to propose practical dynamics with fast convergence; rather, we use simple dynamics as a tool for understanding the inherent relative stability of equilibria. Instead of relying on player coordination to avoid the Nash equilibria with unbounded cost (as is done in the study of strong equilibria), we show that these bad equilibria are inherently unstable in the face of occasional uncoordinated mistakes. We conjecture that the price of stochastic anarchy is closer to the linear lower bound, paralleling the price of strong anarchy.

In light of our results, we believe the techniques in this paper will be useful for understanding the relative stability of Nash equilibria in other games for which the worst equilibria are brittle. Indeed, for a variety of games in the price of anarchy literature, the worst Nash equilibria of the lower bound instances are not stochastically stable.

1.1 Related Work

We give a brief survey of related work in three areas: alternatives to Nash equilibria as a solution concept, stochastic stability, and the unrelated load balancing game.

Recently, several papers have noted that the Nash equilibrium is not always a suitable solution concept for computationally bounded agents playing in a repeated game, and have proposed alternatives. Goemans et al. [15] study players who sequentially play myopic best responses, and quantify the *price of sinking* that results from such play. Fabrikant and Papadimitriou [9] propose a model in which agents play restricted finite automata. Blum et al. [4,3] assume only that players' action histories satisfy a property called *no regret*, and show that for many games, the resulting social costs are no worse than those guaranteed by price of anarchy results.

Although we believe this to be the first work studying stochastic stability in the computer science literature, computer scientists have recently employed other tools from evolutionary game theory. Fisher and Vöcking [13] show that under replicator dynamics in the routing game studied by Roughgarden and Tardos [22], players converge to Nash. Fisher et al. [12] went on to show that using a simultaneous adaptive sampling method, play converges quickly to a Nash equilibrium. For a thorough survey of algorithmic results that have employed or studied other evolutionary game theory techniques and concepts, see Suri [23].

Stochastic stability and its adaptive learning model as studied in this paper were first defined by Foster and Young [14], and differ from the standard game theory solution concept of evolutionarily stable strategies (ESS). ESS are a refinement of Nash equilibria, and so do not always exist, and are not necessarily associated with a natural play dynamic. In contrast, a game always has stochastically stable states that result (by construction) from natural dynamics. In addition, ESS are resilient only to single shocks, whereas stochastically stable states are resilient to persistent noise.

Stochastic stability has been widely studied in the economics literature (see, for example, [24,17,19,5,7,21,16]). We discuss in Sect. 2 concepts from this body of literature that are relevant to our results. We recommend Young [25] for a readable introduction to stochastic stability, its adaptive learning model, and some related results. Our work differs from prior work in stochastic stability in that it is the first to quantify the social utility of stochastically stable states, the *price of stochastic anarchy*.

We also note a connection between the stochastically stable states of the game and the sinks of a game, recently introduced by Goemans et al. as another way of studying the dynamics of computationally bounded agents. In particular, the stochastically stable states of a game under the play dynamics we consider correspond to a subset of the sink equilibria, and so provide a framework for identifying the stable sink equilibria. In potential games, the stochastically stable states of the play dynamics we consider correspond to a subset of the Nash equilibria, thus providing a method for identifying which of these equilibria are stable.

In this paper, we study the price of stochastic anarchy in load balancing. Even-Dar et al. [8] show that when playing the load balancing game on unrelated machines, any turn-taking improvement dynamics converge to Nash. Andelman et al. [1] observe that the price of Nash anarchy in this game is unbounded and they show that the strong price of anarchy is linear in the number of machines. Fiat et al. [11] tighten their upper bound to match their lower bound at a strong price of anarchy of exactly m.

2 Model and Background

We now formalize (from Young [24]) the adaptive play model and the definition of stochastic stability. We then formalize the play dynamics that we consider. We also provide in this section the results from the stochastic stability literature that we will later use to obtain our results.

2.1 Adaptive Play and Stochastic Stability

Let $G = (X, \pi)$ be a game with n players, where $X = \prod_{j=1}^{n} X_i$ represents the strategy sets X_i for each player i, and $\pi = \prod_{j=1}^{n} \pi_i$ represents the payoff functions $\pi_i : X \rightarrow \mathbb{R}$ for each player. G is played repeatedly for successive time periods $t = 1, 2, \ldots$, and at each time step t, player i plays some action $s_i^t \in X_i$. The collection of all players' actions at time t defines a play profile $S^t = (S_1^t, S_2^t, \ldots, S_n^t)$. We wish to model computationally efficient agents, and so we imagine that each agent has some finite memory of size z, and that after time step t, all players remember a history consisting of a sequence of play profiles $h^t = (S^{t-z+1}, S^{t-z+2}, \ldots, S^t) \in (X)^z$.

We assume that each player i has some efficiently computable function $p_i : (X)^z \times X_i \rightarrow \mathbb{R}$ that, given a particular history, induces a sampleable probability distribution over actions (for all players i and histories h, $\sum_{a \in X_i} p_i(h, a) = 1$). We write p for $\prod_i p_i$. We wish to model imperfect agents who make mistakes, and so we imagine that at time t each player i plays according to p_i with probability $1 - \epsilon$, and with probability ϵ plays some action in X_i uniformly at random.[1] That is, for all players i, for all actions $a \in X_i$, $\Pr[s_i^t = a] = (1 - \epsilon) p_i(h^t, a) + \frac{\epsilon}{|X_i|}$. The dynamics we have described define a Markov process $P^{G,p,\epsilon}$ with finite state space $H = (X)^z$ corresponding to the finite histories. For notational simplicity, we will write the Markov process as P^ϵ when there is no ambiguity.

The potential successors of a history can be obtained by observing a new play profile, and "forgetting" the least recent play profile in the current history.

[1] The mistake probabilities need not be uniform random—all that we require is that the distribution has support on all actions in X_i.

Definition 2.1. *For any $S' \in X$, A history $h' = (S^{t-z+2}, S^{t-z+3}, \ldots, S^t, S')$ is a* successor *of history $h^t = (S^{t-z+1}, S^{t-z+2}, \ldots, S^t)$.*

The Markov process P^ϵ has transition probability $p^\epsilon_{h,h'}$ of moving from state $h = (S^1, \ldots, S^z)$ to state $h' = (T^1, \ldots, T^z)$:

$$p^\epsilon_{h,h'} = \begin{cases} \prod_{i=1}^n (1-\epsilon)\, p_i(h, T_i^z) + \frac{\epsilon}{|X_i|} & \text{if } h' \text{ is a successor of h;} \\ 0 & \text{otherwise.} \end{cases}$$

We will refer to P^0 as the unperturbed Markov process. Note that for $\epsilon > 0$, $p^\epsilon_{h,h'} > 0$ for every history h and successor h', and that for any two histories h and $\hat{h}$ not necessarily a successor of h, there is a series of z histories $h_1, \ldots, h_z$ such that $h_1 = h$, $h_z = \hat{h}$, and for all $1 < i \leq z$, h_i is a successor of h_{i-1}. Thus there is positive probability of moving between any h and any $\hat{h}$ in z steps, and so P^ϵ is irreducible. Similarly, there is a positive probability of moving between any h and any $\hat{h}$ in $z+1$ steps, and so P^ϵ is aperiodic. Therefore, P^ϵ has a unique stationary distribution μ^ϵ.

The stochastically stable states of a particular game and player dynamics are the states with nonzero probability in the limit of the stationary distribution.

Definition 2.2 (Foster and Young [14]). *A state h is* stochastically stable *relative to P^ϵ if $\lim_{\epsilon \to 0} \mu^\epsilon(h) > 0$.*

Intuitively, we should expect a process P^ϵ to spend almost all of its time at its stochastically stable states when ϵ is small.

When a player i plays at random rather than according to p_i, we call this a mistake.

Definition 2.3 (Young [24]). *Suppose $h' = (S^{t-z+1}, \ldots, S^t)$ is a successor of h. A* mistake *in the transition between h and h' is any element S_i^t such that $p_i(h, S_i^t) = 0$. Note that mistakes occur with probability $\leq \epsilon$.*

We can characterize the number of mistakes required to get from one history to another.

Definition 2.4 (Young [24]). *For any two states h, h', the* resistance *$r(h, h')$ is the minimum total number of mistakes involved in the transition $h \to h'$ if h' is a successor of h. If h' is not a successor of h, then $r(h, h') = \infty$.*

Note that the transitions of zero resistance are exactly those that occur with positive probability in the unperturbed Markov process P^0.

Definition 2.5. *We refer to the sinks of P^0 as* recurrent classes. *In other words, a* recurrent class *of P^0 is a set of states $C \subseteq H$ such that any state in C is reachable from any other state in C and no state outside C is accessible from any state inside C.*

We may view the state space H as the vertex set of a directed graph, with an edge from h to h' if h' is a successor of h, with edge weight $r(h, h')$.

Observation 2.6. *We observe that the recurrent classes $H_1, H_2, \ldots$, where each $H_i \subseteq H$, have the following properties:*

1. *From every vertex $h \in H$, there is a path of cost 0 to one of the recurrent classes.*
2. *For each H_i and for every pair of vertices $h, h' \in H_i$, there is a path of cost 0 between h and h'.*
3. *For each H_i, every edge (h, h') with $h \in H_i, h' \notin H_i$ has positive cost.*

Let $r_{i,j}$ denote the cost of the shortest path between H_i and H_j in the graph described above. We now consider the complete directed graph $\mathcal{G}$ with vertex set $\{H_1, H_2, \ldots\}$ in which the edge (H_i, H_j) has weight $r_{i,j}$. Let T_i be a directed minimum-weight spanning in-tree of $\mathcal{G}$ rooted at vertex H_i. (An in-tree is a directed tree where each edge is oriented toward the root.) The *stochastic potential* of H_i is defined to be the sum of the edge weights in T_i.

Young proves the following theorem characterizing stochastically stable states:

Theorem 2.7 (Young [24]). *In any n-player game G with finite strategy sets and any set of action distributions p, the stochastically stable states of $P^{G,p,\epsilon}$ are the recurrent classes of minimum stochastic potential.*

2.2 Imitation Dynamics

In this paper, we study agents who behave according to a slight modification of the imitation dynamics introduced by Josephson and Matros [16]. (We note that this modification is of no consequence to the results of Josephson and Matros [16] that we present below.) Player i using imitation dynamics parameterized by $\sigma \in \mathbb{N}$ chooses his action at time $t + 1$ according to the following mechanism:

1. Player i selects a set Y of σ play profiles uniformly at random from the z profiles in history h_t.
2. For each play profile $S \in Y$, i recalls the payoff $\pi_i(S)$ he obtained from playing action S_i.
3. Player i plays the action among these that corresponds to his highest payoff; that is, he plays the ith component of $\text{argmax}_{S \in Y}\, \pi_i(S)$. In the case of ties, he plays a highest-payoff action at random.

The value σ is a parameter of the dynamics that is taken to be $n \leq \sigma \leq z/2$. These dynamics can be interpreted as modeling a situation in which at each time step, players are chosen at random from a pool of identical players, who each played in a subset of the last z rounds. The players are computationally simple, and so do not counterspeculate the actions of their opponents, instead playing the action that has worked the best for them in recent memory.

We will say that a history h is *monomorphic* if the same action profile S has been repeated for the last z rounds: $h = (S, S, \ldots, S)$. Josephson and Matros [16] prove the following useful fact:

Proposition 2.8. *A set of states is a recurrent class of the imitation dynamics if and only if it is a singleton set consisting of a monomorphic state.*

Since the stochastically stable states are a subset of the recurrent classes, we can associate with each stochastically stable state $h = (S, \ldots, S)$ the unique action profile S it contains. This allows us to now define the price of stochastic anarchy with respect to imitation dynamics. For brevity, we will refer to this throughout the paper as simply the price of stochastic anarchy.

Definition 2.9. *Given a game $G = (X, \pi)$ with a social cost function $\gamma : X \to \mathbb{R}$, the* price of stochastic anarchy *of G is equal to* $\max \frac{\gamma(S)}{\gamma(\mathbf{OPT})}$, *where* $\mathbf{OPT}$ *is the play profile that minimizes γ and the* max *is taken over all play profiles S such that $h = (S, \ldots, S)$ is stochastically stable.*

Given a game G, we define the *better response graph* of G: The set of vertices corresponds to the set of action profiles of G, and there is an edge between two action profiles S and S' if and only if there exists a player i such that S' differs from S only in player i's action, and player i does not decrease his utility by unilaterally deviating from S_i to S'_i. Josephson and Matros [16] prove the following relationship between this better response graph and the stochastically stable states of a game:

Theorem 2.10. *If $\mathbb{V}$ is the set of stochastically stable states under imitation dynamics, then $V = \{S : (S, \ldots, S) \in \mathbb{V}\}$ is either a strongly connected component of the better response graph of G, or a union of strongly connected components.*

Goemans et al. [15] introduce the notion of sink equilibria and a corresponding notion of the "price of sinking", which is the ratio of the social welfare of the worst sink equilibrium to that of the social optimum. We note that the strongly connected components of the better response graph of G correspond to the sink equilibria (under sequential better-response play) of G, and so Theorem 2.10 implies that the stochastically stable states under imitation dynamics correspond to a subset of the sinks of the better response graph of G, and we get the following corollary:

Corollary 2.11. *The price of stochastic anarchy of a game G under imitation dynamics is at most the price of sinking of G.*

2.3 Load Balancing: Game Definition and Price of Nash Anarchy

An instance of the load balancing game on unrelated machines is defined by a set of n players and m machines $M = \{M_1, \ldots, M_m\}$. The action space for each player is $X_i = M$. Each player i has some cost $c_{i,j}$ on machine j. Denote the cost of machine M_j for action profile S by $C_j(S) = \sum_{\{i|S_i=M_j\}} c_{i,j}$. Each player i has utility function $\pi_i(S) = -C_{s_i}(S)$. The social cost of an action profile S is $\gamma(S) = \max_{j\in M} C_j(S)$. We define $\mathbf{OPT}$ to be the action profile that minimizes social cost: $\mathbf{OPT} = \mathrm{argmin}_{S\in X}\, \gamma(S)$. Without loss of generality, we will always normalize so that $\gamma(\mathbf{OPT}) = 1$.

The coordination ratio of a game (also known as the price of anarchy) was introduced by Koutsoupias and Papadimitriou [18], and is intended to quantify the loss of efficiency due to selfishness and the lack of coordination among rational agents. Given a game G and a social cost function γ, it is simple to quantify the $\mathbf{OPT}$ game state S: $\mathbf{OPT} = \mathrm{argmin}\, \gamma(S)$. It is less clear how to model rational selfish agents. In most prior work it has been assumed that selfish agents play according to a Nash equilibrium, and the price of anarchy has been defined as the ratio of the cost of the worst (pure strategy) Nash state to $\mathbf{OPT}$. In this paper, we refer to this measure as the price of Nash anarchy, to distinguish it from the price of stochastic anarchy, which we defined in Sect. 2.2.

Definition 2.12. *For a game G with a set of Nash equilibrium states $\mathcal{E}$, the* price of (Nash) anarchy *is* $\max_{S\in\mathcal{E}} \frac{\gamma(S)}{\gamma(\mathbf{OPT})}$.

We show here that even with only two players and two machines, the load balancing game on unrelated machines has a price of Nash anarchy that is unbounded by any function of m and n. Consider the two-player, two-machine game with $c_{1,1} = c_{2,2} = 1$ and $c_{1,2} = c_{2,1} = 1/\delta$, for some $0 < \delta < 1$. Then the play profile $\mathbf{OPT} = (M_1, M_2)$ is a Nash equilibrium with cost 1. However, observe that the profile $S^* = (M_2, M_1)$ is also a Nash equilibrium, with cost $1/\delta$ (since by deviating, players can only increase their cost from $1/\delta$ to $1/\delta + 1$). The price of anarchy of the load balancing game is therefore $1/\delta$, which can be unboundedly large, although $m = n = 2$.

3 Upper Bound on Price of Stochastic Anarchy

The load balancing game is an ordinal potential game [8], and so the sinks of the better-response graph correspond to the pure strategy Nash equilibria. We therefore have by Corollary 2.11 that the stochastically stable states are a subset of the pure strategy Nash equilibria of the game, and the price of stochastic anarchy is at most the price of anarchy. We have noted that even in the two-person, two-machine load balancing game, the price of anarchy is unbounded (even for pure strategy equilibria). In this section we give upper bounds on the price of stochastic anarchy for both the two-player two-machine case and the general n-player m-machine game.

Theorem 3.1. *In the two-player, two-machine load balancing game on unrelated machines, the price of stochastic anarchy is 2.*

The proof of the above theorem can be found in the full version of this paper.

Theorem 3.2. *The general load balancing game on unrelated machines has price of stochastic anarchy bounded by a function Ψ depending only on n and m, and $\Psi(n, m) \leq m \cdot F_{(n)}(nm + 1)$, where $F_{(n)}(i)$ denotes the ith n-step Fibonacci number.*[2]

To prove this upper bound, we show that any solution worse than our upper bound cannot be stochastically stable. To show this impossibility, we take any arbitrary solution worse than our upper bound and show that there must always be a minimum cost in-tree in $\mathcal{G}$ rooted at a different solution that has strictly less cost than the minimum cost in-tree rooted at that solution. We then apply Proposition 2.8 and Theorem 2.7. The proof proceeds by a series of lemmas.

Definition 3.3. *For any monomorphic Nash state $h = (S, \ldots, S)$, let the* Nash Graph *of h be a directed graph with vertex set M and directed edges (M_i, M_j) if there is some player i with $S_i = M_i$ and $\mathbf{OPT}_i = M_j$. Let the* closure *$\bar{M}_i$ of machine M_i, be the set of states reachable from M_i by following 0 or more edges of the Nash graph.*

---

[2] $F_{(n)}(i) = \begin{cases} 1 & \text{if } i \leq n; \\ \sum_{j=i-n}^{i} F_{(n)}(j) & \text{otherwise.} \end{cases}$ $F_{(n)}(i) \in o(2^i)$ for any fixed n.

Lemma 3.4. *In any monomorphic Nash state $h = (S, \ldots, S)$, if there is a machine M_i such that $C_i(S) > m$, then every machine $M_j \in \bar{M}_i$ has cost $C_j(S) > 1$.*

Proof. Suppose this were not the case, and there exists an $M_j \in \bar{M}_i$ with $C_j(S) \leq 1$. Since $M_j \in \bar{M}_i$, there exists a simple path $(M_i = M_1, M_2, \ldots, M_k = M_j)$ with $k \leq m$. Since S is a Nash equilibrium, it must be the case that $C_{k-1}(S) \leq 2$ because by the definition of the Nash graph, the directed edge from M_{k-1} to M_k implies that there is some player i with $S_i = M_{k-1}$, but $\mathbf{OPT}_i = M_k$. Since $1 = \gamma(\mathbf{OPT}) \geq C_k(\mathbf{OPT}) \geq c_{i,k}$, if player i deviated from his action in Nash profile S to $S'_i = M_k$, he would experience cost $C_k(S) + c_{i,k} \leq 1 + 1 = 2$. Since he cannot benefit from deviating (by definition of Nash), it must be that his cost in S, $C_{k-1}(S) \leq 2$. By the same argument, it must be that $C_{k-2}(S) \leq 3$, and by induction, $C_1(S) \leq k \leq m$. □

Lemma 3.5. *For any monomorphic Nash state $h = (S, \ldots, S) \in \mathcal{G}$ with $\gamma(S) > m$, there is an edge from h to some $g = (T, \ldots, T)$ where $\gamma(T) \leq m$ with edge cost $\leq n$ in $\mathcal{G}$.*

Proof. Let $D = \{M_j : C_i(S) \geq m\}$, and define the closure of D, $\bar{D} = \bigcup_{M_i \in D} \bar{M}_i$. Consider the successor state h' of h that results when every player i such that $S_i^t \in \bar{D}$ makes a mistake and plays on their OPT machine $S_i^{t+1} = \mathbf{OPT}_i$, and all other players do not make a mistake and continue to play $S_i^{t+1} = S_i^t$. Note that by the definition of $\bar{D}$, for $M_j \in \bar{D}$, for all players i playing machine j in S, $\mathbf{OPT}_i \in \bar{D}$. Let $T = S^{t+1}$. Then for all j such that $M_j \in \bar{D}$, $C_j(T) \leq 1$, since $C_j(T) \leq C_j(\mathbf{OPT}) \leq 1$. To see this, note that for every player i such that $S_i^t = M_j \in \bar{D}$, $S_i^{t+1} = M_j$ if and only if $\mathbf{OPT}_i = M_j$. Similarly, for every player i such that $S_i^{t+1} = M_j \in \bar{D}$ but $S_i^t \neq M_j$, $\mathbf{OPT}_i = M_j$, and so for each machine $M_j \in \bar{D}$, the agents playing on M_j in T are a subset of those playing on M_j at $\mathbf{OPT}$. Note that by Lemma 3.4, for all $M_j \in \bar{D}$, $C_j(S) > 1$. Therefore, for every agent i with $S_i^t \in \bar{D}$, $\pi_i(T) > \pi_i(S)$, and so for $h'' = (S, \ldots, S, T, T)$ a successor of h', $r(h', h'') = 0$. Reasoning in this way, there is a path of zero resistance from h' to $g = (T, \ldots, T)$. We have therefore exhibited a path between h and g that involves only $|\{i : S_i^t \in \bar{D}\}| \leq n$ mistakes. Finally, we observe that if $M_j \in \bar{D}$ then $C_j(T) \leq 1$, and by construction, if $M_j \notin \bar{D}$, then $C_j(T) = C_j(S) < m$, since as noted above $M_j \notin \bar{D}$ implies that the players playing M_j in S are the same set playing M_j in T. Thus, we have $\gamma(T) \leq m$, which completes the proof. □

Lemma 3.6. *Let $h = (S, \ldots, S) \in \mathcal{G}$ be any monomorphic state with $\gamma(S) \leq m$. Any path in $\mathcal{G}$ from h to a monomorphic state $h' = (S', \ldots, S') \in \mathcal{G}$ where $\gamma(h') > m \cdot F_{(n)}(mn + 1)$ must contain an edge with cost $\geq \sigma$, where $F_{(n)}(i)$ denotes the ith n-step Fibonacci number.*

Proof. Suppose there were some directed path $\mathcal{P}$ in $\mathcal{G}$ $(h = h_1, h_2, \ldots, h_l = h')$ such that all edge costs were less than σ. We will imagine assigning costs to players on machines adversarially: for a player i on machine M_j, we will consider $c_{i,j}$ to be undefined until play reaches a monomorphic state h_k in which he occupies machine j, at which point we will assign $c_{i,j}$ to be the highest value consistent with his path from h_{k-1} to h_k. Note that since initially $\gamma(S) \leq m$, we must have for all $i \in N$, $c_{i,S_i} \leq m = mF_{(n)}(n)$.

There are mn costs $c_{i,j}$ that we may assign, and we have observed that our first n assignments have taken values $\leq mF_{(n)}(n) = mF_{(n)}(1)$. We will assume inductively that our k^{th} assignment takes value at most $mF_{(n)}(k)$. Let $h_k = (T, \ldots, T)$ be the last monomorphic state in $\mathcal{P}$ such that only k cost assignments have been made, and $h_{k+1} = (T', \ldots, T')$ be the monomorphic state at which the $k + 1^{st}$ cost assignment is made for some player i on machine M_j. Since by assumption, fewer than σ mistakes are made in the transition $h_k \to h_{k+1}$, it must be that $c_{i,j} \leq C_{T_i}(T)$; that is, $c_{i,j}$ can be no more than player i's experienced cost in state T. If this were not so, player i would not have continued playing on machine j in T' without additional mistakes, since with fewer than σ mistakes, any sample of size σ would have contained an instance of T which would have yielded higher payoff than playing on machine j. Note however that the cost of any machine M_j in T is at most:

$$C_j(T) \leq \sum_{i:c_{i,j} \neq \text{ undefined}} c_{i,j} \leq \sum_{i=0}^{n-1} mF_{(n)}(k-i) = mF_{(n)}(k+1)$$

where the inequality follows by our inductive assumption. We have therefore shown that the kth cost assigned is at most $mF_{(n)}(k)$, and so the claim follows since there are at most nm costs $c_{i,j}$ that may be assigned, and the cost on any machine in S' is at most the sum of the n highest costs. □

Proof (of Theorem 3.2). Given any state $h = (S, \ldots, S) \in \mathcal{G}$ where $\gamma(S) > m \cdot F_{(n)}(mn+1)$, we can exhibit a state $f = (U, U, \ldots, U)$ with lower stochastic potential than h such that $\gamma(U) \leq m \cdot F_{(n)}(nm+1)$ as follows.

Consider the minimum weight spanning in-tree T_h of $\mathcal{G}$ rooted at h. We will use it to construct a spanning in-tree T_f rooted at a state f as follows: We add an edge of cost at most n from h to some state $g = (T, \ldots, T)$ such that $\gamma(T) \leq m$ (such an edge is guaranteed to exist by Lemma 3.5). This induces a cycle through h and g. To correct this, we remove an edge on the path from g to h in T_h of cost $\geq \sigma$ (such an edge is guaranteed to exist by Lemma 3.6). Since this breaks the newly induced cycle, we now have a spanning in-tree T_f with root $f = (U, U, \ldots, U)$ such that $\gamma(U) \leq m \cdot F_{(n)}(mn+1)$. Since the added edge has lower cost than the removed edge, T_f has lower cost than T_h, and so f has lower stochastic potential than h.

Since the stochastically stable states are those with minimum stochastic potential by Theorem 2.7 and Proposition 2.8, we have proven that h is not stochastically stable. □

4 Lower Bound on Price of Stochastic Anarchy

In this section, we show that the price of stochastic anarchy for load balancing is at least m, the price of strong anarchy. All proofs can be found in the full version of this paper.

Theorem 4.1. *The price of stochastic anarchy of the load balancing game on unrelated machines is at least m.*

We present the lower bound instance for $m = 4$ in Fig. 1(a). Here, the entry corresponding to player i and machine M_j represents the cost $c_{i,j}$. The δs represent some sufficiently small positive value and the ∞s can be any sufficiently large value.

	M_1	M_2	M_3	M_4
1	1	$1-\delta$	∞	∞
2	$2-2\delta$	1	$2-3\delta$	∞
3	$3-4\delta$	∞	1	$3-5\delta$
4	$4-6\delta$	∞	∞	1

(a)

	M_1	M_2	M_3	M_4
1	1	1	∞	$4-3\delta$
2	$2-\delta$	1	$2-\delta$	∞
3	$3-2\delta$	$3-2\delta$	1	$3-2\delta$
4	$4-3\delta$	$5-4\delta$	∞	1

(b)

Fig. 1.

More complicated examples like Fig. 1(b) show that the price of stochastic anarchy is at least $m+1$, and so our lower bound is not tight.

We note the exponential separation between our upper and lower bounds. We conjecture, however, that the true value of the price of stochastic anarchy is $O(m)$. If this conjecture is correct, then the $O(m)$ bound from the strong price of anarchy [1] can be achieved without coordination.

5 Conclusion and Open Questions

In this paper, we propose the evolutionary game theory solution concept of stochastic stability as a tool for quantifying the relative stability of equilibria. We show that in the load balancing game on unrelated machines, for which the price of Nash anarchy is unbounded, the "bad" Nash equilibria are not stochastically stable, and so the price of stochastic anarchy is bounded. We conjecture that the upper bound given in this paper is not tight and the cost of stochastic stability for load balancing is $O(m)$. If this conjecture is correct, it implies that the fragility of the "bad" equilibria in this game is attributable to their instability, not only in the face of player coordination, but also to minor uncoordinated perturbations in play. We expect that the techniques used in this paper will also be useful in understanding the relative stability of Nash equilibria in other games for which the worst equilibria are brittle. This promise is evidenced by the fact that the worst Nash in the worst-case instances in many games (for example, the Roughgarden and Tardos [22] lower bound showing an unbounded price of anarchy for routing unsplittable flow) are not stochastically stable.

Acknowledgments. We would like to thank Yishay Mansour for bringing the unbounded price of anarchy in the load balancing game to our attention, Avrim Blum and Tim Roughgarden for useful discussion, and Alexander Matros for his early guidance. We also are grateful to Mallesh Pai and Sid Suri for helpful discussions about evolutionary game theory.

References

1. Andelman, N., Feldman, M., Mansour, Y.: Strong price of anarchy. In: SODA 2007 (2007)
2. Awerbuch, B., Azar, Y., Richter, Y., Tsur, D.: Tradeoffs in worst-case equilibria. Theor. Comput. Sci. 361(2), 200–209 (2006)
3. Blum, A., Even-Dar, E., Ligett, K.: Routing without regret: On convergence to Nash equilibria of regret-minimizing algorithms in routing games. In: PODC 2006 (2006)

4. Blum, A., Hajiaghayi, M., Ligett, K., Roth, A.: Regret minimization and the price of total anarchy. In: STOC 2008 (2008)
5. Blume, L.E.: The statistical mechanics of best-response strategy revision. Games and Economic Behavior 11(2), 111–145 (1995)
6. Chen, X., Deng, X.: Settling the complexity of 2-player Nash-equilibrium. In: FOCS 2006 (2006)
7. Ellison, G.: Basins of attraction, long-run stochastic stability, and the speed of step-by-step evolution. Review of Economic Studies 67(1), 17–45 (2000)
8. Even-Dar, E., Kesselman, A., Mansour, Y.: Convergence time to Nash equilibria. In: Baeten, J.C.M., Lenstra, J.K., Parrow, J., Woeginger, G.J. (eds.) ICALP 2003. LNCS, vol. 2719, Springer, Heidelberg (2003)
9. Fabrikant, A., Papadimitriou, C.: The complexity of game dynamics: Bgp oscillations, sink equilibria, and beyond. In: SODA 2008 (2008)
10. Fabrikant, A., Papadimitriou, C., Talwar, K.: The complexity of pure nash equilibria. In: STOC 2004 (2004)
11. Fiat, A., Kaplan, H., Levy, M., Olonetsky, S.: Strong price of anarchy for machine load balancing. In: Arge, L., Cachin, C., Jurdziński, T., Tarlecki, A. (eds.) ICALP 2007. LNCS, vol. 4596, Springer, Heidelberg (2007)
12. Fischer, S., Räcke, H., Vöcking, B.: Fast convergence to wardrop equilibria by adaptive sampling methods. In: STOC 2006 (2006)
13. Fischer, S., Vöcking, B.: On the evolution of selfish routing. In: Albers, S., Radzik, T. (eds.) ESA 2004. LNCS, vol. 3221, Springer, Heidelberg (2004)
14. Foster, D., Young, P.: Stochastic evolutionary game dynamics. Theoret. Population Biol. 38, 229–232 (1990)
15. Goemans, M., Mirrokni, V., Vetta, A.: Sink equilibria and convergence. In: FOCS 2005 (2005)
16. Josephson, J., Matros, A.: Stochastic imitation in finite games. Games and Economic Behavior 49(2), 244–259 (2004)
17. Kandori, M., Mailath, G.J., Rob, R.: Learning, mutation, and long run equilibria in games. Econometrica 61(1), 29–56 (1993)
18. Koutsoupias, E., Papadimitriou, C.: Worst-case equilibria. In: 16th Annual Symposium on Theoretical Aspects of Computer Science, Trier, Germany, March 4–6, 1999, pp. 404–413 (1999)
19. Larry, S.: Stochastic stability in games with alternative best replies. Journal of Economic Theory 64(1), 35–65 (1994)
20. Nisan, N., Roughgarden, T., Tardos, E., Vazirani, V.V. (eds.): Algorithmic Game Theory. Cambridge University Press, Cambridge (2007)
21. Robson, A.J., Vega-Redondo, F.: Efficient equilibrium selection in evolutionary games with random matching. Journal of Economic Theory 70(1), 65–92 (1996)
22. Roughgarden, T., Tardos, É.: How bad is selfish routing. J. ACM 49(2), 236–259 (2002); In: FOCS 2000 (2000)
23. Suri, S.: Computational evolutionary game theory. In: Nisan, N., Roughgarden, T., Tardos, É., Vazirani, V.V., Vazirani, V.V. (eds.) Algorithmic Game Theory, Cambridge University Press, Cambridge (2007)
24. Peyton Young, H.: The evolution of conventions. Econometrica 61(1), 57–84 (1993)
25. Peyton Young, H.: Individual Strategy and Social Structure. Princeton University Press, Princeton (1998)

Singleton Acyclic Mechanisms and Their Applications to Scheduling Problems*

Janina Brenner and Guido Schäfer

Institute of Mathematics, Technical University Berlin, Germany
{brenner,schaefer}@math.tu-berlin.de

Abstract. Mehta, Roughgarden, and Sundararajan recently introduced a new class of cost sharing mechanisms called *acyclic mechanisms.* These mechanisms achieve a slightly weaker notion of truthfulness than the well-known Moulin mechanisms, but provide additional freedom to improve budget balance and social cost approximation guarantees. In this paper, we investigate the potential of acyclic mechanisms for combinatorial optimization problems. In particular, we study a subclass of acyclic mechanisms which we term *singleton acyclic mechanisms.* We show that every ρ-approximate algorithm that is *partially increasing* can be turned into a singleton acyclic mechanism that is weakly group-strategyproof and ρ-budget balanced. Based on this result, we develop singleton acyclic mechanisms for parallel machine scheduling problems with completion time objectives, which perform extremely well both with respect to budget balance and social cost.

1 Introduction

We consider the problem of designing truthful mechanisms for *binary demand cost sharing games.* We are given a universe U of players that are interested in a common service, and a cost function $C : 2^U \to \mathbb{R}^+$ that specifies the cost $C(S)$ to serve player set $S \subseteq U$. We require that the cost function C is *increasing*, i.e., $C(T) \leq C(S)$ for every $T \subseteq S \subseteq U$, and satisfies $C(\emptyset) = 0$. In this paper, we assume that C is given implicitly by the cost of an optimal solution to an underlying combinatorial optimization problem $\mathcal{P}$. Every player $i \in U$ has a private, non-negative *valuation* v_i and a non-negative *bid* b_i for receiving the service.

A *cost sharing mechanism* M takes the bid vector $\boldsymbol{b} := (b_i)_{i\in U}$ as input, and computes a binary allocation vector $\boldsymbol{x} := (x_i)_{i\in U}$ and a payment vector $\boldsymbol{p} := (p_i)_{i\in U}$. Let S^M be the subset of players associated with the allocation vector $\boldsymbol{x}$, i.e., $i \in S^M$ iff $x_i = 1$. We say that S^M is the player set that receives service. We require that a cost sharing mechanism complies with the following two standard assumptions: $p_i = 0$ if $i \notin S^M$ and $p_i \leq b_i$ if $i \in S^M$ (*individual rationality*) and $p_i \geq 0$ for all $i \in S^M$ (*no positive transfer*). In

* This work was supported by the DFG Research Center MATHEON "Mathematics for key technologies".

B. Monien and U.-P. Schroeder (Eds.): SAGT 2008, LNCS 4997, pp. 315–326, 2008.

addition, the mechanism has to compute a (potentially suboptimal) feasible solution to the underlying optimization problem $\mathcal{P}$ on the player set S^M. We denote the cost of the computed solution by $\bar{C}(S^M)$. M is *β-budget balanced* if $\bar{C}(S^M) \leq \sum_{i \in S^M} p_i \leq \beta \cdot C(S^M)$. The *social cost* [17] of a set $S \subseteq U$ is defined as $\Pi(S) := \bar{C}(S) + \sum_{i \notin S} v_i$. A mechanism M is said to be *α-approximate* if the social cost of the served set S^M satisfies $\Pi(S^M) \leq \alpha \cdot \Pi^*$, where $\Pi^* := \min_{S \subseteq U}(C(S) + \sum_{i \notin S} v_i)$ denotes the optimal social cost.

We assume that players act strategically and each player's goal is to maximize his own utility. The *utility* u_i of player i is defined as $u_i(\boldsymbol{x}, \boldsymbol{p}) := v_i x_i - p_i$. Since the outcome $(\boldsymbol{x}, \boldsymbol{p})$ computed by the mechanism M solely depends on the bids $\boldsymbol{b}$ of the players, a player may have an incentive to declare a bid b_i that differs from his valuation v_i. We say that M is *strategyproof* if bidding truthfully is a dominant strategy for every player. In this paper, we consider *cooperative cost sharing games*, i.e., players are allowed to form coalitions in order to coordinate their bids. A mechanism is *group-strategyproof* if no coordinated bidding of a coalition $S \subseteq U$ can ever strictly increase the utility of some player in S without strictly decreasing the utility of another player in S.

In recent years, considerable progress has been made in devising truthful mechanisms for cost sharing games. Most notably, Moulin [15] proposed a general framework for designing so-called *Moulin mechanisms* that are truthful and (approximately) budget balanced. The strength of Moulin mechanisms lies in the fact that they achieve one of the strongest notions of truthfulness, i.e., group-strategyproofness. Most of the mechanisms for cooperative cost sharing games that are currently prevailing in literature are Moulin mechanisms (e.g., [1,2,4,9,12,17,18]). However, recent negative results [1,2,10,13,17] show that for several fundamental cost sharing games, Moulin mechanisms can only achieve a very poor budget balance factor, and this effect is further amplified if approximate social cost is desired as additional objective [2,4,17,18].

Very recently, Mehta, Roughgarden, and Sundararajan [14] introduced a more general framework for designing truthful cost sharing mechanisms, termed *acyclic mechanisms*. Acyclic mechanisms implement a slightly weaker notion of truthfulness, called *weak group-strategyproofness*, but therefore leave more flexibility to improve upon the approximation guarantees with respect to budget balance and social cost. A mechanism is *weakly group-strategyproof* [5,14] if no coordinated bidding of a coalition $S \subseteq U$ can ever strictly increase the utility of *every* player in S. Mehta, Roughgarden, and Sundararajan [14] showed that primal-dual approximation algorithms for several combinatorial optimization problems naturally give rise to acyclic mechanisms.

Our Results. In this paper, we investigate the potential of acyclic mechanisms for combinatorial optimization problems. Our contribution is twofold:

1. Singleton Acyclic Mechanisms. We study a subclass of acyclic mechanisms that we call *singleton acyclic mechanisms*. We show that a ρ-approximation algorithm for the underlying optimization problem $\mathcal{P}$ yields a singleton acyclic mechanism that is ρ-budget balanced and weakly group-strategyproof if the cost function $\bar{C}$ induced by the approximation algorithm is increasing. In fact, even a slightly weaker condition suffices, namely that the induced cost function is *partially increasing* (definition will be given in Section 3). Our proof is constructive, i.e., we provide a framework that enables to turn any such approximation algorithm into a corresponding singleton acyclic mechanism. We also provide a means to prove approximate social cost for singleton mechanisms that fulfill an additional *weak monotonicity* property. While previously, most cost sharing mechanisms were developed in case-by-case studies, this is the first attempt to provide a general framework for obtaining cost sharing mechanisms from existing approximation algorithms.

Implications: A direct consequence of this result is that for several problems, lower bounds on the budget balance factor for Moulin mechanisms can be overcome by acyclic mechanisms. We mention three examples from the scheduling context here only: Graham's *largest processing time* rule [8] yields a 4/3-budget balanced acyclic mechanism for $P||C_{\max}$, beating the lower bound of essentially 2 for Moulin mechanisms [1]. Moreover, the *shortest remaining processing time* algorithm gives rise to a 2-budget balanced acyclic mechanism for $P|r_i, pmtn| \sum C_i$ [6] and a 1-budget balanced acyclic mechanism for $1|r_i, pmtn| \sum F_i$ [19], both overcoming the lower bounds of $\Omega(n)$ for Moulin mechanisms [2].

2. Singleton Acyclic Mechanisms for Completion Time Scheduling. We demonstrate the applicability of our singleton acyclic mechanism framework, also when social cost is concerned, by developing acyclic mechanisms for completion time scheduling with and without release dates and preemption. Namely, we achieve 1-budget balance and 2-approximate social cost for $P| | \sum C_i$, 1.21-budget balance and 2.42-approximate social cost for $P| | \sum w_i C_i$, and 1-budget balance and 4-approximate social cost for $1|r_i, pmtn| \sum C_i$. Not only are these the first cost sharing mechanisms to achieve constant social cost approximation factors, but we also outperform the strong lower bound of $\Omega(n)$ on the budget balance factor of any Moulin mechanism for all completion time related objectives [2].

Implications: We remark that every cost sharing mechanism that approximates social cost by a factor of α also is an α-approximation algorithm for the price-collecting variant of the underlying optimization problem. As a by-product of the results mentioned above, we therefore obtain constant approximation algorithms for the respective machine scheduling problems *with rejection* (see also [7] and the references therein).

2 Preliminaries

2.1 Acyclic Mechanisms

We briefly review the definition of acyclic mechanisms introduced by Mehta, Roughgarden, and Sundararajan (see [14] for a more detailed description).

An *acyclic mechanism* is defined in terms of a cost sharing method ξ and an offer function τ. A *cost sharing method* $\xi : U \times 2^U \to \mathbb{R}^+$ specifies for every subset $S \subseteq U$ and every player $i \in S$ a non-negative *cost share* $\xi_i(S)$; we define $\xi_i(S) := 0$ for all $i \notin S$. ξ is *β-budget balanced* if for every subset $S \subseteq U$ we have $C(S) \leq \sum_{i \in S} \xi_i(S) \leq \beta \cdot C(S)$. An *offer function* $\tau : U \times 2^U \rightarrow \mathbb{R}^+$ defines for every subset $S \subseteq U$ and every player $i \in S$ a non-negative *offer time* $\tau(i, S)$.

The acyclic mechanism $M(\xi, \tau)$ induced by ξ and τ receives the bid vector $\boldsymbol{b}$ as input and proceeds as follows:

1. Initialize $S := U$.
2. If $\xi_i(S) \leq b_i$ for every player $i \in S$, then halt. Output the characteristic vector $\boldsymbol{x}$ of S and payments $\boldsymbol{p} := (\xi_i(S))_{i \in U}$.
3. Among all players in S with $\xi_i(S) > b_i$, let i^* be one with minimum $\tau(i, S)$ (breaking ties arbitrarily).
4. Set $S := S \setminus \{i^*\}$ and return to Step 2.

For a given subset $S \subseteq U$ and a player $i \in S$, we partition the player set S into three sets with respect to the offer time of i: let $L(i, S)$, $E(i, S)$ and $G(i, S)$ be the sets of players with offer times $\tau(\cdot, S)$ strictly less than, equal to, or strictly greater than $\tau(i, S)$, respectively. The following definition is crucial to achieve weak group-strategyproofness.

Definition 1. *Let ξ and τ be a cost sharing method and an offer function on U. The offer function τ is* valid *for ξ if the following two properties hold for every subset $S \subseteq U$ and player $i \in S$:*

(P1) $\xi_i(S \setminus T) = \xi_i(S)$ *for every subset* $T \subseteq G(i, S)$*;*
(P2) $\xi_i(S \setminus T) \geq \xi_i(S)$ *for every subset* $T \subseteq G(i, S) \cup (E(i, S) \setminus \{i\})$.

We summarize the main result of Mehta, Roughgarden, and Sundararajan [14] in the following theorem:

Theorem 1 ([14]). *Let ξ be a β-budget balanced cost sharing method on U and let τ be an offer function on U that is valid for ξ. Then, the induced acyclic mechanism $M(\xi, \tau)$ is β-budget balanced and weakly group-strategyproof.*

2.2 Parallel Machine Scheduling

In a parallel machine scheduling problem, we are given a set U of n jobs that are to be scheduled on m identical machines. Every job $i \in U$ has a non-negative *release date* r_i, a positive *processing time* p_i, and a non-negative weight w_i. The release date specifies the time when job i becomes available for execution. The processing time describes the time needed to execute i on one of the machines. Every machine can execute at most one job at a time. In the *preemptive* setting, the execution of a job can be interrupted at any point of time and resumed later; in contrast, in the *non-preemptive* setting, job interruption is not permitted.

Given a scheduling algorithm ALG, we denote by $C_i^{\text{ALG}}(S)$ the completion time of job $i \in S$ in the schedule for the set of jobs $S \subseteq U$ output by ALG. We omit the superscript ALG if it is clear from the context to which schedule we refer. Depending on the underlying application, there are different objectives for machine scheduling problems. Among the most common objectives are the minimization of the total weighted completion time, i.e., $\sum_i w_i C_i$, and the makespan, i.e., $\max_i C_i$, over all feasible schedules.

In our game-theoretic view of scheduling problems, each job is identified with a player who wants his job to be processed on one of the m machines.

3 Singleton Acyclic Mechanisms

In this section, we describe our general framework for converting an approximation algorithm into a weakly group-strategyproof acyclic mechanism.

When thinking about acyclic mechanisms and their offer functions, we like to think of *clusters*. By a cluster we mean a maximal set of players that have the same offer time with respect to a set S, i.e., two players $i, j \in U$ are in the same cluster iff $\tau(i, S) = \tau(j, S)$. With this view, it becomes clear to which extent acyclic mechanisms generalize Moulin mechanisms: To one end, if there is only one cluster that contains all players, Definition 1 reduces to cross-monotonicity (see [15] for a definition), leading to Moulin mechanisms. To the other end, if all clusters are singletons, i.e., every player has a unique offer time, then (P2) of Definition 1 reduces to (P1) and once a cost share is announced to a player, it can never be changed again. Between these two extremes, there is a great range of other acyclic mechanisms. However, in this paper, we concentrate on the subclass of acyclic mechanisms that result from *singleton offer functions*, i.e., offer functions that induce singleton clusters. We call these mechanisms *singleton acyclic mechanisms*, or simply *singleton mechanisms*.

Let τ be a singleton offer function. In the following, we assume that the elements of a subset $S \subseteq U$ are ordered according to non-decreasing offer times, i.e., $S =: \{i_1, \dots, i_q\}$ with $\tau(i_l, S) < \tau(i_k, S)$ for all $1 \leq l < k \leq q$. Moreover, we define $S_k := \{i_1, \dots, i_k\} \subseteq S$ as the set of the first $1 \leq k \leq q$ elements in S. We slightly abuse notation and let for every $i \in S$, S_i refer to the set S_k with $i_k = i$. We are particularly interested in singleton offer functions that satisfy the following *consistency property*.

Definition 2. *A singleton offer function τ is called* consistent *if for all subsets $P \subseteq S \subseteq U$, ordered as $P =: \{j_1, j_2, \dots, j_p\}$ and $S =: \{i_1, i_2, \dots, i_q\}$, the following holds: If k is minimal with $i_k \notin P$, then $i_l = j_l$ for all $l < k$.*

Let ALG be a ρ-approximate algorithm for the underlying optimization problem $\mathcal{P}$ and let $\bar{C}$ denote the cost function induced by ALG, i.e., $\bar{C}(S)$ is the cost of the solution computed by ALG for player set $S \subseteq U$. We say that ALG is *partially increasing with respect to a singleton offer function τ* if for every $S \subseteq U$ and $i \in S$, we have $\bar{C}(S_i) \geq \bar{C}(S_{i-1})$. The main result of this section is the following:

Theorem 2. *Let ALG be a ρ-approximate algorithm. If there exists a consistent singleton offer function τ with respect to which ALG is partially increasing, then there is a singleton acyclic mechanism that is weakly group-strategyproof and ρ-budget balanced.*

A singleton offer function τ together with a partially increasing approximation algorithm ALG naturally induce the following cost sharing method ξ: for every $S \subseteq U$ and every $i \in S$, define

$$\xi_i(S) := \bar{C}(S_i) - \bar{C}(S_{i-1}).$$

Note that these cost shares are non-negative since ALG is partially increasing.

Truthfulness and Budget Balance. The following lemma together with Theorem 1 proves Theorem 2.

Lemma 1. *Let τ be a singleton offer function and let* ALG *be a ρ-approximate algorithm that is partially increasing with respect to τ. Moreover, let ξ be the cost sharing method induced by* ALG *and τ. Then the following holds:*

1. *ξ is ρ-budget balanced.*
2. *If τ is consistent, then τ is valid for ξ.*

Proof. By definition of ξ, we have $\sum_{i \in S} \xi_i(S) = \sum_{i \in S}(\bar{C}(S_i) - \bar{C}(S_{i-1})) = \bar{C}(S) - \bar{C}(\emptyset) = \bar{C}(S)$ for all $S \subseteq U$, proving that ξ is ρ-budget balanced.

We next show that τ is valid for ξ. Fix $S \subseteq U$ and $i \in S$. Since τ is a singleton offer function, $E(i, S) \setminus \{i\} = \emptyset$, and so (P2) of Definition 1 reduces to (P1). To prove (P1), let $P := S \setminus T$ for some subset $T \subseteq G(i, S)$ and consider the ordered sets $S =: \{i_1, i_2, \ldots, i_q\}$ and $P =: \{j_1, j_2, \ldots, j_p\}$. Let k be minimal with $i_k \notin P$. Then, by Definition 2, for all $l < k$, $i_l = j_l$ and hence $P_l = S_l$. Since $T \subseteq G(i, S)$, we have $\tau(i, S) < \tau(i_k, S)$. We conclude that $\xi_i(P) = \bar{C}(P_i) - \bar{C}(P_{i-1}) = \bar{C}(S_i) - \bar{C}(S_{i-1}) = \xi_i(S)$. □

From now on, for a consistent singleton offer function τ and an approximation algorithm ALG that is partially increasing with respect to τ, we call the mechanism $M := M(\xi, \tau)$ the *singleton mechanism induced by* ALG *and* τ. Given an approximation algorithm ALG, we remark that the budget balance factor of M is independent of the consistent singleton offer function used. However, the choice of the singleton offer function may very well influence the social cost of the solution output by the mechanism. Hence, if the cost function $\bar{C}$ induced by ALG is increasing, i.e., $\bar{C}(T) \leq \bar{C}(S)$ for all $T \subseteq S \subseteq U$, we can choose τ solely to achieve a good social cost approximation factor. If not, the *no positive transfer* property restricts the choice of τ to offer functions with respect to which ALG is partially increasing.

Social Cost. The social cost analysis of singleton mechanisms can be alleviated if the induced cost sharing method has the following property: We call a cost sharing method ξ *weakly monotone* if for all subsets $T \subseteq S \subseteq U$, $\sum_{i \in T} \xi_i(S) \geq \bar{C}(T)$.

Theorem 3. *Let $M = M(\xi, \tau)$ be the singleton mechanism induced by* ALG *and a consistent singleton offer function τ. Suppose that ξ is weakly monotone. Then, M approximates social cost by a factor of α if*

$$\frac{\bar{C}(S^M \cup S^*)}{C(S^*) + C(S^M \setminus S^*)} \leq \alpha.$$

Proof. We need to upper bound the ratio between the social cost of the set S^M chosen by the mechanism and a set $S^* := \arg\min_{S \subseteq U}(C(S) + \sum_{i \notin S} v_i)$. We have

$$\frac{\Pi(S^M)}{\Pi^*} = \frac{\bar{C}(S^M) + \sum_{i \in S^* \setminus S^M} v_i + \sum_{i \notin S^M \cup S^*} v_i}{C(S^*) + \sum_{i \in S^M \setminus S^*} v_i + \sum_{i \notin S^M \cup S^*} v_i} \leq \frac{\bar{C}(S^M) + \sum_{i \in S^* \setminus S^M} v_i}{C(S^*) + \sum_{i \in S^M \setminus S^*} v_i}$$
$$\leq \frac{\bar{C}(S^M) + \sum_{i \in S^* \setminus S^M} v_i}{C(S^*) + \sum_{i \in S^M \setminus S^*} \xi_i(S^M)} \leq \frac{\bar{C}(S^M) + \sum_{i \in S^* \setminus S^M} v_i}{C(S^*) + C(S^M \setminus S^*)}.$$

Here, the first inequality follows from the fact that $\frac{a}{b} \leq \frac{a-c}{b-c}$ for arbitrary real numbers $a \geq b > c > 0$. The second inequality holds because $v_i \geq \xi_i(S^M)$ for every player $i \in S^M$. The last inequality follows from weak monotonicity of ξ and the fact that $\bar{C}(S) \geq C(S)$ for every set S.

Without loss of generality, number the players in $S^* \setminus S^M$ in the order in which they were rejected in the course of the mechanism M, i.e., $S^* \setminus S^M =: \{1, \ldots, \ell\}$. For every $i \in S^* \setminus S^M$, let R^i be the subset of players in $S^* \cup S^M$ that were remaining in the iteration in which i was removed, i.e., $R^i := S^M \cup \{i, i+1, \ldots, \ell\}$. Since i rejected, we have $v_i < \xi_i(R^i)$. Moreover, by definition of the sets R^i and weak monotonicity of ξ, we obtain $\bar{C}(R^i) = \sum_{j \in R^i} \xi_j(R^i) = \xi_i(R^i) + \sum_{j \in R^{i+1}} \xi_j(R^i) \geq \xi_i(R^i) + \bar{C}(R^{i+1})$. Summing over all $i \in \{1, \ldots, \ell\}$ yields

$$\sum_{i \in S^* \setminus S^M} v_i \leq \sum_{i=1}^{\ell} \left(\bar{C}(R^i) - \bar{C}(R^{i+1})\right) = \bar{C}(S^M \cup S^*) - \bar{C}(S^M). \qquad \square$$

4 Completion Time Scheduling

In this section, we study the performance of singleton mechanisms for parallel machine scheduling problems with total completion time objectives. We distinguish between the model with weights, in which all jobs arrive at time zero and no preemption is allowed, and the model in which jobs have release dates and may be preempted.

4.1 Weighted Completion Time

We consider the problem $P||\sum w_i C_i$ of scheduling a set of jobs $U := [n]$ non-preemptively on m parallel machines such that the total weighted completion time is minimized. Lenstra proves that this problem is NP-complete [3]. Even for the unweighted version, i.e., $w_i = 1$ for all $i \in U$, no Moulin mechanism can achieve a budget balance factor better than $n/2$ [2]. However, using singleton acyclic mechanisms, we can heavily improve upon this.

Let ρ^{SM} denote the approximation guarantee achieved by Smith's rule [20], which schedules the jobs in non-increasing order of their weight per processing time ratios w_i/p_i. For $P||\sum w_i C_i$, the produced schedule is $(1+\sqrt{2})/2 \approx 1.21$-approximate [11]. In the single machine case, Smith's rule produces an optimal schedule. In the unweighted setting, Smith's rule reduces to the *shortest processing time* policy and also delivers an optimal schedule.

Let $M^{\text{wct}} := M(\xi, \tau)$ be the singleton mechanism induced by Smith's rule and the offer function τ defined as follows:

Singleton offer function for Smith's rule: Let σ be a non-increasing weight per processing time order on $U = [n]$. If two jobs $i, j \in U$ satisfy $w_i/p_i = w_j/p_j$, we define $\sigma(i) < \sigma(j)$ iff $i < j$. For every subset $S \subseteq U$, let $\tau(\cdot, S)$ be the order on S induced by σ.

One easily verifies that τ is a consistent singleton offer function. We have $\xi_i(S) = \bar{C}(S_i) - \bar{C}(S_{i-1}) = w_i C_i(S)$, where $C_i(S)$ is the completion time of job i in the schedule computed by Smith's rule. Since $w_i C_i(S) \geq 0$, Smith's rule is obviously partially increasing with respect to τ.

Theorem 4. *The singleton mechanism $M^{\text{wct}} = M(\xi, \tau)$ induced by Smith's rule and τ is weakly group-strategyproof, ρ^{SM}-budget balanced, and $2\rho^{\text{SM}}$-approximate.*

Proof. It follows from Theorem 2 that M^{wct} is weakly group-strategyproof and ρ^{SM}-budget balanced. It remains to be shown that M^{wct} is $2\rho^{\text{SM}}$-approximate with respect to social cost. To see that the induced cost sharing method ξ is weakly monotone, note that $C_i(S) \geq C_i(T)$ for every $i \in T \subseteq S$. Thus, $\sum_{i \in T} \xi_i(S) \geq \sum_{i \in T} \xi_i(T) = \bar{C}(T)$. The social cost approximation factor now follows from Theorem 3 and Lemma 2 given below. □

Lemma 2. *Let* ALG *be an algorithm for $P||\sum w_i C_i$ with cost function $\bar{C}$. Let A and B be two disjoint sets of jobs. Then, the cost of an optimal schedule for $A \cup B$ can be bounded by $C(A \cup B) \leq 2(\bar{C}(A) + \bar{C}(B))$.*

Proof. We prove the inequality individually for each machine $\hat{M}$. Consider the jobs $\hat{A} \subseteq A$ and $\hat{B} \subseteq B$ scheduled on $\hat{M}$ in the runs of ALG on A and B, respectively. We denote by c_i the completion time of job i in his respective schedule, i.e. $c_i := \bar{C}_i(A)$ for all $i \in \hat{A}$ and $c_i := \bar{C}_i(B)$ for all $i \in \hat{B}$.

Consider the schedule which processes all jobs in $\hat{A} \cup \hat{B}$ on $\hat{M}$ according to non-decreasing c_i. The completion time of a job $i \in \hat{A}$ in this schedule is $c_i + c_{i^*}$, where i^* denotes the last job in $\hat{B}$ that is processed before i. Since i^* is processed before i, we have $c_i + c_{i^*} \leq 2c_i$. By exchanging the roles of A and B, we can show the same for the completion time of every job $i \in \hat{B}$.

Since the cost of an optimal schedule for $A \cup B$ is at most that of the schedule produced by repeating the above procedure for each machine, we have

$$C(A \cup B) \leq \sum_{i \in A \cup B} w_i \cdot 2c_i = 2\Big(\sum_{i \in A} w_i c_i + \sum_{i \in B} w_i c_i\Big) = 2(\bar{C}(A) + \bar{C}(B)). \quad \square$$

We remark that a simple example shows that our social cost analysis is tight, even in the unweighted case (details will be given in the full version of the paper).

4.2 Completion Time with Release Dates and Preemption

Now, consider the problem $1|r_i, pmtn|\sum C_i$ of scheduling a set of jobs $U := [n]$ on a single machine such that the total completion time is minimized. Each job $i \in U$ has a non-negative release date r_i, and preemption of jobs is allowed. The *shortest remaining processing time* (SRPT) policy delivers an optimal schedule for this problem [19].

We introduce some more notation in order to give a formal definition of SRPT. Let $e_i(t)$ be the amount of time that has been spent on processing job i up to time t. The *remaining processing* time $x_i(t)$ of job i at time t is $x_i(t) := p_i - e_i(t)$. We call a job i *active* at time t if it has been released but not yet completed at this time, i.e., $r_i \leq t < C_i$. Let $A(t)$ be the set of jobs that are active at time t. SRPT works as follows: At any time $t \geq 0$, SRPT schedules an active job $i \in A(t)$ with minimum remaining processing time, i.e. $x_i(t) \leq x_k(t)$ for all $k \in A(t)$. In the following, we assume that SRPT uses a consistent tie breaking rule, e.g., if $x_i(t) = x_k(t)$ for two different jobs i and k, then schedule the one with smaller index. Throughout this section, let $C_i(S) := C_i^{\text{SRPT}}(S)$ for all $S \subseteq U$.

Let $M^{pct} := M(\xi, \tau)$ be the singleton mechanism induced by SRPT and the following singleton offer function τ:

Singleton offer function for SRPT: For a given subset $S \subseteq U$, let $\tau(\cdot, S)$ be the order induced by increasing completion times of the jobs in S, i.e., $\tau(i, S) < \tau(j, S)$ iff $C_i(S) < C_j(S)$.

The offer function τ is consistent; we defer the proof to the end of this section. Recall that SRPT is an optimal scheduling policy and thus $\bar{C}(S) = C(S)$. We thus have $\xi_i(S) = C(S_i) - C(S_{i-1}) = C_i(S)$, where the latter follows from Lemma 5. Note that SRPT is partially increasing with respect to τ because $C_i(S) \geq 0$.

Theorem 5. *The singleton mechanism $M^{pct} = M(\xi, \tau)$ induced by SRPT and τ is weakly group-strategyproof, budget balanced, and 4-approximate.*

Proof. It follows from Theorem 2 that M^{pct} is weakly group-strategyproof and budget balanced. To prove that M^{pct} approximates social cost, we first show that ξ is weakly monotone. Fix some set S and let $T \subseteq S$. Consider the SRPT schedule for S. If we remove from this schedule all jobs in $S \setminus T$, we obtain a feasible schedule for T of cost at most $\sum_{i \in S \setminus T} C_i(S) \geq C(T)$. Since $\xi_i(S) = C_i(S)$, we have weak monotonicity. Now, the bound on the social cost approximation factor follows from Theorem 3, using Lemma 3 given below. □

The following lemma is used to prove the social cost approximation factor.

Lemma 3. *Let* ALG *be an algorithm for $P|r_i, pmtn|\sum C_i$ with cost function $\bar{C}$. Let A and B be two disjoint sets of jobs. Then, the cost of an optimal schedule for $A \cup B$ can be bounded by $C(A \cup B) \leq 4(\bar{C}(A) + \bar{C}(B))$.*

Proof. Phillips et al. [16] prove that any preemptive schedule for $P|r_i, pmtn|\sum C_i$ can be turned into a non-preemptive schedule NP with at most twice the cost. With Lemma 2, we obtain $C(A \cup B) \leq 2(C^{\text{NP}}(A) + C^{\text{NP}}(B)) \leq 4(\bar{C}(A) + \bar{C}(B))$. □

Consistency. In order to prove that τ is consistent, we need some more notation. Consider the SRPT schedule for a set $S \subseteq U$. Let $i, j \in A(t)$ be two jobs that are active at time t. We define $i \prec_t j$ iff either $x_i(t) < x_j(t)$ or $x_i(t) = x_j(t)$ and $i \leq j$. Note that at any point of time t, SRPT schedules the job $i \in A(t)$ with $i \prec_t j$ for all $j \in A(t)$. Thus, if $i \prec_t j$ for some t, then $i \prec_{t'} j$ for all $t' \in [t, C_i)$. We therefore simply write $i \prec j$ iff there exists a time t with $i \prec_t j$. Let $\sigma(t)$ denote the job that is executed at time t in the SRPT schedule for S; we define $\sigma(t) = \emptyset$ if $A(t) = \emptyset$.

Let $j \in S$ be an arbitrary job and consider the time interval $[r_j, C_j)$. We define the set $\mathcal{C}_j$ of jobs that are *competing* with j as $\mathcal{C}_j := \{i \in S \setminus \{j\} : [r_i, C_i) \cap [r_j, C_j) \neq \emptyset\}$. Note that $j \notin \mathcal{C}_j$. We partition the jobs in $\mathcal{C}_j$ into a set $\mathcal{W}_j$ of *winning jobs* and a set $\mathcal{L}_j$ of *losing jobs* with respect to j: $\mathcal{W}_j := \{i \in \mathcal{C}_j : i \prec j\}$ and $\mathcal{L}_j := \mathcal{C}_j \setminus \mathcal{W}_j$. Intuitively, suppose i and j are both active at some time t. If i is a winning job, then i prevents j from being executed by SRPT. On the other hand, if i is a losing job, then j prevents i from being executed.

We next investigate the effect of removing a job j from S. We use the superscript T if we refer to the SRPT schedule for $T := S \setminus \{j\}$.

Lemma 4. *Consider the two SRPT schedules on job sets S and $T := S \setminus \{j\}$. For every job $i \in \mathcal{C}_j$ that is active at time $t \in [r_j, C_j)$,*

$$x_i^T(t) = x_i(t) \text{ if } i \in \mathcal{W}_j \quad \text{and} \quad x_i^T(t) \geq x_j(t) \text{ if } i \in \mathcal{L}_j.$$

Proof. We partition the time interval $[r_j, C_j)$ into a sequence of maximal subintervals $I_1, I_2, \ldots, I_f$ such that the set of active jobs remains the same within every subinterval $I_\ell := [s_\ell, e_\ell)$. We prove by induction over ℓ that the claim holds for every $t \in [r_j, e_\ell)$.

Note that both schedules are identical up to time $r_j = s_1$. If $\sigma(s_1) \neq j$, then both schedules process the same job during I_1 and the claim follows. Suppose $\sigma(s_1) = j$. This implies that $A(s_1) \cap \mathcal{W}_j = \emptyset$ and thus all jobs in $A(s_1) \setminus \{j\} = A^T(s_1)$ are losing jobs. If $A^T(s_1) = \emptyset$, the claim follows. Otherwise, let $k := \sigma^T(s_1)$ be the job that is processed in the schedule for T. Since k is a losing job, we have $x_k^T(s_1) = x_k(s_1) \geq x_j(s_1)$. Since k and j receive the same processing time during I_1 in their respective schedules, the claim holds for all $t \in [r_j, e_1)$.

Now, assume that the claim is true for every $t \in [r_j, e_{\ell-1})$ for some $\ell > 1$. We show that it remains true during the time interval I_ℓ. By the induction hypothesis, $x_i^T(t) = x_i(t)$ for every job $i \in \mathcal{W}_j$ that is active at time $t \in [r_j, e_{\ell-1})$. This implies that a job $j \in \mathcal{W}_i$ is executed at time $t \in [r_j, e_{\ell-1})$ in the schedule for S iff it is executed at time t in the schedule for T. We thus have $A^T(s_\ell) \cap \mathcal{W}_j = A(s_\ell) \cap \mathcal{W}_j$. Moreover, $x_i^T(t) \geq x_j(t)$ for every job $i \in \mathcal{L}_j$ that is active at time $t \in [r_j, e_{\ell-1})$. Since $x_j(t) > 0$ for every $t \in [r_j, C_j)$, every job $i \in \mathcal{L}_j$ that is active at time $t \in [r_j, e_{\ell-1})$ in the schedule for S must also be active at time t in the schedule for T. Thus, $A^T(s_\ell) \cap \mathcal{L}_j = A(s_\ell) \cap \mathcal{L}_j$. We now distinguish two cases:

(i) First, assume $\sigma(s_\ell) =: k \in \mathcal{W}_j$. Job k then has smallest remaining processing time, i.e., $x_k(s_\ell) \leq x_i(s_\ell)$ for all $i \in A(s_\ell)$. We conclude that

$$x_k^T(s_\ell) = x_k(s_\ell) \le x_i(s_\ell) = x_i^T(s_\ell) \;\; \forall i \in A(s_\ell) \cap \mathcal{W}_j = A^T(s_\ell) \cap \mathcal{W}_j$$
$$x_k^T(s_\ell) = x_k(s_\ell) \le x_j(s_\ell) \le x_i^T(s_\ell) \;\; \forall i \in A(s_\ell) \cap \mathcal{L}_j = A^T(s_\ell) \cap \mathcal{L}_j.$$

Since we assume that SRPT uses a consistent tie breaking rule, this implies that $\sigma^T(s_\ell) = k$ and the claim follows.

(ii) Now, suppose $\sigma(s_\ell) = j$. (Note that $\sigma(s_\ell) \in \mathcal{L}_j$ is impossible.) Then $x_j(s_\ell) \le x_i(s_\ell)$ for every $i \in A(s_\ell)$ and $A(s_\ell) \cap \mathcal{W}_j = \emptyset$. But then we also have $A^T(s_\ell) \cap \mathcal{W}_j = \emptyset$ and thus $A^T(s_\ell) \subseteq \mathcal{L}_j$. If $A^T(s_\ell) = \emptyset$, the claim follows. Otherwise, let $k := \sigma^T(s_\ell) \in \mathcal{L}_j$ be the job that is executed at time s_ℓ in the schedule for T. Since $x_k^T(s_\ell) \ge x_j(s_\ell)$ and the remaining processing times of k and j in their respective schedules reduce by the same amount during I_ℓ, the claim follows. □

We omit the proof of the following lemma due to lack of space.

Lemma 5. *Let $T \subseteq S \subseteq U$ and consider the* SRPT *schedule for $S \setminus T$. We have:*

1. *$C_i(S \setminus T) = C_i(S)$ for every job $i \in S \setminus T$ with $C_i(S) < C_j(S)$ for all $j \in T$.*
2. *$C_\ell(S \setminus T) \ge \min_{j \in T} C_j(S)$ for every job $\ell \in S \setminus T$ with $C_\ell(S) > C_j(S)$ for some $j \in T$.*

Lemma 6. *The singleton offer function τ is consistent.*

Proof. Consider two sets $P \subseteq S \subseteq U$, ordered by τ as $P =: \{j_1, j_2, \ldots, j_p\}$ and $S =: \{i_1, i_2, \ldots, i_q\}$. Let k be minimal with $i_k \notin P$. Then, for all $l < k$, we have $i_l \in P$ by minimality of k, and $C_{i_l}(S) < C_{i_k}(S)$ by definition of τ. Also by minimality of k, for all other $i \notin P$, we have $C_{i_l}(S) < C_{i_k}(S) < C_i(S)$. Hence, Lemma 5 proves that $C_{i_l}(S) = C_{i_l}(P)$ for all $l < k$.

For all other jobs $j \in P$, we have $C_j(S) > C_k(S)$ and thus by Lemma 5, $C_j(P) \ge C_k(S) > C_{k-1}(S) = C_{k-1}(P)$. Hence, we have $i_l = j_l$ for all $l < k$. □

References

1. Bleischwitz, Y., Monien, B.: Fair cost-sharing methods for scheduling jobs on parallel machines. In: Calamoneri, T., Finocchi, I., Italiano, G.F. (eds.) CIAC 2006. LNCS, vol. 3998, pp. 175–186. Springer, Heidelberg (2006)
2. Brenner, J., Schäfer, G.: Cost sharing methods for makespan and completion time scheduling. In: Thomas, W., Weil, P. (eds.) STACS 2007. LNCS, vol. 4393, pp. 670–681. Springer, Heidelberg (2007)
3. Brucker, P.: Scheduling Algorithms. Springer, New York, USA (1998)
4. Chawla, S., Roughgarden, T., Sundararajan, M.: Optimal cost-sharing mechanisms for Steiner forest problems. In: Proc. of the 2nd Int. Workshop on Internet and Network Economics, pp. 112–123 (2006)
5. Devanur, N., Mihail, M., Vazirani, V.: Strategyproof cost-sharing mechanisms for set cover and facility location games. In: Proc. of the ACM Conference on Electronic Commerce (2003)
6. Du, J., Leung, J.Y.T., Young, G.H.: Minimizing mean flow time with release time constraint. Theoretical Computer Science 75(3), 347–355 (1990)

7. Engels, D., Karger, D., Kolliopoulos, S., Sengupta, S., Uma, R., Wein, J.: Techniques for scheduling with rejection. Journal of Algorithms 49, 175–191 (2003)
8. Graham, R.L.: Bounds on multiprocessing timing anomalies. SIAM Journal on Applied Mathematics 17(2), 416–429 (1969)
9. Gupta, A., Könemann, J., Leonardi, S., Ravi, R., Schäfer, G.: An efficient cost-sharing mechanism for the prize-collecting Steiner forest problem. In: Proc. of the 18th ACM-SIAM Sympos. on Discrete Algorithms, pp. 1153–1162 (2007)
10. Immorlica, N., Mahdian, M., Mirrokni, V.S.: Limitations of cross-monotonic cost sharing schemes. In: Proc. of the 16th ACM-SIAM Sympos. on Discrete Algorithms, pp. 602–611 (2005)
11. Kawaguchi, T., Kyan, S.: Worst case bound of an LRF schedule for the mean weighted flow time problem. SIAM Journal on Computing 15(4), 1119–1129 (1986)
12. Könemann, J., Leonardi, S., Schäfer, G.: A group-strategyproof mechanism for Steiner forests. In: Proc. of the 16th ACM-SIAM Sympos. on Discrete Algorithms, pp. 612–619 (2005)
13. Könemann, J., Leonardi, S., Schäfer, G., van Zwam, S.: From primal-dual to cost shares and back: a stronger LP relaxation for the Steiner forest problem. In: Caires, L., Italiano, G.F., Monteiro, L., Palamidessi, C., Yung, M. (eds.) ICALP 2005. LNCS, vol. 3580, pp. 930–942. Springer, Heidelberg (2005)
14. Mehta, A., Roughgarden, T., Sundararajan, M.: Beyond Moulin mechanisms. In: Proc. of the ACM Conference on Electronic Commerce (2007)
15. Moulin, H.: Incremental cost sharing: Characterization by coalition strategy-proofness. Social Choice and Welfare 16, 279–320 (1999)
16. Phillips, C., Stein, C., Wein, J.: Minimizing average completion time in the presence of release dates. Math. Program 82, 199–223 (1998)
17. Roughgarden, T., Sundararajan, M.: New trade-offs in cost-sharing mechanisms. In: Proc. of the 38th ACM Sympos. on Theory of Computing, pp. 79–88 (2006)
18. Roughgarden, T., Sundararajan, M.: Optimal efficiency guarantees for network design mechanisms. In: Proc. of the 12th Int. Conf. on Integer Programming and Combinatorial Optimization, pp. 469–483 (2007)
19. Schrage, L.: A proof of the optimality of the shortest remaining processing time discipline. Operations Research 16, 687–690 (1968)
20. Smith, W.: Various optimizers for single-stage production. Naval Research Logistics Quarterly 3, 59–66 (1956)

Is Shapley Cost Sharing Optimal?

Shahar Dobzinski[1,⋆], Aranyak Mehta[2], Tim Roughgarden[3,⋆⋆], and Mukund Sundararajan[3,⋆⋆⋆]

[1] The School of Computer Science and Engineering, The Hebrew University of Jerusalem
shahard@cs.huji.ac.il
[2] Google, Inc., Mountain View, CA
tim@cs.stanford.edu
[3] Department of Computer Science, Stanford University, 353 Serra Mall, Stanford, CA 94305
mukunds@cs.stanford.edu

Abstract. We study the best guarantees of efficiency approximation achievable by cost-sharing mechanisms. Our main result is the first quantitative lower bound that applies to all truthful cost-sharing mechanisms, including randomized mechanisms that are only truthful in expectation, and only β-budget-balanced in expectation. Our lower bound is optimal up to constant factors and applies even to the simple and central special case of the public excludable good problem. We also give a stronger lower bound for a subclass of deterministic cost-sharing mechanisms, which is driven by a new characterization of the Shapley value mechanism. Finally, we show a separation between the best-possible efficiency guarantees achievable by deterministic and randomized cost-sharing mechanisms.

1 Introduction

1.1 Approximation in Algorithmic Mechanism Design

Algorithmic mechanism design studies the possibilities and impossibilities of optimization with incomplete information by incentive-compatible mechanisms. The main positive result in the area is, of course, the *VCG mechanisms* [18,3,8], a family of truthful, direct-revelation mechanisms that maximize objective functions of the form

$$\max_{o \in \Omega} \sum_i w_i v_i(o) - C(o), \tag{1}$$

⋆ This work was done while the author was visiting Stanford University. Supported by the Adams Fellowship Program of the Israel Academy of Sciences and Humanities, and by grants from the Israel Science Foundation and the USA-Israel Bi-national Science Foundation.

⋆⋆ Supported in part by NSF CAREER Award CCF-0448664, an ONR Young Investigator Award, and an Alfred P. Sloan Fellowship.

⋆⋆⋆ Supported by NSF Award CCF-0448664 and a Stanford Graduate Fellowship.

B. Monien and U.-P. Schroeder (Eds.): SAGT 2008, LNCS 4997, pp. 327–336, 2008.

where Ω is the outcome space, v_i is a valuation private to a self-interested player i, and the w_i's and $C(o)$'s are known real-valued constants. In other words, affine maximization of private data is always possible by compensating the self-interested participants appropriately.

For many central applications, VCG mechanisms are irrelevant or infeasible, and research has focused on the design and analysis of truthful approximation mechanisms (see e.g. [10]). For example, for some optimization problems different from affine maximization, *no* truthful mechanism can achieve full optimality, even with unbounded computational power (e.g. [15,16]). Another common reason for designing truthful approximation mechanisms is the exponential communication and/or computation required by the VCG mechanism for some affine maximization problems, such as welfare maximization in combinatorial auctions (see e.g. [1]). This paper is motivated by a different flaw with truthful welfare-maximizing mechanisms: no such mechanism achieves non-trivial worst-case revenue guarantees, even if unbounded computation is allowed. Precisely, when the outcome-dependent constant $C(o)$ in (1) represents the production costs for outcome o, then no truthful and individually rational mechanism that maximizes the welfare $\sum_i v_i(o) - C(o)$ guarantees that the revenue obtained is at least a constant fraction of the incurred cost. This impossibility result applies even to extremely simple single-parameter settings [6,7,16]. An important research goal, to which this paper contributes, is to quantify the minimum efficiency loss required to recover non-trivial budget-balance guarantees.

1.2 Randomization in Algorithmic Mechanism Design

A related issue is quantifying the power of randomization in the design of truthful approximation mechanisms. Recall that a randomized mechanism is *truthful in expectation* if truthful revelation is a dominant strategy for a player that wants to maximize its expected payoff, and is *universally truthful* if it is a distribution over truthful deterministic mechanisms. (The second condition effectively assumes that players can predict the outcome of the mechanism's internal randomization and therefore is stronger than the first.) For non-affine problems, universally truthful mechanisms are provably more powerful than deterministic ones [15]. We show, for the first time, an analogous separation between the best-possible performance of deterministic and randomized revenue-constrained mechanisms.

1.3 Our Results

Our main result is the first quantitative lower bound on efficiency loss that applies to all truthful and budget-balanced mechanisms. Our lower bound applies even in the special case of a single-parameter *public excludable good* problem, where the outcome set Ω is the subsets of the participants (the "winners") and the cost $C(o)$ is zero for the empty set and 1 otherwise. The public excludable good problem occupies a central position in the economic cost-sharing literature [5,4]. It is also a special case of nearly all of the cost-sharing problems that have been studied in the theoretical computer science literature, including fixed-tree multicast, uncapacitated facility location, and vertex cover cost-sharing problems (see [2]). Naturally, our lower bound carries over to all of these more general classes of cost-sharing problems. Previous lower bounds for approximate efficiency in cost-sharing mechanisms applied only to subclasses of deterministic mechanisms (to Moulin mechanisms in [17] and to acyclic mechanisms in [11]).

Precisely, we prove the following. Call a truthful and individually rational mechanism for a public excludable good problem *β-budget-balanced* if its revenue is always at least a $1/\beta$ fraction of and no more than the incurred cost. We show that *every β-budget-balanced truthful mechanism is $\Omega(\log k/\beta)$-approximate* in the sense of [17], where k is the number of participants. Our lower bound applies even to randomized mechanisms that are only truthful in expectation, and only β-budget-balanced in expectation. Our lower bound is optimal up to constant factors for all $\beta = O(\sqrt{\log k})$, with the nearly matching upper bound provided by a scaled version of the Shapley value mechanism [14,17]. All of our lower bounds apply to both the social cost approximation measure introduced in [17] and to the additive efficiency loss measure studied earlier by Moulin and Shenker [14].

We also give stronger results for a subclass of deterministic cost-sharing mechanisms. Specifically, we show that the *Shapley value mechanism is optimal* among all deterministic, symmetric, and budget-balanced cost-sharing mechanisms for public excludable good problems. (A similar result of Moulin and Shenker [14] proves only that the Shapley value mechanism is an optimal Moulin mechanism [13].) Here, "symmetric" means that players that submit equal bids are given the same allocations and prices. This proof is based on a new characterization of the Shapley value mechanism, which improves upon a previous characterization of Deb and Razzolini [5].

Finally, we give the first separation between the power of deterministic and randomized cost-sharing mechanisms: we prove a lower bound on the approximation factor of all deterministic mechanisms for the 2-player public excludable good problem, and exhibit a universally truthful randomized mechanism that possesses a strictly better approximation guarantee.

2 Preliminaries

There is a population U of k players and a public cost function C defined on all subsets of U. We always assume that $C(\emptyset) = 0$ and that C is nondecreasing (i.e., $S \subseteq T$ implies that $C(S) \leq C(T)$). Player i has a private value v_i for service. We focus on direct revelation mechanisms; such mechanisms accept a bid b_i from each player i and determine an allocation $S \subseteq U$ and payments p_i for the players.

We discuss only mechanisms that satisfy the following standard assumptions: *individual rationality*, meaning that $p_i = 0$ if $i \notin S$ and $p_i \leq b_i$ if $i \in S$; and *no positive transfers*, meaning that prices are always nonnegative. We also assume that players have quasilinear utilities, meaning that each player i aims to maximize $u_i(S, p_i) = v_i x_i - p_i$, where $x_i = 1$ if $i \in S$ and $x_i = 0$ if $i \notin S$.

A mechanism is *strategyproof*, or *truthful*, if no player can ever strictly increase its utility by misreporting its valuation. Formally, truthfulness means that for every player i, every bid vector b with $b_i = v_i$, and every bid vector b' with $b_j = b'_j$ for all $j \neq i$, $u_i(S, p_i) \geq u_i(S', p'_i)$, where (S, p) and (S', p') denote the outputs of the mechanism for the bid vectors b and b', respectively. When discussing truthful mechanisms, we typically assume that players bid their valuations and conflate the (unknown) valuation profile v with the (known) bid vector b.

In Section 4 we use the following standard fact about truthful mechanisms (see e.g. [12]).

Proposition 1. *Let M be a truthful, individually rational cost-sharing mechanism with the player set U. Then for every $i \in U$ and bid vector b_{-i} for players other than i, there is a* threshold *$t_i(b_{-i})$ such that: (i) if i bids more than $t_i(b_{-i})$, then it receives service at price $t_i(b_{-i})$; (ii) if i bids less than $t_i(b_{-i})$, then it does not receive service.*

A randomized mechanism is, by definition, a probability distribution over deterministic mechanisms. Such a mechanism is *universally truthful* if every mechanism in its support is truthful. Such a mechanism is *truthful in expectation* if no player can ever strictly increase its *expected* utility by misreporting its valuation. Every universally truthful mechanism is truthful in expectation, but the converse need not hold.

We study two kinds of objectives for cost-sharing mechanisms, one for the revenue of the mechanism, and one for its economic efficiency. First, for a parameter $\beta \geq 1$, a mechanism is *β-budget-balanced* if it always recovers at least a $1/\beta$ fraction of and at most the cost incurred. We say that a mechanism is *budget-balanced* if it is 1-budget-balanced.

We measure the efficiency (loss) achieved by a cost-sharing mechanism via the *social cost* objective. The social cost of an outcome S with respect to a cost function C and valuation profile v is, by definition, the service cost $C(S)$ plus the excluded value $v(U \setminus S) = \sum_{i \notin S} v_i$. This objective function is ordinally equivalent to the more standard welfare objective, which is the difference between the value served $\sum_{i \in S} v_i$ and the cost $C(S)$. Moreover, it is, in a precise sense, the "minimal perturbation" of the welfare objective function that admits non-trivial relative approximation guarantees; see [17] for details and additional justification for studying this objective. A cost-sharing mechanism is *α-approximate* if, assuming truthful bids, it is an α-approximation algorithm for the social cost objective. We state all of our lower bounds in terms of this approximation measure, but our proofs immediately yield comparable lower bounds for the additive efficiency loss measure adopted by Moulin and Shenker [14].

For a *public excludable problem*, in which $C(S) = 1$ for every non-empty S, the optimal solution is either U (for valuation profiles v with $v(U) \geq 1$) or $\emptyset$ (otherwise).

We conclude this section by describing a central mechanism [17,5,4,14] for the public excludable good problem. Following [14], we call this the *Shapley value mechanism.* Given a set of bids, the mechanism serves the largest set $S \subseteq U$ such that for each player $i \in S$, $b_i \geq 1/|S|$. (Such sets are closed under union, and hence there is a unique largest such set.) Every player in S pays $1/|S|$ and the other players pay 0; the price that a player in S pays is precisely its Shapley value in the set S with respect to the function $C(\cdot)$. The mechanism is obviously budget-balanced; it is also truthful [14] and $\mathcal{H}_k$-approximate [17], where $k = |U|$ and $\mathcal{H}_k$ is the kth harmonic number. (Recall that $\mathcal{H}_k \approx \ln k$.) We recall here the example that shows that the result is tight.

Example 1. Let ϵ be a small positive number. Consider the truthful bid vector $1-\epsilon, 1/2-\epsilon, 1/3-\epsilon \ldots 1/k-\epsilon$. The solution which optimizes social cost serves all the players and has social cost 1. On the other hand, the Shapley value mechanism serves no players and has social cost $\mathcal{H}_k - k\epsilon$. Since ϵ can be arbitrarily small, the Shapley value mechanism is no better than $\mathcal{H}_k$-approximate.

This paper investigates whether or not there are truthful budget-balanced mechanisms that outperform the Shapley value mechanism.

3 A Lower Bound on Cost-Sharing Mechanisms

In this section we prove that every $O(1)$-budget-balanced cost-sharing mechanism for the public excludable good problem is $\Omega(\log k)$-approximate. This lower bound applies even to randomized mechanisms, and even to mechanisms that are only truthful in expectation.

Theorem 1. *Every cost-sharing mechanism for the public excludable good problem that is truthful in expectation and β-budget-balanced in expectation is $\Omega((\log k)/\beta)$-approximate, where k is the number of players.*

Proof. Fix values for k and $\beta \geq 1$. The plan of the proof is to define a distribution over valuation profiles such that the sum of the valuations is likely to be large but every mechanism is likely to produce the empty allocation. Let $a_1, \ldots, a_k$ be i.i.d. draws from the distribution with density $1/z^2$ on $[1, k]$ and remaining mass $(1/k)$ at zero. Set $v_i = a_i/4k\beta$ for each i and $V = \sum_{i=1}^{k} v_i$. We first note that V is likely to be $\Omega((\log k)/\beta)$. To see why, we have $\mathbf{E}[V] = k\mathbf{E}[v_i] = (\ln k)/4\beta$, $\mathbf{Var}[V] = k\mathbf{Var}[v_i] \leq k\mathbf{E}[v_i^2] = 1/(16\beta^2)$, and $\sigma[V] = 1/4\beta$. By Chebyshev's Inequality, V is at least $(\ln k - 2)/4\beta = \Omega(\log k/\beta)$ with probability at least 3/4.

Let M be a mechanism that is truthful in expectation and β-budget-balanced in expectation, meaning that for every bid vector, the expected revenue of M is at least a β fraction of its expected cost. For a public excludable good problem, the expected cost equals 1 minus the probability that no player is served. We can finish the proof by showing that the expected revenue of M, over both the random choice of valuation profile and the internal coin flips of the mechanism, is at most $1/4\beta$: if true, the expected cost of M is at most 1/4, so no player is served with probability at least 3/4. By the Union Bound, the probability that no player is served and also the sum of the valuations is $\Omega((\log k)/\beta)$ is at least 1/2. Thus, there is a valuation profile for which the optimal social cost is 1 but the expected social cost of M is $\Omega((\log k)/\beta)$.

We next apply a transformation of Mehta and Vazirani [12], originally developed for digital goods auctions, to assist in upper bounding the revenue obtained by M. Given a bid vector b, a *randomized threshold mechanism* chooses a random threshold $t_i(b_{-i})$ for each player i (cf., Proposition 1) from a distribution that is independent of b_i. Such mechanisms are truthful in the universal sense. By Mehta and Vazirani [12], there is a randomized threshold mechanism M' that has the same expected revenue as M on every bid vector.

To upper bound the expected revenue of M', consider a single truthful player i with (random) valuation v_i. Every fixed threshold t extracts expected revenue $t \cdot \mathbf{Pr}[v_i \geq t] \leq 1/4k\beta$ from the player. By the Principle of Deferred Decisions, a randomized threshold that is independent of v_i also obtains expected revenue at most $1/4k\beta$ from player i. Linearity of expectation implies that the expected revenue of M', and hence of M, is at most $1/4\beta$, completing the proof.

Scaling the prices of the Shapley value mechanism down by a $\beta \geq 1$ factor gives a β-budget-balanced, $O(\beta + (\log k)/\beta)$-approximate mechanism [17]. Thus, the lower bound in Theorem 1 is optimal up to constant factors for all $\beta = O(\sqrt{\log k})$.

4 Deterministic, Symmetric Mechanisms: Characterizations and Lower Bounds

In this section we prove a lower bound on the social cost approximation factor of every deterministic, budget-balanced cost-sharing mechanism that satisfies the "equal treatment" property. We derive this lower bound from a new characterization of the Shapley value mechanism, discussed next.

Proposition 1 does not specify the behavior of a truthful mechanism when a player bids exactly its threshold $t_i(b_{-i})$. There are two valid possibilities, each of which yields zero utility to a truthful player: the player is not served (at price 0), or is served and charged its bid. The following technical condition breaks ties in favor of the second outcome.

Definition 1. *A mechanism satisfies* upper semi-continuity *if and only if the following condition holds for every player i and bids b_{-i} of the other players: if player i receives service at every bid larger than b_i, then it also receives service at bid b_i.*

We stress that while our characterization result (Theorem 2) relies on this condition, our lower bound (Corollary 1) does not depend on it.

Our results concern mechanisms satisfying the following symmetry property.

Definition 2. *A mechanism satisfies* equal treatment *if and only if every two players i and j that submit the same bid receive the same allocation and price.*

The Shapley value mechanism (Section 2) satisfies equal treatment and upper semi-continuity. It uses the same threshold function for each player, namely:

$$\forall\ b_{-i}: \qquad t(b_{-i}) = \frac{1}{f(b_{-i})+1}. \tag{2}$$

Here, $f(b_{-i})$ is the size of the largest subset S of $U \setminus \{i\}$ such that $b_j \geq 1/(|S|+1)$ for all $j \in S$. Intuitively, this is precisely the set of other players that the Shapley value mechanism services if player i pays its share and also receives service.

Our characterization theorem is the following.

Theorem 2. *A deterministic and budget-balanced cost-sharing mechanism satisfies equal treatment, consumer sovereignty, and upper-semicontinuity if and only if it is the Shapley value mechanism.*

Proof. Fix such a mechanism M. We first note that all thresholds $t_i(b_{-i})$ induced by M must lie in $[0, 1]$: every threshold is finite by consumer sovereignty, and is at most 1 by the budget-balance condition. We proceed to show that for all players i and bids b_{-i} by the other players, the threshold function t_i has the same value as that for the Shapley value mechanism. We prove this by downward induction on the number of coordinates of b_{-i} that are equal to 1.

For the base case, fix i and suppose that b_{-i} is the all-ones vector. Suppose that $b_i = 1$. Since all thresholds are in $[0, 1]$ and M is upper semi-continuous, all players are served. By equal treatment and budget-balance, all players pay $1/k$. Thus, $t_i(b_{-i}) = 1/k$ when b_{-i} is the all-ones vector, as for the Shapley value mechanism.

For the inductive step, fix a player i and a bid vector b_{-i} that is not the all-ones vector. Set $b_i = 1$ and consider the bid vector $b = (b_i, b_{-i})$. Let S denote the set of players j with $b_j = 1$. Let $R \supseteq S$ denote the output of the Shapley value mechanism for the bid vector b — the largest set of players such that $b_j \geq 1/|R|$ for all $j \in R$.

As in the base case, consumer sovereignty, budget-balance, and equal treatment imply that M serves all of the players of S at a common price p. For a player j outside S, b_{-j} has one more bid of 1 than b_{-i} (corresponding to player i), and the inductive hypothesis implies that its threshold is that of the Shapley value mechanism for the same bid vector b. For players of $R \setminus S$, this threshold is $1/|R|$. For a player outside R, this threshold is some value strictly greater than its bid. Since $b_j \geq 1/|R|$ for all $j \in R$ and M is upper semicontinuous, it serves precisely the set R when given the bid vector b. This generates revenue $|S|p + (|R| - |S|)/|R|$. Budget-balance dictates that the common threshold p for all players of S, and in particular the value of $t_i(b_{-i})$, equals $1/|R|$. This agrees with player i's threshold for the bids b_{-i} in the Shapley value mechanism, and the proof is complete.

Theorem 2 implies that the Shapley value mechanism is the optimal deterministic, budget-balanced mechanism that satisfies the equal treatment property.

Corollary 1. *Every deterministic, budget-balanced cost-sharing mechanism that satisfies equal treatment is at least $\mathcal{H}_k$-approximate.*

We briefly sketch the proof. Let M be such a mechanism. If M fails to satisfy consumer sovereignty, then we can find a player i and bids b_{-i} such that $t_i(b_{-i}) = +\infty$. Letting the valuation of player i tend to infinity shows that the mechanism fails to achieve a finite social cost approximation factor.

Suppose that M also satisfies consumer sovereignty. The proof of Theorem 2 shows that the outcome of the mechanism agrees with that of the Shapley value mechanism except on the measure-zero set of bid vectors for which there is at least one bid equal to $1/i$ for some $i \in \{1, \ldots, k\}$. As in Example 1, bid vectors of the form $1 - \epsilon$, $\frac{1}{2} - \epsilon, \ldots, \frac{1}{k} - \epsilon$ for small $\epsilon > 0$ show that M is no better than $\mathcal{H}_k$-approximate.

Remark 1. Other characterizations of the Shapley value mechanism are known. See Moulin and Shenker [14] and Immorlica, Mahdian, and Mirrokni [9] for related characterizations of *groupstrategyproof* mechanisms that satisfy various properties. (A groupstrategyproof mechanism is robust to coordinated false bids when there are no side payments between players. The Shapley value mechanism satisfies this strong incentive-compatibility condition.) Our Theorem 2 is incomparable to these results because we work with the much richer class of truthful, not necessarily groupstrategyproof, mechanisms. Our characterization is more similar to that of Deb and Razzolini [5], who also show that the Shapley value mechanism is the only one that satisfies certain conditions. We weaken their stand-alone condition to consumer sovereignty and do not require the voluntary non-participation condition. Also, our proof is arguably simpler.

An interesting research problem is to characterize the class of mechanisms obtained after dropping the (admittedly strong) equal treatment condition. There are several mechanisms that satisfy the remaining conditions and appear hard to characterize (e.g. [9, Example 4.1]).

5 The Power of Randomization

Theorem 1 shows that the best-possible approximation guarantee of a randomized cost-sharing mechanism cannot be more than a constant factor smaller than that of the (deterministic) Shapley value mechanism. We now show that randomized mechanisms are in fact strictly more powerful than deterministic ones, even in the two-player public excludable good problem.

Proposition 2. *Let M be a deterministic budget-balanced cost-sharing mechanism for the 2-player public excludable good problem. Then, M is at least* 1.5*-approximate.*

Proof. Consider the bid vector with $b_1 = b_2 = 1$. Every mechanism that provides an approximation ratio better than 2 must serve both players. Suppose this is the case and player 1 pays p while player 2 pays $1 - p$. Without loss of generality, assume that $p \leq 0.5$. By Proposition 1, player 2's threshold function satisfies $t_2(1) = 1 - p$.

Now suppose $b_1 = 1$ and $b_2 = 1-p-\epsilon$ for small $\epsilon > 0$. The optimal social cost is 1, with both players served. Since $t_2(1) = 1-p$, player 2 is not served by M. Whether or not player 1 is served, the incurred social cost is $1+1-p-\epsilon \geq 1.5-\epsilon$.

There is a randomized mechanism with strictly better approximate efficiency.

Proposition 3. *There is a universally truthful, budget-balanced, 1.25-approximate randomized mechanism for the two-player public excludable good problem.*

Proof. The mechanism starts by selecting $\gamma \in [0,1]$ uniformly at random. Then, players 1 and 2 are offered service at prices γ and $1-\gamma$, respectively. A player who refuses is not served. If both players accept, then both are served at their respective prices. If exactly one player accepts, it is served (at price 1) if and only if its bid is at least 1.

The mechanism is clearly universally truthful and budget-balanced with probability 1. To bound its expected social cost, assume truthful bids with $v_1 \geq v_2$ and define $x = v_1 + v_2 - 1$. If $x < 0$ then, with probability 1, neither player is served and this is optimal. If $v_2 \geq 1$, then both players are served with probability 1, which again is optimal.

The most interesting case is when $x, v_1, v_2 \in [0,1]$. The optimal social cost in this case is 1. The mechanism selects a γ such that $v_1 \geq \gamma$ and $v_2 \geq 1-\gamma$ with probability x. In this event, both players are served and the incurred social cost is 1. Otherwise, neither player is served and the social is $1+x$. The expected approximation ratio obtained by the algorithm for this valuation profile is $x \cdot 1 + (1-x) \cdot (1+x)$. Choosing $x = 0.5$ maximizes this ratio, at which point the ratio is 1.25.

Finally, if $v_1 \geq 1$ but $v_2 < 1$, both players are served with probability v_2, and the mechanism serves only player 1 otherwise. The optimal social cost is again 1 and the expected social cost incurred by the mechanism is $v_2 \cdot 1 + (1-v_2)(1+v_2)$. This quantity is maximized when $v_2 = 0.5$, at which point the expected social cost (and hence the expected approximation ratio) is 1.25.

Unfortunately, universally truthful mechanisms cannot help further.

Proposition 4. *Let M be a universally truthful, budget-balanced cost-sharing mechanism for the two-player public excludable good problem. Then M is no better than 1.25-approximate.*

Proof. By Yao's Minimax Principle, we only need to exhibit a distribution over valuation profiles so that the approximate efficiency of every deterministic budget-balanced mechanism is large.

Let M be a deterministic, budget-balanced truthful mechanism. Let t_1 and t_2 denote the threshold functions for M in the sense of Proposition 1. Since M is budget-balanced on the bid vector $(1,1)$, $t_1(1) + t_2(1) = 1$. Fix $\epsilon > 0$ and randomize uniformly between the profiles $v_1 = 1, v_2 = (1/2) - \epsilon$ and $v_1 = (1/2) - \epsilon, v_2 = 1$. The optimal social cost is 1 for both of these profiles. Since either $t_1(1) \geq 1/2$ or $t_2(1) \geq 1/2$, the expected social cost of M is at least $(1/2) \cdot 1 + (1/2) \cdot (1 + (1/2) - \epsilon)$, which tends to $5/4$ as $\epsilon \to 0$.

References

1. Blumrosen, L., Nisan, N.: Combinatorial auctions. In: Nisan, N., Roughgarden, T., Tardos, E., Vazirani, V. (eds.) Algorithmic Game Theory
2. Brenner, J., Schäfer, G.: Cost sharing methods for makespan and completion time scheduling. In: Thomas, W., Weil, P. (eds.) STACS 2007. LNCS, vol. 4393, Springer, Heidelberg (2007)
3. Clarke, E.H.: Multipart pricing of public goods. Public Choice V11(1), 17–33 (1971)
4. Deb, R., Razzolini, L.: Auction-like mechanisms for pricing excludable public goods. Journal of Economic Theory 88(2), 340–368 (1999), `http://ideas.repec.org/a/eee/jetheo/v88y1999i2p340-368.html`
5. Deb, R., Razzolini, L.: Voluntary cost sharing for an excludable public project. Mathematical Social Sciences 37, 123–138 (1999)
6. Feigenbaum, J., Krishnamurthy, A., Sami, R., Shenker, S.: Hardness results for multicast cost sharing. Theoretical Computer Science 304, 215–236 (2003)
7. Green, J., Kohlberg, E., Laffont, J.J.: Partial equilibrium approach to the free rider problem. Journal of Public Economics 6, 375–394 (1976)
8. Groves, T.: Incentives in teams. Econometrica 41(4), 617–631 (1973)
9. Immorlica, N., Mahdian, M., Mirrokni, V.S.: Limitations of cross-monotonic cost-sharing schemes. In: Proceedings of the 16th Annual ACM-SIAM Symposium on Discrete Algorithms (SODA), pp. 602–611 (2005)
10. Lavi, R.: Computationally efficient approximation mechanisms. In: Nisan, N., Roughgarden, T., Tardos, E., Vazirani, V. (eds.) Algorithmic Game Theory
11. Mehta, A., Roughgarden, T., Sundararajan, M.: Beyond Moulin mechanisms. In: EC 2007: Proceedings of the 8th ACM conference on Electronic commerce, pp. 1–10 (2007)
12. Mehta, A., Vazirani, V.V.: Randomized truthful auctions of digital goods are randomizations over truthful auctions. In: ACM Conference on Electronic Commerce, pp. 120–124 (2004)
13. Moulin, H.: Incremental cost sharing: Characterization by coalition strategy-proofness. Social Choice and Welfare 16, 279–320 (1999)
14. Moulin, H., Shenker, S.: Strategyproof sharing of submodular costs: Budget balance versus efficiency. Economic Theory 18, 511–533 (2001)
15. Nisan, N., Ronen, A.: Algorithmic mechanism design. In: STOC 1999 (1999)
16. Roberts, K.: The characterization of implementable choice rules. In: Laffont, J.J. (ed.) Aggregation and Revelation of Preferences, North-Holland, Amsterdam (1979)
17. Roughgarden, T., Sundararajan, M.: New trade-offs in cost-sharing mechanisms. In: Proceedings of the 38th Annual ACM Symposium on the Theory of Computing (STOC), pp. 79–88 (2006)
18. Vickrey, W.: Counterspeculation, auctions, and competitive sealed tenders. Journal of Finance 16(1), 8–37 (1961)

Non-cooperative Cost Sharing Games Via Subsidies

Niv Buchbinder[1], Liane Lewin-Eytan[2], Joseph (Seffi) Naor[1], and Ariel Orda[2]

[1] Computer Science Department, Technion, Haifa, Israel
{nivb,naor}@cs.technion.ac.il
[2] Department of Electrical Engineering, Technion, Haifa, Israel
liane@tx.technion.ac.il, ariel@ee.technion.ac.il

Abstract. We consider a cost sharing system where users are selfish and act according to their own interest. There is a set of facilities and each facility provides services to a subset of the users. Each user is interested in purchasing a service, and will buy it from the facility offering it at the lowest cost. The notion of *social welfare* is defined to be the total cost of the facilities chosen by the users. A central authority can encourage the purchase of services by offering subsidies that reduce their price, in order to improve the social welfare. The subsidies are financed by taxes collected from the users. Specifically, we investigate a non-cooperative game, where users join the system, and act according to their *best response*. We model the system as an instance of a set cover game, where each element is interested in selecting a cover minimizing its payment. The subsidies are updated dynamically, following the selfish moves of the elements and the taxes collected due to their payments. Our objective is to design a *dynamic* subsidy mechanism that improves on the social welfare while collecting as taxes only a small fraction of the sum of the payments of the users. The performance of such a subsidy mechanism is thus defined by two different quality parameters: (i) the *price of anarchy*, defined as the ratio between the social welfare cost of the Nash equilibrium obtained and the cost of an optimal solution; and (ii) the *taxation ratio*, defined as the fraction of payments collected as taxes from the users.

1 Introduction

Individual self-interest is the basis for the modern market system in which a consumer acts in its self-interest when buying goods at lowest prices. A government, or any other central authority, can influence natural market forces in several ways, such as taxation or regulation. In cases where a government wishes to support and encourage the production of a good that is regarded as being in the public interest, it gives out an assistance called a *subsidy* (also called negative taxation). Subsidies are thus a way to influence the state of the market in a world of independent self-interested consumers.

An example where government supervision can be very effective is an urban passenger transportation system. An employee commuting to work in a city usually has many transportation options. He can use a private car, join a car-pool, or use public transportation, e.g., bus or a train. The common choices as to how to travel to work have significant environmental impacts and a major influence on road traffic congestion. It is thus a governmental interest to reduce the number of single occupancy vehicles on the road and encourage people to use public transport when commuting to and from work.

B. Monien and U.-P. Schroeder (Eds.): SAGT 2008, LNCS 4997, pp. 337–349, 2008.

Letting the invisible hand of the free market take its course can sometimes be devastating. Consider, for example, a setting in which a new building is being built. Each new resident can either purchase a private car, or initiate the use of some public transport at a much higher cost. As no bus line is available at the new residence when it is established, the cheapest way for each new resident to commute is to buy his own car, and then no public transport will ever be established. Thus, in this case, it is the role of a central authority to develop public transport by offering subsidies. After public transportation means are established, it is likely that residents will switch from private to public transport, since the latter are cheaper.

Central authorities have limited budgets. Therefore, subsidies are financed by taxes collected from the users. A central authority thus aims to improve on the overall system performance, while using a policy that collects taxes constituting a bounded fraction of the total payments made by the users.

Our Model. We investigate a system where facilities provide services to users. Each user is interested in purchasing a service which is typically provided by only a subset of the facilities. Users naturally buy the service from the facility offering it at the lowest cost. A central authority can encourage the purchase of services by offering subsidies that reduce their price. We investigate settings where users share services and thereby also share their cost. Back to the public transportation example, each transportation option corresponds to a different facility having a different cost. The cost of each facility is essentially the cost of operating the type of transport it represents. The cost of a facility that provides service to several users is shared amongst them, and can be subsidized by the central authority in order to shift market share, e.g., from cars to public transport.

We model the system as an instance of the set cover problem. Let $N = \{1, 2, \ldots, n\}$ be a ground set of n elements (the users), and let $\mathbb{S}$ be a family of subsets of N, $|\mathbb{S}| = m$ (the facilities). A *cover* of $N' \subseteq N$ is a collection of sets such that their union contains N'. In our public transportation example, a cover is a choice of transport types allowing all users belonging to N' to get to work. The notion of *social welfare* of a collection of sets $\mathbb{T}$ is defined to be the total cost of the sets belonging to $\mathbb{T}$. Each subset $s \in \mathbb{S}$ has a non-negative *cost* c_s associated with it. Each subset is also associated with a subsidy value (possibly equal to zero, in case no subsidy is offered to the set). The *effective cost* of a set s, denoted by $\hat{c}_s$, is defined to be c_s minus the subsidy associated with s.

In a feasible cover, each user is assigned to one of the sets in the cover containing it. Users sharing the same set also share its effective cost. We consider an egalitarian cost sharing mechanism, which evenly splits the effective cost of a set among its users. More precisely, if n_s users use set s, then each user pays $\hat{c}_s/n_s$ for this set. This cost sharing mechanism has an intuitive appeal, and satisfy essential properties such as cross monotonicity (the cost share of a user for using a set cannot increase when additional users join the set) and budget balance (the sum of the payments of the users receiving service from a set is equal to its effective cost).

The Non-Cooperative Game. We consider a set cover game with selfish non-cooperative players (also called users, or elements). Each player is interested in selecting a cover that minimizes its payment. Thus, the strategies of the players in the game correspond to the different sets that can provide service to the players. Each player independently chooses a strategy minimizing its payment, i.e., its *best response*. The best response of a player in the set cover game is thus defined as the set(s) that can provide service to the player at minimum cost (with respect to the *current* state of the system). The mutual influence of the players is determined by the egalitarian cost sharing mechanism.

We focus on a dynamic setting, where players follow the natural game course induced by *best-response dynamics*. Each player, in his turn, chooses a cover that minimizes his cost. We start the game from an empty configuration; upon arrival, a user chooses a cover selfishly. As a result, players that have joined the game previously may change their strategy later on by choosing a cover of lower cost. The central authority is allowed to increase the subsidies of the sets in every step of the game in order to improve the social welfare of the final cover. We assume that the game is controlled by an adversarial scheduler that decides which user plays in each step. The order by which the users play is not known beforehand (as it is chosen adversarially) as well as the set of elements (users) $N' \subseteq N$ that actually participates in the game. (Note that N' may be a strict subset of N in general.) However, we assume that the set cover instance, i.e., N and $\mathbb{S}$, is known in advance.

The natural game course continues until Nash equilibrium is reached. A Nash equilibrium of the set cover game corresponds to a choice of covers for all users in N', where no user can unilaterally reduce its payment by choosing a different cover. We note that the set cover game is a special case of the well known class of *congestion games* [9]. Rosenthal [9] showed that a potential function can be defined for each congestion game with the property that it decreases in case a player makes a move that improves his cost, thus establishing convergence to Nash equilibrium. We note that as the subsidies can only lower the cost of the users, the potential function of the set cover game decreases in the presence of a subsidy mechanism as well, and convergence to Nash equilibrium is still guaranteed.

The Nash equilibrium of the set cover game is not unique and the greedy nature of the users could lead to Nash equilibrium points with a very high price of anarchy, even when initializing the game from an empty configuration. We use subsidies in order to guarantee that best-response dynamics will not converge to such bad equilibria. The following example is instructive as to why subsidies are needed for minimizing the cost of the final solution when considering an arbitrary set system. Consider n users where each user can be covered by a unit-cost "private" set containing only herself. There is also a set containing all the users that costs $\sqrt{n}$. The users appear one by one and the best response of each user is to pick the private set covering him. Once a user picks his private set, he will have no incentive to change his strategy. How will the set covering all the users (the social optimal solution) be chosen without subsidies?

Ideally, we would like our subsidy mechanism to spend on subsidies only a bounded fraction of the revenue. However, in a non-cooperative game setting, the users' payments are dynamic, and can vary significantly during the game course due to strategy changes. Consider, for example, a set s shared by many users, who decide to leave it at some point of the game in order to join other subsidized sets. In this case, the revenue that was accrued from the use of s is now reduced to zero. In order to cope with such dynamic scenarios, we propose a natural framework where subsidies are offered via *taxes*. A tax is a non-refundable sum paid to the central authority only in case a user purchases a new set. It is equal to a fixed fraction of the effective cost of the purchased set. The central authority collects as taxes a fraction of the payments made by the users that open new sets, and later on offers the revenue from the taxes as subsidies. The total amount of subsidies offered should always be bounded by the amount of taxes collected. The performance of a subsidy mechanism is a function of two quality parameters:

- **The price of anarchy:** The ratio between the social welfare cost of a Nash equilibrium solution (that is, the sum of the subsidies and the payments of the users) and the cost of an optimal solution.
- **The taxation ratio:** The fraction of the payments collected as taxes from the users.

There is a trade-off between the taxation ratio and the price of anarchy achieved by our subsidy mechanism. The higher the fraction of payments collected as taxes and spent on subsidies is, the lower the cost of the final solution becomes. The taxation ratio is determined by a parameter $\epsilon \leq 1$ which is given as input to the subsidy mechanism. Denoting by P the total payments of the user, the objective is to achieve the best price of anarchy while collecting taxes (and spending on subsidies) at most ϵP.

Compare the set cover game to the multicast game [3,6] in which users (terminals) connect to a source by making a routing decision that minimizes their payment. Chuzhoy et al [6] analyze the price of anarchy of a Nash equilibrium resulting from the best-response dynamics of a game course in which the players first join the game sequentially beginning from an *empty configuration*. Their setting is thus a special case of our model in which there is no central authority intervention. It is shown in [5] that the price of anarchy of this setting is $O(\log^3 n)$. In the multicast setting, unlike the set cover game, an initial empty configuration coupled with best response dynamics does guarantee a low price of anarchy with no need to offer subsidies.

Results and Techniques. We consider two different models: (i) an *integral* model in which sets can only be fully bought (i.e., integrally) and each element is covered by a single set; and (ii) a *fractional* model in which a fraction of a set can be bought and each user can be covered by several sets (provided that their fractions add up to 1). Note that the fraction of coverage an element gets from a set cannot be greater than the fraction associated with the set. The subsidies, similarly to the choices of the sets, can be given either integrally or fractionally, depending on the model. In the fractional model, the central authority is allowed to subsidize only a fraction of a set.

The importance of the fractional model is two-fold. First, a fractional solution is a first step towards obtaining an integral solution. Second, it is interesting in its own right as it captures many practical "fractional" scenarios. In the urban transport system example, a fractional solution can correspond to the case where a user uses different transport options during the week. Then, subsidizing a fraction of a set can be interpreted as subsidizing a transportation mean only during part of the day, or part of the week. Let f denote the maximum *frequency* of an element, that is, the maximum number of sets that an element can belong to. For the fractional model, we prove the following theorem:

Theorem 1 (Fractional Cover). *There exists a subsidy policy such that, for any* $\epsilon \leq 1$*, the price of anarchy is* $O(\frac{\log f}{\epsilon})$ *and the taxation ratio is* ϵ.

Theorem 1 provides a trade-off between the taxation ratio and the price of anarchy: as the central authority collects a higher fraction of the payments as taxes (later on invested in subsidies), the cost of the final solution decreases.

For the integral model, we obtain the following slightly inferior bound.

Theorem 2 (Integral Cover). *There exists a subsidy policy such that, for any $\epsilon \leq 1$, the price of anarchy is $O(\frac{\log f \log(\frac{n}{\epsilon})}{\epsilon})$ and the taxation ratio is ϵ.*

In order to design mechanisms for both the fractional and integral models, we draw on ideas from [1,4]. In [1] Alon et al. considered an online version of the set cover problem, where elements arrive one by one and need to be covered upon arrival. The goal of [1] is to design an online algorithm achieving the best possible competitive ratio with respect to the optimal solution, i.e., optimal social welfare. In [1], a $O(\log m \log n)$-competitive algorithm is presented for this online setting.

In our work, we take into consideration not only the overall system performance, but also the selfish nature of the users who play according to their best response. As the goal of the users is to minimize the payment for their cover, they may change their strategy after joining the system until Nash equilibrium is reached. We thus go beyond the online version of the set cover problem considered in [1], and analyze its non-cooperative game extension. The model investigated in [1] can be seen as a special case of ours, where the central authority pays for the full cost of the cover and users pay nothing. (Also, users join one by one without reaching equilibrium.) Thus, using the algorithm of [1], a central authority will not be able to finance the subsidies from taxes. Bounding the taxation ratio while maintaining a low price of anarchy requires a new algorithm and a different analysis achieved via the primal-dual approach of [4].

We note that for both the integral and fractional models our results are tight, since our subsidy mechanism applies also to the special case of an online setting where users join the system one by one and act according to their *best response* by choosing a cover of minimum cost. In case ϵ is a fixed constant, the resulting price of anarchy almost matches the lower bound of $\Omega(\frac{\log m \log n}{\log \log m + \log \log n})$ shown in [1] for the online setting.

In the fractional model, our subsidy mechanism keeps a bounded taxation ratio by investing money in subsidies only when a user purchases a new set (or a fraction thereof), and pays a tax. The idea is therefore to invest in subsidies in each iteration only a small fraction of the payment of a user (corresponding to the tax paid). The total cost of the taxes is bounded by maintaining an (almost) feasible dual solution during the execution of the algorithm, which also allows us to bound the price of anarchy of the solution. As in [1], we maintain a fractional primal solution, however, in our case, it is not always feasible. Rather, the feasibility of the cover is obtained by the best response of the users joining the system. The primal solution we maintain corresponds to the subsidies given by the central authority.

Developing an integral subsidy mechanism requires several more ideas. As opposed to the fractional case, it is no longer possible to offer in each iteration a fraction of the user's payment. Instead, the algorithm keeps a bounded taxation ratio by giving an integral subsidy only after accumulating the taxes paid by the users over several iterations. In [1], Alon et al. obtained an integral solution for their online setting by maintaining at each iteration a fractional feasible solution and using a potential function that determines which of the sets should be chosen to the integral cover. We design a new potential function and note that the potential function defined in [1] cannot satisfy our needs, as it would lead to a high taxation ratio. The analysis we perform is more delicate and bounds both the price of anarchy and the taxation ratio of the algorithm.

Perspective on Other Approaches that Improve on the Social Welfare. The issue of improving on the overall system performance even in the face of selfish behavior has been considered extensively in the game theory literature, and designing mechanisms to improve the coordination of selfish agents is a well known idea. A central topic in game theory is the notion of mechanism design in which rules of a game are designed to achieve a specific outcome. This is done by setting up a structure where players are paid (or penalized), and thus each player has an incentive to behave as the designer intends. Planning such a mechanism is based on an assumption that the players have private information known only to them and which affects their decisions.

Coordination mechanisms [7,8] is another concept that tries to improve the performance in systems with independent selfish and non-colluding agents. A coordination mechanism attempts to redesign the system by selecting policies and rules of the game (for example, adding delays and priorities to a congestion game [7]). An important aspect of both mechanism design and coordination mechanisms is that the designer must design the system once and for all. In contrast, in our setting, the policy of the central authority is dynamic, changes over time, and is determined by the state of the system.

Extensions. Our fractional subsidy algorithm can be generalized for the game extensions of the wide range of online graph and network optimization problems considered in [2] and which concern connectivity and cut problems in graphs. In a general online connectivity problem, there is a communication network known to the algorithm in advance, where each edge in the network has a nonnegative cost. The connectivity demands, specifying subsets of vertices to be connected, arrive online. The notion of *social welfare* of a subgraph G is defined to be the total cost of the edges belonging to G. Thus, an optimal solution with respect to the overall system performance consists of a minimum cost subgraph satisfying all connectivity demands. The algorithm presented in [2] satisfies each new demand, so as to achieve the best possible competitive ratio.

In the non-cooperative game version of these problems, a user corresponds to a connectivity demand, and is thus interested in choosing a minimum cost subgraph satisfying its own demand. The central authority is allowed to subsidize the costs of some of the edges by collecting taxes, in order to improve on the overall system performance. The game extensions of this range of problems belong to the class of congestion games [9], and thus their natural game course induced by best-response dynamics converges to a Nash equilibrium. Our subsidy algorithm achieves a taxation ratio of ϵ, while maintaining a price of anarchy of $O(\frac{\log m}{\epsilon})$, where m is the number of edges in the graph. Examples of problems belonging to this class are fractional versions of Steiner trees, generalized Steiner trees, and the group Steiner problem. It remains an open question whether an integral solution can be obtained for this set of problems as well.

2 Formal Definitions

In this section we formally describe our model. Let $N = \{1, 2, \ldots, n\}$ be a ground set of n elements (the users), and let $\mathbb{S}$ be a family of subsets of N, $|\mathbb{S}| = m$ (the facilities). Each $s \in \mathbb{S}$ has a non-negative *cost* c_s associated with it. Let f be the maximum frequency of an element, i.e., the maximum number of sets that can contain an element. A *cover* is a collection of sets such that their union is N. The *cost* of a cover is the sum of the costs of the sets that are included in the cover. A *fractional cover* is an assignment of weights, w_s, to each $s \in \mathbb{S}$, such that the total weight of the sets that contain each element is at least 1. The cost of a fractional cover is $\sum_{s\in\mathbb{S}} w_s c_s$. A linear programming formulation of the minimum fractional set cover problem appears in Figure 1. We have a variable w_s for each set $s \in \mathbb{S}$ indicating the fraction of set s that is taken to the cover. For each element, we demand that the sum of the fractions of the sets that contain the element is at least 1. In the dual program (see also Figure 1) we have a variable y_e corresponding to each of the elements. We require that the total sum of variables that correspond to elements that belong to a set s is at most the cost of the set. The integral set cover problem corresponds to the special case where $w_s \in \{0, 1\}$.

Primal		Dual	
Minimize:	$\sum_{s\in\mathbb{S}} c_s w_s$	Maximize:	$\sum_{e\in N} y_e$
Subject to:		Subject to:	
$\forall e \in N$:	$\sum_{s\mid e\in s} w_s \geq 1$	$\forall s \in \mathbb{S}$:	$\sum_{e\in s} y_e \leq c_s$
$\forall s \in \mathbb{S} \mid w_s \geq 0$		$\forall e \in N \mid y_e \geq 0$	

Fig. 1. A primal-dual pair for the set-cover problem

Cost Shares & Subsidies Structures. We turn to define the set cover game more precisely. As our subsidy mechanism works under fairly general assumptions, not all the definitions here are needed for the algorithms and analysis in the next sections. Rather, any setting where the subsidies offered are fully financed from taxes, is sufficient. We provide here a precise and natural definition of the game for completeness.

For simplicity, we assume taxes are collected only when a user purchases a new set (or a fraction thereof). This can happen either when a user joins the system or when it changes its strategy. The cost c_s of a new set s that has not yet been opened (purchased) is called its *opening cost*. When a new set is opened by a user, a fraction equal to ϵ of its opening cost is collected as tax. The *operating* cost of a facility is defined to be its opening cost minus the payment collected as tax. The operating cost of a facility that provides service to several users is shared amongst them. The tax paid by a user is non-refundable, while the remaining part of the payment (the operating cost) is a variable amount which may decrease when additional users join the same set and share its cost. We note that our mechanism can support other settings where both taxes and operating costs are shared by the users, as long as taxes are non-refundable, and can thus cover the subsidies. Each set is associated with a subsidy value (possibly zero, in case no subsidy is offered to the set). The subsidy can be applied either to the opening cost, in case the set has not been opened yet, or to the operating cost in case it is used

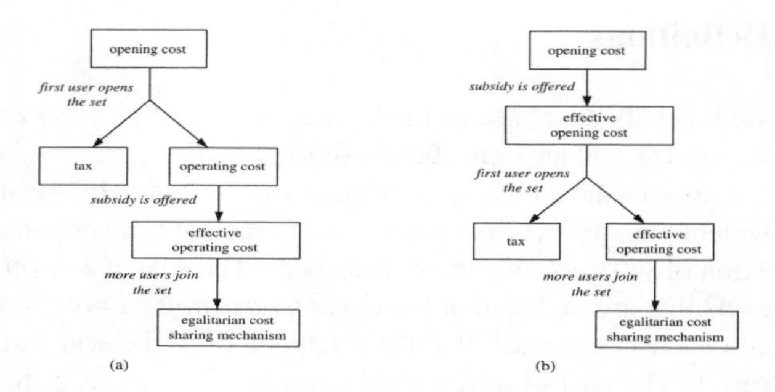

Fig. 2. (a) A user opens a new set that is not subsidized, and pays its opening cost, consisting of a tax and the operating cost. The subsidy offered lowers the operating cost, changing it to effective operating cost. In case more users join the set, they share its effective operating cost. (b) A subsidy is offered to a set that has not been opened yet. The first user joining the set pays its effective opening cost, consisting of a tax and the effective operating cost. In case more users join the set, they share its effective operating cost.

by at least one user (see Figure 2). In the latter case, subsidies are given to a set that has already been purchased so as to lower its cost and encourage more users to join it. The *effective opening cost* $\hat{c}_s$ of a set s, is defined to be its opening cost minus the subsidy associated with s. The *effective operating cost* of s is defined similarly with respect to the operating cost of the set.

In the *integral* model, sets are taken integrally and each element is covered by a single set. The effective opening cost $\hat{c}_s$ that a user will have to pay for purchasing a new set s that is not subsidized, is composed of a non-refundable tax of $\epsilon \cdot \hat{c}_s$, and a variable payment of $(1-\epsilon)\hat{c}_s$ that is equal to the effective operating cost of the set. In case a user joins a set s that is not subsidized, and shared by other users, its payment is equal to $(1-\epsilon)\hat{c}_s/n_s$, where n_s is the number of users sharing s.

In the *fractional* model, each set s is associated with a fraction x_s which is fully subsidized (that is, its effective cost equals zero). The cost of any other fraction of this set, that is, a fraction λ that is not subsidized, is equal to $\lambda \cdot c_s$. Each element can be covered by several fractions of different sets adding up to 1. Denote the fraction of set s used by user i by $\lambda_{s,i}$ and the number of users using set s by n_s. Assume without loss of generality that $\lambda_{s,1} \leq \lambda_{s,2} \leq \ldots \leq \lambda_{s,n_s} \leq 1$. Define $\lambda_{s,0} = x_s$. The cost of each fraction of s is as follows: the interval $[\lambda_{s,j-1}, \lambda_{s,j}]$ is shared by $(n_s - j + 1)$ users, where the variable payment of each user equals $(\lambda_{s,j} - \lambda_{s,j-1}) \cdot (1-\epsilon) \cdot c_s/(n_s - j + 1)$. The first user who opened the interval $[\lambda_{s,j-1}, \lambda_{s,j}]$ will also pay a non-refundable tax equal to $(\lambda_{s,j} - \lambda_{s,j-1}) \cdot \epsilon \cdot c_s$.

Nash Equilibrium Existence & Convergence. For both the fractional and integral models, the set cover game always converges to a Nash equilibrium. This property is established by means of a global potential function Φ on the strategy space. Given a strategy profile $\mathcal{T}$ consisting of the integral cover choices of all players, the potential function $\Phi(\mathcal{T})$ defined for our integral set cover game is the following:

$$\Phi(\mathcal{T}) = \sum_{s \in \mathcal{T}} \epsilon \cdot \hat{c}_s + \sum_{s \in \mathcal{T}} \left(\sum_{j=1}^{n_s} \frac{(1-\epsilon)\hat{c}_s}{j} \right).$$

The potential of the fractional model follows directly, as each fraction λ_s of set s can be considered as a different set, with opening cost $\lambda_s \cdot c_s$ and operating cost $\lambda_s \cdot (1-\epsilon)c_s$.

3 The Fractional Model

In this section we design a fractional subsidy algorithm that is executed by the central authority. The algorithm receives as input a parameter $\epsilon \leq 1$, and generates a solution with price of anarchy $O(\frac{\log f}{\epsilon})$, and taxation ratio ϵ. The subsidy algorithm runs in iterations, where each iteration corresponds to a new set fraction purchased by some user. This can be the case either when a new user joins the system, or when an existing user changes its strategy. In each such case, the user pays as tax an ϵ fraction of its payment. The amount of subsidies given in each iteration is bounded by the amount of collected taxes, thus allowing the subsidies to be fully financed from taxes. The central authority does not determine which sets are chosen to the cover by the users. The only guarantee is that the users act according to their best response by choosing a fractional cover of minimum cost. Each iteration of the algorithm solely consists of an update of the subsidies. Consider an element (user) e that either joins the system or changes its strategy. There are four different types of set fractions that can be chosen by e.

1. Fractions that are fully subsidized. These fractions have zero cost.
2. Fractions that are not subsidized, yet are used by other users. A user joining such fractions does not have to pay any tax. The operating cost of such a fraction is evenly split between its users.
3. Fractions that are not subsidized and are **not** used by other users. A user choosing these fractions will have to pay their full cost (tax and a full operating cost).
4. Fractions that have been previously opened, but are currently **not** used by any user (users left them following strategy changes). A user joining such a fraction does not have to pay any tax (as a tax was paid when opening it for the first time), but has to pay its full operating cost.

The minimum cost cover chosen by element e consists of the lowest cost combination of these four types of fractions adding up to one. In case the element chooses a second-type fraction, its payment will lower the payments of other elements using the fraction, but will not have any effect on the cost of the solution. In addition, we do not consider in our analysis fractions that are "deserted" following strategy changes performed by users. As we do not reduce the cost of such fractions once they are unused, we do not take them into account either when an element chooses a fourth-type fraction. Moreover, as both fractions of type three and four are chosen from the minimum cost feasible set that covers e, any user that already opened a new fraction of a set in the past will always prefer to return to this fourth-type fraction (that requires no tax payment), before opening new third-type fractions (that require tax payment).

Thus, following the best response of user e, the total cost of the solution increases only due to third-type fractions. Note that a third-type fraction is chosen by e in order to "complete" its cover, after choosing all possible first, second, and fourth-type fractions of lower cost. Let ρ be the third-type fraction chosen by e. User e chooses the fraction ρ from the minimal cost set that covers it. Let $c_{\min}$ be the opening cost of this minimal set.

We refer to ρ as the *greedy choice*, or *greedy cover* of the user, and to $\rho \cdot c_{\min}$ as its *greedy cost*. Let x_s be the fraction of set s that is subsidized by the central authority. Initially, $x_s = 0$ for all sets, and the dual variables $y_e = 0$ for all elements. The algorithm that updates the subsidies offered by the central authority is the following:

Fractional Subsidy Algorithm (with input ϵ):
When user e purchases a new (third-type) set fraction:

1. $y_e \leftarrow y_e + \epsilon \cdot \rho \cdot c_{\min}$
2. For each set s that contains e: $x_s \leftarrow x_s \cdot \left(1 + \frac{\epsilon \cdot \rho \cdot c_{\min}}{2c_s}\right) + \frac{\epsilon \cdot \rho \cdot c_{\min}}{f \cdot 2c_s}$

The variables y_e are the variables of the dual linear program of the fractional set cover problem (Figure 1). These variables are used to maintain an (almost) feasible dual solution. The cost of the dual solution allows us to bound both the price of anarchy and the taxation ratio of the algorithm. Note that the value of the primal variables w_s, indicating the fraction of set s that is taken to the cover (Figure 1), is determined both by the third-type fractions chosen by the user, and by the subsidized fractions x_s.

Let Δx_s^i be the change of x_s in the ith iteration (i.e., the additional subsidy given to set s). The amount of subsidies given in the ith iteration is $\sum_{s \in \mathbb{S}} \Delta x_s^i c_s$. We show that this amount is bounded by $\epsilon \cdot \rho \cdot c_{\min}$, which is the tax paid by the user. To do so, we establish a relationship between the fractional greedy cost G, the fractional subsidy cost F and the total profit D of the dual solution we produce. Note that in each iteration the amount of taxes collected is exactly the change in the dual cost ($\epsilon \cdot \rho \cdot c_{\min}$). Let ΔG_i, ΔF_i and ΔD_i be the change of the fractional greedy cost, the fractional subsidy cost, and the dual cost, respectively, in the ith iteration.

Lemma 1. *In each iteration i, $\Delta G_i = \frac{1}{\epsilon} \Delta D_i$, and $\Delta F_i \leq \Delta D_i$. Thus, $\Delta F_i / \Delta G_i \leq \epsilon$.*

Proof. In each iteration the greedy cost $\Delta G_i = \rho \cdot c_{\min}$, and $\Delta D_i = \epsilon \cdot \rho \cdot c_{\min}$. Thus, $\Delta G_i = \frac{1}{\epsilon} \cdot \Delta D_i$. In case $\sum_{s|e \in s} x_s > 1$, element e is covered by fully subsidized set fractions and thus $\rho = 0$. Thus, we get that in each iteration, the subsidy cost, ΔF_i, is:

$$\sum_{s|e \in s} c_s \frac{\epsilon \cdot \rho \cdot c_{\min}}{2c_s} \left(x_s + \frac{1}{f}\right) \leq \epsilon \cdot \rho \cdot c_{\min} \leq \Delta D_i. \tag{1}$$

□

As ΔD_i equals the amount of new taxes collected in the ith iteration, ΔF_i is the subsidy cost in the ith iteration, and ΔG_i is the opening cost of the new set fraction purchased in the same iteration, the next corollary follows directly.

Corollary 1. *The taxation ratio of the fractional subsidy algorithm is ϵ. Moreover, the subsidy cost is bounded by the amount of taxes collected.*

Lemma 2. *The dual solution D produced is feasible up to factor of $O(\log f)$.*

Theorem 3. *The price of anarchy of the final solution is $O(\frac{\log f}{\epsilon})$, and the taxation ratio is ϵ.*

Proof. The taxation ratio follows by Corollary 1. Let D' be a feasible dual solution obtained from D by dividing it by $O(\log f)$. The total cost of the solution is bounded by the sum of the subsidies given by the central authority and the total greedy cost. By Lemmas 1 and 2 we get that the total cost of the solution is then at most:

$$G+F=O\left((1+\frac{1}{\epsilon})\log f\right)D'=O\left(\frac{\log f}{\epsilon}\mathrm{OPT}\right). \qquad \square$$

4 The Integral Model

In this section we show how to obtain a subsidy algorithm for the integral version of the problem, which requires several more ideas and a careful analysis. The algorithm receives as input a parameter $\epsilon \leq 1$, and generates a solution with price of anarchy $O\left(\frac{1}{\epsilon}\log f \log\left(\frac{n}{\epsilon}\right)\right)$, and taxation ratio ϵ.

Let OPT be the cost of an optimal integral solution. We design a subsidy algorithm that computes a solution with the properties stated above, given the value of OPT. Note that we can assume (using doubling) that the value of OPT is known up to a factor of 2. The complete subsidy algorithm runs in phases, as follows. We start by guessing $\alpha = \min_{s\in\mathbb{S}} c_s$. If it turns out that the total cost of the solution exceeds $\Theta(\alpha\frac{\log f \log(\frac{n}{\epsilon})}{\epsilon})$, we update the value of α by doubling it, and start a new phase by restarting the algorithm from the current event. Since the success of our algorithm is guaranteed whenever $\alpha \geq$ OPT, then it holds in the last phase that $\alpha \leq 2\mathrm{OPT}$. Therefore, the total cost of the solution is the sum of a geometric sequence which is at most twice the bound on the cost of the last phase of our algorithm. Moreover, this does not influence the taxation ratio, that is ϵ in each such phase separately. Note that as we guess the value of the optimum solution, we can ignore all sets with cost greater than α, since these sets cannot belong to an optimal solution.

The algorithm maintains a variable $x_s \geq 0$ for each $s \in \mathbb{S}$, and updates it as in the fractional case. Unlike the fractional case, these variables do not denote (fractional) subsidies. Rather, the value of x_s is used in order to determine whether the set s should be (fully) subsidized. Let $x_j = \sum_{s\in\mathbb{S}_j} x_s$ for each element $j \in N$, where $\mathbb{S}_j$ denotes the collection of sets containing element j. We define $\mathbb{C}$ to be the family of sets in $\mathbb{S}$ that are chosen to the cover, either by the greedy choices of the users, or by the central authority, and define $\tilde{\mathbb{C}} \subseteq \mathbb{C}$ to be the family of sets that are (fully) subsidized. We denote by C and $\tilde{C}$ the set of all elements covered by the members of $\mathbb{C}$ and $\tilde{\mathbb{C}}$, respectively. The following potential function is used throughout the algorithm:

$$\Phi(\epsilon)=\sum_{j\notin\tilde{C}}\exp\left((x_j-1)\cdot\ln\left(\frac{e\cdot n}{\epsilon}\right)\right)+\exp\left(\frac{1}{2\alpha}\sum_{s\in\mathbb{S}}\left[c_s\cdot\mathcal{I}_{\tilde{\mathbb{C}}}(s)-\frac{3}{2}x_s c_s\cdot\ln\left(\frac{e\cdot n}{\epsilon}\right)\right]-\epsilon\right).$$

The function $\mathcal{I}_{\tilde{\mathbb{C}}}$ above is the characteristic function of $\tilde{\mathbb{C}}$, that is, $\mathcal{I}_{\tilde{\mathbb{C}}}(s) = 1$ if $s \in \tilde{\mathbb{C}}$, and $\mathcal{I}_{\tilde{\mathbb{C}}}(s) = 0$ otherwise. The potential function is used to determine whether a set s should be subsidized. More specifically, after increasing the value x_s, the set s is added to the cover $\tilde{\mathbb{C}}$ (that is, s is subsidized), only if as a result the potential function decreases. Throughout the analysis of the algorithm, the first term of the potential function ensures that whenever the fraction assigned to an element j is at least 1 (that is,

$x_j \geq 1$), then j is covered by a fully subsidized set. The second term is used to bound the cost of the subsidized sets.

Consider a user e that either joins the system, or performs a best-response move. In either case, e chooses an integral cover of minimum cost, i.e., it chooses a min cost set covering it. If this set is either subsidized, or used by users that joined the system previously, then the total solution cost does not increase and the subsidy algorithm does nothing. In addition, similarly to the fractional model, we do not consider in our analysis sets that are "deserted" following strategy changes performed by users. In case the user chooses a new set, that is, a set that is neither subsidized, "deserted", nor used by other users, we implement the following subsidy algorithm:

Integral Subsidy Algorithm (with input ϵ):
Let $\epsilon'' = \frac{1}{16}\epsilon$ and let $\epsilon' = \frac{\epsilon}{3\ln\left(\frac{e \cdot n}{\epsilon''}\right)}$.
Let s' be the new set chosen by the user and let $c_{\min}$ be the cost of the set:

1. $y_e \leftarrow y_e + \epsilon' \cdot c_{\min}$
2. For each set s that contains e:
   (a) $x_s \leftarrow x_s \cdot \left(1 + \frac{\epsilon' c_{\min}}{2c_s}\right) + \frac{\epsilon' \cdot c_{\min}}{f \cdot 2c_s}$.
   (b) Subsidize the full cost of set s (add it to $\tilde{\mathbb{C}}$) if by doing so the value of the potential function $\Phi(\epsilon'')$ is at most its value before the increment of x_s.

The algorithm updates the variables x_s each time a user purchases a new set. In each such iteration the tax collected from the user is $\epsilon c_{\min}$. We will show that the total subsidy given by the algorithm is at most the amount of tax that was collected until that time. The analysis of our algorithm's performance is based on the following lemma.

Lemma 3. *For any value $\epsilon \leq 1$, $\Phi(\epsilon)$ satisfies the following properties:*

1. *At start $\Phi(\epsilon) \leq 1$, and at any time during the execution of the algorithm $\Phi(\epsilon) > 0$.*
2. *Each time the fraction x_s of a set s is increased by the algorithm, then either adding it to $\tilde{\mathbb{C}}$, or not adding it, does not increase the value of $\Phi(\epsilon)$.*

By Lemma 3 the algorithm is well defined throughout the execution of the algorithm, and it follows that $\Phi(\epsilon)$ is monotonically non-increasing. Using Lemma 3, we prove our main Theorem 4.

Theorem 4. *For any $\epsilon \leq 1$, the price of anarchy of the solution is $O\left(\frac{\log f \log\left(\frac{n}{\epsilon}\right)}{\epsilon}\right)$, and the taxation ratio is ϵ.*

References

1. Alon, N., Awerbuch, B., Azar, Y., Buchbinder, N., Naor, J.: The online set cover problem. In: Proc. of the 35th Annual ACM Symposium on the Theory of Computation, pp. 100–105 (2003)
2. Alon, N., Awerbuch, B., Azar, Y., Buchbinder, N., Naor, J.: A general approach to online network optimization problems. ACM Transactions on Alg. 2, 640–660 (2006)

3. Anshelevich, E., Dasgupta, A., Kleinberg, J., Tardos, É., Wexler, T., Roughgarden, T.: The price of stability for network design with fair cost allocation. In: Proc. of the 45th Annual IEEE Symposium on Foundations of Computer Science, pp. 295–304 (2004)
4. Buchbinder, N., Naor, J.: Online primal-dual algorithms for covering and packing problems. In: Proc. of the 13th Annual European Symposium on Algorithms, pp. 689–701 (2005)
5. Charikar, M., Karloff, H., Mathieu, C., Saks, M., Naor, J.: Online multicast with egalitarian cost sharing (manuscript, 2007)
6. Chekuri, C., Chuzhoy, J., Lewin-Eytan, L., Naor, J., Orda, A.: Non-Cooperative multicast and facility location games. IEEE Journal on Selected Areas in Communication (Special Issue on Non-Cooperative Behavior in Networking) 25, 1193–1206 (2007)
7. Christodoulou, G., Koutsoupias, E., Nanavati, A.: Coordination mechanisms. In: Proc. of the 31st Inter. Colloq. on Automata, Languages and Programming, pp. 345–357 (2004)
8. Immorlica, N., Li, L., Mirrokni, V., Schulz, A.: Coordination mechanisms for selfish scheduling. In: Proc. of the 1st Inter. Workshop on Internet and Network Econonomics, pp. 55–69 (2005)
9. Rosenthal, R.W.: A class of games possessing pure strategy Nash equilibria. International Journal of Game Theory 2, 65–67 (1973)

Group-Strategyproof Cost Sharing for Metric Fault Tolerant Facility Location*

Yvonne Bleischwitz[1] and Florian Schoppmann[1,2]

[1] Faculty of Computer Science, Electrical Engineering and Mathematics, University of Paderborn, Fürstenallee 11, 33102 Paderborn, Germany
{yvonneb,fschopp}@uni-paderborn.de
[2] International Graduate School of Dynamic Intelligent Systems

Abstract. In the context of *general demand* cost sharing, we present the first group-strategyproof mechanisms for the *metric fault tolerant uncapacitated facility location problem.* They are $(3L)$-budget-balanced and $(3L \cdot (1 + \mathcal{H}_n))$-efficient, where L is the maximum service level and n is the number of agents. These mechanisms generalize the seminal *Moulin mechanisms* for *binary demand.* We also apply this approach to the *generalized Steiner problem in networks.*

1 Introduction and Model

Satisfying agents' connectivity requirements at minimum cost is a major challenge in network design problems. In many cases, these problems are NP-hard. The situation gets even more intricate when we consider these problems in the context of *cost sharing.*

In cost-sharing scenarios, a *service provider* offers a common *service* (e.g., connectivity within a network) to agents. Based on *bids* that indicate the agents' willingness to pay, the provider determines a service allocation and payments. Moreover, he computes a solution to provide the service according to the allocation (e.g., to meet the connectivity requirements).

The decision-making of the provider is governed by a commonly known *cost-sharing mechanism.* Essential properties of these mechanisms are *group-strategyproofness*, preventing collusion by guaranteeing that rational agents communicate bids equal to their *true valuations*; *budget-balance*, ensuring recovery of the provider's cost as well as competitive prices in that the generated surplus is always relatively small; and *economic efficiency*, providing a reasonable trade-off between the provider's cost and the valuations of the excluded agents. Finally, practical applications demand for polynomial-time computability (in the size of the problem).

Most research assumes *binary demand*, where agents are "served" or "not served". In contrast, we consider the *general demand* setting, providing *service levels* ranging from 0 to some maximum number. This is of particular interest

* This work was partially supported by the IST Program of the European Union under contract number IST-15964 (AEOLUS).

B. Monien and U.-P. Schroeder (Eds.): SAGT 2008, LNCS 4997, pp. 350–361, 2008.

when agents require different *qualities of service*. For connectivity problems, the service level of an agent is the number of her (distinct) connections. More connections correspond to a higher quality of service, for reasons including throughput and resistance to link failure.

1.1 The Model

Notation. For $n \in \mathbb{N}$, let $[n] := \{1, \dots, n\}$ and $[n]_0 := [n] \cup \{0\}$. Let $\mathcal{H}_n := \sum_{i=1}^{n} \frac{1}{i} \in (\log n, 1 + \log n)$. For $\boldsymbol{x}, \boldsymbol{y} \in \mathbb{Q}^n$, we write $\boldsymbol{x} \geq \boldsymbol{y}$, if for all $i \in [n]$, $x_i \geq y_i$. Let $\mathbf{0}, \mathbf{1}, \boldsymbol{e}_i$ be the zero, one, and i-th standard basis vector. We say that $\boldsymbol{x} \in \{0,1\}^n$ *indicates* $X \subseteq [n]$ if $x_i = 1 \Leftrightarrow i \in X$.

We consider a finite set $[n]$ of agents. Each agent $i \in [n]$ has a *maximum level of service* $L_i \in \mathbb{N}$ she can receive. Let $\mathcal{Q} := [L_1]_0 \times \dots \times [L_n]_0$, $\mathcal{L} := \mathbb{Q}^{L_1} \times \dots \times \mathbb{Q}^{L_n}$, $\mathcal{L}_{\geq 0} := \mathbb{Q}_{\geq 0}^{L_1} \times \dots \times \mathbb{Q}_{\geq 0}^{L_n}$, and $L := \max_{i \in [n]}\{L_i\}$.

An *allocation* is a vector $\boldsymbol{q} \in \mathcal{Q}$ representing the service level given to each agent. Given $\boldsymbol{q} \in \mathcal{Q}$ and $l \in [L]$, we define $\boldsymbol{q}^{\leq l}$ by $q_i^{\leq l} := \min_i\{q_i, l\}$, and let $\boldsymbol{q}^l \in \{0,1\}^n$ indicate the set $Q_l := \{i \in [n] \mid q_i \geq l\}$.

The *valuation vector* $\boldsymbol{v}_i \in \mathbb{Q}^{L_i}$ of agent i consists of the *marginal valuations* $v_{i,l}$ of receiving level l additionally to level $l-1$. Agent i's total valuation for level m is thus $\sum_{l=1}^{m} v_{i,l}$. We call $\boldsymbol{V} = (\boldsymbol{v}_1, \dots, \boldsymbol{v}_n) \in \mathcal{L}$ the *valuation matrix*; accordingly, $\boldsymbol{B} = (\boldsymbol{b}_1, \dots, \boldsymbol{b}_n) \in \mathcal{L}$ denotes a *bid matrix*.

Assumption: We assume that $v_{i,1} \geq \dots \geq v_{i,L_i}$ for all $i \in [n]$.

Definition 1. *A* cost-sharing mechanism $M = (q, x) : \mathcal{L} \to \mathcal{Q} \times \mathbb{Q}_{\geq 0}^n$ *is a function that gets as input a bid matrix $\boldsymbol{B} \in \mathcal{L}$. It outputs an allocation $q(\boldsymbol{B}) \in \mathcal{Q}$ and a vector of cost shares $x(\boldsymbol{B}) \in \mathbb{Q}_{\geq 0}^n$.*

We always require three standard properties of cost-sharing mechanisms:

- *No positive transfers* (NPT): For all $\boldsymbol{B} \in \mathcal{L}$, it is $x(\boldsymbol{B}) \geq 0$.
- *Voluntary participation* (VP): Agents are never charged more than they bid, i.e., for all $\boldsymbol{B} \in \mathcal{L}$ and all $i \in [n]$, it is $x_i(\boldsymbol{B}) \leq \sum_{l=1}^{q_i(\boldsymbol{B})} b_{i,l}$.
- *Consumer sovereignty* (CS): An agent can always ensure to receive a certain service level, i.e., for all $i \in [n]$ and all $l \in [L_i]_0$, there is a bid vector $\boldsymbol{b}_i^{+l} \in \mathbb{Q}^{L_i}$ such that for all $\boldsymbol{B} \in \mathcal{L}$ with $\boldsymbol{b}_i = \boldsymbol{b}_i^{+l}$, it is $q_i(\boldsymbol{B}) = l$.

An agent i aims to submit a bid vector $\boldsymbol{b}_i$ such that her utility is maximized, where her utility is given as $u_i(\boldsymbol{B}) := \sum_{l=1}^{q_i(\boldsymbol{B})} v_{i,l} - x_i(\boldsymbol{B})$.

Definition 2. *A cost-sharing mechanism $M = (q, x)$ is* group-strategyproof *(GSP) if for every true valuation matrix $\boldsymbol{V} \in \mathcal{L}$ and any coalition $K \subseteq [n]$ there is no bid matrix $\boldsymbol{B} \in \mathcal{L}$ with $\boldsymbol{b}_i = \boldsymbol{v}_i$ for all $i \notin K$, such that $u_i(\boldsymbol{B}) \geq u_i(\boldsymbol{V})$ for all $i \in K$ with at least one strict inequality.*

Cost shares computed by a GSP mechanism only depend on the computed allocation [Mou99]. This gives rise to the following definition:

Definition 3. *A cost-sharing method $\xi : \mathcal{Q} \to \mathbb{Q}^n_{\geq 0}$ maps each allocation $\boldsymbol{q} \in \mathcal{Q}$ to a vector of cost shares. If $q_i = 0$, we require $\xi_i(\boldsymbol{q}) = 0$.*

For *binary demand*, *cross-monotonic* cost-sharing methods $\xi : \{0,1\}^n \to \mathbb{Q}^n_{\geq 0}$ are of particular interest for achieving GSP. For all allocations $\boldsymbol{p}$ and all agents $i \in [n]$ and $j \in [n] \setminus \{i\}$ with $p_j = 0$, they fulfill $\xi_i(\boldsymbol{p}) \geq \xi_i(\boldsymbol{p} + \boldsymbol{e}_j)$. Using such a ξ, the (binary demand) mechanism $Moulin_\xi$ is GSP [Mou99]:

Algorithm 1 (computing $Moulin_\xi = (q, x)$**).**
Input: $\boldsymbol{B} \in \mathbb{Q}^n$ $\triangleright$ $\boldsymbol{B} = (b_{1,1}, \ldots, b_{n,1})$
Output: assignment $q(\boldsymbol{B}) \in \{0,1\}^n$, cost-share vector $x(\boldsymbol{B}) \in \mathbb{Q}^n_{\geq 0}$

1: $\boldsymbol{p} := \boldsymbol{1}$;
2: **while** $\boldsymbol{p} \neq \boldsymbol{0}$ and there exists i with $b_{i,1} < \xi_i(\boldsymbol{p})$ **do**
3: $\quad p_j := 0$ for an arbitrary j with $b_{j,1} < \xi_j(\boldsymbol{p})$
4: **return** $(\boldsymbol{p}, \xi(\boldsymbol{p}))$

Our general demand mechanisms use *marginal* cost-sharing methods:

Definition 4. *A* marginal cost-sharing method *is a function $\chi : \mathcal{Q} \to \mathcal{L}_{\geq 0}$, where $\chi_{i,l}(\boldsymbol{q})$ is the marginal cost-share of agent i for additionally receiving service level l to level $l-1$. If $l > q_i$, we require $\chi_{i,l}(\boldsymbol{q}) = 0$.*

We now turn to the service cost. A *cost-sharing problem* is specified by a *cost function* $C : \mathcal{Q} \to \mathbb{Q}_{\geq 0}$ mapping each $\boldsymbol{q}$ to the cost of providing each agent i with service level q_i. Typically, costs stem from solutions to a combinatorial optimization problem and are defined only implicitly. We let $C(\boldsymbol{q})$ be the value of a minimum-cost solution for the instance induced by $\boldsymbol{q}$. This cost can in general not be recovered exactly due to restrictions placed by the GSP requirement. Further difficulties arise when problems are hard. We denote the cost of an approximate solution by $C'(\boldsymbol{q})$ and require the total charge of a mechanism to lie within reasonable bounds:

Definition 5. *A general demand cost-sharing mechanism $M = (q, x)$ is β-*budget-balanced *(β-BB, for $\beta \geq 1$) with respect to $C, C' : \mathcal{Q} \to \mathbb{Q}_{\geq 0}$ if for all $\boldsymbol{B} \in \mathcal{L}$ it holds that*

$$C'(q(\boldsymbol{B})) \leq \sum_{i=1}^{n} x_i(\boldsymbol{B}) \leq \beta \cdot C(q(\boldsymbol{B})) \ .$$

As a quality measure for the computed allocation, we use *optimal social costs* $SC_{\boldsymbol{V}}(\boldsymbol{q}) := C(\boldsymbol{q}) + \sum_{i=1}^{n} \sum_{l=q_i+1}^{L_i} \max\{0, v_{i,l}\}$ and *actual social costs* $SC'_{\boldsymbol{V}}(\boldsymbol{q}) := C'(\boldsymbol{q}) + \sum_{i=1}^{n} \sum_{l=q_i+1}^{L_i} \max\{0, v_{i,l}\}$. The cost incurred and the valuations of the rejected agents should be traded off as good as possible:

Definition 6. *A general demand cost-sharing mechanism $M = (q, x)$ is γ-*efficient *(γ-EFF, for $\gamma \geq 1$) with respect to $C, C' : \mathcal{Q} \to \mathbb{Q}_{\geq 0}$ if for all true valuations $\boldsymbol{V} \in \mathcal{L}$ it holds that $SC'_{\boldsymbol{V}}(q(\boldsymbol{V})) \leq \gamma \cdot \min_{\boldsymbol{q} \in \mathcal{Q}}\{SC_{\boldsymbol{V}}(\boldsymbol{q})\}$.*

The efficiency of $Moulin_\xi$ can be analyzed via the *summability* of ξ:

Definition 7. *A binary demand cost-sharing method* $\xi : \{0,1\}^n \to \mathbb{Q}^n_{\geq 0}$ *is* α*-summable* (α*-SUM, for* $\alpha \geq 1$*) with respect to* $C : \mathcal{Q} \to \mathbb{Q}_{\geq 0}$ *if for every* $\boldsymbol{s} \in \{0,1\}^n$ *and every ordering* $s_1, \ldots, s_{|S|}$ *of* $S := \{i \in [n] \mid s_i = 1\}$*, it is* $\sum_{i=1}^{|S|} \xi_{s_i}(\boldsymbol{s}^i) \leq \alpha \cdot C(\boldsymbol{s})$*, where* $\boldsymbol{s}^i \in \{0,1\}$ *indicates* $S_i := \{s_1, \ldots, s_i\}$*.*

If ξ is β-BB and α-SUM, then $Moulin_\xi$ is $(\alpha + \beta)$-EFF [RS06].

1.2 The Problems

FAULTTOLERANTFL (Metric Fault Tolerant Uncapacitated Facility Location Problem): An instance of this problem is given by a set of agents $[n]$, a set F of facilities, an opening cost $o_f \in \mathbb{N}$ for each $f \in F$, and a non-negative cost function $c : ([n] \cup F) \times ([n] \cup F) \to \mathbb{N}$ that satisfies the triangle inequality. Given $\boldsymbol{q} \in \mathcal{Q}$ with $\max_i\{q_i\} \leq |F|$, the aim is to open a set of facilities and connect each agent i to q_i distinct open facilities, such that the total opening and connection cost is minimized. For $k \in \mathbb{N}$, let $\mathcal{F}_k := \{F' \subseteq F \mid |F'| \geq k\}$. For $F' \in \mathcal{F}_{\max_i\{q_i\}}$, let F'_i be a set of q_i closest facilities in F' to $i \in [n]$. Then the optimal cost is $C(\boldsymbol{q}) := \min_{F' \in \mathcal{F}_{\max_i\{q_i\}}} \{\sum_{f \in F'} o_f + \sum_{i \in [n]} \sum_{f \in F'_i} c(i, f)\}$.

GENERALIZEDSTEINER (Generalized Steiner Problem in Networks): An instance of this problem is given by a set of agents $[n]$, an undirected graph $G := (V, E)$ with edge costs $c : E \to \mathbb{N}$, and a pair of nodes (s_i, t_i) for each agent $i \in [n]$. For a requirement vector $\boldsymbol{q}$, the aim is to determine a minimum-cost subgraph with cost $C(\boldsymbol{q})$ that has q_i edge-disjoint paths between s_i and t_i. We consider a simplification allowing to use multiple edge copies. The cost of such an edge copy is equal to the cost of the edge.

1.3 Related Work

The only general design technique for binary demand GSP mechanisms is applying $Moulin_\xi$ [Mou99, JV01]. Particularly, cross-monotonic cost-sharing methods were designed by Pál et al. [PT03] for metric uncapacitated facility location and by Könemann et al. [KLS05] for Steiner forests. A non-Moulin and only SP mechanism for facility location was introduced by Devanur et al. [DMV05].

Prior to this work, incremental cost-sharing mechanisms were the only known GSP mechanisms for general demand cost sharing. They simply consider an ordering that specifies which agent's level is incremented next and make the agent pay for the corresponding marginal cost. Incremental mechanisms are only known to be GSP for certain costs, and are essentially the only GSP and 1-BB mechanisms for these costs [Mou99]. However, for costs induced by FAULTTOLERANTFL and GENERALIZEDSTEINER, incremental mechanisms are not GSP. The interested reader can find examples in the extended version of this paper.

The only other general demand mechanisms we are aware of are *acyclic mechanisms* introduced by Mehta et al. [MRS07], which were originally designed for binary demand to overcome the limitations of Moulin mechanisms [IMM05, RS06].

The main drawback of acyclic mechanisms is that they are only *weak* GSP, meaning that coalitions are only assumed to be successful if *every* agent strictly improves her utility. Technically, *all* inequalities in Definition 2 are strict. For FaultTolerantFL, [MRS07] give a $O(L^2)$-BB and $O(L^2 \cdot (1 + \log n))$-EFF acyclic mechanism. In the full version of their paper, they present a $\mathcal{H}_n$-BB and $(2\mathcal{H}_n \cdot (1 + \mathcal{H}_L))$-EFF acyclic mechanism for the non-metric case.[1]

To the best of our knowledge, the best approximation algorithm for FaultTolerantFL yields an appproximation factor of 2.076 [SSL03], while the best approximation factor for GeneralizedSteiner is 2, with and without the simplification of allowing edge copies [Jai01].

1.4 Contribution

The point of departure for this work is a rather obvious idea for generalizing Moulin mechanisms: Start with the maximum allocation and iteratively reduce service levels until every agent can afford her remaining levels. The cost shares are extracted from marginal cost-sharing methods χ. These mechanisms, termed $MoulinGD_\chi$, are stated in Section 2.

- We identify three properties of marginal cost-sharing methods that are sufficient for $MoulinGD_\chi$ to be GSP. It comes as no surprise that a generalization of binary-demand cross-monotonicity is among them.
- We give marginal cost-sharing methods χ^{FL} for every instance of FaultTolerantFL in Section 3 and show that $MoulinGD_{\chi^{FL}}$ is GSP, $(3L)$-BB and $(3L \cdot (1+\mathcal{H}_n))$-EFF. These are the first GSP mechanisms for this problem. Method χ^{FL} is a natural generalization of the binary demand cost-sharing method for facility location in [PT03]. In contrast, the generalization used within acyclic mechanisms in [MRS07] does not guarantee GSP.
- We give the first GSP mechanisms for GeneralizedSteiner in Section 4.

Our work adapts the common assumption that marginal valuations are nonincreasing in the service level. Omitted proofs are given in the extended version.

2 Generalized Moulin Mechanisms

Given a marginal cost-sharing method χ, we propose to generalize Moulin mechanisms as in Algorithm 2:

Algorithm 2 (computing $MoulinGD_\chi := (q, x)$**).**
Input: bid matrix $\boldsymbol{B} \in \mathcal{L}$
Output: allocation $q(\boldsymbol{B}) \in \mathcal{Q}$; cost-share vector $x(\boldsymbol{B}) \in \mathbb{Q}^n_{\geq 0}$
1: $\boldsymbol{q} := (L_1, \ldots, L_n)$;
2: **while** there exists i with $b_{i,q_i} < \chi_{i,q_i}(\boldsymbol{q})$ **do**
3: $\quad q_j := q_j - 1$ for an arbitrary j with $b_{j,q_j} < \chi_{j,q_j}(\boldsymbol{q})$
4: **return** $(\boldsymbol{q}, \boldsymbol{x})$ with $x_i := \sum_{l=1}^{q_i} \chi_{i,l}(\boldsymbol{q})$

[1] Note that these results are adjusted to our notion of β-BB.

We state three properties of χ that are sufficient for $MoulinGD_\chi$ to be GSP. The first is a generalization of binary demand cross-montonicity:

Definition 8. *A marginal cost-sharing method is* cross-monotonic *if for all allocations $\boldsymbol{q} \in \mathcal{Q}$, all agents $i \in [n]$ and $j \in [n] \setminus \{i\}$ with $q_j < L_j$, and all service levels $l \in [L_i]$, it holds that $\chi_{i,l}(\boldsymbol{q}) \geq \chi_{i,l}(\boldsymbol{q} + \boldsymbol{e}_j)$.*

The second property ensures that the marginal cost-share $\chi_{i,l}(\boldsymbol{q})$ of agent i with $q_i \geq l$ is exactly the marginal cost-share $\chi_{i,l}(\boldsymbol{q}^{\leq l})$.

Definition 9. *A marginal cost-sharing method is* level-restricted *if for all allocations $\boldsymbol{q} \in \mathcal{Q}$, for all agents $i \in [n]$, and for all service levels $l \in [L_i]$, it holds that $\chi_{i,l}(\boldsymbol{q}) = \chi_{i,l}(\boldsymbol{q}^{\leq l})$.*

The third property together with cross-monotonicity implies that the marginal cost-share of an agent is non-decreasing in the number of levels:

Definition 10. *A marginal cost-sharing method is* non-decreasing *if for all allocations $\boldsymbol{q} \in \mathcal{Q}$, for service level $l := \max_i\{q_i\}$, and for all agents $i \in [n]$ with $q_i = l < L_i$, it holds that $\chi_{i,l}(\boldsymbol{q}) \leq \chi_{i,l+1}(\boldsymbol{q} + \sum_{j \in [n]: q_j = l < L_j} \boldsymbol{e}_j)$.*

Lemma 1. *If χ is non-decreasing and cross-monotonic, it holds for all allocations $\boldsymbol{q} \in \mathcal{Q}$, for all service levels $l \in [L]$, and for all agents $i \in [n]$ with $q_i > l$ that $\chi_{i,l}(\boldsymbol{q}^{\leq l}) \leq \chi_{i,l+1}(\boldsymbol{q}^{\leq l+1})$.*

If χ is level restricted, agent i's utility is the sum over the *marginal utilities* $v_{i,l} - \chi_{i,l}(\boldsymbol{q}^{\leq l})$. If χ is additionally cross-monotonic and non-decreasing, by Lemma 1 and non-increasing marginal valuations, these marginal utilities are non-increasing in l. The proof of Theorem 1 heavily relies on non-increasing marginal utilities and uses the main idea from [MS01], which shows that $Moulin_\xi$ is GSP if ξ is cross-monotonic.

Theorem 1. *$MoulinGD_\chi$ is GSP given any level-restricted, cross-monotonic, and non-decreasing marginal cost-sharing method χ. Furthermore, it satisfies NPT, VP and CS.*

Due to space limitations we only show that if exactly one of the three properties required for χ does not hold, $MoulinGD_\chi$ is not GSP anymore. It remains an open problem whether there are cost functions for which all three properties need to be fulfilled at once in order to obtain GSP.

For all examples, let $n = 2$, $L_1 = L_2 = 2$, and $\chi_{2,l}(\boldsymbol{q}) := 2$ for all $l \in [2]$ and all $\boldsymbol{q} \in \mathcal{Q}$ with $q_2 \geq l$. We always assume that $\boldsymbol{v}_2 = (2, 2)$.

Example 1. Consider χ with $\chi_{1,l}(\boldsymbol{q}) := 1$ for all $l \in [2]$ and all $\boldsymbol{q} \in \mathcal{Q}$ with $q_1 \geq l$, with the only exception that $\chi_{1,2}(2, 2) := 2$. Obviously, χ is level-restricted and non-decreasing, but not cross-monotonic since $\chi_{1,2}(2, 1) < \chi_{1,2}(2, 2)$. For the case that $\boldsymbol{v}_1 = (2, 2)$, both agents get service level 2, where $u_1(\boldsymbol{v}_1, \boldsymbol{v}_2) = 1$ and $u_2(\boldsymbol{v}_1, \boldsymbol{v}_2) = 0$. Agent 2 may then bid $\boldsymbol{b}_2 = (-1, -1)$ in order to not receive the service with the result that agent 1 receives level 2 with utility $u_1(\boldsymbol{v}_1, \boldsymbol{b}_2) = 2$.

Example 2. Consider χ with $\chi_{1,1}(\boldsymbol{q}) := 1$ for all $\boldsymbol{q} \in \mathcal{Q}$ with $q_1 = 1$, $\chi_{1,1}(\boldsymbol{q}) := 2$ for all $\boldsymbol{q} \in \mathcal{Q}$ with $q_1 = 2$ and $\chi_{1,2}(\boldsymbol{q}) := 3$ for all $\boldsymbol{q} \in \mathcal{Q}$ with $q_1 = 2$. Here, χ is cross-monotonic and non-decreasing but fails to be level-restricted due to $\chi_{1,1}(2,2) > \chi_{1,1}(1,1)$. If now $\boldsymbol{v}_1 = (3,3)$, then agent 1 receives level 2 and $u_1(\boldsymbol{v}_1, \boldsymbol{v}_2) = 1$. However, for $\boldsymbol{b}_1 = (2,2)$, agent 1 receives only level 1, and $u_1(\boldsymbol{b}_1, \boldsymbol{v}_2) = 2$.

On the other hand, we get the situation $\chi_{1,1}(2,2) < \chi_{1,1}(1,1)$ when we change χ such that $\chi_{1,1}(\boldsymbol{q}) = 2$ for all $\boldsymbol{q} \in \mathcal{Q}$ with $q_1 = 1$ and $\chi_{1,1}(\boldsymbol{q}) = 1$ for all $\boldsymbol{q} \in \mathcal{Q}$ with $q_1 = 2$. If $\boldsymbol{v}_1 = (3, 3-\varepsilon)$, then agent 1 receives only one level and $u_1(\boldsymbol{v}_1, \boldsymbol{v}_2) = 1$. However, she may bid $\boldsymbol{b}_1 = (3,3)$ to receive both levels such that $u_1(\boldsymbol{b}_1, \boldsymbol{v}_2) = 2 - \varepsilon$.

Example 3. Consider χ with $\chi_{1,1}(\boldsymbol{q}) := 2$ for all $\boldsymbol{q} \in \mathcal{Q}$ with $q_1 \geq 1$ and $\chi_{1,2}(\boldsymbol{q}) := 1$ for all $\boldsymbol{q} \in \mathcal{Q}$ with $q_1 \geq 2$. Now we have the case that χ is cross-monotonic and level-restricted, but not non-decreasing. For $\boldsymbol{v}_1 = (1,1)$, agent 1 receives level 2 and has a utility of $u_1(\boldsymbol{v}_1, \boldsymbol{v}_2) = -1$. However, bidding $\boldsymbol{b}_1 = (-1,-1)$ ensures a utility of zero.

3 Metric Fault Tolerant Uncapacitated Facility Location

We explain how to define $\chi^{FL} : \mathcal{Q} \to \mathcal{L}_{\geq 0}$ and how to construct a solution in polynomial time. For $\boldsymbol{q} \in \{0,1\}^n$, both reduces to the method and solution by Pál and Tardos [PT03] for binary demand facility location.

Fix $\boldsymbol{q} \in \mathcal{Q}$ and $l \in [L]$. We only need to determine $\chi_{i,l}^{FL}(\boldsymbol{q})$ for all $i \in Q_l := \{j \in [n] \mid q_j \geq l\}$. Simultaneously, every agent i in Q_l uniformly grows a ball with i at its center to infinity. This ball, the *ghost* of i, has radius t at time t. We say that the ghost of i *touches* facility f at time t if $c(i,f) \leq t$. After the ghost of i touches f it starts *filling* f, contributing $t - c(i,f)$ at time $t \geq c(i,f)$. Facility f is said to be *full* if all such contributions sum up to its opening cost o_f. Let $t(f)$ denote the time when f becomes full, and $S_f := \{i \in [n] \mid c(i,f) < t(f)\}$ the set of agents that contributed to filling f. It holds that $\sum_{i \in S_f}(t(f) - c(i,f)) = o_f$.

We define $\chi_{i,l}^{FL}(\boldsymbol{q})$ to be the time that it takes for the ghost of i to touch l full facilities. Note that $\chi_{*,l}^{FL}(\boldsymbol{q})$ only depends on $Q_l := \{j \in [n] \mid q_j \geq l\}$. For $\boldsymbol{q}^l \in \{0,1\}^n$ indicating Q_l, it is $\chi_{i,l}^{FL}(\boldsymbol{q}) := \chi_{i,l}^{FL}(l \cdot \boldsymbol{q}^l)$. This is even stronger than level-restriction.

Lemma 2. *χ^{FL} is level-restricted, cross-monotonic, and non-decreasing.*

We construct a solution with cost $C'(\boldsymbol{q})$ iteratively by computing $\chi_{i,l}^{FL}(\boldsymbol{q})$ for all $i \in Q_l$ for $l = 1, \ldots, \max_i\{q_i\}$. Fix an iteration l. Let $t(f)$ and S_f be the values obtained for all $f \in F$ by growing the ghosts of Q_l.

Facilities are opened in iteration l according to the following rule: Let F_{l-1} be the set of the already opened facilities in iterations $1, \ldots, l-1$. If in iteration l, a facility $f \notin F_{l-1}$ becomes full, we open f if and only if conditions O_1 and O_2 hold:

O_1) There is no facility g that was already opened in iteration l and for which $c(g, f) \leq 2 \cdot t(f)$.

O_2) There are no l distinct facilities $g_1, \ldots, g_l \in F_{l-1}$ for which $c(g_k, f) \leq 2 \cdot t(f)$ for all $k \in [l]$.

In order to simplify the analysis, we connect every agent $i \in Q_l$ to one (more) open facility in each iteration l, rather than connecting her with the q_i closest facilities in the final facility set. This is how we do it:

C_1) If $i \in S_f$ for an f opened in iteration l, we connect i to f. (It can be shown that for f, g with $f \neq g$ opened in iteration l, $S_f \cap S_g = \emptyset$.)

C_2) Otherwise, if at time $\chi_{i,l}^{FL}(\boldsymbol{q})$ the ghost of i touches an arbitrary open facility f to which i is not connected yet, we connect i to f.

C_3) Otherwise, let f be an arbitrary full but closed facility that the ghost of i touches at time $\chi_{i,l}^{FL}(\boldsymbol{q})$; f was not opened in iteration l, because O_1 or O_2 do not hold:
  a) If O_1 does not hold because of facility g, connect i to g.
  b) If O_2 does not hold because of facilities $g_1, \ldots, g_l$, connect i to a $g \in \{g_1, \ldots, g_l\}$ to which i is not connected yet.

If there are ties in C_3a and C_3b, or when facilities become full simultaneously, break them arbitrarily. It is straightforward to see that for every instance of FaultTolerantFL, χ^{FL} and the solution constructed above can be computed in polynomial time (in the size of the input). Deleting O_2 and C_3b, we get the rules of [PT03]. However, these two rules are crucial for a reasonable BB approximation. We refer to the extended version for the details.

Example 4. Facilities are illustrated as houses, where the roofs are labeled with the opening cost. Agents are circles labeled with the agent's identity. Edges (i, j) are labeled with $c(i, j)$. If i and j are not directly linked, $c(i, j)$ is defined as the cost of a shortest path between i and j in the network.

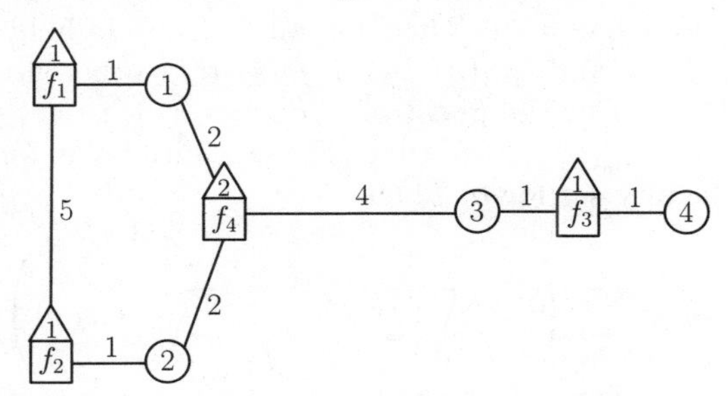

We look at allocation $\boldsymbol{q} = (2, 2, 2, 1)$. The marginal cost shares for level 1 are determined by growing the ghosts of agent set $Q_1 = \{1, 2, 3, 4\}$, where $\boldsymbol{q}^1 = (1, 1, 1, 1)$. For level 2, we grow the ghosts of agent set $Q_2 = \{1, 2, 3\}$, where $\boldsymbol{q}^2 = (1, 1, 1, 0)$. We write $i \circ f$ if i touches f, but f is not full yet; $\mathbf{f}$ if f becomes full; $i \bullet \mathbf{f}$ if i touches a full facility f. Then for timesteps t:

t	Events for Level 1
1	$1 \circ f_1, 2 \circ f_2, 3 \circ f_3, 4 \circ f_3$
$\frac{3}{2}$	$\mathbf{f_3}, 3 \bullet \mathbf{f_3}, 4 \bullet \mathbf{f_3}$
2	$\mathbf{f_1}, \mathbf{f_2}, 1 \bullet \mathbf{f_1}, 2 \bullet \mathbf{f_2}, 1 \circ f_4, 2 \circ f_4$

t	Events for Level 2
1	$1 \circ f_1, 2 \circ f_2, 3 \circ f_3$
2	$\mathbf{f_1}, \mathbf{f_2}, \mathbf{f_3}, 1 \bullet \mathbf{f_1}, 2 \bullet \mathbf{f_2}, 3 \bullet \mathbf{f_3}$ $1 \circ f_4, 2 \circ f_4$
3	$\mathbf{f_4}, 1 \bullet \mathbf{f_4}, 2 \bullet \mathbf{f_4}$
4	$3 \bullet \mathbf{f_4}$

Note that $i \in S_f$ iff event $i \circ f$ occurs at a strictly smaller time step than event $i \bullet \mathbf{f}$. The cost shares are $\chi_{*,1}(\boldsymbol{q}) = \chi_{*,1}(1,1,1,1) = (2, 2, \frac{3}{2}, \frac{3}{2})$ and $\chi_{*,2}(\boldsymbol{q}) = \chi_{*,2}(2,2,2,0) = (3,3,4,0)$. The final cost shares are thus $(5, 5, \frac{11}{2}, \frac{3}{2})$. For Level 1, we open f_3, f_1, and f_2. Due to C_1, we connect agents 3 and 4 to f_3, agent 1 to f_1, and agent 2 to f_2. For Level 2, f_4 stays closed due to O_2. All agents in $\{1,2,3\}$ are connected due to C_3b; 1 is connected to f_2, 2 is connected to f_1, and 3 is connected to f_1 or f_2.

Theorem 2. *For any* $\boldsymbol{q} \in \mathcal{Q}$, $X(\boldsymbol{q}) := \sum_{l=1}^{L} \sum_{i \in Q_l} \chi_{i,l}^{FL}(\boldsymbol{q}) \leq L \cdot C(\boldsymbol{q})$. *Furthermore, there exists a solution for allocation* $\boldsymbol{q}$ *with cost* $C'(\boldsymbol{q})$, *such that* $\frac{1}{3} \cdot C'(\boldsymbol{q}) \leq X(\boldsymbol{q})$.

Proof. Let $\chi := \chi^{FL}$. Fix $\boldsymbol{q} \in \mathcal{Q}$. We show the upper bound; the lower bound is obtained by modifying the proof in [PT03]. Consider an arbitrary facility set $F' \subseteq \mathcal{F}_{\max_i\{q_i\}}$. Fix $l \in [\max_i\{q_i\}]$ and $i \in Q_l$. Let $F'_i \subseteq F'$ be a set of q_i distinct closest facilities in F' to i. We show:

$$\exists f \in F'_i : \chi_{i,l}(\boldsymbol{q}) \leq \begin{cases} t(f) & \text{if } i \in S_f \\ c(i,f) & \text{otherwise} . \end{cases} \tag{1}$$

Assume that (1) does not hold. Then for all $f \in F'_i$ it holds that $\chi_{i,l}(\boldsymbol{q}) > t(f) > c(i,f)$ if $i \in S_f$ and $\chi_{i,l}(\boldsymbol{q}) > c(i,f) \geq t(f)$ otherwise. Thus, at time $t := \max_{f \in F'_i}\{t(f), c(i,f)\}$, the ghost of i touches at least $q_i \geq l$ full facilities, a contradiction to $t < \chi_{i,l}(\boldsymbol{q})$. Note that (1) especially holds for $F' = F^*$, when F^* is an optimal facility set for $\boldsymbol{q}$. Then,

$$\begin{aligned}
\sum_{i \in Q_l} \chi_{i,l}(\boldsymbol{q}) &\leq \sum_{i \in [n]} \left(\sum_{f \in F_i^* : i \in S_f} t(f) + \sum_{f \in F_i^* : i \notin S_f} c(i,f) \right) \\
&= \sum_{f \in F^*} \sum_{i \in S_f : f \in F_i^*} (t(f) - c(i,f)) + \sum_{i \in [n]} \sum_{f \in F_i^*} c(i,f) \\
&\leq \sum_{f \in F^*} o_f + \sum_{i \in [n]} \sum_{f \in F_i^*} c(i,f) = C(\boldsymbol{q}) .
\end{aligned}$$

Finally, $X(\boldsymbol{q}) = \sum_{l=1}^{L} \sum_{i \in Q_l} \chi_{i,l}(\boldsymbol{q}) \leq L \cdot C(\boldsymbol{q})$. □

Theorem 3. *MoulinGD*$_{\chi^{FL}}$ *is* $(3L \cdot (1 + \mathcal{H}_n))$*-EFF with respect to* C *and* C'.

Whereas we refer to the full version for the proof of Theorem 3, we show a property of χ^{FL} similar to binary demand summability (see e.g. [RS06]) in Lemma 3, which constitutes the main part of the proof.

Lemma 3. *For any* $\boldsymbol{q} \in \mathcal{Q}$ *and any ordering* $s_1, \ldots, s_{|S|}$ *of the set* $S := \{i \in [n] \mid q_i > 0\}$*, where* $\boldsymbol{s}^j \in \{0,1\}^n$ *indicates* $S_j := \{s_1, \ldots, s_j\}$*, it is*

$$\sum_{j=1}^{|S|} \chi^{FL}_{s_j, q_{s_j}}(q_{s_j} \cdot \boldsymbol{s}^j) \leq \mathcal{H}_n \cdot C(\boldsymbol{q}) \ .$$

Proof. Let $\chi := \chi^{FL}$. Roughly speaking, the main idea of the proof is a "reduction" to the summability of a (binary demand) cost-sharing method $\xi^{FL} : \{0,1\}^n \to \mathbb{Q}^n_{\geq 0}$ that we define according to Pál and Tardos [PT03] with a new facility location instance: It has agent set $[n]$, facility set G, and a new cost function $d : ([n] \cup G) \times ([n] \cup G) \to \mathbb{Q}_{\geq 0}$. Let $D := \{0,1\}^n \to \mathbb{Q}_{\geq 0}$ be the optimal cost function for the new instance.

Fix $\boldsymbol{q} \in \mathcal{Q}$ and an ordering $s_1, \ldots, s_{|S|}$ of $S := \{i \in [n] \mid q_i > 0\}$. Fix $j \in [|S|]$ and look at $\chi_{s_j, q_{s_j}}(q_{s_j} \cdot \boldsymbol{s}^j)$, computed for the original instance. For all $f \in F$, let $t(f)$ and S_f be the corresponding values for growing the ghosts of set S_j. In the original instance, let F^* be an optimal facility set for $\boldsymbol{q}$, and $F^*_{s_j}$ be the facilities that s_j is connected to in an optimal solution. It is $F^* \subseteq \mathcal{F}_{\max_i\{q_i\}}$. In the proof of Theorem 2, we have already shown that there exists $f_j \in F^*_{s_j}$, such that $\chi_{s_j, q_{s_j}}(q_{s_j} \cdot \boldsymbol{s}^j)$ is at most $t(f_j)$ if $i \in S_{f_j}$, or $c(s_j, f_j)$ otherwise.

Let the new facilities be $G := \{f_1, \ldots, f_{|S|}\}$. For a each pair in $\{(s_j, f_j)\}_{j \in [|S|]}$, let $d(s_j, f_j) := c(s_j, f_j)$. Furthermore, for all j, j' such that $f_j = f_{j'}$, we define $d(s_j, s_{j'}) := c(s_j, s_{j'})$. All other costs are defined to be sufficiently large, while ensuring that d satisfies the triangle inequality. The networks below illustrate the old (left) and the new (right) facility instance. The grey parts correspond to the unchanged distances.

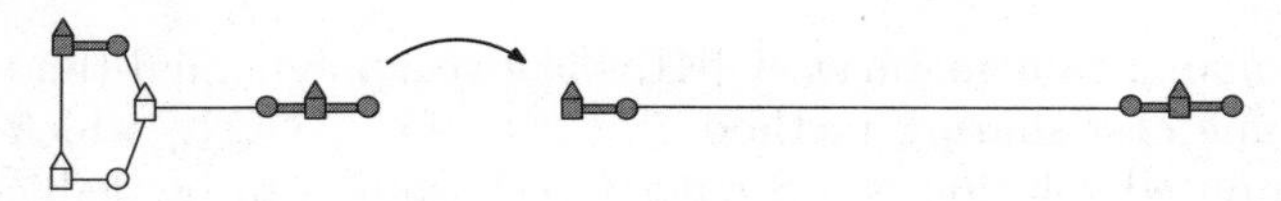

By construction of the new instance, it is $\chi_{s_j, q_{s_j}}(q_{s_j} \cdot \boldsymbol{s}^j) \leq \xi^{FL}_{s_j}(\boldsymbol{s}^j)$ for all $j \in [|S|]$, where $\xi^{FL}_{s_j}(\boldsymbol{s}^j)$ is computed on the new instance. Additionally, $D(\boldsymbol{s}) \leq C(\boldsymbol{q})$, with $\boldsymbol{s} \in \{0,1\}^n$ indicating S. We further use the fact that ξ^{FL} is $\mathcal{H}_n$-SUM [RS07] in order to obtain

$$\sum_{j=1}^{|S|} \chi^{FL}_{s_j, q_{s_j}}(q_{s_j} \cdot \boldsymbol{s}^j) \leq \sum_{j=1}^{|S|} \xi^{FL}_{s_j}(\boldsymbol{s}^j) \leq \mathcal{H}_n \cdot D(\boldsymbol{s}) \leq \mathcal{H}_n \cdot C(\boldsymbol{q}) \ . \qquad \square$$

Corollary 1. *There is a marginal cost-sharing method* χ *such that MoulinGD*$_\chi$ *is* $(3L)$*-BB and* $(3L \cdot (1 + \mathcal{H}_n))$*-EFF with respect to* C *and* C'.

Proof. Define χ by multiplying χ^{FL} by 3. Adjusting the proofs of Theorem 2 and Theorem 3 leads to the stated BB and EFF guarantees.

We shortly describe the marginal cost-sharing method χ^{AFL} used in the acyclic mechanism introduced by Mehta et al. [MRS07]. The mechanism itself is essentially Mechanism *MoulinGD*$_\chi$, where line 2 is replaced by "while there exists i with $b_{i,l} < \chi_{i,l}(\boldsymbol{q})$ for an $l \in [q_i]$". The main difference between χ^{AFL} and χ^{FL} is that $\chi^{FL}_{i,l}(\boldsymbol{q})$ is independent of connections computed in iterations 1 to $l-1$.

The marginal cost shares $\chi^{AFL}_{i,l}(\boldsymbol{q})$ for all $i \in Q_l$ are computed iteratively for $l = 1, \dots, L$. In each iteration l, the cross-monotonic cost-sharing method of Pál and Tardos [PT03] is invoked for all agents with $q_i \geq l$. The instance is changed in such a way that opening costs are set to 0 for already opened facilities. In order to ensure that each agent i is connected to q_i distinct facilities, the distance $c(i,f)$ for already existing connections between i and f is set to infinity. All other distances stay the same. Consider the network below, given by [MRS07]:

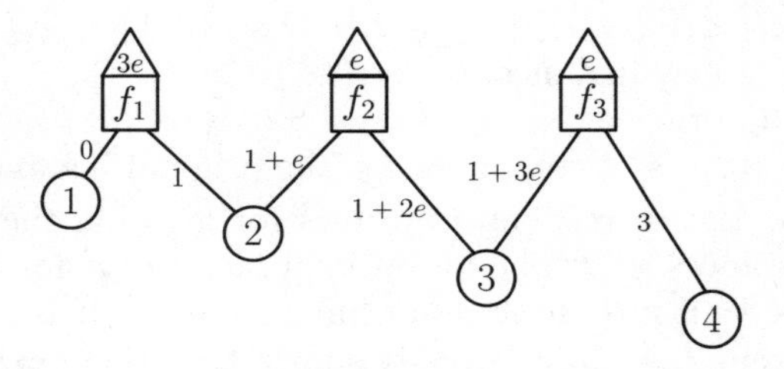

Here, $\chi^{AFL}_{4,2}((2,2,2,2)) = 5 + 5\varepsilon > 3 = \chi^{AFL}_{4,2}(0,2,2,2)$ obviously violates cross-monotonicity. Essentially, this happens due to that fact that for $(2,2,2,2)$, $c(4,3)$ is set to infinity in the first iteration, making her ghost grow longer in the second iteration. However, for $(0,2,2,2)$ it is $c(4,3) = 3$ in both iterations.

4 Generalized Steiner Problem in Networks

For binary demand, Könemann et al. [KLS05] give a polynomial-time computable cross-monotonic cost-sharing method $\xi^{GS} : \{0,1\}^n \to \mathbb{Q}^n_{\geq 0}$ which is $(2 - \frac{1}{n})$-BB. The computed solution is a Steiner forest which can be deduced from the cost-share computation. Additionally, it is known that ξ^{GS} is $O(\log^2 n)$-SUM [CRS06]. The cost-sharing method can essentially be computed by the algorithm by Agrawal et al. [AKR95] with only a small modification which is crucial for cross-monotonicity.

We combine the cost-sharing method from [KLS05] with a straightforward solution construction by Goemans and Bertsimas [GB93]. The marginal cost-sharing method $\chi^{GS} : \mathcal{Q} \to \mathcal{L}_{\geq 0}$ is simply defined by letting $\chi^{GS}_{i,l}(\boldsymbol{q}) := \xi^{GS}_i(\boldsymbol{q}^l)$ for all $\boldsymbol{q} \in \mathcal{L}$, where $\boldsymbol{q}^l \in \{0,1\}$ indicates $Q_l := \{i \in [n] \mid q_i \geq l\}$. Our computed solution is simply the union of the Steiner forests from each round (as implied by ξ^{GS}), where multiple edges count as copies.

Theorem 4. *There is a marginal cost-sharing method* χ^{GS} *and a solution with cost* $C'(\boldsymbol{q})$ *for each* $\boldsymbol{q} \in \mathcal{Q}$*, such that* $MoulinGD_{\chi^{GS}}$ *is* $((2-\frac{1}{n})\cdot\mathcal{H}_L)$*-BB,* $((2-\frac{1}{n}+\log^2 n)\cdot\mathcal{H}_L)$*-EFF, and GSP with respect to* C *and* C'*. Furthermore, it satisfies NPT, VP, and CS.*

References

[AKR95] Agrawal, A., Klein, P., Ravi, R.: When trees collide: An approximation algorithm for the generalized steiner problem in networks. SIAM Journal on Computing 24(3), 445–456 (1995)

[CRS06] Chawla, S., Roughgarden, T., Sundararajan, M.: Optimal cost-sharing mechanisms for Steiner forest problems. In: Spirakis, P.G., Mavronicolas, M., Kontogiannis, S.C. (eds.) WINE 2006. LNCS, vol. 4286, pp. 112–123. Springer, Heidelberg (2006)

[DMV05] Devanur, N.R., Mihail, M., Vazirani, V.V.: Strategyproof cost-sharing mechanisms for set cover and facility location games. Decision Support Systems 39(1), 11–22 (2005)

[GB93] Goemans, M., Bertsimas, D.: Survivable networks, linear programming relaxations and the parsimonious property. Mathematical Programming 60, 145–166 (1993)

[IMM05] Immorlica, N., Mahdian, M., Mirrokni, V.: Limitations of cross-monotonic cost sharing schemes. In: Proceedings of the 16th Annual ACM-SIAM Symposium on Discrete Algorithms, pp. 602–611 (2005)

[Jai01] Jain, K.: A factor 2 approximation algorithm for the generalized steiner network problem. Combinatorica 21(1), 39–60 (2001)

[JV01] Jain, K., Vazirani, V.: Applications of approximate algorithms to cooperative games. In: Proceedings of the 33th Annual ACM Symposium on Theory of Computing, pp. 364–372 (2001)

[KLS05] Könemann, J., Leonardi, S., Schäfer, G.: A group-strategyproof mechanism for Steiner forests. In: Proceedings of the 16th Annual ACM-SIAM Symposium on Discrete Algorithms, pp. 612–619 (2005)

[Mou99] Moulin, H.: Incremental cost sharing: Characterization by coalition strategy-proofness. Social Choice and Welfare 16(2), 279–320 (1999)

[MRS07] Mehta, A., Roughgarden, T., Sundararajan, M.: Beyond Moulin mechanisms. In: Proceedings of the 8th ACM Conference on Electronic Commerce, pp. 1–10 (2007), `http://theory.stanford.edu/~tim/papers/bmm.pdf`

[MS01] Moulin, H., Shenker, S.: Strategyproof sharing of submodular costs: budget balance versus efficiency. Economic Theory 18, 511–533 (2001)

[PT03] Pál, M., Tardos, É.: Group strategyproof mechanisms via primal-dual algorithms. In: Proceedings of the 44th Annual IEEE Symposium on Foundations of Computer Science, pp. 584–593 (2003)

[RS06] Roughgarden, T., Sundararajan, M.: New trade-offs in cost-sharing mechanisms. In: Proceedings of the 38th ACM Symposium on Theory of Computing, pp. 79–88 (2006)

[RS07] Roughgarden, T., Sundararajan, M.: Optimal efficiency guarantees for network design mechanisms. In: Proceedings of the 12th Conference on Integer Programming and Combinatorial Optimization, pp. 469–483 (2007)

[SSL03] Shmoys, D.B., Swamy, C., Levi, R.: Fault-tolerant facility location. In: Proceedings of the 14th Annual ACM-SIAM Symposium on Discrete Algorithms, pp. 735–736 (2003)

Author Index

Printing: Mercedes-Druck, Berlin
Binding: Stein + Lehmann, Berlin

Lecture Notes in Computer Science

Sublibrary 3: Information Systems and Application, incl. Internet/Web and HCI

For information about Vols. 1–4566
please contact your bookseller or Springer

Vol. 4803: R. Meersman, Z. Tari (Eds.), On the Move to Meaningful Internet Systems 2007: CoopIS, DOA, ODBASE, GADA, and IS, Part I. XXIX, 1173 pages. 2007.

Vol. 4802: J.-L. Hainaut, E.A. Rundensteiner, M. Kirchberg, M. Bertolotto, M. Brochhausen, Y.-P.P. Chen, S.S.-S. Cherfi, M. Doerr, H. Han, S. Hartmann, J. Parsons, G. Poels, C. Rolland, J. Trujillo, E. Yu, E. Zimányie (Eds.), Advances in Conceptual Modeling – Foundations and Applications. XIX, 420 pages. 2007.

Vol. 4801: C. Parent, K.-D. Schewe, V.C. Storey, B. Thalheim (Eds.), Conceptual Modeling - ER 2007. XVI, 616 pages. 2007.

Vol. 4797: M. Arenas, M.I. Schwartzbach (Eds.), Database Programming Languages. VIII, 261 pages. 2007.

Vol. 4796: M. Lew, N. Sebe, T.S. Huang, E.M. Bakker (Eds.), Human–Computer Interaction. X, 157 pages. 2007.

Vol. 4794: B. Schiele, A.K. Dey, H. Gellersen, B. de Ruyter, M. Tscheligi, R. Wichert, E. Aarts, A. Buchmann (Eds.), Ambient Intelligence. XV, 375 pages. 2007.

Vol. 4777: S. Bhalla (Ed.), Databases in Networked Information Systems. X, 329 pages. 2007.

Vol. 4761: R. Obermaisser, Y. Nah, P. Puschner, F.J. Rammig (Eds.), Software Technologies for Embedded and Ubiquitous Systems. XIV, 563 pages. 2007.

Vol. 4747: S. Džeroski, J. Struyf (Eds.), Knowledge Discovery in Inductive Databases. X, 301 pages. 2007.

Vol. 4744: Y. de Kort, W. IJsselsteijn, C. Midden, B. Eggen, B.J. Fogg (Eds.), Persuasive Technology. XIV, 316 pages. 2007.

Vol. 4740: L. Ma, M. Rauterberg, R. Nakatsu (Eds.), Entertainment Computing – ICEC 2007. XXX, 480 pages. 2007.

Vol. 4730: C. Peters, P. Clough, F.C. Gey, J. Karlgren, B. Magnini, D.W. Oard, M. de Rijke, M. Stempfhuber (Eds.), Evaluation of Multilingual and Multi-modal Information Retrieval. XXIV, 998 pages. 2007.

Vol. 4723: M. R. Berthold, J. Shawe-Taylor, N. Lavrač (Eds.), Advances in Intelligent Data Analysis VII. XIV, 380 pages. 2007.

Vol. 4721: W. Jonker, M. Petković (Eds.), Secure Data Management. X, 213 pages. 2007.

Vol. 4718: J. Hightower, B. Schiele, T. Strang (Eds.), Location- and Context-Awareness. X, 297 pages. 2007.

Vol. 4717: J. Krumm, G.D. Abowd, A. Seneviratne, T. Strang (Eds.), UbiComp 2007: Ubiquitous Computing. XIX, 520 pages. 2007.

Vol. 4715: J.M. Haake, S.F. Ochoa, A. Cechich (Eds.), Groupware: Design, Implementation, and Use. XIII, 355 pages. 2007.

Vol. 4714: G. Alonso, P. Dadam, M. Rosemann (Eds.), Business Process Management. XIII, 418 pages. 2007.

Vol. 4704: D. Barbosa, A. Bonifati, Z. Bellahsène, E. Hunt, R. Unland (Eds.), Database and XML Technologies. X, 141 pages. 2007.

Vol. 4690: Y. Ioannidis, B. Novikov, B. Rachev (Eds.), Advances in Databases and Information Systems. XIII, 377 pages. 2007.

Vol. 4675: L. Kovács, N. Fuhr, C. Meghini (Eds.), Research and Advanced Technology for Digital Libraries. XVII, 585 pages. 2007.

Vol. 4674: Y. Luo (Ed.), Cooperative Design, Visualization, and Engineering. XIII, 431 pages. 2007.

Vol. 4663: C. Baranauskas, P. Palanque, J. Abascal, S.D.J. Barbosa (Eds.), Human-Computer Interaction – INTERACT 2007, Part II. XXXIII, 735 pages. 2007.

Vol. 4662: C. Baranauskas, P. Palanque, J. Abascal, S.D.J. Barbosa (Eds.), Human-Computer Interaction – INTERACT 2007, Part I. XXXIII, 637 pages. 2007.

Vol. 4658: T. Enokido, L. Barolli, M. Takizawa (Eds.), Network-Based Information Systems. XIII, 544 pages. 2007.

Vol. 4656: M.A. Wimmer, J. Scholl, Å. Grönlund (Eds.), Electronic Government. XIV, 450 pages. 2007.

Vol. 4655: G. Psaila, R. Wagner (Eds.), E-Commerce and Web Technologies. VII, 229 pages. 2007.

Vol. 4654: I.-Y. Song, J. Eder, T.M. Nguyen (Eds.), Data Warehousing and Knowledge Discovery. XVI, 482 pages. 2007.

Vol. 4653: R. Wagner, N. Revell, G. Pernul (Eds.), Database and Expert Systems Applications. XXII, 907 pages. 2007.

Vol. 4636: G. Antoniou, U. Aßmann, C. Baroglio, S. Decker, N. Henze, P.-L. Patranjan, R. Tolksdorf (Eds.), Reasoning Web. IX, 345 pages. 2007.

Vol. 4611: J. Indulska, J. Ma, L.T. Yang, T. Ungerer, J. Cao (Eds.), Ubiquitous Intelligence and Computing. XXIII, 1257 pages. 2007.

Vol. 4607: L. Baresi, P. Fraternali, G.-J. Houben (Eds.), Web Engineering. XVI, 576 pages. 2007.

Vol. 4606: A. Pras, M. van Sinderen (Eds.), Dependable and Adaptable Networks and Services. XIV, 149 pages. 2007.

Vol. 4605: D. Papadias, D. Zhang, G. Kollios (Eds.), Advances in Spatial and Temporal Databases. X, 479 pages. 2007.

Vol. 4602: S. Barker, G.-J. Ahn (Eds.), Data and Applications Security XXI. X, 291 pages. 2007.

Vol. 4601: S. Spaccapietra, P. Atzeni, F. Fages, M.-S. Hacid, M. Kifer, J. Mylopoulos, B. Pernici, P. Shvaiko, J. Trujillo, I. Zaihrayeu (Eds.), Journal on Data Semantics IX. XV, 197 pages. 2007.

Vol. 4592: Z. Kedad, N. Lammari, E. Métais, F. Meziane, Y. Rezgui (Eds.), Natural Language Processing and Information Systems. XIV, 442 pages. 2007.

Vol. 4587: R. Cooper, J. Kennedy (Eds.), Data Management. XIII, 259 pages. 2007.

Vol. 4577: N. Sebe, Y. Liu, Y.-t. Zhuang, T.S. Huang (Eds.), Multimedia Content Analysis and Mining. XIII, 513 pages. 2007.

Vol. 4568: T. Ishida, S. R. Fussell, P. T. J. M. Vossen (Eds.), Intercultural Collaboration. XIII, 395 pages. 2007.